Sources and Studies in the History of Mathematics and Physical Sciences

Sources and Studies in the History of Mathematics and Physical Sciences was inaugurated as two series in 1975 with the publication in Studies of Otto Neugebauer's seminal three-volume History of Ancient Mathematical Astronomy, which remains the central history of the subject. This publication was followed the next year in Sources by Gerald Toomer's transcription, translation (from the Arabic), and commentary of Diocles on Burning Mirrors. The two series were eventually amalgamated under a single editorial board led originally by Martin Klein (d. 2009) and Gerald Toomer, respectively two of the foremost historians of modern and ancient physical science. The goal of the joint series, as of its two predecessors, is to publish probing histories and thorough editions of technical developments in mathematics and physics, broadly construed. Its scope covers all relevant work from pre-classical antiquity through the last century, ranging from Babylonian mathematics to the scientific correspondence of H. A. Lorentz. Books in this series will interest scholars in the history of mathematics and physics, mathematicians, physicists, engineers, and anyone who seeks to understand the historical underpinnings of the modern physical sciences. All works are reviewed to meet the highest standards of scientific literature.

More information about this series at http://www.springer.com/series/4142

Fabio Acerbi

The Logical Syntax of Greek Mathematics

Springer

Fabio Acerbi
CNRS, UMR8167 Orient et Méditerranée
Paris, France

ISSN 2196-8810 ISSN 2196-8829 (electronic)
Sources and Studies in the History of Mathematics and Physical Sciences
ISBN 978-3-030-76958-1 ISBN 978-3-030-76959-8 (eBook)
https://doi.org/10.1007/978-3-030-76959-8

This Springer imprint is published by the registered company Springer Nature Switzerland AG
The registered company address is: Gewerbestrasse 11, 6330 Cham, Switzerland

My torment is memory

L'analyse logique est secondaire
par rapport à l'analyse linguistique
sur laquelle elle est fondée

TABLE OF CONTENTS

LIMINALIA

The aim of this book is to describe Greek mathematics as a literary product. This means studying its style from a logico-syntactic point of view and setting parallels with logical and grammatical doctrines developed in antiquity. In this way, major philosophical themes such as the expression of mathematical generality and the selection of criteria of validity for arguments can be treated without anachronism. My strategic goal is to show that Greek mathematics is better identified by the stylistic codes it employs than by its contents. This interpretive framework allows reassessing branches of Greek mathematics which scholarship usually deems debased if not simply childish, and getting a renewed look, in the perspective of register-crossing, at late mathematical elaborations.

The approach I am advocating raises several issues. First, it is a fact that some syntactic structures among those I shall be studying are fairly common in ordinary Greek prose style. However, in the mathematical style they are adhered to with remarkable consistency, sometimes almost exclusively, even in cases in which the general Greek prose style allows for alternative structures. Now, the resources of ordinary language cannot by themselves univocally determine the expressions to be used in a strictly codified idiolect as the one of mathematics, where the shades of sense may make a difference as regards the mathematics involved. Therefore, the evidence about Greek mathematical style requires in many instances an explanation suited to its own expressive purposes, independently of contemporary linguistic practice and of contemporary grammatical conceptions.

Second, no ancient discussions are known of the Greek mathematical idiolect:[1] the standard style appears to be the result of a self-regulating and self-reproducing practice. This means that such a practice was perceived as imposing implicit stylistic constraints. As a consequence, on the one hand, investigations into this domain can be based on statistical evidence only, but, on the other hand, sufficiently clear-cut data can be safely read as the consequence of regulating syntactic criteria. After all, the so-called syntactic, metrical, and philological "laws" and "rules" were first constituted on the basis of statistical predominance and of shared practices, acquiring a normative character only later.[2] The limits of such an approach were already clear to the Pyrrhonists:[3] on the one hand, this is the only approach grounded on sound scientific method once factual data must be analyzed; on the other hand, and exactly for the same reason, this approach is nearly tautological.

These facts raise the major methodological problem of what "explanation" may mean in this context, and in fact whether there is anything to explain at all. To anticipate some of our major problematic points, it is not clear what might count for an "explanation" of some choices of word order that are so remarkably adhered to in the Greek mathematical style, or for an "explanation" of the use of the aorist stem instead of the present stem in specific pieces of mathematical writing. If we are interested in such issues, traditional, descriptive grammars are useless. What these traditional grammars lack—they hide this lacuna behind a thicket of (sub)typologies and myriads of examples—is a theoretical framework: without such a framework, "solutions" to problems of word order, of use of the article, or of aspectual choice are simply out of reach. These are in fact problems that pertain to pragmatics, and they happen to arise quite naturally when Greek mathematical style is studied. The paradigm of functional grammar appears to offer satisfactory solutions to all these problems; for this reason, I shall formulate most of my grammatical "explanations" in the idiolect

[1] A notable exception are some passing remarks in Galen's *Institutio logica*, in particular in chapters XII and XVI.
[2] See for instance the observations in Pasquali 1998, 49–66. Similar observations, specifically about the difficulty of finding principles that can "explain" Greek word order, are in Dover 1960, 1, 33–34, 41.
[3] Read the criticisms in Sextus Empiricus, *M* I.221–227.

of this paradigm—but the reader must know that competing paradigms exist, and they may happen to use the same technical terms with a different meaning. It remains that any theory-laden approach to Greek grammar has a major limit: the linguistic structures any approach of this kind systematizes usually belong to the archaic period (Homer, Herodotus) or to the classical period (historians, orators, Plato), whereas Greek mathematics is obviously a Hellenistic literary product.

The absence of ancient discussions of Greek mathematical style does not entail that the Greek mathematicians worked in a vacuum chamber. As we shall see, specific syntactic constraints suggest that cross-fertilization has occurred to some extent between mathematical style and some prescriptions of Stoic logic, whereas homologies with Peripatetic logic appear to be post-factum elaborations of ancient and modern commentators. On the other hand, we shall also see that only a part of the logical structures that emerge in mathematical treatises fit prescriptions typical of Stoic logic.[4] However, we may extract from the mathematical texts a consistent, and peculiar, approach to almost all logical notions and practices studied or recommended either by the Stoic school or by the Peripatetic school. To state it tersely: mathematics constitutes a self-contained and independent logical system, a "third logical school",[5] whose logical syntax is very well suited to a structural analysis as the one I shall be developing in this book.

Setting a comparison with ancient grammatical doctrines is much more problematic. First, a fully-fledged *ars grammatica* was developed well after Greek mathematical style had reached a stable format: among the extant grammatical treatises, a systematic account of connectors cannot be found before Apollonius Dyscolus' *On connectors*, dating to the 2nd century CE. Second, grammatical theories appear to depend on the linguistic doctrines of the Stoic,[6] even if to an extent which it is not easy to determine.[7] In principle, then, parallels between mathematical style and grammatical categorizations could be explained through the intermediation of Stoic linguistic—and ultimately of Stoic logic. It is also clear that the borderline between logic and grammar is rather thin, and even thinner if the substratum is Greek mathematical style: for linguistic regularity there replaces formalism, and such a regularity is the linguistic counterpart of mathematical "rigour" insofar as this compels the approval of peers and thereby necessitates the mathematical facts established by means of regular linguistic patterns.

I shall show that Greek mathematical style comprises three main codes, which interacted with one another through the remarkable metadiscursive resource of "validation". The three codes are strictly incompatible, insofar as each of them is optimized to express mathematical contents that cannot, or hardly can, be expressed by any of the others. My main source for the "demonstrative code" is the *Elements*; when appropriate, I shall also present evidence drawn from Apollonius or Archimedes, or from Euclid's *Data*.[8] The reason for such a choice is obvious: the *Elements* is the reference work of Greek mathematics, as regards both contents and style.[9] This fact undercuts the difficulties arising from the unwelcome accident that almost all Greek mathematical treatises, and

---

[4] The best general accounts of Stoic logic are Bobzien 1999 and the excellent synthesis Crivelli, forthcoming. Other surveys, sometimes with a less technical focus, include Mates 1960, Frede 1974, Brunschwig 1978a, Gourinat 2000.

[5] Dissatisfaction with the two major logical schools (which he found "useless for establishing proofs") is what made Galen "steer clear of the arguments of these people, while emulating the model provided by geometrical proof", as he explains in *Libr. Propr.*, 116.16 and 117.14–16.

[6] Read, for instance, the introductory remarks Apollonius Dyscolus prefaces to his *On connectors*, in particular at *GG* II.1.1, 214.1–3. Even if Apollonius declares that he will depart from Stoic doctrines (a discussion of the issue is in Barnes 2007, 181–182), his statement entails that these constituted the standard approach to grammatical issues.

[7] See the overview in Blank, Atherton 2003.

[8] The *Data* has been less affected by revisions than the *Elements*; it displays the demonstrative code in its purest form.

[9] There does not exist such a reference work for the other two stylistic codes, to be named just below.

those of the main authors more than the others, have been revised in late antiquity.[10] Very simply, the point is that we are interested in the *canonical* work for Greek demonstrative style, and what has been transmitted to us is by definition the canonized version of the *Elements*. However, we shall sometimes need to go deeper into philological issues, for some syntactic peculiarities of the *Elements* can be shown—for instance by comparison with the indirect tradition of the treatise—to be the result of late, large-scale campaigns of revision.

My book is divided into five parts; its ordering principle is the (decreasing) size of the linguistic units involved. I shall first describe the three stylistic codes of Greek mathematics, which I shall call "demonstrative code", "procedural code", and "algorithmic code". The second part of the book expounds in detail the mechanism of "validation", both within the demonstrative code and as a major form of interaction between the three codes. From the third part on, I shall focus my attention on the demonstrative code, for the simple reasons that it is lexically and syntactically richer than the others and that it was more thoroughly practised by the most renowned Greek mathematicians. The third part will deal exclusively with the status of mathematical objects and with the problem of mathematical generality; I shall explain in detail the "Greek solution" to this problem. The remaining two parts are an extensive study of all logico-syntactic structures that figure in a mathematical proposition. The fourth part will analyse the main features of the "deductive machine", namely, the suprasentential logical system dictated by the traditional division of a mathematical proposition into specific parts: enunciation, setting-out, construction, and proof. As for the proof, I shall pay particular attention to the logic of relations. The fifth part deals with the sentential logical system of a mathematical proposition, with special emphasis on quantification, modalities, and connectors. Three appendices offer complementary material. The five parts of this book are strictly intertwined, for identical linguistic units can be studied from different perspectives; frequent cross-references between the sections of the book make these links explicit.

As said, the strategic goal of this book is to show that the way the Greek mathematicians have said the things they had to say is much more important than the things they have actually said: after all, most of Greek mathematics is "trivial" to modern eyes. I have also tried to make my book as interpretation-free as possible; notable exceptions—in a traditional historiographic perspective, these are the highlights of my study—are the notion of validation and the "Greek solution" to the problem of mathematical generality. However, these two exegetic complexes must be regarded as narratives that try to put a part of the documentary record into a consistent order: I do not claim that validation was consciously practised by the Greek geometers, or that they concocted such a supremely subtle way to phrase their mathematics out of conscious and well-informed rejection of generic objects. Thus, the reader will not find here neither lucubrations about poorly documented periods nor extensive descriptions of mathematical contents. In a sense, my book is an exercise in what can or cannot be said about Greek mathematics—and in how to say it—without indulging in either of these approaches and in the associated rhetoric. This is the real reason for its title.

This book presents a fair number of Greek texts with associated translation. As I am interested in linguistic issues, diagrams are attached to complete propositions only, and only if they are required as a visual aid. The translations are often idiosyncratic; my conventions (some of which are explained below) will be obvious to any perceptive reader; their aim is to preserve as much of the structure of the original as possible. I have tacitly operated editorial choices whenever this was required. All Greek texts are punctuated anew; there is a dedicated section showing how to do that. I shall not explain the mathematics behind the texts I present, nor

[10] The textual tradition of the *Elements* is outlined in Sect. 1.5.

shall I systematically provide historical information. The reader is assumed to be acquainted with Greek, with mathematics, and with Greek mathematics.

In order to avoid confusion with the established mathematical lexicon, I shall use "statement" as a synonym of "proposition" in the technical sense the latter term has in logic; its linguistic expression is a "(declarative) sentence"; my "statement" also translates the Stoic ἀξίωμα, which is currently (and inaccurately as to its being a ματ-stem noun) rendered with "assertible". I shall carefully distinguish between a unit of meaning and its linguistic expression (the latter "formulates" the former). I hope I shall be forgiven for a major exception: I shall use the same denomination for a specific statement and for its linguistic expression as a sentence; for instance, I shall write in both cases that "a conditional is made of an antecedent and of a consequent". The linguistic expression of a "supposition" has the verb in the imperative. An "assumption" is any mathematical fact featuring as an independent premise in a deduction, independently of the linguistic form in which it is formulated. The "reduction assumption" (or "reduction supposition") is the one driven to contradiction in a proof by reduction to the impossible. I shall say that a linguistic item "opens" or "introduces" a larger linguistic item if the former is in a liminal position in the latter and if it is necessary for the latter to be the kind of linguistic item it is (for instance, to be a coassumption, or to be the antecedent of a conditional); according to Greek syntax, linguistic items can be said to be in "liminal" position in a sentence even if they do not occupy the first position in the word sequence. Words supplied in translation are put between simple angular brackets. In translation, I shall frequently abbreviate the two formulae τὸ ὑπό "the ‹rectangle contained› by" and τὸ ἀπό "the ‹square described› on" to "that by" and "that on"; however, every translation of a self-contained Greek text will exhibit the longer forms, possibly bracketed, on their first occurrence.

The titles of the writings of classical authors are usually abbreviated as in the Liddell-Scott-Jones lexicon; the page(s) referred to are those of the canonical editions cited in the *Index fontium* of the present book. Propositions in mathematical writings are referred to by book and number, like for instance "*Con.* IX.15". Pappus' *Collectio* is cited by book and chapter as in Hultsch's edition. If the indication of the work is missing, this is the *Elements*. The sign "def" stands for "definition", "post" for "postulate", "cn" for "common notion", *alt* for "alternative (proof)"; a sign like "*alt*X.23" signifies item 23 of the Appendix to Heiberg's edition of Book X. A porism (= corollary) to proposition *x* is denoted *x*por; a lemma between propositions *x* and *y* is denoted *x*/*y* (the *Elements* ends with a lemma; it is denoted 18/19 even if there is no proposition XIII.19). If there are several groups of definitions in the same book, several lemmas between two propositions, or several alternative proofs of the same proposition, these are numbered by means of Roman figures in the former two cases, by means of Arabic figures in the latter. Frequently used sources are cited according to the *sigla* listed at the beginning of the bibliography, followed by "volume, page.line(s)", as in *AOO* II, 528.12–16. The boldface sigla that denote some Greek manuscripts are explained in the *Index fontium*.

I shall cite a fair amount of secondary literature, and a fair amount of my own studies. In general, I shall not cite literature that is either irrelevant to the subject of my book—which is not a history of Greek mathematics, but a manual for learning how to write Greek mathematics—or which, albeit relevant, I regard as scholarly ballast (example: previous literature on analysis and synthesis, en bloc), or that represents dead historiographic fashions (example: the "archaeological" reading of the *Elements*), or that is just an empty exercise (example: the debate about "geometric algebra") or sheer speculation (examples: most of what has been written on pre-Euclidean mathematics and on loans from other mathematical cultures). However, I will make an exception for some studies that belong to one of the previous categories, and will assess them as they deserve. Why shall I be so trenchant in my assessments? Because, to paraphrase a supreme scholar who was much more entitled to be acerbic than me (and whom of course I will not cite), I feel sure that the reader's hope of extracting a low enjoyment from scurrilous assessments of other scholars' studies may partly compensate for his or her natural disrelish for such a combination of a tedious subject with an odious author as the one exhibited in my book.

The final form of this study owes much to discussions with the participants in a course co-held with Jonathan Beere within the Doctoral Program of the Research Training Group "Philosophy, Science and the Sciences" of the Humboldt-Universität zu Berlin. This book was first written in Italian in 2007 and remained a draft; I have radically rewritten it several times since then. Now it has found a publisher and at least five readers: Jonathan Beere, David Ebrey, Paolo Fait, Ramon Masià, and especially Jan von Plato, who supported me in manyfold ways. I am relieved to say farewell to this ball and chain and to ancient Greek mathematics. *Verum tempus est desinere*.

1. THE THREE STYLISTIC CODES OF GREEK MATHEMATICS

Greek mathematics is a complex literary genre: it comprises three main subgenres and a number of hybrids of them.[1] I shall call the three main subgenres the "demonstrative code", the "procedural code", and the "algorithmic code".[2] Each of the three codes is a *Kunstsprache* (and not simply a *Fachsprache*) exhibiting a limited lexicon and highly regimented syntactic features, which in some cases may well be termed "extreme".[3]

I shall summarily describe these three stylistic codes in the next three Sections. However, it is an obvious fact that one of the three codes, namely, the demonstrative code, is both lexically and syntactically richer than the others. For this reason, the remaining parts of this book will expound in detail the logico-syntactic structures of the demonstrative code.

The demonstrative code is the stylistic environment we are used to associate with Greek mathematics. It is practised by such celebrated authors as Archimedes, Euclid, Apollonius, and Diophantus, in such landmarks of mathematics as the *Method*, the *Elements*, the *Conics*, or the *Arithmetics*. Its deductively-marked discursive progression has become synonymous with mathematical rigour. It makes extensive and characteristic use of diagrams and of abstract designators.

The procedural and the algorithmic code were systematically applied in doing a kind of mathematics which modern scholarship has always considered debased. These are two stylistic resources that formulate chains of operations on geometric or numeric entities, and such that the output of an operation is taken as the input of the subsequent operation: these two codes are the ancient counterpart of our computer programmes. In particular, the procedural code was used to express in words operational sequences we would summarize in an algebraic "formula". As the examples I shall give will show, any of the three codes was devised to formulate mathematical states of affairs that could not find a satisfactory expression—or even any expression at all—in the other two.

My analysis of each of the three stylistic codes of Greek mathematics will first propose a short description, corroborated by an example of a self-contained piece of mathematical discourse written in the intended code. A more detailed description, which specifically refers to the example set out, will follow; this description focuses on four kinds of stylistic marker:

- Discursive arrow, including coordination and subordination.
- Verb forms, with emphasis on the use of stems and moods.
- Reference to objects in the form of designations.
- Mathematical generality: how the self-contained piece of mathematical discourse applies to the most general class of intended objects.

I shall point out in each case what part of speech has the function of making the said self-contained piece of mathematical discourse a connected whole.

---

[1] The hybrids are studied in Acerbi 2012b—where I have first proposed the approach developed in this Section—and in Acerbi, Vitrac 2014, 411–427.

[2] This basic tripartition was already obvious in late antiquity: Theon carefully distinguishes between the ψιλαὶ ἔφοδοι "bare methods" (= my "algorithms") of his "small commentary" on Ptolemy's *Handy Tables* (Tihon 1978, 199.8–9), intended for those who πρὸς τὴν τῆς τοιαύτης διδασκαλίας μάθησιν μετὰ τοῦ μηδὲ τοῖς πολλαπλασιαμοῖς ἢ μερισμοῖς τῶν ἀριθμῶν ἱκανῶς παρακολουθεῖν δύνασθαι, ἔτι καὶ ἀμύητοι παντάπασι καὶ τῶν γραμμικῶν δείξεων τυγχάνουσιν "follow this teaching while being not only unable adequately to follow the multiplications and divisions of numbers, but even completely ignorant of linear proofs" (ibid., 199.5–7), the λογικαὶ ἔφοδοι (= my "procedures") of his "great commentary" on the *Handy Tables* (ibid., 199.2, and Mogenet, Tihon 1985–99 I, 93.4), which aim to teach τοὺς λόγους τοὺς κατὰ τῶν ψηφοφοριῶν "the reasons of the calculations" (ibid., 93.14–15), and the γραμμικαὶ ἔφοδοι "linear methods" (= geometric proofs), to be found only in his commentary on the *Almagest* (ibid., 94.5–6).

[3] But every written expression in ancient Greek has a *Kunstcharacter*: see for instance Des Places 1934.

F. Acerbi, *The Logical Syntax of Greek Mathematics*, Sources and Studies in the History of Mathematics and Physical Sciences,
https://doi.org/10.1007/978-3-030-76959-8_1

1.1. THE DEMONSTRATIVE CODE

Short description and example

The syntactic structure of a self-contained mathematical unit of the demonstrative code, namely, of a "proposition", is an argumental or operational coordination.[4] Every fully-fledged proposition can be divided into self-contained segments, called "specific parts". These parts are canonized in Proclus' well-known isagogic scheme:[5] they are πρότασις "enunciation"; ἔκθεσις "setting-out"; διορισμός "determination"; κατασκευή "construction"; ἀπόδειξις "proof"; συμπέρασμα "conclusion".[6] The logical import of each of the specific parts is variable. Enunciation and conclusion (Sect. 4.1 below) are deductively inert, since they comprise a single statement. The setting-out and the construction (Sects. 4.2 and 4.3) have a purely conjunctive structure: they only contain sentences formulated with the verb form in the imperative, and these sentences are coordinated by the conjunction καί "and". Assumptions of disparate kinds are supposed in the setting-out and in the construction, among which the "constructive acts".[7] These assumptions are used in the proof (Sect. 4.5), which is a complex argument that comprises connected chains "conclusion / assumption + coassumption → conclusion / assumption".[8] The very beginning of a proof can be isolated as a self-standing syntactic structure with a crucial logical import; this structure is introduced by the subordinant ἐπεί "since" and has recently been baptized "anaphora" (Sect. 4.4). As we shall see throughout this book (see also the table at the end of this Section), the specific parts of a mathematical proposition are sharply demarcated by a series of linguistic markers; the only (partial) exception—and despite the presence of a specific linguistic marker at the beginning of each proof—is the boundary between the construction and the proof, whose steps may intermingle (see Sect. 4.3).

Our sources for the demonstrative code are the most celebrated mathematical authors and corpora of Greek antiquity: Archimedes, Euclid, Apollonius, the so-called "little astronomy", Pappus'

---

[4] There are two main types of hypotactic structure in a mathematical proposition: conditional statements in the enunciations and in clauses that initialize a reduction to the impossible, παρασυνημμένα "paraconditional" clauses, which characterize the "anaphora" (see just below). The occurrences of genitive absolute are fairly rare (an example is in the enunciation of proposition I.16, cited in III.2, to be read below).

[5] At *iE*, 203.1–207.25. Proclus' scheme takes an idealized geometric theorem as its model. This is a mathematical proposition that is not a problem nor concerns number theory, in which only one result is proved, and whose construction is not interspersed with deductive steps. As we shall see in Sect. 4.1, and contrary to Proclus' contention, just a strict minority of theorems in the *Elements* do have a general conclusion, namely, a sentence that exactly repeats the enunciation.

[6] Proclus' terminology is inspired—with some mismatches: mathematical πρότασις and συμπέρασμα always coincide, whereas this is impossible in a syllogism—on supposedly analogous notions in Aristotle's dialectic. At least three denominations, however, are traditional in the mathematical field: these are πρότασις, ἀπόδειξις, and συμπέρασμα; for the former two, just read the opening sentence of Archimedes' *Method*, at *AOO* II, 426.4–7; for the latter, read the sentence closing *Meth.* 1, which Heiberg misplaced at the beginning of *Meth.* 2, ibid., 438.16–21 (on the meaning of πρότασις "premise" in Aristotle's *APr.* see Crivelli, Charles 2011). Κατασκευή is a technical term of Aristotle's dialectic and denotes the act of "establishing" a thesis by means of an argument; it is a weaker notion than "proof"; the verb that denotes the opposite action is ἀνασκευάζω "to refute" a thesis, normally by means of a counterexample: see for instance *Top.* II.3, 110a32–b7; in this passage, Aristotle explicitly contrasts the general case of "establishing" a result, where some form of preliminary agreement on a general axiom is normally required, with a geometric proof, where the form of the argument secures full generality within a given domain of objects. The verb κατασκευάζω in the meaning "to perform a construction" is already attested in Archimedes (18 occurrences of the verb, 2 of the associated noun: see *AOO* III, 364 *sub voce*): the associated noun is thus a well-established denomination of a specific part of a proposition. Only ἔκθεσις and διορισμός cannot be found before Proclus with this meaning (cf. Sect. 4.2).

[7] To avoid confusions, I shall henceforth use the term "construction" for the specific part of a proposition or in a generic meaning (as in "problem of construction"), "constructive act" for a specific construction such as "let a straight line be drawn that joins points A and B". See Sect. 4.3 for a more detailed typology.

[8] The slash means that the conclusion of an argument directly serves as the (main) assumption of the subsequent argument. No premise is stated twice in a chain of mathematical arguments. Likewise, well-formed Aristotelian or Stoic (chains of) syllogisms do not allow for doubled premises (for the latter, cf. Bobzien 1999, 150–151, and 2019, 251–252).

Collectio. The stylistic unity of this code shows conspicuous cracks in such later champions of register-crossing as Ptolemy, Hero, and Diophantus.[9]

The example of proposition we shall read is III.2; it proves, by reduction to the impossible, that a circle is a convex figure. The specific parts of III.2 are separated by inner borders in the transcription + translation below. The bracketed references in the translation are to principles or propositions of the *Elements*, or to other geometric facts; no constructive acts must be invoked to justify the assumptions in the setting-out; I have underlined the two clauses that make the "anaphora".

The text of III.2 (*EOO* I, 168.17–170.17) is followed by two diagrams (see next page):

<table>
<tr><td>ἐὰν κύκλου ἐπὶ τῆς περιφερείας ληφθῇ δύο τυχόντα σημεῖα, ἡ ἐπὶ τὰ σημεῖα ἐπιζευγνυμένη εὐθεῖα ἐντὸς πεσεῖται τοῦ κύκλου.</td><td>If on the circumference of a circle two random points be taken, the straight line joined at the points will fall within the circle.</td></tr>
<tr><td>ἔστω κύκλος ὁ ΑΒΓ, καὶ ἐπὶ τῆς περιφερείας αὐτοῦ εἰλήφθω δύο τυχόντα σημεῖα τὰ Α Β.</td><td>Let there be a circle, ΑΒΓ, and let two random points, Α, Β, be taken on its circumference.</td></tr>
<tr><td>λέγω ὅτι ἡ ἀπὸ τοῦ Α ἐπὶ τὸ Β ἐπιζευγνυμένη εὐθεῖα ἐντὸς πεσεῖται τοῦ κύκλου.</td><td>I claim that the straight line joined from Α to Β will fall within the circle.</td></tr>
<tr><td>μὴ γάρ, ἀλλ' εἰ δυνατόν, πιπτέτω ἐκτὸς ὡς ἡ ΑΕΒ, καὶ εἰλήφθω τὸ κέντρον τοῦ ΑΒΓ κύκλου καὶ ἔστω τὸ Δ, καὶ ἐπεζεύχθωσαν αἱ ΔΑ ΔΒ, καὶ διήχθω ἡ ΔΖΕ.</td><td>In fact not, but if possible, let it fall outside as ΑΕΒ, and let the centre of circle ΑΒΓ be taken [III.1] and let it be Δ, and let ‹straight lines›, ΔΑ, ΔΒ, be joined [I.post.1], and let a ‹straight line›, ΔΖΕ, be drawn through.</td></tr>
<tr><td>ἐπεὶ οὖν ἴση ἐστὶν ἡ ΔΑ τῇ ΔΒ, ἴση ἄρα καὶ γωνία ἡ ὑπὸ ΔΑΕ τῇ ὑπὸ ΔΒΕ. καὶ ἐπεὶ τριγώνου τοῦ ΔΑΕ μία πλευρὰ προσεκβέβληται ἡ ΑΕΒ, μείζων ἄρα ἡ ὑπὸ ΔΕΒ γωνία τῆς ὑπὸ ΔΑΕ· ἴση δὲ ἡ ὑπὸ ΔΑΕ τῇ ὑπὸ ΔΒΕ· μείζων ἄρα ἡ ὑπὸ ΔΕΒ τῆς ὑπὸ ΔΒΕ· ὑπὸ δὲ τὴν μείζονα γωνίαν ἡ μείζων πλευρὰ ὑποτείνει· μείζων ἄρα ἡ ΔΒ τῆς ΔΕ· ἴση δὲ ἡ ΔΒ τῇ ΔΖ· μείζων ἄρα ἡ ΔΖ τῆς ΔΕ ἡ ἐλάττων τῆς μείζονος, ὅπερ ἐστὶν ἀδύνατον· οὐκ ἄρα ἡ ἀπὸ τοῦ Α ἐπὶ τὸ Β ἐπιζευγνυμένη εὐθεῖα ἐκτὸς πεσεῖται τοῦ κύκλου. ὁμοίως δὴ δείξομεν ὅτι οὐδὲ ἐπ' αὐτῆς τῆς περιφερείας· ἐντὸς ἄρα.</td><td>Then since ΔΑ is equal to ΔΒ [setting-out, construction, I.def.15], therefore angle ΔΑΕ is also equal to ΔΒΕ [I.5]. And since one side ΑΕΒ of a triangle ΔΑΕ turns out to be produced [constr.], therefore angle ΔΕΒ is greater than ΔΑΕ [I.16]; and ΔΑΕ is equal to ΔΒΕ [proved above]; therefore ΔΕΒ is greater than ΔΒΕ; and the greater side extends under the greater angle [I.19]; therefore ΔΒ is greater than ΔΕ; and ΔΒ is equal to ΔΖ [I.def.15]; therefore ΔΖ is greater than ΔΕ, the less than the greater [form of design. ἡ ΔΖΕ], which is really impossible; therefore it is not the case that the straight line joined from Α to Β will fall outside the circle. Very similarly we shall prove that ‹it will› not ‹fall› on the circumference itself either: therefore ‹it will fall› within.[10]</td></tr>
<tr><td>ἐὰν ἄρα κύκλου ἐπὶ τῆς περιφερείας ληφθῇ δύο τυχόντα σημεῖα, ἡ ἐπὶ τὰ σημεῖα ἐπιζευγνυμένη εὐθεῖα ἐντὸς πεσεῖται τοῦ κύκλου, ὅπερ ἔδει δεῖξαι.</td><td>Therefore if on the circumference of a circle two random points be taken, the straight line joined at the points will fall within the circle, which it was really required to prove.</td></tr>
</table>

[9] I have thoroughly analyzed the style of Hero's *Metrica* in Acerbi, Vitrac 2014, 59–81 and 363–427, Diophantus' style in Acerbi 2011e, 57–113.

[10] The case in which the straight line joined from τὸ Α to τὸ Β coincides with the circumference is similar but not identical to the case proved in full in the theorem: the construction should be rewritten, since point τὸ Ζ is no longer necessary; the proof should end with "therefore ΔΒ is greater than ΔΕ; but they are also equal, which is really impossible". In metacontextual mentions like "joined from τὸ Α to τὸ Β" in the first line of the present footnote, the article in front of the denotative letters must be retained, because it is an integral part of the designation (see Sect. 2.2). Thus, I have written, and shall write, "joined from τὸ Α to τὸ Β" and not "joined from A to B".

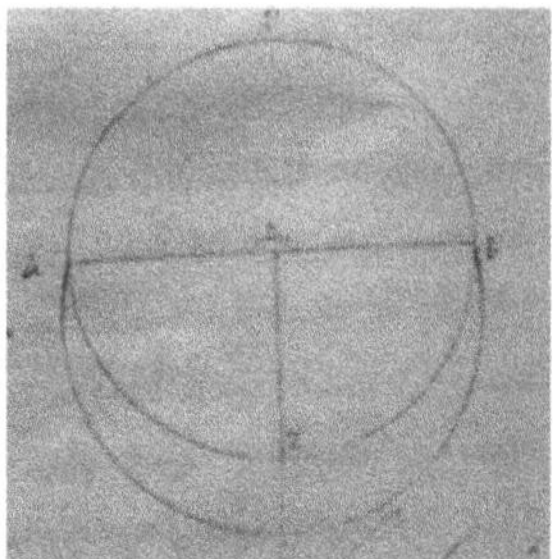

Bodl. Dorv. 301, f. 44v
a highly oversymmetric diagram[11]

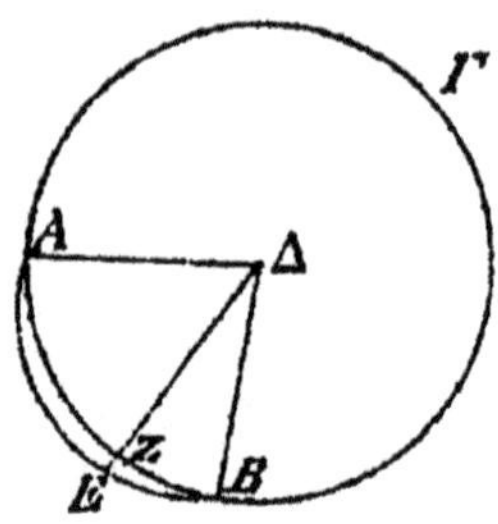

August-Heiberg's diagram for *El.* III.2
duly de-oversymmetrized[12]

My choices of translation will be fully justified below and in Sects. 2.2 and 3.1; for obvious reasons, here as elsewhere I translate every lexical item contained in the Greek text, even if some particles are more naturally rendered with punctuation (weak boundary markers like δέ), with paragraphs (the scope particles), or with graphic emphasis like italics (δή, περ). I shall normally translate identical Greek syntagms with identical English syntagms, regardless of their meaning and of their syntactic function, and different Greek syntagms with different English syntagms. Unfortunately, there is no real alternative to translating both the boundary marker δέ and the conjunction καί with "and".[13] Therefore, it will not be possible in every case to recover the original from my translations; in principle, however, "and" preceded by a comma translates καί, preceded by a semicolon or by no punctuation sign it translates δέ.

Before going on, an obvious yet capital fact must be highlighted, as well as its consequences in all three codes: there is no temporality in a mathematical argument, nor is there a narrative structure. As a consequence, the temporal arrow is replaced by other discursive arrows, which, depending on the piece of mathematics involved, may be termed "constructive", "deductive", or "operational" arrow. These arrows are expressed by various lexical items: these are the invariant parts of speech called "particles", which act as subordinants and coordinants; the verb forms, which operate on the mathematical objects; and the mode of reference to these objects.

Within the demonstrative code, the sentences that formulate constructive acts are conjoined, a kind of coordination that in principle is logically and time-symmetric: the ordering in the sequence of constructive acts points to the fact that any act within the sequence produces an object that is apt to be operated upon by the subsequent constructive act: the temporal arrow is replaced by a constructive arrow. Likewise, within the procedural and the algorithmic codes, the sentences that formulate the atomic operational steps are conjoined:[14] the ordering in the sequence of steps points to the fact that an operation produces an object which the subsequent operation accepts as input. In

[11] For the phenomenon of "oversymmetrization" of diagrams in Greek manuscripts see Sect. 3.5. In a nutshell: in medieval manuscripts, generic triangles are frequently represented as isosceles, a point taken at random on a segment actually bisects it, parallelograms are represented as rectangles, etc. In the case of III.2, all manuscripts make straight line ἡ ΔΖΕ bisect both arc ἡ ΑΒ and the straight line joined from τὸ Α to τὸ Β (which is unnecessarily represented as an arc of a circumference), place letter Γ exactly opposite to Z, and make the whole diagram left-right symmetric. The main diagram of manuscript **B** (see figure above) even makes radii ἡ ΔΑ and ἡ ΔΒ belong to the same diameter.

[12] As a matter of fact, Heiberg simply passed the clichés of the diagrams in August's edition *in usum tironum* (1826–29) on to his typesetters. It is easy to check that all figures are identical. A further confirmation of this comes from the very diagram of III.2: in both editions the letter Γ is omitted (see *EOO* I, 171). I have added this letter, suitably "aged".

[13] An exception to the other arm of the one-to-one correspondence is γάρ, translated "in fact" or "for".

[14] Procedures also exhibit a subordinative structure, which I shall discuss in Sect. 1.2, point (B).

these two codes (see Sects. 1.2–3), the operational arrow is mainly secured by the mode of reference to the objects, either in the form of "objectual overlap" (in procedures: a named output of a step is also the input of the subsequent step) or of a nearest-neighbour "referential range" (in algorithms: a named output of a step is referred to by a demonstrative or relative pronoun at the beginning of the subsequent step). Again within the demonstrative code, the proof is framed as an argumental coordination, in which the deductive arrow is secured by the parts of speech called "connectors".

Modern mathematics, in which mathematical discourse is mainly reduced to proofs, lives in an eternal present. Greek mathematics, instead, fully availed itself of a remarkable feature of the Greek verb system: the so-called "aspect" or semantic value, which in a sense describes the internal structure (or the lack of internal structure) of the action expressed by a verb, independently of the temporal relationships such an action bears with other actions or with the moment of the utterance.[15] The aspectual connotations are carried by the tense stems (present, aorist, perfect, future, future perfect); they are organized in a system of polar oppositions, with the proviso that an assigned polar opposition can be undetectable in specific linguistic patterns: in this case, the opposition is said to be "neutralized".[16] Let me clarify this; I shall also add an example that is important in our perspective. The aspectual values of the first three verb stems mentioned above can be characterized as follows: the present stem signifies that a state of affairs is being carried out and is thereby non-completed (imperfective or progressive value); the aorist stem signifies that a state of affairs is completed (confective or punctual value); the perfect stem signifies both that a state of affairs is completed and that a state exists as a result (stative-confective value).[17] In particular, the aspectual values are exhibited by such moods as do not carry a temporal value, that is, by all moods with the exception of the indicative (in our perspective, the most relevant non-indicative mood is the imperative). Clearly, the three characterizations above can be organized as a hierarchy of oppositions: perfect vs. {present,aorist}, according to the presence / absence of the stative value; aorist vs. present, according to the presence / absence of the progressive value. Each polarity has a marked term and an unmarked term, according to its presenting or not the value that defines the opposition; the unmarked term may also be "indifferent" to the opposition, if the opposition is irrelevant to, or not expressed by, the context. The neat result of all of this is a hierarchy in which the aorist stem denotes a state of affairs as an undivided whole, and in fact it is the only stem that is apt to express an operation or a constructive act "in itself".[18] Now, in specific linguistic patterns, a given opposition can, as said, be "neutralized". A major, and quite natural, source of neutralization is whenever a verb is defective, that is, some of its tense stems do not occur; in this case, the occurring stem(s) may carry the aspectual value of the non-occurring stems. Two defective verbs are central to the demonstrative code: one of them is of course εἰμί "to be", the other is κεῖμαι "to be set", mainly in the prefixed form ἔκκειμαι "to be set out". Thanks to neutralization, these present-stem verbs are apt to carry, while of course occurring only in the present stem, a punctual or a stative-confective value. In particular, these verbs may occur in linguistic contexts in which other stems predominate, and in these contexts they carry the same aspectual value as the other verb stems.

[15] The best introductions to the Greek verb system are Duhoux 2000 and Rijksbaron 2006; see also Rijksbaron 1989. For the temporal values of the Greek verb see also Ruijgh 1985, 1991, 1995, and 2000, the latter with a valuable discussion of the ancient grammarians' opinions on temporality insofar as this is expressed in the verb system. Thus, if temporality is evacuated from any piece of Greek mathematical discourse, aspectuality is not: Greek mathematics is an ideal segment of Greek prose where to study aspectual features in a pure form.

[16] The term "neutralization" originates in phonology. A very good summary of neutralization phenomena in the Greek verb system is in Duhoux 2000, 58–64, but see also Ruipérez 1982, 26–35.

[17] I have quoted here, without marks for simplicity's sake, from Rijksbaron 2006, 1.

[18] See the clear exposition in Ruipérez 1982, sects. III and IV, and below in this Section.

The last remark in the previous paragraph is crucial. For it is a fact, on which we shall repeatedly return, that each of the three codes, or each of the specific parts of a proposition within the demonstrative code, carries out its mathematical operations within a well-defined "aspectual environment". As for a proposition, the enunciation and the conclusion are irrelevant, insofar as they are deductively structureless. The aspectual environment proper to the setting-out and to the construction (in which states of affairs are all formulated in the imperative mood) is the one of the perfect stem: the present stem verbs that figure in these two specific parts, most notably εἰμί "to be" and ἔκκειμαι "to be set out" mentioned above, carry a stative value, and as such cannot be copular verbs.[19] The aspectual environment proper to the proof is the one of the present stem: very much like modern mathematics, Greek proofs live in an eternal present. Constructive acts in the perfect stem as the one we have read above in the *proof* of III.2 may only occur within the "anaphora"; their presence is justified by the fundamental principle of citation by conformity to a template I shall discuss in Sect. 2.[20] Conversely, we must expect that, within the proof, the indicative mood is only represented by the present stem or by the perfect stem; preterites such as the imperfect or the aorist indicative are only allowed to occur in the metadiscourse (see Sect. 4.5.2).

I now describe the main stylistic markers of the demonstrative code according to the quadripartite typology presented at the beginning of Sect. 1; I shall also frequently add references to other, relevant Sections of this book.

A) *Discursive arrow*

The cohesion of a demonstrative argument is secured by coordinants and connectors.[21] The abundance of connective particles in the demonstrative code reflects a general feature of Greek language;[22] however, mathematical style operates a sharp selection among particles,[23] and eliminates non-logical connotations of the selected items. Particles may be arranged in three, nested, logico-syntactic levels, according to the range of their scope: supra-argumental, argumental, and sentential. Some particles, like ἄρα "therefore", δή "thus", and καί (adverbial "also" vs. conjunctive "and") may belong to different levels, but any assigned occurrence of them always has one single function: no ambiguities occur. The levels may superpose to one another; the associated particles are in this case paired: their co-occurrence is routinized. As we shall repeatedly see, particles in a specific position within a clause also have the function of linguistic markers; in this function, however, they can be replaced by items that belong to other lexical types, like the verb forms.

---

[19] The last remark will be further developed in Sect. 3.1.

[20] As is well known, the perfect stem refers to the *present* time of the agent (as is witnessed to by its primary endings), while stressing that the present-time condition is the accomplished result of some past history. See for instance the discussion in Duhoux 2000, 421–426.

[21] In using "connector" instead of "connective" I endorse the lexical choice and the underlying justification in Barnes 2007, 173. I shall normally drop the distinction between connectors and connective particles.

[22] The literature on Greek particles is huge. The standard overview of the use of particles in classical writers is Denniston 1954; in the Hellenistic period, the reference is Blomqvist 1969. Among the relevant titles in our perspective, see Bakker 1988 for a comprehensive theoretical framework, and Bakker 1993 on particles (in particular δέ) as boundary markers; Sicking, van Ophuijsen 1993 for particles in Lysias and Plato (for the latter, see also des Places 1929); De Jong 1997 and Bakker 2009b for γάρ, and in general the studies edited in Rijksbaron 1997a. The statistical survey in Duhoux 1997 refutes the belief that Plato employs so many particles because his dialogues try to mimic some features of spoken Greek. See also Hellwig 1974 for particles in New testament Greek.

[23] A particle erased from the standard lexicon is οὐκοῦν "and then", frequently used by Plato (Des Places 1929, sect. VI), by Aristotle with a strong logical connotation, and as a fossil in Euclid's optical treatises (see Sect. 5.3.6). Discussing possible reasons why some particles and not others were chosen will only end in conjecture.

- The first level comprises the scope particles (cf. Sect. 5.3.6).[24] These have range over complexes of deductions and mainly mark a specific part of a proposition or a fresh start of an argument after a hiatus. In III.2, these particles are γάρ "in fact", which introduces the construction,[25] οὖν "then", which opens the proof, ἄρα "therefore", which introduces the general conclusion. There are two major exceptions to the rule that a scope particle is required at every "paragraph". First, the imperative ἔστω "let there be" with "presential" value that initializes the setting-out and in fact the whole proposition (Sects. 3.1 and 4.2) is a stylistic trait and therefore also serves as a marker of a specific part of a proposition; for this reason, it is not followed by any scope particle. Second, λέγω "I claim" and δείξομεν "we shall prove" are personal forms and hence they are stylistic traits too: this entails that they also have the function of markers of specific parts of a proposition, namely, the determination and a "potential" fully-fledged proof (Sects. 4.2.1 and 4.5.2). Such markers, as well as the impersonal verb form δεῖ "it is required" that opens the determination of a problem, solve a non-trivial logico-syntactic problem: to place an "assertion sign" where no conclusion can be drawn, so that ἄρα "therefore" cannot be used. Some such markers are accompanied by δή, which in this case is simply emphatic: "very".
- The second level comprises the inferential particles; these are characteristic of the proof. They mark the assumption–coassumption–conclusion structure of any self-contained atomic inference in a proof: coassumptive particles are δέ "and", ἀλλά "but", emphatic καί "also"; conclusive particles are ἄρα "therefore", consecutive ὥστε "so that", δή "thus"—the latter marks the statement of a fact that is obvious from what precedes (Sect. 5.3.6). Explicative γάρ "for" introduces postposed explanations (Sect. 4.5.3).
- The third level comprises coordinants and subordinants, which form sentences formulating non-simple statements from sentences formulating simple statements: these are conjunctive καί "and", dis-junctive ἤ(τοι) "either" / "or", conjunctive [μέν …] δέ …, conditional εἰ "if" and ἐάν "if", "paraconditional" ἐπεί "since" (Sects. 5.3.4–5; 4.1, 5.3.1; 4.4, 5.3.2). The particle ἄρα "therefore" is also included in this category, for it may introduce the consequent of a paraconditional clause in the "anaphora" ("apodotic" ἄρα; Sects. 4.4 and 5.3.2). Such particles may figure in (almost) any specific parts.

This tripartite structure is summarized in the following two tables. The first table sets out the sequence of particles extracted from III.2:

enunciation	ἐάν	if
setting-out	καί	and
determination	ὅτι	that
construction	γάρ, ἀλλ', εἰ, καί, καί, καί, καί	in fact, but, if, and, and, and, and
proof	ἐπεὶ οὖν … ἄρα καί, καὶ ἐπεί … ἄρα, δέ, ἄρα, δέ, ἄρα, δέ, ἄρα, οὐκ ἄρα, δή, ὅτι, οὐδέ, ἄρα	then since … therefore … also, and since … therefore, and, therefore, and, therefore, and, therefore, therefore it is not the case that, very, that, not … either, therefore
conclusion	ἐὰν ἄρα	therefore if

[24] "Liminal" means here "according to Wackernagel's law": such postpositive particles having sentence-range scope, enclitic in origin, occupy the second position in a given colon; see Ruijgh 1990.
[25] This γάρ also has the function of introducing one of the standard linguistic complexes that initialize a reduction to the impossible (see Sect. 5.2.1).

The second table exhibits the particles that occur in the proof of III.2; they are arranged according to their level as described above (I exclude ὅτι):

scope	οὖν											
infer.	καί	καί	δέ	ἄρα	δέ	ἄρα	δέ	ἄρα	οὐκ ἄρα			ἄρα
conn.	ἐπεί ... ἄρα	ἐπεί ... ἄρα									οὐδέ	
emph.										δή		

B) *Verb forms*

The verb structure of a proposition is rich and variegated. The stems (and the associated "aspectual environment") and the moods are characteristically distributed among the specific parts.[26]

Present stem. In the indicative, the present stem is ubiquitous in the proof, mainly as forms of the verb "to be". This verb simply marks each step of a logically ordered sequence of states of affairs that follow from what is supposed in the setting-out or in the construction. The present indicative also figures in declarative enunciations, in the consequent of enunciations in conditional form (Sect. 4.1), and in the antecedent of the conditional that, within a proof, initializes a reduction to the impossible (Sect. 5.2.1). The future indicative sometimes replaces the present indicative in the consequent of conditional clauses,[27] as it does in the enunciation of III.2, but this variant is not systematic. The present subjunctive is used in the antecedent of an enunciation in conditional form, when the verb "to be" is required (Sects. 4.1 and 5.3.1). The present imperative is mainly represented by forms of the stative-presential verbs "to be" and "to be set out": these are ἔστω(σαν) "let there be", which opens the setting-out, and ἐκκείσθω "let it be set out",[28] which figures in the construction (Sect. 4.2).

Aorist stem. The presence of the aorist subjunctive in a conditional enunciation is mandatory for verbs different from "to be" (see Sects. 4.1 and 5.3.1). Since in this case it expresses operations undergone by the mathematical objects, the aorist simply intimates absence of temporal or aspectual connotations.[29] Finite forms of the aorist are virtually absent in the other specific parts of a proposition, with the exception of particular verbs, as for instance ἐμπίπτω "to fall".[30] Participial

[26] As for the voice, middle-passive voice verbs with object do not have a reflexive connotation, but intimate that the object that undergoes the action is of special interest to the subject. A good mathematical example is provided by the verb ἐφάπτομαι "to be tangent", used in the middle voice because a straight line is tangent to "its own" circle (the active voice has a causative value). On the middle voice in ancient Greek, see Bakker 1994, Allan 2003 (which I will follow in using "voice" instead of "diathesis": cf. his 16 n. 19). For the ancient grammarians on the middle voice, see Rijksbaron 1986 and the texts and commentaries in the excellent editions—with excellent indices—Lallot 1998 (Dionysius Thrax' *Ars grammatica*) and Lallot 1997 (Apollonius Dyscolus' *Syntax*).

[27] The future emphasizes the idea that the condition stated in the consequent of a conditional necessarily follows from what is stated in the antecedent.

[28] The verb ἔκκειμαι "to be set out" serves as the passive of ἐκτίθημι "to set out" (see Sect. 4.2).

[29] All occurrences of the aorist stem may carry a zero-grade aspectual connotation since this stem is the less connotated pole of the system of aspectual oppositions (Ruipérez 1982, sect. IV; cf. Humbert 1960, 133–181 passim). The literature on aspect is huge; the modern seminal study is Ruipérez 1982 (Spanish original 1954). On aspectual choice, see first and foremost Duhoux 1995, and also Brunel 1939, Amigues 1977, and Sicking, Stork 1996, Part I; for aspectual differences between the present and aorist imperatives, see Bakker 1966 and Sicking 1991; on aspect and middle-passive voice, see Bakker 1994; on the aspectual usage of infinitive, Stork 1982. Aspect in Plato is studied in Jacquinod 2000; in New Testament Greek, it is studied in Fanning 1990, and see also the approach in Pang 2016 (mainly interesting as a very mild specimen of the present-day formalistic drift of linguistics).

[30] Six occurrences, in I.44–46, II.10, VI.3 (*bis*), all in citations of I.post.5. This is a (bewildering) variant, for a perfect stem ἐμπέπτωκεν is to be expected, as is confirmed by all other citations of the same postulate (12 occurrences). The occurrence of ἔπεσεν in proposition III.13, in another context, has an obvious metamathematical connotation.

forms are instead fairly frequent, in particular when they identify the objects that are "given" (Sects. 2.4.1 and 5.1.5), for what is of interest is the state of affairs that these objects are given, no matter how and when: given objects are not "given" as the result of some construction. The enunciation of a problem of construction is characterized by an active or middle aorist infinitive with directive connotation and no temporal or aspectual value (again, the pure action signified by the verb is intended).[31] In the ("instantiated", that is, carrying denotative letters) conclusion of a problem and in all citations of this conclusion, the aorist that figures in the enunciation is transformed into a passive perfect indicative (see Sect. 4.3).[32]

Perfect stem.[33] This stem characterizes the demonstrative code. It is used in three different contexts, almost always in the passive voice and with terminative verbs.[34] First, a passive perfect indicative figures in the (instantiated) conclusions of problems and in their citations within the proof, like the citation of I.16 in the "anaphora" of III.2 (see Sects. 2.3, 4.3, and 4.5.1). The perfect stem has in this case a standard confective-stative value: "such-and-such an object has been *ed in the deductive / operational past and it is in the state of being *ed since then"; translation "it turns out to be *ed" (cf. Sect. 4.3). Second and third, in the setting-out and in the construction, a passive perfect imperative predominates (translation "let it be *ed"), with the sole exceptions, still in the imperative, of presential "to be" and of some verbs (for instance πίπτω "to fall" and ἐφάπτομαι "to be tangent") whose traditional form is the *praesens pro perfecto*.[35] In the Greek verb system there is nothing more impersonal than a passive perfect imperative: the agent is unnaturally attached to an imperative; the aspectual value of the perfect stem presents any constructive act as accomplished and "ready for use"; the passive voice indicates that the action is undergone by the mathematical object. The mathematician disappears in this way behind the propositions.[36]

---

[31] On imperatival infinitive see most recently Allan 2010, whose analysis fits mathematical practice particularly well—but on the other hand, his complementary characterization of non-infinitival imperative as "deriv[ing] its directive force directly from the speaker's will" (205), fits it badly. See Bakker 1966 and Amigues 1977 for the aspectual choice of imperative. Imperatival infinitives are frequently met in inscriptions that report decrees or laws (Moreschini Quattordio 1970–71), in treaties (Neuberger-Donath 1980 for the Homeric examples), and in medical recipes (almost everywhere in suitable treatises of the Hippocratic corpus). All *topoi* in Aristotle's *Topica* are enunciated with an imperatival infinitive (I owe this remark to B. Wilck). The imperatival infinitive does not exist in English, contrary to some romance languages as French or Italian (in Italian, it is mandatory for negative prescriptions in the singular); therefore, standard translations like "to construct such-and-such a triangle" are ungrammatical. For this reason, one must translate with an imperative: "construct such-and-such a triangle".

[32] Thus, in III.2, the "same" operation of taking a point is formulated in the enunciation with a passive aorist subjunctive (ληφθῇ), in the construction with a passive perfect imperative (εἰλήφθω). The aorist infinitive is otherwise seldom used; a small sample is within διὰ τό + infinitive clauses, to be generally regarded as spurious (Sect. 4.5.3).

[33] The classical study of the perfect stem is Chantraine 1926, whose theory on the evolution of the so-called "resultative" perfect has been first refuted in McKay 1965 and 1980; see also Rijksbaron 1984, Sicking, Stork 1996, Part II, and Duhoux 2000, 426–431. On the explosion of transitive active perfect in the 5th century BCE, see Willi 2003, 126–133. My translations of perfect stems are sometimes (and purposely) over-emphatic.

[34] Terminative verbs express actions that have an inherent end point; constructive acts quite naturally have an end point. In the mathematical lexicon, there are some fossil perfect stems of intransitive verbs with pure connotation of state. These are βέβηκα "to happen to stand", said of an angle upon an arc (III.def.9), ἀντιπέπονθα "to happen to be in inverse relation", said of magnitudes (VI.def.2), γεγονέτω "let it happen to come to be", which initializes an analysis (see Sect. 2.4.1; but in the *Elements* this verb form applies to establishing a proportion or to finding a fourth proportional), κέκλιμαι "to happen to be inclined" (a middle voice), said of a straight line to a plane (XI.def.7).

[35] But these verbs present geometric states of affairs that are not governed by constructive acts licensed by a postulate or by a problem (see Sect. 4.3): a straight line "falls on" a straight line and it is not "made to fall" on it. Note also that πίπτω is an *activum tantum* and that, as seen just above, ἐφάπτομαι is only used in the middle voice.

[36] This stylistic choice might be read as an implicit reply to Plato's criticism in *Rsp.* VII, 527A6, addressed to the mathematicians who λέγουσι μέν που μάλα γελοίως τε καὶ ἀναγκαίως "speak in a most ridiculous yet necessary way" (the reference is to the active-voice formulation of constructive acts such as "squaring", "applying", "adding"). Recall that personal forms in formulae such as λέγω ὅτι and ὁμοίως δὴ δείξομεν are mere stylistic traits.

The following table summarizes the distribution of verb forms among the specific parts of a proposition; an asterisk marks passive voices (marginal occurrences are in parentheses):

mood/stem	present	future	aorist	perfect
indicative	enunciation proof	enunciation proof	(setting-out) (proof)	proof*
subjunctive	proof		enunciation*	
imperative	setting-out		enunciation (construction*)	setting-out* construction*
participle			enunciation* proof*	

C) *Reference to objects*

References to objects are operated by means of designations in forms of noun phrases or by means of prepositional expressions that include letters; these may simply reduce to complexes of letters (Sect. 2.2). The presence of the article in such expressions gives rise to subtle issues (Sects. 3.2–3). The referential range of the designations by "denotative letters" is the single mathematical proposition.[37] As no premise is stated twice in a chain of mathematical arguments (this is the "deductive overlap": the conclusion of an argument directly serves as the first premise of the subsequent argument), there is no objectual overlap between consecutive inferential steps. Let us also anticipate here some interesting features of III.2, which we shall discuss in Sect. 3.2:

- the circle ὁ ΑΒΓ is designated by three letters but the letter Γ is never assigned to any point;
- the "real" straight line ἡ ΑΒ is never joined: for this reason, its fictitious avatars are designated by the periphrasis ἡ ἀπὸ τοῦ Α ἐπὶ τὸ Β ἐπιζευγνυμένη εὐθεῖα (setting-out and proof) or by ἡ ΑΕΒ (construction and proof), and not by ἡ ΑΒ;
- that the letters E and Z are names of points becomes clear only within the proof, even if these letters exhibit their first occurrence in the construction.

D) *Generality*

Full generality of a proposition is warranted by the indefinite structure we shall discuss in detail in Sect. 3.3. Summarizing the results of that discussion, no particular object is ever mentioned in a mathematical proposition: the *enunciation* and the *conclusion* are fully general statements—even if use is seldom made of quantification (see Sect. 5.1.1)—because the geometric species are there designated, in their first occurrence, by indefinite noun phrases; the *setting-out* and the *construction* first "present", then refer to mathematical objects only by means of indefinite noun phrases or by lettered designations, which are in fact the names of (the primary, indefinite designations of) these objects; in the *proof*, lettered designations that are neutral as to the opposition definite / indefinite predominate; references to previous results are formulated so as to secure full generality. The denotative letters do not carry any particularizing value.

[37] The referential range of a designation is the longest segment of text in which the referent is unique.

I end this Section with a summary of the markers that allow an exact identification of the specific parts of a proposition. These markers are the presence or absence of denotative letters, the type and token of the linguistic item that opens a specific part, the verb moods. As seen above—and despite the existence of well-defined aspectual environments—the verb stems are not sufficiently polarized to act as markers of the specific parts of a proposition. As seen above again, the boundary between the construction and the proof may not be so sharply defined: deductive steps may be inserted in the middle of a construction, so as to pinpoint features of the arrived at configuration that make the subsequent constructive act possible; inversely, constructive acts may be inserted in the middle of a proof, if this is required by the argument.

The markers of the specific parts of a proposition are as follows:

enunciation	absence of denotative letters
setting-out	introduced either by a scope particle, usually γάρ, or by incipitary ἔστω; verbs in the imperative; denotative letters are first assigned
determination	introduced by λέγω ὅτι (theorems) or δεῖ δή (problems); verbs in the indicative
construction	introduced by a scope particle, usually γάρ; verbs in the imperative
proof	introduced by a scope particle, usually οὖν or γάρ; verbs in the indicative; it begins with a series of "paraconditional" clauses
conclusion	introduced by ἄρα serving as a scope particle; word-for-word repetition of the enunciation in the case of theorems; in the case of problems, the denotative letters normally remain ("instantiated" conclusion) and the verb is transformed into a perfect indicative; the conclusion is frequently absent

In general, a stylistic or linguistic marker is any linguistic item that regularly occurs in a self-contained text type and that singles out such a text as a type among other, possibly nearby, self-contained, and related pieces of discourse. Markers can serve their function as types or as tokens within a well-defined type. For instance, scope particles are a marker type that can single out the text type "specific part of a proposition", whereas the particles γάρ, οὖν, and ἄρα are tokens of the marker type "scope particle" that single out—within the text type "specific part of a proposition"—the text tokens "construction", "proof", and "conclusion", respectively.

As seen throughout this Section, and as summarized in the above table, (1) different marker types may co-occur—and in fact usually co-occur—thereby contributing to single out one and the same text token; (2) it is not said that tokens of an assigned marker type can single out every text token (the "enunciation" and the "determination" are not—and in fact cannot be—singled out by any scope particle); (3) tokens that belong to different marker types may be used in alternative to mark the same text token in exactly the same position: in our perspective, the most important example of this phenomenon is the mutually exclusive use of the liminal scope particle γάρ and of the liminal verb form ἔστω as markers of the setting-out (but "liminal" has different meanings!).

Time and again in this book, I shall also pinpoint stylistic markers of Books of the *Elements*, or of Archimedes' style as a whole; in general, these are marker tokens that, because of their frequency, single out a self-contained piece of discourse. For instance, the inferential particle ὥστε "so that" is a stylistic marker of Book X because its frequency in this Book is appreciably higher than in the other Books of the *Elements*. Linguistic items are "negative" markers if they serve as markers because of their low frequency or of their absence. For instance, the specific part "conclusion" is a negative marker of the arithmetic Books of the *Elements*.

1.2. THE PROCEDURAL CODE

Short description and example

The procedural code formulates its prescriptions as a sequence of coordinated principal clauses with the verb in the imperative or in the first person plural, in the present or future stem; one or more participial clauses coordinated with one another are subordinated to each principal clause; the participle is a satellite and performs the function of modifier of the operating subject. There necessarily are, moreover, an initializing clause, which feeds the initial input into the procedure, and an end clause, which identifies the result of the chain of operations as the quantity to be calculated; this quantity is usually declared in a clause that precedes the entire procedure. This code is used to formulate operatory prescriptions in the most general way; the verb forms—either finite or participial—represent the operations, and each verb form corresponds to one and only one operation; the involved mathematical objects—the "operands"—are the complements of the verb forms and are identified by (sometimes extremely long) definite descriptions. The operations may be unary or binary, that is, they may accept one or two operands as input.

As for sources, one single example of a procedure can be found in the *Elements*, still an important one: it is the technical core of the so-called "method of exhaustion" (see Sect. 4.5.2). We find procedures in Hero's *Metrica* and in the geometric metrological corpus.[38] Concerning number theory, Diophantus once but crucially sets out a procedure (see just below); Nicomachus formulates in this way the most general prescriptions in his *Introductio arithmetica*; he stresses the gain in generality that results from adopting this format.[39] Procedures prominently figure in the astronomical corpus; they expound how to use numeric tables for computing quantities relevant to astronomy. Thus, we frequently find procedures in Ptolemy, both in the *Almagest*[40] and in the instruction manual to the *Handy Tables* (see below), in Pappus' and Theon's commentaries thereon, in the anonymous *Prolegomena to the Almagest*—a late antiquity primer to the elementary arithmetic operations in the sexagesimal system, which has been transmitted as a preliminary to the *Almagest*—in Stephanus' commentary on the *Handy Tables*, and in all similar Byzantine primers and commentaries. In the Byzantine texts, procedures precede paradigmatic examples presented in algorithmic form (Sect. 1.3), and are intended to validate them (see Sect. 2).

The best example of a procedure I know of is in Diophantus' treatise *On polygonal numbers*. Diophantus starts with a definition of polygonal numbers that is formulated as a relation between the side and the multiplicity of the angles of any such number:[41]

---

[38] The prescriptions of medical recipes are often formulated in procedural form; particularly well-developed are the procedures in the Hippocratic *Internal affections*.

[39] See for instance *Ar.* II.1.2 and II.2.4; Nicomachus calls procedures μέθοδοι or ἔφοδοι.

[40] See *Alm.* II.9, III.8 (to be read below), III.9, V.9, V.19, VI.9–10, XI.12, XIII.6.

[41] Acerbi 2011e, 197.2–4. There are infinitely many instances of polygonal numbers of any assigned species (pentagonal, hexagonal, etc.): the multiplicity of the angles identifies the species of polygonal (and it figures in its name: <u>penta</u>gonal, <u>hexa</u>gonal, etc.), the side identifies a specific polygonal number within the sequence of numbers that belong to a species (for instance, 5, 12, 22, 35 are pentagonal numbers of sides 2, 3, 4, 5, respectively). With the exception of the unit (which is "potentially" a polygonal number of any species) and of the dyad, any number is polygonal for some species and side (I also include triangular and square numbers, which strictly speaking are not polygonal: see my parenthetical remark just below). As a consequence, polygonal numbers cannot be defined by means of a standard *genus cum differentia* statement. On why Diophantus' definition "hides" the square on the right-hand side of the expression to be read in the next page, see Acerbi 2011e, 44. Polygonal numbers are treated at length by Nicomachus, *Ar.* II.6–12.

πᾶς πολύγωνον πολλαπλασιασθεὶς ἐπὶ τὸν ὀκταπλασίονα τοῦ δυάδι ἐλάσσονος τοῦ πλήθους τῶν γωνιῶν καὶ προσλαβὼν τὸν ἀπὸ τοῦ τετράδι ἐλάσσονος τοῦ πλήθους τῶν γωνιῶν ποιεῖ τετράγωνον.

Every polygonal ‹number› multiplied by the octuple of the ‹number› less by a dyad than the multiplicity of the angles and taking in addition the ‹number› less by a tetrad than the multiplicity of the angles makes a square.

If we develop the "square" according to Diophantus' argument, the relation that defines a polygonal number P in terms of its side s and of the multiplicity of its angles v can be written as follows (note the second addendum in the left-hand side of the equality):

$$8P(v-2)+(v-4)^2=[2+(v-2)(2s-1)]^2$$

Now, Diophantus explains how to find, once v is fixed, a polygonal number P whose side s is given, and vice versa. The description of what he himself calls μέθοδος "procedure" is as follows:[42]

λαβόντες γὰρ τὴν πλευρὰν τοῦ πολυγώνου ἀεὶ διπλασιάσαντες ἀφελοῦμεν μονάδα, καὶ τὸν λοιπὸν πολλαπλασιάσαντες ἐπὶ τὸν δυάδι ἐλάσσονα τοῦ πλήθους τῶν γωνιῶν τῷ γενομένῳ προσθήσομεν ἀεὶ δυάδα, καὶ λαβόντες τὸν ἀπὸ τοῦ γενομένου τετράγωνον ἀφελοῦμεν ἀπ' αὐτοῦ τὸν ἀπὸ τοῦ τετράδι ἐλάσσονος τοῦ πλήθους τῶν γωνιῶν, καὶ τὸν λοιπὸν μερίσαντες εἰς τὸν ὀκταπλασίονα τοῦ δυάδι ἐλάσσονος τοῦ πλήθους τῶν γωνιῶν εὑρήσομεν τὸν ζητούμενον πολύγωνον.
πάλιν δὲ αὐτοῦ τοῦ πολυγώνου δοθέντος εὑρήσομεν οὕτως τὴν πλευράν· πολλαπλασιάσαντες γὰρ αὐτὸν ἐπὶ τὸν ὀκταπλασίονα τοῦ δυάδι ἐλάσσονος τοῦ πλήθους τῶν γωνιῶν καὶ τῷ γενομένῳ προσθέντες τὸν ἀπὸ τοῦ τετράδι ἐλάσσονος τοῦ πλήθους τῶν γωνιῶν τετράγωνον εὑρήσομεν τετράγωνον – ἐάνπερ ᾖ ὁ ἐπιταχθεὶς πολύγωνος – τούτου δὲ τοῦ τετραγώνου ἀπὸ τῆς πλευρᾶς ἀφελόντες ἀεὶ δυάδα τὸν λοιπὸν μερίσομεν ἐπὶ τὸν δυάδι ἐλάσσονα τοῦ πλήθους τῶν γωνιῶν, καὶ τῷ γενομένῳ προσθέντες μονάδα καὶ τοῦ γενομένου λαβόντες τὸ ἥμισυ ἕξομεν τὴν τοῦ ζητουμένου πολυγώνου πλευράν.

In fact, taking the side of the polygonal always doubling we shall subtract a unit, and multiplying the remainder by the ‹number› less by a dyad than the multiplicity of the angles we shall always add a dyad to the result, and taking the square on the result we shall subtract from it the ‹square› on the ‹number› less by a tetrad than the multiplicity of the angles, and dividing the remainder by the octuple of the ‹number› less by a dyad than the multiplicity of the angles we shall find the sought polygonal.
And again, the polygonal itself being given, we shall find the side as follows: multiplying it by the octuple of the ‹number› less by a dyad than the multiplicity of the angles and adding to the result the square on the ‹number› less by a tetrad than the multiplicity of the angles we shall find a square—whenever the assigned ‹number› be really polygonal—and always subtracting from the side of this square a dyad we shall divide the remainder by the ‹number› less by a dyad than the multiplicity of the angles, and adding to the result a unit and taking half of the result we shall have the side of the sought polygonal.

Let us see the main features of this extraordinary piece of mathematics.

[42] Acerbi 2011e, 197.18–30. Transitions from one principal clause to the subsequent principal clause take place where an algebraic transcription "puts brackets". Diophantus introduces the procedure, contrasting it with the just preceding "validation" by means of a proof framed in the "language of the givens" (see Sect. 2.4.2), as follows: διδασκαλικώτερον δὲ ὑποδείξομεν καὶ τοῖς βουλομένοις εὐχερῶς ἀκούειν τὰ ζητούμενα διὰ μεθόδων "we shall provide a more instructional description, for those who want easily to learn what is sought by means of procedures" (Acerbi 2011e, 197.16–17).

A) *Discursive arrow*

A procedure contains two operational flows: the flow of the principal clauses and the flow of the participial subordinates. Only the principal flow includes metadiscursive pointers—they are not operators—that identify the function of the final output: these are εὑρήσομεν "we shall find" and ἕξομεν "we shall have". Conversely, only the participial flow includes a metadiscursive pointer that identifies the primary input: this is λαβόντες "taking". The several conjunctions καί "and" in a procedure may belong either to the subordinate or to the principal flow; there are no other coordinants or subordinants, unless in metadiscursive clauses. The discursive arrow is put to effect by means in two ways: subordination (an operation in a participial subordinate must precede the operation in the principal clause) and merging of the operands. In fact, there is no distinction between the output of one operation and the input of the subsequent operation; no operand is retrieved later in one and the same procedure [see also point (C) below]. For this reason, a procedure cannot admit of hiatuses; if this happens, the result is two independent procedures.[43] Merging of the operands also shows that the several καί "and" that give a procedure its structure are not simply coordinative but carry an operational arrow; this operational arrow replaces the temporal arrow. Very much as in a construction within the demonstrative code, the ordering in the sequence of procedural steps points to the fact that an operation yields an object which the subsequent operation accepts as input.

The mixed (coordination / subordination) structure of the two Diophantine procedures is made apparent in the following tables, where I have only retained the operations:

Direct procedure

λαβόντες	ἀεὶ διπλασιάσαντες	πολλαπλασιάσαντες ἐπί	λαβόντες τὸν ἀπὸ [...] τετράγωνον	μερίσαντες εἰς
ἀφελοῦμεν		προσθήσομεν ἀεί	ἀφελοῦμεν ἀπ’	εὑρήσομεν

Inverse procedure

πολλαπλασιάσαντες [...] ἐπί	προσθέντες	ἀπὸ [...] ἀφελόντες ἀεί	προσθέντες	λαβόντες τὸ ἥμισυ
εὑρήσομεν		μερίσομεν ἐπί	ἕξομεν	

Direct procedure

taking	always doubling	multiplying [...] by	taking the square on	dividing [...] by
we shall subtract		we shall always add [...] to	we shall subtract from	we shall find

Inverse procedure

multiplying [...] by	adding to	always subtracting from	adding to	taking half
we shall find		we shall divide [...] by	we shall have	

Since the two Diophantine procedures are one the inverse of the another, the operations in them should map into one another, inversely and in inverse order. And this is exactly what happens:

Direct procedure	Inverse procedure
ἀεὶ διπλασιάσαντες	λαβόντες τὸ ἥμισυ
ἀφελοῦμεν	προσθέντες (*sec.*)
πολλαπλασιάσαντες ἐπί	μερίσομεν ἐπί
προσθήσομεν ἀεί	ἀπὸ [...] ἀφελόντες ἀεί
λαβόντες τὸν ἀπὸ [...] τετράγωνον	*
ἀφελοῦμεν ἀπ’	προσθέντες (*pr.*)
μερίσαντες εἰς	πολλαπλασιάσαντες [...] ἐπί

[43] See also Ptolemy's procedure (*Alm.* III.8) at the end of this Section.

Direct procedure	Inverse procedure
always doubling	taking half
we shall subtract	adding to (*sec.*)
multiplying [...] by	we shall divide [...] by
we shall always add [...] to	always subtracting from
taking the square on	*
we shall subtract from	adding to (*pr.*)
dividing [...] by	multiplying [...] by

However, the correspondence by inversion is not syntactically rigid: only one operation of the subordinated flow in the direct procedure is transferred to the principal flow of the inverse procedure, while all operations of the principal flow of the direct procedure pass to the subordinated flow of the inverse procedure. There are border-effect constraints that justify this discrepancy. First, since both procedures must end with a metadiscursive pointer and begin with an operation in the subordinated flow, the four extremal operations cannot figure in the principal flow. Second, an operation "disappears" in the inverse procedure, namely, the operation corresponding to "taking the square on"—and rightly so: a formulation like "multiplying by ... we shall add to ..., and taking a side of the result we shall always subtract ..." runs into the difficulty that the third operation in the sequence might not be performed on any result of the former two, for not every number has a rational square root. The way out to this difficulty consists in anticipating, in the inverse procedure, the numeric species that results from the first addition: for this reason we find the metadiscursive pointer εὑρήσομεν, whose object is "a square", and the postposed antecedent ἐάνπερ ᾖ ὁ ἐπιταχθεὶς πολύγωνος "whenever the assigned ‹number› be really polygonal".

B) *Verb forms*

The verb forms are the operators of a procedure. The formulation is prescriptive or involves personal forms, which convey a directive connotation: the verb in the principal clause is in the imperative or in the first person plural, in the present or future stem.[44] The subordinated flow is made of a sequence of conjoined circumstantial participles that qualify the operating subject. The aorist stem of such participles deprives the action of any aspectual or temporal connotation: thus, an operation has no internal structure. Since an operational arrow replaces the temporal arrow, the systematic use of the aorist stem in the subordinated flow may be read in this perspective; it is not necessary—albeit legitimate—to suppose that the aorist stem is used because a string of operations in the subordinated flow cannot but be temporally prior to the operation in the principal flow to which they are subordinated.[45] In principle, to any operation is associated one typical verb form; the preposition that possibly accompanies the verb form may instead change, as εἰς vs. ἐπί governed by μερίζω "to divide" above.[46] As noted, the operation of taking a square root is included in the designation τούτου τοῦ τετραγώνου ἡ πλευρά. Other operations are embedded in designations of objects.[47] For instance, in τὸν ὀκταπλασίονα τοῦ δυάδι ἐλάσσονος τοῦ πλήθους τῶν γωνιῶν "the octuple of the

[44] Recall that the future strengthens the idea of necessity. No other stem can be used in the principal flow.

[45] To show that this reading is not necessary, just consider that a procedure may in principle entirely consist in a principal flow, possibly with the sole exception of the initializing clause.

[46] The preposition that characterizes division in a "geometric" framework is in fact παρά.

[47] Of such a kind are items 5, 8, 10, and 12 in point (C) below.

‹number› less by a dyad than the multiplicity of the angles" a dyad must first be subtracted from the multiplicity of the angles; the octuple of the result can be taken only after this subtraction. Such embedded operations might well have been made explicit by means of verb forms.

C) *Reference to objects*

The operands are designated by definite descriptions; their referential range is the single procedure. Each of the operands is mentioned just once in a procedure (this is the "merging of the operands"); objectual overlap is rigidly adhered to.[48] Let us list the objects of the two Diophantine procedures, in the order in which they are introduced. Unlike operations, the parallelism between the designations in the two procedures is rigid: the same wording is adopted in both procedures and the objects are mentioned in inverse order:

1	τὴν πλευρὰν τοῦ πολυγώνου	τὴν τοῦ ζητουμένου πολυγώνου πλευράν
2	*	τοῦ γενομένου
3	μονάδα	μονάδα
4	τὸν λοιπὸν	τῷ γενομένῳ
5	τὸν δυάδι ἐλάσσονα τοῦ πλήθους τῶν γωνιῶν	τὸν δυάδι ἐλάσσονα τοῦ πλήθους τῶν γωνιῶν
6	τῷ γενομένῳ	τὸν λοιπὸν
7	δυάδα	δυάδα
8	τοῦ γενομένου	τούτου τοῦ τετραγώνου τῆς πλευρᾶς
9	αὐτοῦ	τετράγωνον
10	τὸν ἀπὸ τοῦ τετράδι ἐλάσσονος τοῦ πλήθους τῶν γωνιῶν	τὸν ἀπὸ τοῦ τετράδι ἐλάσσονος τοῦ πλήθους τῶν γωνιῶν τετράγωνον
11	τὸν λοιπὸν	τῷ γενομένῳ
12	τὸν ὀκταπλασίονα τοῦ δυάδι ἐλάσσονος τοῦ πλήθους τῶν γωνιῶν	τὸν ὀκταπλασίονα τοῦ δυάδι ἐλάσσονος τοῦ πλήθους τῶν γωνιῶν
13	τὸν ζητούμενον πολύγωνον	αὐτὸν

1	the side of the polygonal	the side of the sought polygonal
2	*	the result
3	a unit	a unit
4	the remainder	the result
5	the ‹number› less by a dyad than the multiplicity of the angles	the ‹number› less by a dyad than the multiplicity of the angles
6	to the result	the remainder
7	a dyad	a dyad
8	the result	the side of this square
9	it	a square
10	the ‹square› on the ‹number› less by a tetrad than the multiplicity of the angles	the square on the ‹number› less by a tetrad than the multiplicity of the angles
11	the remainder	the result
12	the octuple of the ‹number› less by a dyad than the multiplicity of the angles	the octuple of the ‹number› less by a dyad than the multiplicity of the angles
13	the sought polygonal	it

[48] The presence of the anaphoric pronoun αὐτοῦ in item 9 is necessary because the noun τετράγωνον "square" to which it refers is part of the designation of an operation.

If we replace the operands with numbers, it is apparent that any of them figures just once in each procedure: the way of reference replaces syntactic links.

λαβόντες γὰρ (1) ἀεὶ διπλασιάσαντες (*) ἀφελοῦμεν (3), καὶ (4) πολλαπλασιάσαντες ἐπὶ (5) (6) προσθήσομεν ἀεὶ (7), καὶ λαβόντες τὸν ἀπὸ (8) τετράγωνον ἀφελοῦμεν ἀπ' (9) (10), καὶ (11) μερίσαντες εἰς (12) εὑρήσομεν (13).
πάλιν δὲ αὐτοῦ τοῦ πολυγώνου δοθέντος εὑρήσομεν οὕτως τὴν πλευράν· πολλαπλασιάσαντες γὰρ (13) ἐπὶ (12) καὶ (11) προσθέντες (10) εὑρήσομεν (9) – ἐάνπερ ᾖ ὁ ἐπιταχθεὶς πολύγωνος – ἀπὸ (8) ἀφελόντες ἀεὶ (7) (6) μερίσομεν ἐπὶ (5), καὶ (4) προσθέντες (3) καὶ (2) λαβόντες τὸ ἥμισυ ἕξομεν τὴν (1).

In fact, taking (1) always doubling (*) we shall subtract (3), and multiplying (4) by (5) we shall always add (6) to (7), and taking the square on (8) we shall subtract from (9) (10), and dividing (11) by (12) we shall find (13).
Again, the polygonal itself being given, we shall find the side as follows: multiplying (13) by (12) and adding to (11) (10) we shall find (9)—whenever the assigned ‹number› be really polygonal—and always subtracting from (8) (7) we shall divide (6) by (5), and adding to (4) (3) and taking half of (2) we shall have (1).

Designations of known objects alternate with designations of the result of a performed operation, such as ὁ γενόμενος "the result" (all operations, subtraction excepted), ὁ λοιπός "the remainder" (subtraction). All objects, indeterminate as to their numeric species, are immediately accepted as input in the subsequent step. The liminal input is directly assumed as an operand in the direct procedure (this is the side of the polygonal); in the inverse procedure, reference is made to the liminal input (this is the polygonal itself) by means of the anaphoric pronoun αὐτός.

Designations may become extremely cumbersome. As said, Ptolemy frequently formulates procedures: this occurs when he describes how to use the tables, both in the instruction manual to the *Handy Tables* and in the *Almagest*. Let us read the whole of *Alm*. III.8 (*POO* I.1, 257.12–258.10; Toomer's translation, modified). The operations or the metadiscursive pointers are in boldface, the objects are underlined, the connectors are in italics. Note that "entering into a table with" a numeric value is an operation.

ὁσάκις οὖν ἂν ἐθέλωμεν τὴν καθ' ἕκαστον τῶν ἐπιζητουμένων χρόνων τοῦ ἡλίου πάροδον ἐπιγιγνώσκειν, τὸν συναγόμενον ἀπὸ τῆς ἐποχῆς χρόνον μέχρι τοῦ ὑποκειμένου πρὸς τὴν ἐν Ἀλεξανδρείᾳ ὥραν **εἰσενεγκόντες εἰς τὰ τῆς ὁμαλῆς κινήσεως κανόνια** τὰς παρακειμένας τοῖς οἰκείοις ἀριθμοῖς μοίρας **ἐπισυνθήσομεν μετὰ** τῶν τῆς ἀποχῆς σξε ιε´ μοιρῶν, *καὶ* **ἀπὸ** τῶν γενομένων **ἐκβαλόντες** ὅλους κύκλους τὰς λοιπὰς **ἀφήσομεν ἀπὸ** τῶν ἐν τοῖς Διδύμοις μοιρῶν ε λ´ εἰς τὰ ἑπόμενα τῶν ζῳδίων, *καί*, ὅπου ἂν ἐκπέσῃ ὁ ἀριθμός, ἐκεῖ τὴν μέσην τοῦ ἡλίου πάροδον **εὑρήσομεν**. ἑξῆς δὲ τὸν αὐτὸν ἀριθμόν (τουτέστιν τὸν ἀπὸ τοῦ ἀπογείου μέχρι τῆς μέσης παρόδου) **εἰσενεγκόντες εἰς τὸ τῆς ἀνωμαλίας κανόνιον** τὰς παρακειμένας τῷ ἀριθμῷ μοίρας ἐν τῷ γ´ σελιδίῳ – κατὰ μὲν τὸ πρῶτον σελίδιον τοῦ ἀριθμοῦ πίπτοντος (τουτέστιν ἕως ρπ μοιρῶν ὄντος)—**ἀφελοῦμεν ἀπὸ** τῆς κατὰ τὴν μέσην πάροδον ἐποχῆς – κατὰ δὲ τὸ β´ σελίδιον τυχόντος τοῦ ἀριθμοῦ (τουτέστιν ὑπερπεσόντος ρπ μοίρας)—**προσθήσομεν** τῇ μέσῃ παρόδῳ, *καὶ* οὕτως τὸν ἀκριβῆ καὶ φαινόμενον ἥλιον **εὑρήσομεν**.

So whenever we want to know the Sun's position for any required time, **entering with** the time from epoch to the given moment reckoned with respect to the local time at Alexandria **into the table of mean motion we shall add up** the degrees corresponding to the various arguments **to** the elongation [from apogee at epoch], 265;15 degrees, *and* **subtracting** complete revolutions **from** the result **we shall count** the remainder **from** Gemini 5;30° rearwards through the signs, *and*, wherever we come to, there **we shall find** the mean position of the Sun. Next **entering with** the same number (namely, the distance from apogee to the Sun's mean position) **into the table of anomaly**—if the argument falls in the first column (that is, if it is less than 180°)—**we shall subtract** the corresponding amount in the third column **from** the mean position—but if the argument falls in the second column (that is, if it is greater than 180°)—**we shall add** ‹it› **to** the mean position, *and* in this way **we shall find** the true or apparent [position of the] Sun.

Note the final metadiscursive οὕτως "in this way" and the bifurcation Ptolemy aptly formulates with a correlative μέν … δέ …

D) *Generality*

Generality is explicit in a procedure, which applies directly to the most general class of intended objects. No denotative or numeral letters are present; therefore, definite descriptions of the intended mathematical objects must be used; as seen, this may make the formulation particularly cumbersome (but not much more cumbersome than what we get by reading aloud an algebraic expression). Generality is sometimes re-affirmed by the presence of the adverb ἀεί "always", which in the demonstrative code has the different meaning "continually" and marks iterative steps (see Sect. 4.5.2). The adverb ἀεί "always" has in a procedure the specific function of marking steps that contain numeric parameters that do not depend on the liminal input: in Diophantus' procedures above, these steps are the initial doubling, and adding or subtracting a dyad. However, not every such step is so marked: the final halving is not.

I end this Section by showing what kind of syntactic tension may arise when a piece of mathematics partly suited to a stylistic code is entirely expressed within another code. Let us read the Euclidean algorithm of reciprocal subtractions applied to numbers, as it is formulated in the enunciation of proposition VII.1 (*EOO* II, 188.13–18)—compare the analogous yet less strained enunciations of X.1–2, to be read in Sect. 4.1:

δύο ἀριθμῶν ἀνίσων ἐκκειμένων, ἀνθυφαιρουμένου δὲ ἀεὶ τοῦ ἐλάσσονος ἀπὸ τοῦ μείζονος, ἐὰν ὁ λειπόμενος μηδέποτε καταμετρῇ τὸν πρὸ ἑαυτοῦ ἕως οὗ λειφθῇ μονάς, οἱ ἐξ ἀρχῆς ἀριθμοὶ πρῶτοι πρὸς ἀλλήλους ἔσονται.	Two unequal numbers being set out, and the lesser being continually subtracted in turn from the greater, if the remaining ‹number› never measure out the ‹number› before itself until a unit has remained, the original numbers will be prime to one another.

This enunciation is a *unicum* in the *Elements*: it features two genitive absolutes, the one subordinated to the other, formulating conditions that normally would be included into the antecedent of the subsequent conditional; it contains the exceedingly rare adverb μηδέποτε "never" (cf. Sect. 1.5). It is no surprise that in VII.2 (at *EOO* II, 192.6–8), where it is required to find the GCD of two non-coprime numbers, we read a pure, even if very short, procedure formulating reciprocal subtractions. On the other hand, the reader is urged to rewrite the entire enunciation of VII.1 (possibly combining it with VII.2) as a procedure for testing relative primality and / or for finding the GCD of two numbers: complementary stylistic tensions will arise (cf. Nicomachus, *Ar.* I.13.11–13).

1.3. The algorithmic code

Short description and example

The algorithmic code processes paradigmatic examples that feature specific numeric values. After the initializing clause, the algorithms are expressed by a sequence of principal clauses coordinated by asyndeton; each clause formulates exactly one step of the algorithm and comprises a verb form in the imperative (this is the operation) and a system of one or two complements[49]—a direct and an indirect complement—in the form of demonstrative or (cor)relative pronouns or of numerals (these are the operands). The operation is often expressed by means of the preposition that introduces the indirect complement, without any verb form: "these by 3" instead of "multiply these by 3". The result of each operation is identified in a dedicated clause, with the verb in the present indicative (forms of γίγνομαι "to yield", "to result"),[50] sometimes replaced by an adjective with predicative value (mainly λοιπός "as a remainder" after a subtraction);[51] both syntactic structures are equivalent to our equality sign.[52] The oscillations between singular γίνεται and plural γίνονται in the identification of the result of each step are not significant, for this term was always written in abbreviated form in origin.[53] An end clause identifies the last output as the quantity to be computed.

This code figures prominently in Hero's *Metrica*, and exclusively in the geometric metrological corpus. In the *Metrica*, proofs that use the "language of the givens" precede paradigmatic examples of computations in algorithmic form, and are intended to validate them (see Sect. 2.4.2). In all astronomical primers mentioned in the previous Section, paradigmatic examples presented in algorithmic form are very frequent; they are systematically preceded by procedures that are intended to validate them (see again Sect. 2.4.2). In these texts, algorithms are frequently replaced—or accompanied—by tabular arrangements of the performed operations; as a matter of fact, the tabular arrangements are nothing but an evolution of the algorithms in a more perspicuous format.[54] Let us read a part of Hero, *Metr.* I.8 (Acerbi, Vitrac 2014, 174.3–7) as an example of an algorithm—this is "Hero's formula" for finding the area of a triangle once its sides are numerically given:

οἷον ἔστωσαν αἱ τοῦ τριγώνου πλευραὶ μονάδων ζ η θ.	For instance, let the sides of the triangle be of 7, 8, 9 units.
σύνθες τὰ ζ καὶ τὰ η καὶ τὰ θ· γίγνεται κδ·	**Compose** the 7 and the 8 and the 9: it yields 24;
τούτων λαβὲ τὸ ἥμισυ· γίγνεται ιβ·	take half of these: it yields 12;
ἄφελε τὰς ζ μονάδας· λοιπαὶ ε.	subtract the 7 units: 5 as a remainder.
πάλιν ἄφελε ἀπὸ τῶν ιβ τὰς η· λοιπαὶ δ.	**Again**, subtract the 8 from the 12: 4 as a remainder.
καὶ ἔτι τὰς θ· λοιπαὶ γ.	**And further** the 9: 3 as a remainder.
ποίησον τὰ ιβ ἐπὶ τὰ ε· γίγνονται ξ·	**Do** the 12 by the 5: they yield 60;
ταῦτα ἐπὶ τὰ δ· γίγνονται σμ·	these by the 4: they yield 240;
ταῦτα ἐπὶ τὰ γ· γίγνεται υκ·	these by the 3: it yields 720;
τούτων λαβὲ πλευράν,	take a side of these,
καὶ ἔσται τὸ ἐμβαδὸν τοῦ τριγώνου.	and it will be the area of the triangle.

[49] Accordingly, the operation is unary or binary, respectively.

[50] "To yield" must be used to translate finite verb forms, "to result" for participial forms.

[51] The adjectives λοιπός and ὅλος—frequently used in all three codes—are canonically treated as adjuncts of state; in this case, they must be invariantly translated "as a remainder" and "as a whole", respectively. The objects they determine are the result of removing or adding items (straight lines, regions, numbers) from or to homogeneous items, respectively.

[52] In mathematical papyri, γίνεται is sometimes replaced by a vertical stroke, |: see e.g. PMich. III.145, in Winter 1936, 34–52. This shows that the verb form is equivalent to our equality sign in a strong sense.

[53] In rare instances, εἰμί "to be" replaces γίνομαι in this function. See for instance *Metr.* I.26, Acerbi, Vitrac 2014, 212.18–19 and 214.1; II.6, ibid., 264.8; III.7, ibid., 324.23, and the formulation of the result of taking a fourth proportional.

[54] In the computational primer included in the *Three Books on Astronomy* of the 14th-century Byzantine polymath Theodorus Meliteniotes, each operation is described three times: by means of a procedure (called μέθοδος), of an algorithm (called ὑπόδειγμα "example"), and of a tabular set-up (called ἔκθεσις τῶν ἀριθμῶν "setting-out of the numbers"). On Byzantine mathematics, as fascinating as it is neglected, see Acerbi 2020b.

A) *Discursive arrow*

The main feature of an algorithm is the systematic and exclusive use of parataxis: no coordinants, (almost) no connectors, no subordination. The algorithmic cross-section of the mathematical corpus certainly constitutes the larger region of coordination by asyndeton consistently and systematically practised within the entire Greek literary production. Thus, the discursive arrow can only be provided by anaphora between the designations: such anaphoric devices mainly comprise demonstrative pronouns (forms of οὗτος "this") and numeral letters, crucially preceded or not by an article [see point (C) below]. Longer-range anaphorae are sometimes found, in the form of citations of algorithms employed as subroutines. The steps of initialization of the algorithm and of identification of the final output as the quantity to be computed receive specific formulations (forms of "to be").

The algorithmic flow is usually one-step: any step (1) accepts a number that is the output of the immediately preceding step as input and (2) feeds in new data by means of the second operand. Operations in which neither operand is the output of the immediately preceding step are less frequent. Such operations induce a hiatus in the algorithmic flow; the hiatus is often syntactically marked by the presence of particles or of liminal verb forms (both in boldface above).[55] Typical particles that mark a hiatus are πάλιν "again" and καὶ ἔτι "and further", but also εἶτα "next" or a simple καί "and" are met; καί, however, sometimes pleonastically precedes a pronoun.

Two operational flows can be singled out in an algorithm: the first flow lists the operations themselves, the second lists the results of the operations. The second flow appears to have the stylistic function of breaking the asyndeton within the chain of imperatives, thereby separating the operations from one another and providing the referent of the subsequent pronoun. The tables below set out the two operational flows in the algorithm of *Metr.* I.8:

σύνθες	λαβὲ τὸ ἥμισυ	ἄφελε	πάλιν ἄφελε ἀπό	καὶ ἔτι	ποίησον [...] ἐπὶ	ἐπί	ἐπί	λαβὲ πλευράν	
γίγνεται	γίγνεται	λοιπαί	λοιπαί	λοιπαί	γίγνονται	γίγνονται	γίγνεται		καὶ ἔσται

com-pose	take half of	subtract	again, subtract [...] from	and further	do [...] by	by	by	take a side of	
it yields	it yields	as a re-mainder	as a remainder	as a re-mainder	they yield	they yield	it yields		and it will be

B) *Verb forms*

The operations are formulated by verb forms in the aorist imperative. The verb forms are very often understood:[56] they are replaced by characteristic prepositions, such as ἐπί (multiplication) and παρά (division), or by adverbs or nouns that univocally determine the operation, such as ὁμοῦ "together" for addition or πλευρά "side" for the square root.[57] An exception is the verb ἄφελε "subtract", which

[55] See, as extreme examples, the isolated verb form σύνθες "compose" in *Metr.* I.14, or the adverb ὁμοῦ "together" in *Metr.* I.16 (Acerbi, Vitrac 2014, 182.1 and 186.15, respectively).

[56] Taking a square is sometimes subsumed under the designation, as in *Metr.* I.14, in Acerbi, Vitrac 2014, 182.4. The same phenomenon occurs in the procedural code, as we have seen in Sect. 1.2.

[57] The reason is that the verb λαμβάνω "to take" that governs the square root operation is not sufficiently characterized, and this induces the presence of the noun.

is almost always expressed.[58] Very elliptical steps are sometimes found, like ταῦτα δίς "these twice" in *Metr.* III.3,[59] or algorithms in which all verb forms are omitted.

The initial input is introduced by resources typical of the demonstrative code, and in particular by the liminal imperative of "to be" typical of the setting-out. The final result is identified by the future of copulative "to be", possibly enriched by a demonstrative pronoun, as in τοσούτου/ων ἔσται "of that much it will be". It must be stressed that such final identifications are not operations.

The steps listing the result of each operation are invariably formulated in the present indicative: this can be explained by the fact that the result of any operation is something taken for automatically yielded once the operands are made expicit. The subsequent numeral cannot be preceded by an article, and for two reasons: it is the nominal complement of the predicate and it is in its first occurrence—the latter a fact of the utmost importance, as we shall see presently.

C) *Reference to objects*

The reference to the objects is secured by demonstrative pronouns (forms of οὗτος "this") and, most frequently, by numeral letters taken as paradigmatic; there is no objectual overlap. Pronouns normally have a very short referential range and point to the immediately preceding clause; this feature, along with an appropriate choice of the numbers assigned as paradigmatic, usually forestalls ambiguities possibly arising whenever the same numeral letter denotes two different magnitudes; if residual ambiguities persist, the numeral letter is accompanied by a definite description of the object referred to. Pronouns are preferred when the verb form is understood, so that the operation is simply identified by a characteristic preposition. If the verb form is present, it usually accepts the last linguistic item mentioned as input.

It seldom occurs that a result is identified by the participle ὁ γενόμενος "resulting":[60] if this is the case, the step in which the result is made explicit is of course omitted.

The operands that correspond to one and the same object are mentioned more than once; since the example processed in the algorithm is paradigmatic, in their first occurrence the operands are not preceded by an article: they invariably acquire it from their second occurrence. The designations of the operands vary according to a fixed scheme: numeral letter → the same numeral letter preceded by a strongly anaphoric article → a short-range demonstrative pronoun. The successive layers of complements of designation in the algorithm read above are set out in the following table:

	a	b	c	d	e	f	g	h	i	l	m
1st mention	ζ	η	θ	κδ	ιβ	ε	δ	γ	ξ	σμ	ψκ
2nd mention	τὰ ζ	τὰ η	τὰ θ	τούτων	τὰ ιβ	τὰ ε	τὰ δ	τὰ γ	ταῦτα	ταῦτα	τούτων
3rd mention	τὰς ζ μονάδας	τὰς η	τὰς θ								

	a	b	c	d	e	f	g	h	i	l	m
1st mention	7	8	9	24	12	5	4	3	60	240	720
2nd mention	the 7	the 8	the 9	of these	the 12	the 5	the 4	the 3	these	these	of these
3rd mention	the 7 units	the 8	the 9								

[58] But see *Metr.* II.8 and 9, in Acerbi, Vitrac 2014, 272.1 and 274.18, respectively, for two steps that do not contain the verb that characterizes subtraction.
[59] Check Acerbi, Vitrac 2014, 314.15.
[60] We find this participle in the retrieval of the algorithm read above at the end of *Metr.* I.8: Acerbi, Vitrac 2014, 168.3. This retrieval is also characterized by the verb συνάγονται "they are collected", a hapax in the *Metrica*: ibid., 168.4.

The article that precedes a numeral letter can be either in the plural neuter or in the plural feminine; in the latter case, μονάδες "units" must be understood; singular masculine is also employed (ἀριθμός "number" is understood), but this happens infrequently.

If, in our paradigmatic algorithm, we replace the operands with the letters that identify them in the previous table, we find many instances of paired identical operands; this replacement gives rise to a skeleton algorithm that represents well the absence of objectual overlap. At the points where there are no paired identical operands and a hiatus is not occurring, the operand "looks at both directions": the way of reference replaces syntactic links.

οἷον ἔστωσαν αἱ τοῦ τριγώνου πλευραὶ μονάδων (a) (b) (c). σύνθες (a) καὶ (b) καὶ (c)· γίγνεται (d)· (d) λαβὲ τὸ ἥμισυ· γίγνεται (e)· ἄφελε (a)· λοιπαὶ (f). πάλιν ἄφελε ἀπὸ (e) (b)· λοιπαὶ (g). καὶ ἔτι (c)· λοιπαὶ (h). ποίησον (e) ἐπὶ (f)· γίγνονται (i)· (i) ἐπὶ (g)· γίγνονται (l)· (l) ἐπὶ (h)· γίγνεται (m)· (m) λαβὲ πλευράν, καὶ ἔσται τὸ ἐμβαδὸν τοῦ τριγώνου.

For instance, let the sides of the triangle be of (a) (b) (c) units. Compose (a) and (b) and (c): it results (d); take half of (d): it results (e); subtract (a): (f) as a remainder. Again, subtract (b) from (e): (g) as a remainder. And further (c): (h) as a remainder. Do (e) by (f): they result (i); (i) by (g): they result (l); (l) by (h): it results (m); take a side of (m), and it will be the area of the triangle.

D) *Generality*

Our discussion above has highlighted two main ways of expression of the mathematical generality of algorithms:[61]

- they are formulated as paradigmatic examples that feature specific yet generic numeric values;
- on their first occurrence, these numeric values do not carry an article; this is added in the second occurrence, and has a strong anaphoric value.

The second linguistic device is as necessary for the expression of generality as the first is: an algorithm really applies to "such numbers as are introduced at the beginning", and hence to any possible choice of these numbers.

The following synoptic table summarizes the markers of the three stylistic codes (the demonstrative code is represented only by proofs):

	structure	discursive arrow	verb forms	reference to objects	generality
proofs	coordination	coordinating particles	third person indicative	denotative letters	indefinite structure
procedures	mixed	designations subordination	first and second person indicative	definite descriptions	explicit, adverbs
algorithms	parataxis	designations, result markers	imperative, third person indicative	numerals, demonstrative pronouns	paradigmatic example

[61] Explicit statements about the generality of an algorithm are very infrequent; likewise, generalizing adverbs like "always" are usually absent.

1.4. Punctuating Greek Mathematical Texts

A fresh approach to punctuating Greek mathematical texts is required by two simple facts, which are apparent to anyone who skims the standard editions of strictly technical texts:[62]

- For reasons that pertain to the history of modern scholarship, a German-style punctuation is very often adopted, which abounds in commas: for instance, commas are found before every relative clause and before every that-clause; a case in point is the comma that separates the sequence λέγω, ὅτι "I claim that" that opens a "theorematic" determination (cf. Sect. 4.2).
- For reasons I am unable to ascertain, punctuation is mostly inconsistent: there is simply no regularity in putting a full stop or an upper point at the end of the setting-out, or after the conclusion of any three-step deduction.

To see this, let us read how Heiberg punctuates our paradigmatic proposition III.2 (paragraphs are included—the reader is urged to check other propositions in Heiberg's text):

Ἐὰν κύκλου ἐπὶ τῆς περιφερείας ληφθῇ δύο τυχόντα σημεῖα, ἡ ἐπὶ τὰ σημεῖα ἐπιζευγνυμένη εὐθεῖα ἐντὸς πεσεῖται τοῦ κύκλου.

Ἔστω κύκλος ὁ ΑΒΓ, καὶ ἐπὶ τῆς περιφερείας αὐτοῦ εἰλήφθω δύο τυχόντα σημεῖα τὰ Α, Β· λέγω, ὅτι ἡ ἀπὸ τοῦ Α ἐπὶ τὸ Β ἐπιζευγνυμένη εὐθεῖα ἐντὸς πεσεῖται τοῦ κύκλου.

Μὴ γάρ, ἀλλ' εἰ δυνατόν, πιπτέτω ἐκτὸς ὡς ἡ ΑΕΒ, καὶ εἰλήφθω τὸ κέντρον τοῦ ΑΒΓ κύκλου, καὶ ἔστω τὸ Δ, καὶ ἐπεζεύχθωσαν αἱ ΔΑ, ΔΒ, καὶ διήχθω ἡ ΔΖΕ.

Ἐπεὶ οὖν ἴση ἐστὶν ἡ ΔΑ τῇ ΔΒ, ἴση ἄρα καὶ γωνία ἡ ὑπὸ ΔΑΕ τῇ ὑπὸ ΔΒΕ· καὶ ἐπεὶ τριγώνου τοῦ ΔΑΕ μία πλευρὰ προσεκβέβληται ἡ ΑΕΒ, μείζων ἄρα ἡ ὑπὸ ΔΕΒ γωνία τῆς ὑπὸ ΔΑΕ. ἴση δὲ ἡ ὑπὸ ΔΑΕ τῇ ὑπὸ ΔΒΕ· μείζων ἄρα ἡ ὑπὸ ΔΕΒ τῆς ὑπὸ ΔΒΕ. ὑπὸ δὲ τὴν μείζονα γωνίαν ἡ μείζων πλευρὰ ὑποτείνει· μείζων ἄρα ἡ ΔΒ τῆς ΔΕ. ἴση δὲ ἡ ΔΒ τῇ ΔΖ. μείζων ἄρα ἡ ΔΖ τῆς ΔΕ ἡ ἐλάττων τῆς μείζονος· ὅπερ ἐστὶν ἀδύνατον. οὐκ ἄρα ἡ ἀπὸ τοῦ Α ἐπὶ τὸ Β ἐπιζευγνυμένη εὐθεῖα ἐκτὸς πεσεῖται τοῦ κύκλου. ὁμοίως δὴ δείξομεν, ὅτι οὐδὲ ἐπ' αὐτῆς τῆς περιφερείας· ἐντὸς ἄρα.

Ἐὰν ἄρα κύκλου ἐπὶ τῆς περιφερείας ληφθῇ δύο τυχόντα σημεῖα, ἡ ἐπὶ τὰ σημεῖα ἐπιζευγνυμένη εὐθεῖα ἐντὸς πεσεῖται τοῦ κύκλου· ὅπερ ἔδει δεῖξαι.

Because of the above facts, I have punctuated anew all mathematical texts—and in fact, all texts, tout court—transcribed in this book. I have followed a set of rules that privilege the "algorithmic" features of any chain of mathematical units of meaning, no matter whether they figure in constructions, proofs, algorithms, or procedures. The rules are rigid and have been rigidly adhered to: of course, this is only possible in the case of strictly technical texts (most of the *Almagest* is already an exception), in which the shades of meaning are programmatically eliminated and the general syntactic frame is likewise rigid. My basic choice is to propose a Greek text for reading that might "sound" like the original: this simply means restoring particles to their function of markers of the progression of an argument—and a part of this function, in the modern, scholarly approach to ancient Greek language, has been replaced by punctuation.

---

[62] On ancient theories of punctuation and on the medieval practice see Blank 1983, Gaffuri 1994, Noret 1995, Mazzucchi 1997, and Geymonat 2008. The punctuation system adopted in one of the most authoritative manuscripts of the *Almagest* is presented in Acerbi 2020c, 252–254.

Thus, my punctuation is rigid and light: the reader will learn to look at particles with sharper attention than it is usually done. The main criterion I adopt is that the comma separates subsentential items, the upper point separates sentential items, the lower point (or "full stop") separates argumental items. Still, my feeling is that three signs are not enough; I see that, adding other signs, semantic issues can be loaded on punctuation, but after all a text should be perspicuous and its punctuation should reflect its multilayered segmental architecture. My conventions are as follows—they entail similar conventions for my English translations:

- *Demonstrative code*. In setting-outs and in constructions: a comma separates independent suppositions or constructive acts. In proofs and enunciations: an upper point separates the deductive steps, even when they are introduced by inferential ὥστε "so that"; a comma separates the antecedent from the consequent of a (para)conditional, digressive relative clauses, and identifications of objects (namely, clauses introduced by τουτέστι).[63] Groups of denotative letters or short-range μέν … δέ … correlatives are not separated by commas.[64] Likewise, no comma separates correlations that are not realized by a relative pronoun; in particular, no comma is inserted in the standard correlative formula of proportionality ὡς … οὕτως … "as …, so …". Again, that-clauses and restrictive relative clauses are never preceded by a comma. A full stop should only be put before scope particles or their homologues, or whenever a deductive hiatus occurs, for instance before "potential" or "analogical" proofs (cf. Sect. 4.5.2). In the former case, the full stop is followed by a paragraph. Postposed explanations (cf. Sect. 4.5.3) should be included in parentheses or between en dashes. The letter that follows a full stop is not capitalized, not even after a paragraph.
- *Procedural code*. A comma separates principal clauses from one another; no comma separates the conjoined participial clauses subordinated to each principal clause; identifications of objects (usually introduced by ἤτοι "namely") are put in parentheses.
- *Algorithmic code*. An upper point separates both the consecutive steps of a continuous algorithmic flow and the operations from their result. Identifications of objects are included in parentheses. A full stop is inserted when a hiatus occurs, and before the identification of the final result, whenever this is opened by a form of τοσοῦτος "that much".

It is a welcome fact that the above rules punctuate a strictly technical text very much in the same way as it is punctuated in medieval manuscripts. The rules also make it clear that the demonstrative code has far richer a syntactic structure than the other codes.

As said, the rest of this book will analyse specific features of the demonstrative code. Before doing this, however, it is necessary to present the main component of my database: Euclid's *Elements*. I shall quickly survey the structure of the treatise and some of its characteristics, focusing then on its lexicon; this will be studied by means of basic tools of computational linguistics. I shall in particular provide complete lists of the linguistic units included in most morphosyntactic categories; each of these units is completed by its translation, by the number of its occurrences in the *Elements*, and by the indication of where poorly represented units occur. I shall finally address the issue of the frequently (and unduly) praised optimization of the Greek mathematical lexicon.

---

[63] There are 114 such clauses in the whole of the *Elements*, out of 135 occurrences of τουτέστι. None of these clauses is a priori suspect of authenticity.

[64] On the unreasonable practice of putting commas between the denotative letters see De Morgan 1915 I, 234–236.

1.5. The *Elements* and its lexical content

The structure of the *Elements* is well known: it is a treatise in 13 sections called "Books", each of which is strongly polarized as to its subject-matter:[65] Book I: triangles, parallelograms, and parallel straight lines; Book II: cut-and-paste geometric lemmas about quadrangles and triangles; Book III: the circle; Book IV: inscribed and circumscribed triangles, squares, and polygons; Book V: general proportion theory for magnitudes; Book VI: similar plane figures; Books VII–IX: number theory (primality, proportion theory for numbers, geometric progressions, perfect numbers); Book X: irrational lines and regions; Book XI: basic solid geometry; Book XII: application of the method of exhaustion to plane and solid figures; Book XIII: construction of the five regular polyhedra. Each Book comprises principles (but they may be absent) and propositions, namely, mathematical results supported by a proof. The principles are definitions, postulates, and "common notions" (the latter two only in Book I). The propositions can be theorems (which prove that an assigned geometric or number-theoretical configuration has such-and-such a property) or problems (which perform the construction of a mathematical object); the two categories are sharply differentiated from the stylistic point of view (see Sect. 4.1). Other material includes results stated but not proved because regarded as obvious (the "porisms"), results that are proved but have the sole function of filling a gap in a subsequent proposition (the "lemmas"), alternative proofs, inauthentic propositions said *vulgo*, variegated adjuncts that do not belong to any of the previous categories.

Most of these "other" items do not belong to the original *Elements*: they are inauthentic or "spurious". For, like any mathematical treatise from Greek antiquity, the *Elements* has been studied and revised by generations of scholars; the most conspicuous output of such revisions is the additional material just listed, simply because deductive completeness has always been the main aim of any reviser. Heiberg relegated part of this additional material in Appendices, but he also kept a good deal of it in his main text, possibly bracketed as spurious; Heiberg's Greek text without the Appendices I shall call "the main text of the *Elements*"; if the Appendices are included, this text is "the whole of the *Elements*". For his edition, Heiberg only used the most ancient manuscripts; he identified two versions of the text, the one witnessed to only in the early 9th-century manuscript Vat. gr. 190, the other—which is the result of a light revision authored by the 4th-century scholar Theon of Alexandria—witnessed to by all other Greek manuscripts, which are thereby called "Theonine". The Arabic and Arabo-Latin translations of the *Elements* give access to a version of the text that is on the whole (that is, disregarding the revisions in the Arabic line of transmission) less adulterated than the version we read in Greek; this fact normally allows settling issues of authenticity.[66] In this book, I shall occasionally use the evidence of the indirect tradition.[67]

The table in the following page sets out in order the following items: the size of each Book in percentage of signs with respect to the whole of the *Elements*, the number of propositions, porisms, lemmas, alternative proofs, propositions *vulgo*, adjuncts of each Book:

---

[65] The deductive structure of the *Elements* is masterly discussed in Mueller 1981. See also Vitrac 1990–2001 passim.

[66] On this renewed approach to the textual tradition of the *Elements*, see Knorr 1996; Vitrac 1990–2001 III–IV; Rommevaux, Djebbar, Vitrac 2001; Acerbi 2003a. Note that much of the additional material predates Theon's revision.

[67] The boldface sigla I shall use to denote some Greek manuscripts are explained in the *Index fontium*. A single manuscript of the Theonine class, Bonon. A 18–19, carries a recension of XI.36–XII that is "aberrant" in the same sense as the indirect tradition is. This recension is identified by the siglum **b** and is edited in Appendix 2 of *EOO* IV; it is not included in "the whole of the *Elements*".

	I	II	III	IV	V	VI	VII	VIII	IX	X	XI	XII	XIII	tot.
% signs	7.6	3.3	6.9	3.4	4.9	7.6	5.7	5	5.1	26.1	9	8.3	6.9	100
prop.	48	14	37	16	25	33	39	27	36	115	39	18	18	465
por.	1	1	3	2	2	4	1	1	1	7	2	3	2	30
lemmas	/	/	/	/	/	1	/	/	/	11	1	2	3	18
alt.	/	1	5	/	/	4	1	/	1	7	1	2	7	29
vulgo	/	/	/	/	/	/	2	/	/	2	1	/	/	5
adj.	/	/	/	/	1	1	/	/	/	11	1	/	2	16

The ὅροι "definitions" contained in the *Elements* are not numbered in the manuscripts. In many cases, an independent definitory item is a cluster of statements; the individual items in the cluster are coordinated by δέ "and".[68] The definitions open a Book, with the remarkable exception of Book X, in which each of the 13 irrational lines is defined within a suitable proposition and the 6 + 6 definitions of the subspecies of the binomial (apotome) are located after propositions X.47 and X.84, respectively (cf. Sect. 4.2). Some Books are not opened by definitions. The following table sets out bookwise, according to the partition adopted in Heiberg's edition, the number of definitions, defined objects, defined relations, and defined operations in the whole of the *Elements*:[69]

	I	II	III	IV	V	VI	VII	VIII	IX	X	XI	XII	XIII	tot.
def.	23	2	11	7	18	5	23	/	/	29	28	/	/	146
obj.	35	1	3	/	/	1	15	/	/	32	22	/	/	109
rel.	1	1	8	7	11	3	7	/	/	4	9	/	/	51
oper.	/	/	/	/	7	1	1	/	/	/	/	/	/	9

The definitions of the *Elements* present conspicuous textual problems: some definitions are obviously spurious (as VI.def.5);[70] most of them are deductively inert (as the definition of "point" in I.def.1) but some are not (as "right angle" in I.def.10 and "proportion" in V.def.5); some define mathematical objects or relations never to be mentioned in the sequel or that are only mentioned in a subsequent definition,[71] whereas some important mathematical objects are never defined (the "parallelogram"—it does not even coincide with the "rhomboid" of I.def.22).

Famously, there are five αἰτήματα "postulates". The first three of them license basic constructive acts: to join any two points; to produce any straight line in a straight line; to describe a circle with any centre and radius.[72] The last two postulates have a theorematic form; they state that all right angles are equal to one another, and (contrapositively) the celebrated necessary condition for two straight lines to be parallel, respectively.[73]

[68] Heiberg's edition splits these clusters in segments that normally "define" one item only, but this may result in statements that do not define anything: see the next-to-last note of this Section.

[69] As a general rule, definitions of objects are marked by the verb forms ἐστί "is" or καλεῖται "is called" / καλείσθω "let it be called"; definitions of relations and of predicates are marked by λέγεται "is said"; see also Sect. 2.4.1. The term "relation" is here meant in a generic sense; in principle, this notion does not coincide with the one introduced in Sect. 4.5.1. On definitions in Greek mathematics, see Mueller 1991 (with an in-depth discussion of principles in Plato and Aristotle); Asper 2007, sects. B1 (lists of definitions) and B2 ("elements" as literary genre); Netz 1999, 91–94; Wilck 2020 (these two offer elaborate typologies). On definition in ancient philosophy see Charles 2010.

[70] Dye, Vitrac 2009 have shown that Sextus Empiricus cannot be regarded as a reliable source for the textual status of the definitions in the *Elements*. This fact refutes Russo 1998, who held that I.def.1–7 are spurious mainly on the grounds of Sextus' reliability as a source.

[71] The former category includes 1 object out of 3 in I.def.19, 3 objects out of 5 in I.def.22, or the 3 relations defined in XI.def.5–7; the latter category includes the remaining 2 objects defined in I.def.19.

[72] None of the magnitudes operated upon in the postulates is said to be "given" (cf. Sects. 2.4 and 4.3).

[73] The names of some Greek mathematicians who tried to prove the fifth postulate are listed, on the authority of Simplicius, by the late 9th-century Persian commentator of the *Elements* an-Nayrīzī (Tummers 1994, 55.20–25): among them is a Diodorus who is probably to be identified with the renowned gnomonist. A proof based on a definition of parallelism

The κοιναὶ ἔννοιαι "common notions" are general axioms: transitivity of equality, stability of (in)equality under addition and removal of equal items, stability of equality under doubling and halving; or general facts about magnitudes (coinciding items are also equal; the whole is greater than any of its parts) and straight lines (two of them cannot contain a region). Ancient debate on the common notions, and accretions originating in revisions, left conspicuously spurious items as their traces.[74] Proclus (who lists only five of the nine common notions witnessed to by the Greek manuscripts) reports the opinions on the subject advocated by Hero (who retained only three common notions) and by Pappus (who added others), as well as his own (*iE*, 196.15–198.15).

Aristotle extensively discussed mathematical principles. At the beginning of *APo.*, he repeatedly presents a tripartite division of the principles of any deductive science, for instance "one ‹principle› is what is proved, the conclusion (this is what applies to some genus per se), one are the axioms (the axioms are from which), the third is the underlying genus, the properties and the per se attributes of which the proof makes clear".[75] Two other divisions, possibly grounded on different ordering criteria, are introduced in *APo*. One is at I.10, 76b23–34: Aristotle distinguishes between ὑποθέσεις "suppositions" and αἰτήματα, the postulates being "something not in accordance with the opinion of the learner". In the subsequent paragraph (76b35–77a4), ὑποθέσεις are instead opposed to ὅροι: ὑποθέσεις have a propositional structure, ὅροι do not, and simply need to be grasped.[76] The same distinction had surfaced earlier as a sub-specification of the classification at I.2, 72a14–22: ἀξιώματα "axioms" are separated from θέσεις "positions", on the basis of whether or not they need to be grasped by anyone who is to learn anything;[77] the θέσεις are further specified as ὑποθέσεις—which have a propositional structure and state existence—and ὁρισμοί (76b35–39).

Before Aristotle, the nature of principles was the subject of an Academic debate (Menaechmus on one side, Speusippus and Amphinomous on the other)[78] about the opportunity of ranging (and hence formulating) all principles that are not definitions among the postulates or among the common notions.[79] A parallel debate among the same philosophers focused on the nature of geometric propositions, whether they are (and hence have to be formulated as) problems or theorems.[80]

Likewise, the 1st-century BCE Stoic philosopher Posidonius reportedly proposed a reform of the enunciations of theorems and of problems with the goal of of making the existential connotation of problems explicit (note that the two examples of theorematic enunciation are quantified):[81]

---

in terms of equidistance and ascribed to a mysterious Aganiz, a mathematician and friend of Simplicius, is then reported (ibid., 56.1–62.14). We may also recall Ptolemy's proofs compiled by Proclus (*iE*, 365.5–367.27). A sophisticated proof attested only in Arabic is to be ascribed to Apollonius, as I have shown in Acerbi 2010b, 165–168 (on Sabra 1968). The renowned Renaissance philosopher George Gemistos Pletho, probably following Geminus (the source is Proclus, *iE*, 181.5–9), relocated the last two postulates among the common notions of the *Elements* by modifying a manuscript that was later used by Grynaeus for the 1536 *editio princeps* of the treatise: see Acerbi, Martinelli Tempesta, Vitrac 2016.

[74] Tannery 1884 argued that all common notions of the *Elements* are spurious. De Risi 2020 is only useful as a survey of ancient and modern literature.

[75] *APo.* I.7, 75a39–b2, and cf. for instance *APo.* I.10, 76b11–16, where the ἀξιώματα are qualified κοινά.

[76] On Aristotle stating that (mathematical) definitions need not prove or state that the defined objects exist see *APo.* II.7, with a most interesting yet cryptic allusion to οἱ νῦν τρόποι τῶν ὅρων "the current methods of definition" (92b19).

[77] That is, the difference is between general principles and assumptions typical of a specific discipline.

[78] Amphinomous also proposed a classification of problems according to the number of their solutions (*iE*, 220.9–12).

[79] The reader will easily realize that every problem can be formulated as a theorem and vice versa, even if every mathematical proposition can be taken to have a "natural" formulation, and an "unnatural" formulation normally looks highly contrived. To see constructions formulated in theorematic form, just read the enunciations of the propositions that "construct" the 13 irrational lines of Book X, or the constructions of the conic sections in Apollonius, *Con.* I.11–13. In all these cases, a definition must figure at the very end of the enunciation, a feature that may well be termed "contrived" on ancient standards. The reader is urged to transform some theorems of *El.* I into problems.

[80] The source is Proclus, *iE*, 181.16–24 and 75.27–78.20, respectively.

[81] Proclus, *iE*, 80.20–81.1 = part of fr. 195 Edelstein-Kidd = fr. 464 Theiler. Posidonius himself and Zeno of Sidon debated about the completeness of the axioms of the *Elements* (Proclus, *iE*, 199.3–200.6 and 214.15–218.11 = frs. 46–

Hence Posidonius distinguished on the one hand the enunciation in which it is investigated the "whether it is possible or not" [τὸ εἰ ἔστιν ἢ μή], on the other hand the enunciation in which it is investigated what or what sort of thing it is [τί ἔστιν ἢ ποῖόν τι], and maintained that the theorematic enunciation should be put in declarative form, for instance, "every triangle has two ‹sides› greater than the remaining one", and "the ‹angles› at the base of every isosceles are equal", the problematic ‹enunciation›, as if one were investigating: "whether it is possible [εἰ ἔστιν] to construct a triangle on this straight line".

These doctrines are not mere verbalism. They hinge upon the nature of mathematical objects, they reflect a sharp divide between a realist and a constructivist approach to mathematics (further discussion in Sects. 3 and 4.3). But there is more. Even if these doctrines were taken to involve only stylistic issues, they would nevertheless retain a conspicuous technical relevance, since in Greek mathematics, which is and was also conceived as a literary genre, the form of expression replaces the formalism, and therefore linguistic and stylistic conventions have a direct mathematical import. These stylistic conventions will be the object of the remaining parts of this book.

Before doing this, however, I shall explain, by applying some basic tools of computational linguistics and with the help of Ramon Masià, the lexical content of the *Elements*.[82] Let us start with some definitions. Every independent sequence of signs in a text is said "occurrence" (or "word-token"); the identical occurrences are a "form" (or "word-type");[83] the set of forms referred to one and the same term is a "lemma" (forget the mathematical meaning in part of this Section). A lemma is identified by writing one of the forms without diacritics; τομαὶ and τομῆς are forms of the lemma τομη; the headwords of any lexicon follow the same principle, while keeping diacritics.

The overall number of lemmas, forms and occurrences in the whole of the *Elements*,[84] in Archimedes' *Sph. cyl.*, in Hero's *Metrica*, and in Pappus' *Collectio* are set out in the following table; the first numerical row exhibits the figures for the prefaces, the second row, the figures for the mathematical text proper, the third row, for the entire treatise:

Elements			*Sph. cyl.*			*Metrica*			*Collectio*		
lem.	for.	occ.	lem.	for.	occ.	lem.	for.	occ.	lem.	for.	occ.
/	/	/	187	351	921	284	503	1047	/	/	/
524	2340	163267	313	1543	24741	505	1930	19360	1796	7260	137539
524	2340	163267	381	1701	25662	612	2179	20407	1796	7260	137539

We may compare the above numbers of occurrences with those in some Aristotelian treatises: *APr.* + *APo.* = 60232; *Ph.* = 55533; *Cael.* = 29763. The sum total is 145528, which is near to the number of occurrences in the *Elements*. The table shows that the universe of discourse of the *Elements* is made of a bit more than 500 different words; *Sph. cyl.* and the *Metrica* have nearly the same lexical content, whereas the *Collectio*, owing to its varied subject-matter, is far richer.

Actually, the operative lexical content of the *Elements* is much more limited than this, as is shown by the following list of the 10 most frequent lemmas in the *Elements*, in *Sph. cyl.*, in the

47 Edelstein-Kidd = fr. 463 Theiler). The foundational interests of Apollonius and of the 1st-century BCE polymath Geminus are studied in Acerbi 2010b.

[82] See the discussion in Acerbi, Vitrac 2014, 59–73, from which I draw some tables and a plot. Compare the discussion, ranging over the entire Greek mathematical corpus, in Netz 1999, sect. 3. On the geometric lexicon see also Mugler 1958. For a detailed lexical analysis of medieval scientific texts, see Roelli 2021, sects. III.1–4.

[83] Variants of accent and elision count as different forms. Thus, ἀπ' is a form of the lemma απο.

[84] This choice aims a giving a faithful picture of what has been transmitted by the main lines of the Greek manuscript tradition; to repeat: "the whole of the *Elements*" is Heiberg' main text, the atheteses included, plus the Appendices of *EOO* I–IV, with the exclusion of Appendix II of *EOO* IV, which contains recension **b** of XI.36–XII. If we exclude the Appendices, Heiberg's main text exhibits 496 lemmas, 2259 forms, and 152701 occurrences.

Metrica, and in the *Collectio*—the lemma "o" is the article, "A" are the denotative letters, "num." are the numeral letters (only relevant in the *Metrica*):

Elements			*Sph. cyl.*			*Metrica*			*Collectio*		
lemma	occ.	%	lemma	occ.	%	lemma	occ.	%	lemma	occ.	%
ο	35740	21.89	ο	6223	24.25	ο	4028	19.27	ο	30348	22.07
A	30723	18.82	A	3535	13.77	A	2263	10.83	A	24421	17.76
ειμι	8394	5.14	ειμι	955	3.72	num.	1227	5.87	και	5841	4.25
και	6715	4.11	προς	855	3.33	και	968	4.63	ειμι	5407	3.93
προς	4078	2.50	και	819	3.19	ειμι	944	4.52	προς	4045	2.94
αρα	4042	2.48	κυκλος	550	2.14	δε	460	2.20	υπο	3468	2.52
ισος	3424	2.10	δε	523	2.04	αρα	420	2.01	δε	2464	1.79
απο	2740	1.68	ισος	488	1.90	προς	384	1.84	ισος	2348	1.71
υπο	2672	1.64	κωνος	423	1.65	μονας	364	1.74	απο	2284	1.66
δε	2655	1.63	επιφανεια	422	1.64	διδωμι	353	1.69	αρα	2188	1.59
tot.	101183	61.97	tot.	14793	57.65	tot.	11361	55.67	tot.	82814	60.21

Thus, the first 10 lemmas in the *Elements* consume nearly 62% of its lexical content; the figures for the other three treatises are lower, still greater than 50%. What is more, articles and denotative letters make over 40% of the lexical content of the *Elements*; this is no surprise, for designations by means of denotative letters are ubiquitous in the demonstrative code, and at least one article must figure in each designation (see Sect. 2.2). As is to be expected, forms of ειμι "to be" are very common. The other most frequent lemmas are also typical of the demonstrative code and of the geometric substratum of most of the *Elements*: αρα "therefore" and δε "and" are the main inferential particles; the adjective ισος "equal" and the conjunction και "and" speak for themselves; the three prepositions in the list are found in such a frequent formula as the one expressing a ratio (προς), and in the designations of angles and rectangles (υπο), and of squares (απο). The different lemmas that figure in the top-ten list for *Sph. cyl.* and the *Metrica* are readily explained on the basis of the subject-matter of these treatises:[85] κωνος "cone", κυκλος "circle", and επιφανεια "surface" attest to a more marked nominal content of *Sph. cyl.*, as well as to a less constraining inferential structure of Archimedes' argument; μονας (the universal "unit" of measurement adopted by Hero) and διδωμι "to give" prominently feature in the *Metrica*, the latter because of validation (see Sect. 2.4.2). The eleventh most frequent lemma in the *Metrica* is γιγνομαι "to yield", "to result", as is natural on account of the fact that the *Metrica* is a prototypical text (partly) written in the algorithmic code. As for the *Elements*, the subsequent positions in the list of the most frequent lemmas are held, in the indicated order, by the multi-purpose pronoun αυτος, by the subordinant ως "as" (key element in the formula for a proportion), by the numeral adjective δυο "two" (this may be surprising, but this numeral is a key ingredient of the formulation of the congruence criteria of triangles), by the subordinant επει "since" (which opens the "anaphora"), by the adverb ουτω "so" (again in the formula for a proportion), and, finally, by two nouns: γωνια "angle" and ευθεια "straight line".

The plot in the following page exhibits the frequency of a lemma (ordinates, suitably rescaled) as a function of its rank in the list of lemmas (abscissae), ordered by decreasing number of occurrences; the lemmas with the same number of occurrences hold the same rank. Treatises correspond to colours as follows: *Elements*, red; *Sph. cyl.*, blue; *Con. sph.*, light blue; *Metrica*, black.[86]

---

[85] The top-ten lemmas of the *Elements* and of the *Collectio* are the same, in different orders.

[86] The data from the Greek mathematical treatises are not best fitted by power functions, as is instead required by Zipf's law: Masià, unpublished typescript [for Zipf(-Mandelbrot)'s law in computational linguistics see Powers 1998].

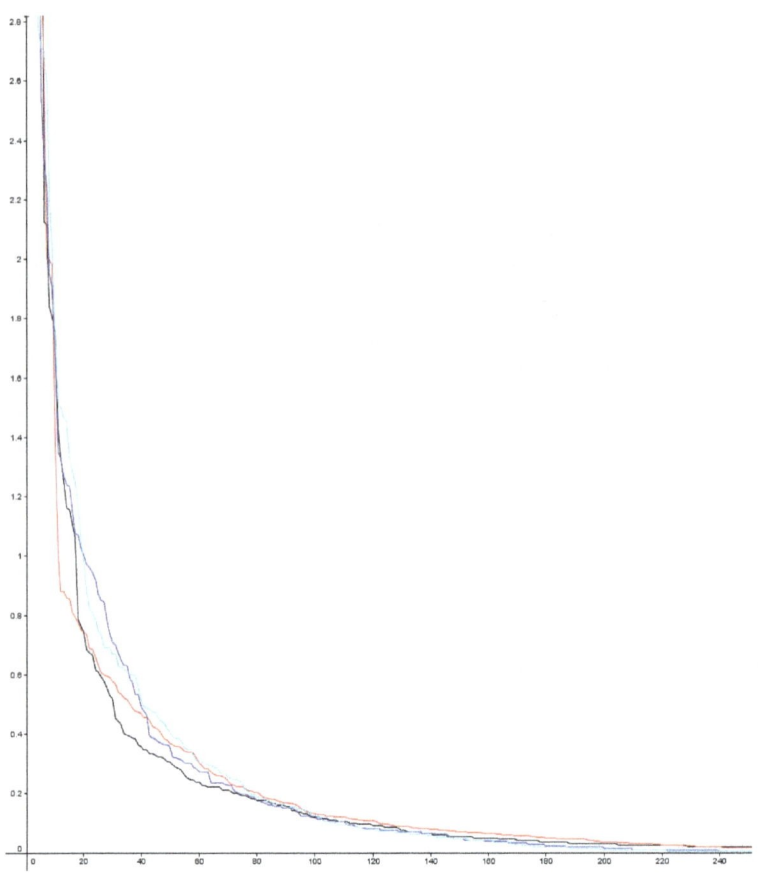

Other lexicographic data are set out in the following tables. Let us see first the 10 lemmas with the highest number of forms in the *Elements*, in *Sph. cyl.*, in the *Metrica*, and in the *Collectio*:

Elements		*Sph. cyl.*		*Metrica*		*Collectio*	
lemma	# forms	lemma	# forms	lemma	# forms	lemma	# forms
ειμι	42	ειμι	37	ειμι	30	λαμβανω	62
τεμνω	38	αυτος	23	διδωμι	23	γιγνομαι	61
γιγνομαι	31	ο	22	μετρεω	22	γραφω	58
περιεχω	29	αγω	20	ποιεω	21	ειμι	57
ποιεω	28	εχω	18	αυτος	19	διδωμι	55
αυτος	27	ος	17	ο	19	ποιεω	54
εχω	26	διδωμι	17	γιγνομαι	17	τεμνω	52
μετρεω	26	ερω	17	διαιρεω	17	αγω	44
ο	25	γιγνομαι	16	ερω	17	εχω	41
διαιρεω / επιζευγνυμι	22	περιγραφω	14	εχω	17	δεικνυμι	39

As for the *Elements*, here is the list of the hapax legomena (namely, lemmas with one single occurrence in the treatise), distributed by morphosyntactic categories (or "PoS" = Parts of Speech); they are 79 items (15.08% of the lemmas), a very low figure, as we shall see in a moment (the hapax occurring in Heiberg's Appendices—that is, in certainly spurious material—are underlined).

adjective

απλατης, αριθμητικος, ασυμπτωτος, δεκαπεντε, δεκατος, διμοιρος, εικοσιπενταπλασιος, ενδεκατος, ενναπλασιος, εννεα, ιδιος, ισαριθμος, κυριος, μακροτερος, μικρος, ομογενης, ομοιοπληθης, παρατελευτος, πενταπλους, πεντεκαιδεκατος, ποιος (encl.), πολυπλευρος, σκαληνος, συντομωτερος

adverb

εντευθεν, εξωθεν, ηδη, ισοπληθως, καθο, μοναχως, ολως, τυχοντως

noun

αναστροφη, απαγωγη, βαθος, δειξις, επισκεψις, ετερομηκες, ευρεσις, θεσις, λημματιον, οικειοτης, ονομασια, παραβολη, πραγματεια, προυπαρξις, ρομβοειδης, ρομβος, σχεσις, υπεξαιρεσις

particle	preposition
ειπερ, μηδε, τοινυν, τοτε (orth.)	ενεκα

verb

αιτεω, αναπληροω, αντιστρεφω, αποτιθημι, αρκεω, δευτερευω, εξακολουθεω, επακουω, επιδεικνυμι, επιλογιζομαι, επινοεω, ζητεω, κατακολουθεω, καταλαμβανω, κλαω, κλινω, μανθανω, μηκυνω, περιαγω, προβαινω, προεκτιθημι, προσαναπληροω, προχωρεω, συγκρινω

Here is also a list of the one-form lemmas that are not hapax or indeclinable (61 items). We may recall that *Sph. cyl.* contains 105 hapax (27.63%) and 66 one-form lemmas that are not hapax, *Con. sph.* 47 hapax (16.04%) and 61 one-form lemmas, the *Metrica*, 218 hapax (35.62%) and 55 one-form lemmas, the *Collectio*, 664 hapax (36.97%).

adjective / pronoun*

ακολουθος, αληθης, αρτιοπλευρος, ατοπος, δηλος, εβδομος, εξαπλασιος, καταλληλος, κοιλος, μετεωροτερος, οποσοσδηποτουν, οσαπλασιων*, οστισουν, πενταπλασιων, πολλαπλασιων, πολυς, προχειροτερος, συνδυο, τετραπλασιων, τοσαυταπλασιων*, τριπλους

noun

αναλυσις, αποτμημα, αρχη, βιβλιον, δεκαγωνον, διαστημα, διχοτομια, δυναμις, ελλειμμα, λημμα, ληψις, ομοιοτης, πεντεκαιδεκαγωνον, πολλαπλασιασμος, πορισμα, προβλημα, προτασις, συνθεσις, τοπος, υποθεσις

verb

αναστρεφω, ανθυφαιρεω, αποκαθιστημι, επισκεπτομαι, ηκω, ιστημι, καταγραφω, καταντα ω, παρακειμαι, περατοω, περιλειπομαι, προστασσω, προστιθημι, προτερεω, στρεφω, συναποδεικνυμι, ταρασσω, υπολειπω, φερω, φημι

The following table sets out, by morphosyntactic categories, the distribution of the occurrences in the whole of the *Elements*; some particles, as for instance καί "and" / "also" and ἤ "or" / "than", may belong to more than one category; denotative letters count as 1 lemma and 1 form; proposition numbers in metadiscursive items count as 1 lemma and 1 form, category "counter"; adjectives as ἥμισυς "one half" are categorized as cardinal numbers. A mildly semantic characterization is introduced for nouns and for some adjectives and pronouns (see just below).

	lemmas	%	forms	%	f/l	occ.	%	o/l	o/f
adjective	138	26.34	637	27.22	4.62	17635	10.80	127.79	27.68
indefinite	6	1.15	14	0.60	2.33	89	0.05	14.83	6.36
indefinite pronominal	13	2.48	84	3.59	6.46	1095	0.67	84.23	13.04
numeral cardinal	14	2.67	56	2.39	4.00	2002	1.23	143.00	35.75
numeral ordinal	11	2.10	56	2.39	5.09	1093	0.67	99.36	19.52
qualifier	75	14.31	369	15.77	4.92	12601	7.72	168.01	34.15
qualifier multiplicative	17	3.24	56	2.39	3.29	723	0.44	42.53	12.91
verbal	2	0.38	2	0.09	1.00	32	0.02	16.00	16.00
adverb	53	10.11	59	2.52	1.11	3560	2.18	67.17	60.34
numeral	8	1.53	8	0.34	1.00	539	0.33	67.38	67.38
others	45	8.59	51	2.18	1.13	3021	1.85	67.13	59.24
article	1	0.19	25	1.07	25.00	35740	21.89	35740.00	1429.60
counter	1	0.19	1	0.04	1.00	34	0.02	34.00	34.00
denotative letter	1	0.19	1	0.04	1.00	30723	18.82	30723.00	30723.00
noun	112	21.37	449	19.19	4.01	15482	9.48	138.23	34.48
action	7	1.34	12	0.51	1.71	39	0.02	5.57	3.25
metadiscursive	21	4.01	28	1.20	1.33	136	0.08	6.48	4.86
object	70	13.36	372	15.90	5.31	13820	8.46	197.43	37.15
relation	14	2.67	37	1.58	2.64	1487	0.91	106.21	40.19
particle	36	6.87	57	2.44	1.58	23179	14.20	643.86	406.65
coordinant	9	1.72	15	0.64	1.67	7682	4.71	853.56	512.13
inferential	10	1.91	16	0.68	1.60	10457	6.40	1045.70	653.56
modal	1	0.19	2	0.09	2.00	30	0.02	30.00	15.00
negative	2	0.38	7	0.30	3.50	612	0.37	306.00	87.43
subordinant	14	2.67	17	0.73	1.21	4398	2.69	314.14	258.71
preposition	22	4.20	45	1.92	2.05	13371	8.19	607.77	297.13
pronoun	20	3.82	150	6.41	7.50	4771	2.92	238.55	31.81
correlative	2	0.38	8	0.34	4.00	53	0.03	26.50	6.63
correlative multiplicative	2	0.38	3	0.13	1.50	26	0.02	13.00	8.67
demonstrative	5	0.95	56	2.39	11.20	1964	1.20	392.80	35.07
demonstrative multiplicative	2	0.38	4	0.17	2.00	23	0.01	11.50	5.75
indefinite	2	0.38	27	1.15	13.50	548	0.34	274.00	20.30
interrogative	1	0.19	2	0.09	2.00	4	0.00	4.00	2.00
personal	1	0.19	1	0.04	1.00	3	0.00	3.00	3.00
reciprocal	1	0.19	7	0.30	7.00	504	0.31	504.00	72.00
reflexive	1	0.19	14	0.60	14.00	247	0.15	247.00	17.64
relative	2	0.38	25	1.07	12.50	1387	0.85	693.50	55.48
relative indefinite	1	0.19	3	0.13	3.00	12	0.01	12.00	4.00
verb	140	26.72	916	39.15	6.54	18772	11.50	134.09	20.49
total	524	100.00	2340	100.00	4.47	163267	100.00	311.58	69.77

As we shall see in Sects. 2.2 and 3.2, the complexes "article + denotative letter(s)" must be counted as nouns (this is a lexicalization mechanism typical of Greek language): therefore, the data set out in this table have a strong bias against nouns. However, no "objectual content" is lost, since such articular complexes are simply denotative shortcuts of the nouns listed later in this Section.

The non-obvious categories in the table are:[87]

a) Indefinite adjective: οποιοσουν, οποσοσδηποτουν, οποσοσουν, οσοσδηποτουν, οστισουν, ποιος (encl.).
b) Pronominal indefinite adjective: αμφοτερος, απας, εκαστος, εκατερος, ετερος, μηδεις, μηδετερος, οποτερος, ουδεις, ουδετερος, πας, συμπας, συναμφοτερος.
c) Particles. Coordinant: η, ηπερ, ητοι, και, μηδε, μητε, ουδε, ουτε, τε; inferential: αλλα, αρα, γαρ, δε, δη, μεν, ουν, τοινυν, τοτε (orth.), ωστε; modal: αν; negative: μη, ου; subordinant: εαν, ει, ειπερ, ειτε, επει, επειδηπερ, επειπερ, εως, ινα, οταν, οτε (orth.), οτι, πλην, ως.
d) Pronouns. Correlative, οιος, οσος; demonstrative, αυτος (subcategory "anaphoric"), εκεινος, ουτος, τοιουτος, τοσουτος; indefinite, αλλος, τις (encl.); interrogative, τις (orth.); multiplicative correlative οσαπλασ*; multiplicative demonstrative τοσαυταπλασ*; personal, εγω; reciprocal, αλληλων; reflexive, εαυτος; relative, ος, οσπερ; indefinite relative, οστις.

Finally, I present a complete list, with sparse comment, of some lexical items that figure in the whole of the *Elements*; most verbs will be listed and discussed in Sect. 4.3. Each item is translated and is followed by the number of its occurrences.

Adverbs are obviously well represented; the high frequency of ἀνάλογον "in proportion" is explained by the fact that it designates a proportion; of οὕτω "so", by its presence in the standard, instantiated formulation of a proportion; of ἀπεναντίον "opposite", by its presence in the enunciation of the key theorems of the theory of parallel lines, I.27–29. Again, δίς "twice" has a notable frequency in the geometric lemmas of Book II and in their applications (formulae like "twice the rectangle contained by"); δίχα figures in the widely used expression τετμήσθω δίχα "let it be bisected"; ἐναλλάξ "by alternation" formulates a standard manipulation of ratios (the same for ἀνάπαλιν "by inversion"); ἑξῆς "successively" expresses iteration; ἰσάκις "equi-" is the key adverb in the theory of equimultiples (Sect. 4.5.1.4); ὁμοίως "similarly" characterizes potential proofs (Sect. 4.5.2); πάλιν "again" introduces a fresh start within a proof; τουτέστι "that is" introduces identifications of objects. The adverb ἀεί "continually" has its standard iterative sense, sometimes within arguments with a marked procedural character (Sect. 4.5.2). Here is the list; the adverbs of manner are underlined, items with a metadiscursive connotation are in italics, items that formulate a multiplication (*–ακις X = * times X) are in boldface: *ἀεί* (34), <u>*ἀκολούθως*</u> "as a consequence of" (6),[88] <u>ἄλλως</u> "otherwise" (21),[89] ἅμα "together" (16), <u>ἀνάλογον</u> (402), ἀνάπαλιν (19), **ἅπαξ** "once" (3),[90] <u>ἀπεναντίον</u> (106), <u>ἁπλῶς</u> "in general" (2), ἀπώτερον "farther" (13), **ἀρτιάκις** "even times" (24),[91] *αὐτόθεν* "immediately" (2),[92] *δηλαδή* "clearly" (10), *δηλονότι* "it is clear that" (3), **δίς** (237), δίχα (157), ἔγγιον "nearer" (13),[93] *ἔμπροσθεν* "previously" (5), ἐναλλάξ (93), *ἐντεῦθεν* "therefrom" (1),[94] ἑξῆς (121), *ἔξωθεν* "from outside" (1),[95] *ἐπάνω* "above" (5), <u>ἑτέρως</u> "otherwise" (3),[96] ἔτι

[87] I use some of the categorizations in van Emde Boas et al. 2019. See Sect. 5.1.4 below for the first category.
[88] Occurrences in IV.15, X.83, X.109–111 (*bis* in 111).
[89] Disregarding the inscriptions of the alternative proofs (18) and the occurrence within XIII.18*alt*, ἄλλως introduces alternative definitions, in V.def.17, XI.def.11.
[90] Occurrences in IX.15, X.60, XIII.2*alt*, always "once the rectangle contained by".
[91] Occurrences in VII.def.8–9, IX.30, 32 (*quinquies*), 33 (*sexties*), 34 (*decies*).
[92] Spurious nexus αὐτόθεν φανερόν "immediately manifest", in III.31 and X.16.
[93] Ἔγγιον and ἀπώτερον are exact correlates and occur in the same formula, in III.7 (*ter*), 8 (*sexties*), 15 (*quater*).
[94] Spurious nexus ἐντεῦθεν δῆλον "clear therefrom", in X.18por.
[95] Isolated, but possibly authentic occurrence, in XI.34.
[96] Occurrences in VI.20*alt*, *alt*X.27, XII.17*alt*.

"and further" (88), ἐφεξῆς "consecutive" (16),[97] ἤδη "previously" (1), **ἰσάκις** (230), ἰσοπληθῶς "with equal multiplicity" (1),[98] *καθάπερ* "really as" (2), *καθό* "as" (1), *καθόλου* "in general" (4),[99] μηδέποτε "never" (4),[100] μήν "of course" (38), μοναχῶς "uniquely" (1),[101] *ὅθεν* "from where" (7),[102] ὅλως "in general" (1),[103] ὁμοίως (205), **ὁσάκις** "how many times" (26), οὐκέτι "never" (3),[104] οὕτως (1146), πάλιν (223), πάντῃ "however" (19),[105] πάντως "in every instance" (13), **περισσάκις** "odd times" (2),[106] *ποτέ* "eventually" (5), πότε "when" (2), *πρότερον* "first" (63), **τετράκις** "four times" (10), **τοσαυτάκις** "so many times" (7), τουτέστι (135), τυχόντως "randomly" (1), *ὡσαύτως* "likewise" (9).[107]

Prepositions;[108] apart from ἀπό "on" / "from" and ὑπό "by", the geometric substratum is well highlighted by the frequency of διά "through", ἐκ "from", ἐν "in", ἐπί "on", κατά "at"; less engaged are εἰς "to", μετά "with" (σύν "with" is obviously normalized out), παρά "to", περί "about". Here is the list: ἄνευ "without" (2),[109] ἀντί "instead of" (2),[110] ἀπό (2740), διά (471), εἰς (293), ἐκ (565), ἐκτός "outside" (71), ἐν (622), ἕνεκεν "for the sake of" (1),[111] ἐντός "inside" (74), ἐπί (630), κατά (476), μετά (202), μεταξύ "between" (42), παρά (243), παρέξ "except" (11),[112] περί (153), πρό "before" (21), πρός "to" (4078), σύν (5),[113] ὑπό (2672), χωρίς "except" (6).[114]

Particles are treated in detail in Sects. 1.1, 3.1, 4.5.2, 5.1.3, and 5.3. Their list is as follows (subordinants are underlined, coordinants are in italics, inferential and scope particles are unmarked, modal and negative particles are in boldface):[115] ἀλλά "but" (451), **ἄν** (30), ἄρα "therefore" (4042), γάρ "in fact" / "for" (1055), δέ "and" (2655), δή "thus" (735), ἐάν "if" (571), εἰ "if" (336), εἴπερ "since … really" (1), εἴτε "either" / "or if" (10), ἐπεί "since" (1148), ἐπειδήπερ "since … really quite" (40), ἐπείπερ "since … really" (2), ἕως "until" (12),[116] *ἤ* "or" / "than" (259), *ἤπερ* "than" (116), *ἤτοι* "either" (85), ἵνα "in order for" (2),[117] *καί* "and" / "also" (6715), μέν (952), **μή** "not" (221), *μηδέ* "nor" (1), *μήτε* "neither" / "nor" (7), ὅταν "whenever" (60), ὅτε "when" (3), ὅτι "that" (863), **οὐ** "not" (391), *οὐδέ* "neither" / "nor" "not … either" (113), οὖν "then" (261), *οὔτε* "neither" / "nor" (10), πλήν "with the exception of" (11), *τε* "both" / "and" (376), τοίνυν "now then" (1), τότε "then" (1),[118] ὡς "as" (1339), ὥστε "so that" (304).

---

[97] All occurrences are in the definition of a right angle in I.def.10, or in its applications.
[98] Isolated, but possibly authentic, occurrence in XII.5.
[99] Occurrences of καθάπερ in XII.17, V.19por, of καθό in X.23cor, of καθόλου in VI.20 (*bis*), X.23, X.10cor.
[100] Certainly authentic occurrences, in VII.1 and X.2, both twice.
[101] Isolated occurrence, in X.41/42, spurious as the entire lemma.
[102] Certainly authentic occurrences, in the definitory clause of the solids of revolution, in XI.def.14, 18, 21, XIII.13–16.
[103] Isolated, but possibly authentic, occurrence in XIII.18.
[104] Occurrences in V.19por and X.9, 28.
[105] All occurrences are in the enunciations of I.22 and of XI.23, or in their applications (see Sects. 4.2 and 4.2.1).
[106] Occurrences in VII.def.10–11.
[107] For ἁπλῶς, see Sect. 4.5.5 (the other occurrence is in XII.4*alt*); for δηλαδή and δηλονότι, Sect. 4.5.2; for ἔμπροσθεν, Sects. 4.5.2 and 4.5.5; for ἐπάνω, Sect. 4.5.5 (the other occurrences are in V.8; VIII.13; X.35, 44, 96); for ἔτι, Sect. 5.3.5; for ἤδη, Sect. 4.5.5; for μήν, Sect. 4.5.5; for πάντως, Sect. 5.1.1; for ποτέ and πότε, Sect. 5.3.1; for τυχόντως, Sect. 5.1.1; for ὡσαύτως, Sect. 4.5.2 [the other occurrences are in V.8, 15 (*bis*), 16, 23, VI.1, X.23, XII.15].
[108] On the sometimes "aberrant" use of prepositions in Aristotle see Stevens 1936.
[109] Occurrences in XIII.1–2*alt*, both in spurious segments of text.
[110] Both occurrences in *alt*X.27, in a spurious segment of text.
[111] Obviously a spurious occurrence, in XII.5.
[112] Certainly authentic occurrences, in IX.13 (*ter*), 14 (*ter*), 32, 36 (*quater*).
[113] Certainly authentic occurrences, in II.def.2, II.6 (*bis*), 10 (*bis*).
[114] All occurrences in IX.10, certainly authentic.
[115] As said, some particles, as for instance καί "and" / "also" and ἤ "or" / "than", may belong to more than one category.
[116] Certainly authentic occurrences, in V.8, VII.1 (*bis*), VIII.3, 9, IX.36 (*bis*), X.1–2 and X.1*alt*, XII.5, within arguments exhibiting a marked procedural character.
[117] Occurrences in X.28, both of them spurious.
[118] The unique occurrences of τοίνυν and of τότε are in I.21 and in V.def.7, respectively.

Nouns provide the most interesting data. Let us first list nominalized states of affairs or relations (mainly –σις nouns); even the items characterized by a very low frequency are certainly authentic: ἀναλογία "proportion" (17), ἀναστροφή "conversion" (1), ἁφή "point of contact" (12), διαίρεσις "division" (12), διχοτομία "bisection" (3), δύναμις "power" (only the relational dative δυνάμει "in power") (355), ἐπαφή "point of tangency" (10), θέσις "position" (1), κλίσις "inclination" (5), λῆψις "taking" (8), λόγος "ratio" (654), μέρος "part" (359), ὁμοιότης "similarity" (13), παραβολή "application" (1), πολλαπλασιασμός "multiplication" (3), συναφή "point of tangency" (3), σύνθεσις "synthesis" (13), σχέσις "relation" (1), τάξις "order" (34), ὑπεξαίρεσις "removing" (1), and finally ὑπεροχή "excess" (20).[119]

Let us also see the nouns that have an obvious metadiscursive connotation; they are all characterized by a low frequency and most of them are certainly spurious: ἀνάλυσις "analysis" (7), ἀπαγωγή "reduction" (1), ἀπόδειξις "proof" (2), ἀρχή "beginning" (25),[120] βιβλίον "book" (2), δείξις "proof" (1), ἐπίσκεψις "investigation" (1), εὕρεσις "finding" (1), θεώρημα "theorem" (8), καταγραφή "diagram" (11), λῆμμα "lemma" (8), λημμάτιον "little lemma" (1), οἰκειότης "property" (1), ὀνομασία "denomination" (1), πόρισμα "porism" (1), πραγματεία "treatise" (1), πρόβλημα "problem" (2), πρότασις "enunciation" (2), προύπαρξις "coming first" (1), τρόπος "way" (7), and finally ὑπόθεσις "supposition" (2).[121]

What remains is a number of count nouns that is small if compared to an exceedingly rich verb system (detail in Sect. 4.3); the entire ontology of the *Elements* is listed in the following lines (nominalized adjectives are underlined; number-theoretical objects are in italics):[122] ἄκρος "extreme" (126),[123] ἄξων "axis" (49), ἀπότμημα "segment" (5), ἀποτομή "apotome" (242), *ἀριθμός* "number" (1012), βάθος "depth" (1), βάσις "base" (768), γνώμων "gnomon" (48), γραμμή "line" (74), γωνία "angle" (1118), δεκάγωνον "decagon" (22), διαγώνιος "diagonal" (3), διάμετρος "diameter" / "diagonal" (171), διάστημα "radius" (30), *δυάς* "dyad" (10), δωδεκάεδρον "dodecahedron" (15), εἶδος "form" (74), εἰκοσάεδρον "icosahedron" (21), ἔλλειμμα "defect" (3), ἑξάγωνον "hexagonon" (36), ἐπίπεδον "plane" (459), ἐπιφάνεια "surface" (18), ἑτερόμηκες "oblong" (1), εὐθεῖα "straight line" (1084), ἡμικύκλιον "semicircle" (61), κέντρον "centre" (267), κορυφή "vertex" (92), κύβος "cube" (162), κύκλος "circle" (713), κύλινδρος "cylinder" (133), κῶνος "cone" (136), μέγεθος "magnitude" (188), μέτρον "measure" (55), μῆκος "length" (432), *μονάς* "unit" (280), ὀκτάεδρον "octahedron" (16), ὄνομα "name" (204), ὅρος "term" (4), παραλληλεπίπεδον "parallelepiped" (79), παραλληλόγραμμον "parallelogram" (340), παραπλήρωμα "complement" (13), πεντάγωνον "pentagon" (92), πεντεκαιδεκάγωνον "pentadecagon" (5), πέρας "boundary" (18), περιφέρεια "arc" / "circumference" (218), *πηλικότης* "value" (2), πλάτος "width" (119), πλευρά "side" (514), *πλῆθος*

[119] Note the rich lexicon of intersection / tangency. Location of the occurrences of poorly represented items: ἀναστροφή, V.def.16; διχοτομία, X.41/42, 43, 44; θέσις, XII.17; κλίσις, I.def.8, XI.def.5–7, 11; λῆψις, V.def.12–17, XIII.1*alt* (*bis*); παραβολή, X.16/17; πολλαπλασιασμός, V.def.5, V.4 (*bis*); συναφή, III.11 (*ter*); σχέσις, V.def.3; ὑπεξαίρεσις, V.def.17.

[120] Always in the expression ἐξ ἀρχῆς "original".

[121] Location of the occurrences of poorly represented items: ἀνάλυσις, XIII.1–5*alt*; ἀπαγωγή, X.13 *vulgo*; ἀπόδειξις, III.31*alt*, V.8; βιβλίον, XII.2, XIII.17; δείξις, IV.16; ἐπίσκεψις, VII.31; εὕρεσις, X.90*alt*; λημμάτιον, X.41/42; οἰκειότης, X.39*alt*; ὀνομασία, X.39*alt*; πόρισμα, X.4por (I disregard the occurrences in mere inscriptions, and the same for θεώρημα and λῆμμα; these are 31, 1, and 18 occurrences, respectively); πραγματεία, X.28/29II; πρόβλημα, XI.23*alt* (*bis*); πρότασις, XI.35, 37; προύπαρξις, X.40*alt*; ὑπόθεσις, X.44, 47.

[122] Some of these nominalized adjectives (most frequently in the neuter) are also used as real adjectives (masculine or feminine); a case in point are the τριγώνους βάσεις "triangular bases" in XII.3–9, 12.

[123] We may doubt that the nouns ἄκρος, ὅρος, and πέρας designate objects, despite the fact that straight lines may be drawn from or through them. In a strict geometric context (for the "extremes" are also terms in a proportion), the first noun is mainly used in the adverbial formula ἄκρον καὶ μέσον λόγον "extreme and mean ratio"; the third characterizes both points and lines as boundaries of lines and of surfaces, respectively (in I.def.3, 6: but these are not definitions: this shows that splitting the clusters of definitions, as Heiberg does in his edition, may lead to incongruities).

"multiplicity" (98), πολύγωνον "polygon" (95), πολύεδρον "polyhedron" (25), πρίσμα "prism" (124), πυραμίς "pyramid" (291), ῥομβοειδές "rhomboid" (1), ῥόμβος "rhombus" (1), σημεῖον "point" (500), στερεόν "solid" (424), σφαῖρα "sphere" (151), σχῆμα "figure" (70), τετράγωνον "square" (954), τετράπλευρον "quadrilateral" (27), τμῆμα "segment" (235), τομεύς "sector" (27), τομή "section" (49), τόπος "place" (4), τραπέζιον "trapezium" (3), τρίγωνον "triangle" (875), τρίπλευρον "trilateral" (3), ὕψος "height" (172), χωρίον "region" (158).[124]

Contrary to what happens in the mathematical verb system (see Sect. 4.3), there is a one-to-one correspondence *in context* between these nouns and their referents (many of the above terms are in fact *defined*, even if some of them are *only* defined). An example is provided by the three nouns τμῆμα, τομή, τομεύς. A τομή is any object generated as a section of higher-dimensional objects, typically a point as a section of two lines or a straight line as a section of two planes; a τομεύς rigidly designates a circular sector (III.def.10, never used in the main text of the *Elements*, 26 occurrences elsewhere, all in Theon's *additamentum* VI.33*alt*—this shows that a fair number of occurrences may originate in a single proposition; the proposition proves that circular sectors are proportional to the arcs that identify them); a τμῆμα designates both a segment of a straight line and a segment of a circle (III.def.6), but the referent is always well-defined in context.

I shall not enter the sea of the 92 qualifiers; a few remarks will suffice. First, I shall discuss the function of indefinite adjectives like ὁποσοιδηποτοῦν "as many as we please" (always in the plural) in Sect. 5.1.4, where I shall call them "generalizing qualifiers" (89 occurrences in all). Second, even if this is prima facie surprising, there are 14 cardinal adjectives and 11 ordinal adjectives in the whole of the *Elements*, for a total of 3095 occurrences. Third, the most frequent qualifiers are ἀδύνατος "impossible" (118), ἄρτιος "even" (101), ἐλάσσων "less" (514), ἐλάχιστος "least" (162), ἰσογώνιος "equiangular" (132), ἰσόπλευρος "equilateral" (111), ἴσος "equal" (3424), κοινός "common" (218), λοιπός "remaining" (488), μείζων "greater" (1039), μόνος mainly adverbial μόνον "only" (264), ὅλος "whole" (444), ὁμόιος "similar" (285), ὀρθογώνιος "rectangular" (161), ὀρθός "right" (649), παράλληλος "parallel" (282) and—pertaining to the theory of irrational lines—ἄλογος "irrational" (156), ἀσύμμετρος "incommensurable" (532), μέσος "medial" / "mean" (801), ῥητός "expressible" (904), σύμμετρος "commensurable" (789).

Finally, a clear case of redundancy must be noted: three different "multiplicative" adjectives are strictly synonymous and mean "double" / "duplicate" (διπλάσιος, διπλασίων, διπλοῦς), three mean "quintuple" (πενταπλάσιος, πενταπλασίων, πενταπλοῦς), two mean just "multiple" (πολλαπλάσιος, πολλαπλασίων), three mean "quadruple" (τετραπλάσιος, τετραπλασίων, τετραπλοῦς), three mean "triple" / "triplicate" (τριπλάσιος, τριπλασίων, τριπλοῦς). Likewise, there are two multiplicative demonstrative pronouns "such a multiple" (τοσαυταπλάσιος, τοσαυταπλασίων) and two multiplicative correlative pronouns "whichever multiple" (ὁσαπλάσιος, ὁσαπλασίων). Happily, only one lemma is found for εἰκοσιπενταπλάσιος "decaquintuple", ἐνναπλάσιος "ennuple", and ἑξαπλάσιος "sextuple" (thus, 21 lemmas in all). The only plausible explanation of this redundancy is that such words were systematically written in abbreviated form in early majuscule manuscripts, as follows: ΤΡΙΠ$^{\Lambda}$. The extant forms just reflect the several ways later copyists resolved these abbreviations. Be that as it may, the redundancies in the supply of verbs and deverbals (Sect. 4.3 and n. 119 above) and adjectives show that the mathematical lexicon is far from being optimized.

---

[124] Location of the occurrences of poorly represented items: ἀπότμημα, XII.12, 10 (*bis*), 11, 12; βάθος, XI.def.1; διαγώνιος, XI.28 (*bis*), 38; δυάς, IX.32 (*ter*), 34 (*quater*), X.36, *alt*X.27 (*bis*); ἔλλειμμα, VI.27, 28 (*bis*); ἑτερόμηκες, I.def.22; ὅρος, I.def.13–14, V.def.8, X.20*alt*; πηλικότης, V.def.3, VI.def.5; ῥομβοειδές and ῥόμβος, I.def.22; τόπος, III.16 (*quater*); τραπέζιον, I.def.22, I.35 (*bis*); τρίπλευρον, I.def.19–21.

2. VALIDATION AND TEMPLATES

Aristotle repeatedly held that the Iliad is "one saying [λόγος] by connectors".[1] Bewildering as this claim may be, it applies to all Aristotelian treatises, and in fact to many Greek writings: ancient Greek language—which did not use extensive punctuation—availed itself of connectors to structure every consistent and sustained complex of thoughts, from subsentential units to suprasentential architectures like a treatise or a poem. Get a look at the connectors that open the non-liminal books of the Iliad or of any Aristotelian treatise; issues of authenticity might even be settled on this sole basis.[2] There is only one major exception to this rule: mathematical treatises. The exception is a glaring one. No connectors open any non-liminal book of any mathematical treatise. No connectors open the self-contained mathematical units of a mathematical treatise: the theorems and the problems in the *Elements* are not linked with one another by connectors—they are not linked by any linguistic token at all. Still, every self-contained mathematical unit of the *Elements* bristles with connectors. What is it, then, that links such units with one another? And what is it that makes Greek mathematical treatises, and in fact Greek mathematics as a whole, a unique λόγος?

My one-word answer is: validation. What does it mean "validation"? It means assessing the pertinency and the validity of a linguistic carrier of meaning by its conformity to a "template" set as a standard. Here, "conformity" means "linguistic conformity": the criterion of validity is a formal criterion, and in a strong sense. Thus, what I shall show in this Section is that Greek mathematics is a net held together not only (and only locally) by connectors, but, locally as well as globally, by a generalized principle of anaphoric conformity to a template. Validation occurs at several levels: subsentential formulaic expressions, sentences, simple and less simple arguments, self-contained mathematical units such as theorems and problems—and, finally, entire strings of such units, in a scholarly and metadiscursive approach.[3] The general principle of anaphora just stated is put to effect, with specific mechanisms, for all the linguistic units just listed, and in addition by means of the standard lexical and syntactic links: the bare designations and the connectors.

Validation might seem to find a prominent application in the *geometric* method of analysis and synthesis: well, we shall see that this is not the case, unless we persist in looking at Greek mathematics by using a wrong historiographic paradigm.[4] This paradigm regards the stylistic code of which the Ancients said that it operates γραμμικῶς "by means of lines"[5]—that is, the demonstrative approach typical of geometry—as their only serious achievement from the logical and technical points of view. Such an obsession for rigour and for mathematical content has made scholars blind to the fact that Greek mathematics is far more complex as a literary genre, for, as we have just seen, it comprises three main, independent, and perfectly legitimate, areas of mathematical discourse.

---

[1] See *APo.* II.10, 93b35–37; *Metaph.* I.6, 1045a12–14; *Po.* 20, 1457a28–30. On this Aristotelian *dictum*, read the bewilderment in Barnes 2007, 180, 231.

[2] For instance, the presence of a liminal δέ shows that the *Sophistici elenchi*, in the form we read it and despite some contrary evidence, is the ninth book of the *Topica*. As explained in Sect. 1.1, "opening" and "liminal" said of a connector must be taken in the sense of Wackernagel's law.

[3] I have shown that Apollonius' and Menelaus' approaches to foundational themes were of such a kind: see Acerbi 2010a and 2010b, and for instance Sect. 5.2.1.

[4] I have argued this point in Acerbi 2011a and 2011b: within the geometric paradigm, analysis could not, and in origin was not intended to, validate the associated synthesis: constructive issues are a later appendage to an original, systematic research program carried out by means of pure analyses. See Sect. 2.4.1 for a sketch of my argument.

[5] The approach "by means of lines" was also adopted in number theory: see *El.* VII–IX and, for an explicit statement using the above adverb, Pappus, *Coll.* II.4 and 15. See also Ptolemy, *Alm.* I.9, in *POO* I.1, 31.3–6, who in order to explain τὴν πραγματείαν τῆς πηλικότητος τῶν ἐν τῷ κύκλῳ εὐθειῶν ἅπαξ γε μελλήσοντες ἕκαστα γραμμικῶς ἀποδεικνύειν "the issue of the chords in a circle, will prove the whole topic by means of lines once and for all".

F. Acerbi, *The Logical Syntax of Greek Mathematics*, Sources and Studies
in the History of Mathematics and Physical Sciences,
https://doi.org/10.1007/978-3-030-76959-8_2

Validation is a solution to two problems. First, pertinence: in which way can a result be used in an argument so as to make it evident that exactly *that* result is used? Second, validity: what is it that conveys validity to a given step of an argument, or to an argument as a whole? To take again our paradigmatic proposition III.2 as an example: what is it that makes us readily concede that in the second "paraconditional" in the proof of III.2—namely, the one formulated by the sentence "and since one side AEB of a triangle ΔAE turns out to be produced, therefore angle ΔEB is greater than ΔAE"—a supposition made in the construction of III.2 is used, and theorem I.16 is applied? And what does it mean "to apply" a theorem within an argument?

The solution to these problems provided by validation assesses the pertinence and the validity of a step in an argument—or of an argument as a whole—by means of its linguistic conformity to a template set as a standard: as said, this is a formal criterion, and in a strong sense. Of course, the point in establishing validation as a criterion is the meaning of "linguistic conformity". For this reason, anyone testing validity by conformity to a linguistic template must be aware of the general linguistic rules that allow adapting syntactic constructs to another context. Greek mathematics adds a specific principle of its own: a general strategy of abbreviation of linguistic complexes, which is in particular carried out by ellipsis of the nominal forms that figure in the templates (cf. Sect. 3.3).

As said, validation operates at several levels: subsentential, sentential, and suprasentential. In each of these cases, a primary linguistic instance, possibly after undergoing a standard sequence of transformations, is explicitly or implicitly referred to as a justification of sorts for writing, *exactly in the form in which it is written*, any perceivedly related linguistic item that has a mathematical import. The linguistic item that carries the reference can be a designation, a formulaic expression, a statement, a constructive act, or a computation. The form of the reference is linguistically stable, the referent univocally identifiable. After all, in natural-language mathematics as Greek mathematics is, the regularity of the adopted linguistic patterns is the only means for shaping arguments whose validity is independent of the singular terms that figure in them—that is, arguments that can be said to be valid in virtue of their form.[6]

I shall first discuss three philosophically-oriented texts—by Aristotle, Galen, and Alexander of Aphrodisias (for templates in Stoic logic see Sects. 2.3 and 5.2.2)—that can be read in the perspective of validation (Sect. 2.1). I shall then pass to mathematical practice. My discussion is organized in three Sections, according to the size of the logico-linguistic units involved: subsentential units like designations and relations (Sect. 2.2), sentential units like constructive acts and deductive steps (Sect. 2.3), and suprasentential units (Sect. 2.4). Suprasentential validation mainly consists in the geometric method of analysis and synthesis (Sect. 2.4.1; an application of the method bears close resemblance to Stoic "analysis", as we shall see) and in poorly-known avatars of it (Sect. 2.4.2). Late mathematicians systematically applied analysis and synthesis to make the three stylistic codes described in Sect. 1 interact, thereby producing the most spectacular application of the general principle of validation, but at the same time clearly showing its limitations. The main features of such an interaction I shall outline, along with such shorter-range forms of validation by conformity to a template as operate on subsentential and sentential units, in this Section. Discussions of specific forms of validation adopted within the demonstrative code will also be found time and again in the relevant subsections of Sects. 4 and 5. The evidence I shall present will, I hope, corroborate the general interpretive framework to the effect that validation innervates the whole Greek mathematical corpus, thereby shaping it as a unique λόγος.

[6] This crucial point will be further discussed in Sect. 4.5.1.4.

2.1. Aristotle and Galen on Linguistic Templates

At least two philosophical texts suggest that a formal principle of validation by conformity to a linguistic template was taken to be operative in mathematics, if not assumed as operative, by thinkers other than mathematicians.[7] Let us first read Aristotle, *APr.* I.24, 41b13–22; the context is a discussion of the fact that a general premise must figure in every valid syllogism:[8]

μᾶλλον δὲ γίνεται φανερὸν ἐν τοῖς διαγράμμασιν, οἷον ὅτι τοῦ ἰσοσκελοῦς ἴσαι αἱ πρὸς τῇ βάσει. ἔστωσαν εἰς τὸ κέντρον ἠγμέναι αἱ Α Β. εἰ οὖν ἴσην λαμβάνοι τὴν ΑΓ γωνίαν τῇ ΒΔ μὴ ὅλως ἀξιώσας ἴσας τὰς τῶν ἡμικυκλίων, καὶ πάλιν τὴν Γ τῇ Δ μὴ πᾶσαν προσλαβὼν τὴν τοῦ τμήματος, ἔτι δ' ἀπ' ἴσων οὐσῶν τῶν ὅλων γωνιῶν καὶ ἴσων ἀφῃρημένων ἴσας εἶναι τὰς λοιπὰς τὰς Ε Ζ, τὸ ἐξ ἀρχῆς αἰτήσεται, ἐὰν μὴ λάβῃ ἀπὸ τῶν ἴσων ἴσων ἀφαιρουμένων ἴσα λείπεσθαι.

It becomes clearer in diagrams, for instance that ‹the angles› at the base of the isosceles are equal. Let there be ‹straight lines›, A, B, drawn to the centre. Then if one assumed (1) angle ΑΓ equal to ΒΔ without asserting in general that (1*a*) the ‹angles› of the semicircles are equal, and again (2) Γ to Δ without further assuming that (2*a*) any ‹angle› of the segment is, and further that, (3) equal ‹angles› being also removed from the whole angles, which are equal, the remaining E, Z are equal, the question will be begged, if one does not assume that, (3*a*) equal items being removed from equal items, equal items remain.

Thus, Aristotle points out that the proof he reports is invalid, and it is actually a circular proof, unless three premises of increasing generality are made explicit, intended to validate particular premises that are either instantiations of these general premises or their applications to specific objects. The general premises are (1*a*) that "the angles of the semicircles are equal", (2*a*) that the two angles of a circular segment are equal in every instance, and finally, (3*a*) the "equals-from-equals" premise, one of Aristotle's mathematical hobby-horses.[9] If validation by conformity to a linguistic template does not seem at issue here, then we might wonder how exactly could Aristotle think that assuming (3*a*) licenses one safely to assume (3).

An interesting parallel can be set with a passage in Galen that has been studied in the context of the so-called "relational syllogisms". Galen provides some examples of these syllogisms and tries to show that they validly conclude (detail in Sect. 4.5.1.1).[10] What is at issue is exactly the way in which particular deductions are validated by general axioms. Let us read Galen's argument in *Inst. Log.* XVI.5–10, a passage that requires emendations, as is frequently the case with this treatise:

[7] The τόποι "commonplaces" in Aristotle's *Topica* appear to have exactly the function of logico-syntactic templates: see e.g. Brunschwig 1967, XXXVIII-XLV.

[8] For the way angles are designated see Sect. 2.2. In non-technical contexts, I shall not translate the particle that usually opens any extract. After all, (in translation) this is frequently the punctuation sign that closes the previous sentence.

[9] Aristotle includes the most general principle (3*a*) here mentioned among the "common axioms" (*APo.* I.10, 76b20–21; I.11, 77a26–31); recall that we read it in the *Elements* as I.cn.3. If we accept that the three premises are put on the same logical footing by Aristotle, and since he apparently considers the proof he expounds to be both representative of the mathematical style and fallacious, we must conclude that none of the general premises was made explicit as a principle in any mathematical writing, a hypothesis that is by no means in contradiction with the plain fact that Aristotle did recognize the importance of the "equals-from-equals" premise as a starting point of so many mathematical deductions. By the way, if Aristotle's example is to reflect the mathematical practice of his times, the impression one gets as for rigour and denotative consistency is rather chilling. Granted, the argumentative and denotative drawbacks of this "proof" can of course be fixed; nevertheless, it must be remarked that the present text of the *Elements* does not allow framing a proof without deductive gaps, essentially because it does not treat non-rectilinear angles. Curvilinear angles are an essential ingredient in the study of homeomeric lines: see my reconstruction in Acerbi 2010a, 27–34.

[10] See also the discussion in Barnes 2007, 419–447, to be referred to again in Sect. 4.5.1.1. My "template" performs the task which Barnes calls "underwriting".

5 πολὺ δὲ πλῆθός ἐστιν, ὡς ἔφην, ἐν ἀριθμητικοῖς τε καὶ λογιστικοῖς τοιούτων συλλογισμῶν ὧν ἁπάντων ἐστὶ κοινὸν ἔκ τινων ἀξιωμάτων τὴν αἰτίαν ἴσχειν συστάσεως [...] **6** ὄντος γὰρ ἀξιώματος τοῦδε καθόλου τὴν πίστιν ἔχοντος ἐξ ἑαυτοῦ "τὰ τῷ αὐτῷ ἴσα καὶ ἀλλήλοις ἐστὶν ἴσα", συλλογίζεσθαί τε καὶ ἀποδεικνύναι ἔστιν ὥσπερ Εὐκλείδης ἐν τῷ πρώτῳ θεωρήματι τὴν ἀπόδειξιν ἐποιήσατο τὰς τοῦ τριγώνου πλευρὰς ἴσας δεικνύων· ἐπεὶ γὰρ τὰ τῷ αὐτῷ ἴσα καὶ ἀλλήλοις ἴσα ἐστίν, δέδεικται δὲ τὸ πρῶτόν τε καὶ τὸ δεύτερον τῷ τρίτῳ ἴσον, ἑκατέρῳ αὐτῶν ἴσον ἂν εἴη οὕτω τὸ τρίτον. **7** ὄντος δὲ πάλιν ἀξιώματος ἐξ ἑαυτοῦ πιστοῦ τοῦδε "ἂν ἴσοις ἴσα προστεθῇ, καὶ τὰ ὅλα ἴσα ἔσται", ἐὰν ὡμολογημένων ἴσων ἀλλήλοις εἶναι τοῦ πρώτου καὶ δευτέρου προτεθῇ τι καθ' ἑκάτερον ἴσον ἴσον, ἔσται καὶ τὸ ὅλον τῷ ὅλῳ ἴσον, ὡδί πως λεγόντων ἡμῶν "ἐπεὶ τὸ πρῶτον ἴσον ἐστὶ τῷ δευτέρῳ, πρόσκειται δὲ τῷ μὲν πρώτῳ τὸ γ^{ον} τῷ δὲ δευτέρῳ τὸ δ^{ον} ἴσα ὄντα καὶ αὐτά, γενηθήσεται καὶ τὸ ὅλον τῷ ὅλῳ ἴσον". **8** ὡσαύτως δὲ κἀπειδὰν ἀπό τινων ἴσων ἴσα ἀφαιρεθῇ, δυνησόμεθα λέγειν "ἐπεὶ τὸ ὅλον τῷ ὅλῳ ἴσον, ἀφαιρεῖται δὲ ἀφ' ἑκατέρου αὐτῶν ἴσα τάδε, καὶ τὸ λοιπὸν τόδε τῷδε τῷ λοιπῷ ἴσον ἔσται". **9** οὕτως δὲ καὶ τὸ τοῦ διπλασίου διπλάσιον τετραπλάσιον ἔσται· ἐὰν δή τινος [ἕτερον] διπλάσιον ληφθῇ κἀκείνου δὲ πάλιν διπλάσιον ληφθῇ, ἔσται τοῦτο τὸ γ^{ον} τοῦ α^{ον} τετραπλάσιον. **10** ὁμοίως δὲ κἀπὶ τῶν ἄλλων ἁπάντων ἡ σύστασις τῶν ἀποδεικτικῶν συλλογισμῶν κατὰ δύναμιν ἀξιώματος ἔσται συνημμένου ἐπὶ ἀριθμῶν ἐπί τε ἄλλων πραγμάτων ἐν τῷ πρός τι γένει καὶ αὐτῶν ὑπαρχόντων.

5 As I have said, in arithmetic and logistic there are plenty of such syllogisms, whose common trait is to get the reason for their formulation from some specific axioms [...] **6** For, there being a general self-validating axiom, this one: "items equal to a same item are also equal to one another", it is possible to argue by proof exactly in the way Euclid made his proof in the first theorem, showing that the sides of the triangle are equal; for since items equal to a same item are also equal to one another, and both the first and the second turn out to be proved equal to the third, the third[11] would in this way be equal to each of them. **7** There being again a self-validating axiom, this one: "if equal items be added to equal items, the wholes will also be equal", if, the first and the second being agreed to be equal to one another, something equal be added to each equal item, the whole will also be equal to the whole, and we might say something like this: "since the first is equal to the second, and the 3rd turns out to be added to the first and the second to the 4th, which are also equal, the whole will also result equal to the whole". **8** Likewise, even whenever equal items be removed from some specific equal items, we might say "since the whole is equal to the whole, and these equal items are removed from each of them, this remainder will also be equal to this remainder". **9** So, the double of the double will also be quadruple; thus if [another] double of something be taken and again the double of that be taken, this 3rd will be quadruple of the 1st. **10** Similarly, for all the others the formulation of the demonstrative syllogisms will also result in virtue of a conjoined axiom, for numbers and for all other things as also belong to the relational genus.

The crucial point in this argument is the meaning of the term σύστασις, which I have translated "formulation" and which Galen uses again, within expressions analogous to the two sentences just read, in *Inst. Log.* XVI.11–12 and XVII.1–2. Σύστασις is the *nomen actionis* related to the verb συνίστημι, which canonically denotes the formation of simple and non-simple statements in Stoic logic:[12] the σύστασις amounts to "compounding" a statement from more elementary components, like a connector or a negation and simple statements. Linguistically, the σύστασις coincides with

[11] This correction of the attested πρῶτον "first" is certain, even if Galen's argument requires something like "the first and the second would in this way be equal to one another".

[12] See for instance D.L. VII.70 and Sextus Empiricus, *M* VIII.108–112. See also the Chrysippean title περὶ τρόπων συστάσεως πρὸς Στησαγόραν "On the formulation of modes to / against Stesagoras" (D.L. VII.194). For the Stoic classification of statements see Goulet 1998.

the actual formulation of a sentence.[13] If we accept this reading, Galen's position amounts to contending that some inferences strictly typical of mathematics are validated by a general axiom insofar as (1) these inferences have a stable linguistic form, which (2) derives from an axiom assumed as a linguistic template. Thus, the steps of an inference are in a strong sense governed by a primary non-instantiated form, which determines their σύστασις: such an anaphora is a kind of validation that operates by identification of, and reference to, a stable linguistic format. Of course, infinite regress must be stopped, so that there must be a "self-validating" axiom (litt. an axiom "drawing credibility from itself"), as Euclid's I.cn.1 in Galen's text.

This approach also allows clarifying the difference between Galen's and Alexander's strategies aimed at showing that relational syllogisms of the kind "*A* is equal to *C*; but *B* is equal to *C*: therefore *A* and *B* are equal", which have particular premises and hence cannot be Aristotelian syllogisms, do validly conclude. Alexander's strategy consists in "forcing"[14] these relational syllogisms into a predicative format:[15] he introduces a general axiom as an additional premise and "merges" the two particular premises of the original syllogism into a single coassumption (cf. Sect. 4.5.1.4):

ὅμοιον τούτῳ καὶ τὸ λαβόντας τὸ Α μεῖζον εἶναι τοῦ Β καὶ τὸ Β τοῦ Γ ἡγεῖσθαι συλλογιστικῶς δείκνυσθαι τὸ καὶ τὸ Α τοῦ Γ μεῖζον εἶναι, ἐπεὶ ἀναγκαίως τοῦτο ἕπεται. ἀλλ' οὔπω συλλογισμὸς τοῦτο, ἂν μὴ προσληφθῇ καθόλου πρότασις ἡ λέγουσα "πᾶν τὸ τοῦ μείζονός τινος μεῖζον καὶ τοῦ ἐλάττονος ἐκείνου μεῖζόν ἐστι", τὰ δὲ κείμενα δύο πρότασις γένηται μία ἡ ἐλάττων ἐν τῷ συλλογισμῷ λέγουσα "τὸ δὲ Α τοῦ Β μείζονος ὄντος τοῦ Γ μεῖζόν ἐστιν"· συναχθήσεται γὰρ οὕτως τὸ καὶ τὸ Α τοῦ Γ μεῖζον εἶναι κατὰ συλλογισμόν.	Similar to this is thinking that taking "A is greater than B and B than Γ" one also proves syllogistically "A is greater than Γ", on the grounds that this follows necessarily. But this is not a syllogism unless a general premise saying "everything greater than something greater is also greater than what is less than that" is coassumed and the two items assumed become one premise—the minor in the syllogism—saying "and A is greater than B, which is greater than Γ"; for in this way it will be deduced by syllogism that A is also greater than Γ.

If I am right in my reading, Galen does not recommend assuming a general axiom as an additional premise: he considers instead the reference to the general axiom a form of linguistic validation, possibly but not necessarily made explicit by the actual presence of that very axiom as an additional premise. Such additional premises, when we find them in Galen's discussion,[16] simply reflect his concern with closely adhering to the argumentative scheme of I.1, where the transitivity-of-equality axiom expressly figures as a coassumption (cf. the discussion in Sect. 4.5.1.1).[17] In all other relevant passages in Galen, a relational syllogism is a particular argument that "derives the reason for its formulation from some axioms"; these syllogisms also draw "the credibility of their formulation and of their demonstrative force from a generic axiom" (*Inst. Log.* XVI.12).[18]

[13] The same point is made in Barnes 2007, 429–430.

[14] At *Inst. Log.* XVI.1, Galen uses the verb βιάζω "to force" to describe Alexander's procedure, which he generically ascribes to "Aristotelians".

[15] See *in APr.*, 21.28–22.23, 68.21–69.4, 344.9–346.6; *in Top.*, 14.18–15.14. The quote is from *in APr.*, 344.20–27.

[16] They can be found at *Inst. Log.* I.3 and XVI.6, read above.

[17] The occurrence in *Inst. Log.* XVI.11 falls into a lacuna whose integration is open to debate. Hankinson 1994, 69, proposes a less general axiom than the one integrated by Kalbfleisch (Hankinson is not willing to introduce premises that display multiple generality, as the example in XVI.11 would instead require; on multiple generality in Stoic logic see Bobzien, Shogry 2020). The possibility that Galen did not intend to introduce an additional premise, but to employ the axiom as an inference rule, is discussed and dismissed, with arguments I find unconvincing, in Barnes 1993, 184–185.

[18] In the perspective of a correct linguistic formulation aiming at validation by means of an axiom can also be read Galen's discussion of the ἰσοδυναμουσῶν προτάσεων "equivalent premises" in *Inst. Log.* XVII.4–9.

2.2. SUBSENTENTIAL VALIDATION: FORMULAIC TEMPLATES

Before entering the issue of formulaic templates, I summarize the pragmatics of the most frequent lettered designations.[19] To see these designations in context, let us read the portion of I.5 (*EOO* I, 20.6–10) in which the denotative letters are introduced:[20]

ἔστω τρίγωνον ἰσοσκελὲς τὸ ΑΒΓ ἴσην ἔχον τὴν ΑΒ πλευρὰν τῇ ΑΓ πλευρᾷ, καὶ προσεκβεβλήσθωσαν ἐπ' εὐθείας ταῖς ΑΒ ΑΓ εὐθεῖαι αἱ ΒΔ ΓΕ. λέγω ὅτι ἡ μὲν ὑπὸ ΑΒΓ γωνία τῇ ὑπὸ ΑΓΒ ἴση ἐστὶν ἡ δὲ ὑπὸ ΓΒΔ τῇ ὑπὸ ΒΓΕ.

Let there be an isosceles triangle, ΑΒΓ, having side ΑΒ equal to side ΑΓ, and let straight lines, ΒΔ, ΓΕ, be produced in a straight line with ΑΒ, ΑΓ. I claim that angle ΑΒΓ is equal to ΑΓΒ and ΓΒΔ to ΒΓΕ.

We thus see that some geometric objects are assigned two letters (straight lines and sides, the latter certainly qua straight lines), others are assigned three letters (triangles and angles); that some of the complexes of letters are preceded by a preposition (angles), while others are not; that all such syntagms are preceded by an article. A natural typology of the most frequent lettered designations takes the number of assigned letters as the type parameter, as follows.

One letter is necessarily assigned to points. However, virtually any figure—and very frequently numbers—can be assigned just one letter, when they are not operated upon.[21] The following table provides a list of one-letter designations of mathematical objects; any such letter is necessarily preceded by an article in the same gender as the noun that gives the object its name (last column):

article	prepos.	article	letter	object
τὸ			Α	point
ἡ			Α	straight line
ἡ			Α	angle
ὁ			Α	circle
τὸ			Α	any figure
ὁ			Α	number
ἡ	πρὸς	τῷ	Α	angle

The one-letter designations of an angle deserve a short discussion,[22] for they provide good evidence for thinking that the standard correspondence "one letter" – "one point" is the outcome of an internal dynamics. In the Euclidean treatises on optics, in fact, we find one-letter designations of

[19] See below and Sect. 3.2 for the fact that the denotative letters are necessarily preceded by an article.

[20] The letters are first introduced in the "setting-out" (cf. Sect. 1.1). The translation of the imperative ἔστω as "let there be" will be fully justified in Sect. 3.1. Earlier discussions of the pragmatics of lettered designations in Greek mathematics include Netz 1999, sect. 2 and 133–135.

[21] For instance, a straight line in VI.17; angles in I.42, 44, 45, VI.7, X.32/33; a circle in Archimedes, *Sph. cyl.* I.5; rectilinear figures in *El.* VI.21 and 28–29; regions and solids frequently in Book XII. It remains that plane and solid figures are usually denoted by letters attached to their boundaries. The designation ἡ πρὸς τῷ Α is met about 100 times in the *Elements*, and occurs in Books I–IV, VI, XI, and XIII. I shall use "figure" as a synonym of "two- or three-dimensional geometric object". No reference is intended to graphic entities, for which I shall always use "diagram". In Greek, the two terms are σχῆμα and καταγραφή, respectively.

[22] In ἡ πρὸς τῷ Α, τὸ Α is the name of the vertex of the angle.

angles without a preposition—namely, designations of the kind ἡ A—much more frequently than in the *Elements*. The data are set out in the following table:[23]

treatise	proposition
Optica **A**	9, 38, 42*alt*, 43*alt*, 48, 53
Optica **B**	4, 7–8, 18, 24, 29, 34, 41–44, 48, 52
Catoptrica	1–6, 13–15, 20–21, 24–25, 28, 30

Actually, the range of designations for angles used in *Optica* **B** and in the *Catoptrica* is wider than this.[24] The typology is as follows—of course, and to repeat, every complex of letters (+ possibly a preposition) is preceded by an article:

1) two different letters, one for the vertex and one for the angle: *Optica* **B** 4, 7–8, 34, 41–44; *Catoptrica* 1–6, 13–15, 21, 24–25, 28, 30;
2) only one letter, which coincides with the name of the vertex: *Optica* **B** 9, 18, 24, 29, 43–44, 48, 52; *Catoptrica* 20;
3) one letter but with πρός: *Optica* **B** 19–20, 23–24, 28, 31, 36, 39, 58; *Catoptrica* 5, 13;
4) three letters, preceded by ὑπό (this is the canonical designation, see below): *Optica* **B** 6, 20, 23, 26, 35–36, 40, 55; *Catoptrica* 19, 20–22, 30;
5) angles are mentioned but they are not named: *Optica* **B** 5, 38, 45–47; *Catoptrica* does not exhibit occurrences.

In the optical treatises and contrary to the *Elements*, types (1) and (2) are locally compositional: angle ἡ AB is the sum of angles ἡ A and ἡ B.[25] Of course, the angle resulting from the composition of two angles named with one letter is no longer named with one letter only. As a consequence, this notation, which coincides with the notation assumed in Aristotle's passage read in Sect. 2.1, is inconsistent. However, this drawback is partly harmless because in our texts these angles are not operated upon except by composing them. Type (1) is also obviously redundant; still, its presence shows that the stylistic practice of specific branches of Greek mathematics had no qualms in admitting of such a redundancy, which was regularized away in more formal treatises.[26]

Two letters. Mainly straight lines, thus: ἡ AB. Numbers are assigned two letters when they result from addition or when operated upon. Quadrilaterals are often named after the letters assigned to two opposite vertices, thus: τὸ AΓ, if the vertices of the quadrilateral are cyclically assigned the letters A, B, Γ, Δ. A square is univocally determined by its side, so that its designation names the letters assigned to two adjacent vertices, thus: τὸ ἀπὸ AB. The fact that this is a square and not a generic quadrilateral is indicated by the preposition ἀπό.

[23] There are two recensions of Euclid's *Optica*, called **A** and **B**, neither of which can straightforwardly be dismissed as less genuine than the other; the way lettered designations are assigned is exactly one of the features that sharply demarcate them: see Jones 1994 and Knorr 1994.
[24] As for *Optica* **A**, most of the designations of angles are standard one- or three-letter names.
[25] In the *Catoptrica*, only propositions 13, 24, 28, 30 do not contain sums of angles designated by juxtaposition of letters.
[26] Other examples can be found in Diocles, *On burning mirrors* 1–3; Hero, *De speculis* 6–8 and 10; and most notably Aristotle, for instance *APr.* I.24, 41b13–22 (read in the previous Section), *APo.* II.11, 94a28–34, and *Mete.* III.5, 376a29. On this basis (the argument in *Meteor.* III.5 pertains to optics), Knorr 1994, 25, surmised that the one-letter designations of type (2) are a lexical fossil that only surfaces in optics.

Three letters. The main designations are set out in the following table:

article	prepos.	letters	object
ἡ		ΑΒΓ	straight line
τὸ		ΑΒΓ	triangle
ὁ		ΑΒΓ	circle
ὁ		ΑΒΓ	gnomon
ὁ		ΑΒΓ	number
ἡ	ὑπὸ	ΑΒΓ	angle
τὸ	ὑπὸ	ΑΒΓ	rectangle

Straight lines named by three letters are found in our paradigmatic proposition III.2. The competing designations of a circle and of a gnomon are never found within the same proposition.[27] The diagrams in the manuscripts (Book II is a case in point) suggest that a gnomon is associated with an arc of a circle, whence the designation. However, nothing in the text corroborates such a reading; a gnomon is likely to be assigned three letters because it is made of three parallelograms (II.def.2).

Four letters. Typically quadrilaterals, but also circles in specific configurations (see for instance propositions III.4 and III.13–15).

More than four letters. Polygons, solid figures, and straight lines (see Ptolemy's text in Sect. 2.4.2) and circles in specific configurations.

Let us come to formulaic templates. The formation of the standard lettered complexes that name angles, rectangles, squares, and products of numbers is explained by adherence to primary, "long", nested designations, according to the following scheme:

article	prepos.	sub-designation			participle	noun
		article	letters	noun		
ἡ	ὑπὸ	τῶν	ΑΒ ΒΓ	εὐθειῶν	περιεχομένη	γωνία
ἡ	ὑπὸ		ΑΒΓ			
τὸ	ὑπὸ	τῶν	ΑΒ ΒΓ	εὐθειῶν	περιεχόμενον	παραλληλόγραμμον ὀρθογώνιον
τὸ	ὑπὸ		ΑΒ ΒΓ			
τὸ	ἀπὸ	τῆς	ΑΒ	εὐθείας	ἀναγραφὲν	τετράγωνον
τὸ	ἀπὸ		ΑΒ			
ὁ	ἐκ	τῶν	ΑΒ ΒΓ	ἀριθμῶν	γενόμενος	ἀριθμός
ὁ	ἐκ		Α Β			

The first row in each sub-table (identified by double separation lines) sets out the "long" expressions, namely, those having all of their linguistic elements made explicit; each of these

[27] The name of a circle requires three letters because a circle is univocally determined by three points on it; however, we must not be thereby deluded into thinking that any of these three letters designates a point on the circle: see Sect. 3.2. More generally, it is clear that this system of designations is unable to prevent ambiguities—so what? Context and mathematical common sense will disambiguate.

expressions includes a sub-designation of a more basic entity: a pair of straight lines (twice), a single straight line, or a pair of numbers, in this order.[28] On their first occurrence in the *Elements*, the designations are attested in the long form, either instantiated or not;[29] no such archetypal occurrence exists for "rectilinear angle", but the long designation provided above is certainly right.[30] All or most of the subsequent occurrences of an assigned designation display "short" forms that naturally derive from the "long" syntagms, most frequently in the compact forms listed in the second row of each sub-table above.[31] The "short" standard forms lack all nominal elements, namely, the nouns and the participles. However, the "short" syntagms *are* noun phrases as they stand: we need not regard them as mere abbreviations of the "long" syntagms (see also Sect. 3.2); a sign of this is that all articles in the sub-designations are usually omitted.[32] I formulate this fact by saying that the "long" designations are the *templates* of the "short" designations.

The archetypal, long designations do not necessarily occur in definitions. Let us read the archetypal occurrences for three of the mathematical objects in question. I give here three passages that relate to "square" in order to show that the stem of the participle ἀναγραφέν in the designation is not univocally determined by the first two occurrences of the clause in I.46;[33] to find what stem must be used one must wait until the first application of the clause within the proof of I.47. Here are the enunciation and the conclusion of I.46 and the first occurrence in I.47 (square), in II.def.1 (rectangle), and in VII.16 (*EOO* I, 108.11–12, 110.7–8, 112.21–22; 118.2–4; and II, 222.10–12):[34]

ἀπὸ τῆς δοθείσης εὐθείας τετράγωνον ἀναγράψαι.	Describe a square on a given straight line[35].
τετράγωνον ἄρα ἐστίν· καί ἐστιν ἀπὸ τῆς AB εὐθείας ἀναγεγραμμένον.	Therefore it is a square; and it turns out to be described[36] on straight line AB.
καί ἐστι τὸ μὲν ΒΔΕΓ τετράγωνον ἀπὸ τῆς ΒΓ ἀναγραφέν.	And ΒΔΕΓ is a square described on ΒΓ.
πᾶν παραλληλόγραμμον ὀρθογώνιον περιέχεσθαι λέγεται ὑπὸ δύο τῶν τὴν ὀρθὴν γωνίαν περιεχουσῶν εὐθειῶν.	Every rectangular parallelogram is said to be contained by two of the straight lines containing the right angle.

[28] The pairs are designated as a whole: see the discussion in Sect. 3.2.

[29] That is, they either carry or do not carry denotative letters, respectively.

[30] This is proved by I.def.9; see also II.def.1 quoted just below. The definition of "rectilinear angle" in I.def.9 does not *directly* provide the template, however: ὅταν δὲ αἱ περιέχουσαι τὴν γωνίαν γραμμαὶ εὐθεῖαι ὦσιν, εὐθύγραμμος καλεῖται ἡ γωνία "and whenever the lines containing the angle are straight lines, the angle is called rectilinear" (*EOO* I, 2.14–15).

[31] The three-letter designation of a rectangle τὸ ὑπὸ ΑΒΓ is less common than the four-letter designation τὸ ὑπὸ AB ΒΓ.

[32] This is not an exception to the rule that denotative letters must be preceded by an article: the syntagm ὑπὸ AB ΒΓ is included in the designation of a parallelogram and is thereby uniquely denotative; the designation itself must be articular.

[33] What is left undecided in the first two occurrences is the fact that the participle must be in the passive aorist; the participles in the other three designations are in the present stem. The definition of "square" is I.def.22.

[34] The occurrence in VII.def.17 (the very definition of a product of numbers) is incomplete.

[35] The polarity definite / indefinite in the noun phrase τῆς δοθείσης εὐθείας is neutralized: therefore, the meaning is indefinite even if τῆς δοθείσης εὐθείας is an articular expression: see further examples, with a discussion, in Sect. 3.3. In τῆς δοθείσης εὐθείας, the modifier is articular because it is required to specify the reference (= "among all straight lines, select the one that is given"); the modifier is prepositive because it is more salient than the head noun; on these issues see Bakker 2009a. The order of the words in the enunciation itself is a standard Topic-Focus-predicate sequence: see Dik 1995. A classical reference for Greek word order, still not within the framework of functional grammar, is Dover 1960.

[36] The strongly marked periphrastic discontinuous construction ἐστίν ... ἀναγεγραμμένον emphasizes both the presential import of the liminal verb form of "to be" (see Sect. 3.1) and the aspectual value of the perfect stem. One must decide between the two in translation, but either of them must be expressed. I have emphasized the latter feature; had I wished to emphasize the former, I would have translated "and here it is, described on straight line AB". The context shows that the subsequent quote, where we read an analogous discontinuous construction, has a different meaning. On periphrastic constructions in ancient Greek, see Aerts 1965 and, more recently, Bailey 2009, sect. 4.7, Bentein 2016; see also the synthesis in Rijksbaron 2006, 126–131. A grammatical study of periphrastic constructions in mathematical propositions is in Federspiel 2010, 104–107. On discontinuous syntax in ancient Greek, see most recently Devine, Stephens 2000.

ἐὰν δύο ἀριθμοὶ πολλαπλασιάσαντες ἀλλήλους ποιῶσί τινας, οἱ γενόμενοι ἐξ αὐτῶν ἴσοι ἀλλήλοις ἔσονται.	If two numbers multiplying one another make some ‹numbers›, those resulting from them will be equal to one another.

Whenever a square or a rectangle is simply used as a figure, it receives a four-letter designation, as the following passage in II.5 (*EOO* I, 130.21–25) shows:

ἀλλὰ ὁ ΜΝΞ γνώμων καὶ τὸ ΛΗ ὅλον ἐστὶ τὸ ΓΕΖΒ τετράγωνον, *ὅ ἐστιν ἀπὸ τῆς ΓΒ*· τὸ ἄρα ὑπὸ τῶν ΑΔ ΔΒ περιεχόμενον ὀρθογώνιον μετὰ τοῦ ἀπὸ τῆς ΓΔ τετραγώνου ἴσον ἐστὶ τῷ ἀπὸ τῆς ΓΒ τετραγώνῳ.	But gnomon ΜΝΞ and ΛΗ are the square ΓΕΖΒ as a whole, *which is on ΓΒ*; therefore the rectangle contained by ΑΔ, ΔΒ with the square on ΓΔ is equal to the square on ΓΒ.

The relative clause (in italics above) identifies the square as "described" on straight line ἡ ΓΒ.[37] This step is necessary to word the subsequent sentence in conformity with the standard formulation provided in the enunciation of II.5 itself (this I call "alignment", see Sects. 4.5.1.3–4).[38]

More complex formulae are used to express relations (see Sect. 4.5.1). Let us take the path leading to the standard form of the relation "being in a straight line with" as an example. The path goes from the first, non-instantiated formulation in I.post.2, through the first application of this postulate as a constructive act in I.2, to the first occurrence of the relation as an instantiated statement in the "determination" of I.14 (*EOO* I, 8.9–10, 12.26–14.1, and 38.12–13):

καὶ πεπερασμένην εὐθεῖαν κατὰ τὸ συνεχὲς ἐπ' εὐθείας ἐκβαλεῖν.	And to produce[39] a bounded straight line continuously in a straight line.
ἐκβεβλήσθωσαν ἐπ' εὐθείας ταῖς ΔΑ ΔΒ εὐθεῖαι αἱ ΑΕ ΒΖ.	Let straight lines, ΑΕ, ΒΖ, be produced in a straight line with ΔΑ, ΔΒ.
λέγω ὅτι ἐπ' εὐθείας ἐστὶ τῇ ΓΒ ἡ ΒΔ.	I claim that ΒΔ is in a straight line with ΓΒ.

Postulate I.post.2 does not mention the adjoined line, whereas its application in I.2 omits the qualifications about the boundedness of the straight line(s) (πεπερασμένην εὐθεῖαν) and about the continuity of the assigned line(s) and of the adjoined line(s) (κατὰ τὸ συνεχές) that we read in the postulate. These are replaced by the evidence that comes either from the context (ΔΑ and ΔΒ have been constructed as bounded straight lines),[40] or from the lettered designations: the letter shared by the assigned line and by the adjoined line (Α or Β) entails continuity. The object πεπερασμένην εὐθεῖαν "bounded straight line" of the verb ἐκβαλεῖν "to produce" in the postulate is transformed, in the application in I.2, into a complement (ταῖς ΔΑ ΔΒ) of the determiner ἐπ' εὐθείας "in a straight line" of the same verb ἐκβαλεῖν. In its turn, the verb is transformed from active aorist infinitive into passive perfect imperative (ἐκβεβλήσθωσαν). A grammatical subject is required, and this function is assumed by the adjoined line(s) εὐθεῖαι αἱ ΑΕ ΒΖ, which, as seen, do(es) not figure in the postulate. The linguistic transformations just described leave some degrees of freedom in the

[37] As for the meaning of the admittedly awkward clause ὅ ἐστιν ἀπὸ τῆς ΓΒ, the article in front of ἀπὸ τῆς ΓΒ might also be absent because this expression is the nominal complement of the copula: "which is a ‹square› on ‹straight line› ΓΒ".

[38] The standard, instantiated, formulation figures in the determination of II.5 (*EOO* I, 130.1–3). Cf. Sect. 4.5.1.3.

[39] The infinitives in the postulates are governed by the finite verb form ᾐτήσθω, so that they must be translated as infinitives, not as directive imperatives.

[40] The context is here crucial since not every straight line denoted by two letters is bounded (a counterexample is in I.12), but those resulting from applications of I.post.2 necessarily are.

(re)arrangement of the terms. The transition from the standard formulation of the constructive act to the standard formulation of the relation in I.14 is just a matter of changing the verb form to ἐστί and of shifting it after the prepositional determiner ἐπ' εὐθείας.[41]

Relations that are canonized by a long-standing tradition may not be formulated according to a template that figures in the *Elements*, if not in a debased sense. Cases in point are the strictly related relations of ratio and of proportion. Let us read the characterization of a ratio of magnitudes and the mathematical definition of "having a ratio" for magnitudes in V.def.3–4 (*EOO* II, 2.6–9), the beginning of the definition of proportion in V.def.5 (*EOO* II, 2.10–11), and the definition of numerical proportion in VII.def.21 (*EOO* II, 188.5–7)—if we exclude the obviously spurious VI.def.5 for the other occurrence of πηλικότης, there are 4 hapax in a row in V.def.3, which is explicit about the categorial status of "ratio" as a relation but intrudes the category of "quality" by means of the indefinite adjective ποια "of a certain kind" that qualifies the σχέσις "relation"; note also the ordinals in V.def.5, on which we shall return in Sect. 5.1.6:

λόγος ἐστὶ δύο μεγεθῶν ὁμογενῶν ἡ κατὰ πηλικότητά ποια σχέσις.	A ratio of two homogeneous magnitudes is a relation of a certain kind according to value.
λόγον ἔχειν πρὸς ἄλληλα μεγέθη λέγεται ἃ δύναται πολλαπλασιαζόμενα ἀλλήλων ὑπερέχειν.	Magnitudes are said to have a ratio to one another that can, if multiples are taken, exceed one another.
ἐν τῷ αὐτῷ λόγῳ μεγέθη λέγεται εἶναι πρῶτον πρὸς δεύτερον καὶ τρίτον πρὸς τέταρτον [...]	Magnitudes are said to be in a same ratio, first to second and third to fourth [...]
ἀριθμοὶ ἀνάλογόν εἰσιν, ὅταν ὁ πρῶτος τοῦ δευτέρου καὶ ὁ τρίτος τοῦ τετάρτου ἰσάκις ᾖ πολλαπλάσιος ἢ τὸ αὐτὸ μέρος ἢ τὰ αὐτὰ μέρη ὦσιν.	Numbers in proportion are whenever the first of the second and the third of the fourth be equimultiple or the same part or be the same parts.

and let us compare them with the most widespread wording of a proportion, in which the standard formulation of a ratio is of course embedded (I purposely do not use quotes): (ἔστιν) ὡς τὸ Α πρὸς τὸ Β οὕτως τὸ Γ πρὸς τὸ Δ "as A is to B, so Γ is to Δ". The only predictable element in this formulation is the expression of a ratio τὸ Α πρὸς τὸ Β; the definition of numerical proportionality does not even mention ratios, if not within the adverb ἀνάλογον, viz. ἀνὰ λόγον, litt. "as for ratio".

However, V.def.4 and V.def.5 act as templates for two other formulations of proportion (the second of them is strictly sectorial, however), which we read for instance in V.13 and in V.20 (*EOO* II, 38.18–22 and 56.6–10), respectively (note again the ordinals):

ἐὰν πρῶτον πρὸς δεύτερον τὸν αὐτὸν ἔχῃ λόγον καὶ τρίτον πρὸς τέταρτον τρίτον δὲ πρὸς τέταρτον μείζονα λόγον ἔχῃ ἢ πέμπτον πρὸς ἕκτον, καὶ πρῶτον πρὸς δεύτερον μείζονα λόγον ἕξει ἢ πέμπτον πρὸς ἕκτον.	If first to second have the same ratio as third to fourth and third to fourth have a greater ratio than fifth to sixth, first to second will also have a greater ratio than fifth to sixth.
ἐὰν ᾖ τρία μεγέθη καὶ ἄλλα αὐτοῖς ἴσα τὸ πλῆθος σύνδυο λαμβανόμενα καὶ ἐν τῷ αὐτῷ λόγῳ δι' ἴσου δὲ τὸ πρῶτον τοῦ τρίτου μεῖζον ᾖ, καὶ τὸ τέταρτον τοῦ ἕκτου μεῖζον ἔσται, κἂν ἴσον, ἴσον, κἂν ἔλαττον, ἔλαττον.	If there be three magnitudes and others equal to them in multiplicity taken two and two together and in a same ratio and the first be greater than the third through an equal, the fourth will also be greater than the sixth, and if equal, equal, and if less, less.

[41] See also Sect. 4.5.1.3 for the main features of the standard formulation of a relation.

2.3. Sentential Validation: Syntactic Templates

Sentential validation consists in a generalized anaphora to sentential items assumed as templates.[42] As we shall see in the dedicated sections, such a generalized principle of anaphora may operate between the specific parts of a proposition: allowance being made for well-defined linguistic transformations, the two-sentence complex setting-out + determination assumes the enunciation as a template (Sect. 4.2); the "anaphora" systematically referes, in its causal subordinate clause, to assumptions supposed in the setting-out or in the construction, which are thereby "discharged" or "closed", whereas the "anaphora" itself may take the form of an instantiated reference to a theorem (Sect. 4.4, and see also below); the general conclusion of a theorem or of a problem, when it is present, is simply identical to the enunciation (Sect. 4.1), whereas the instantiated conclusion of a problem is a strictly codified transform of the enunciation (Sect. 4.3).

Within a proof, validation is mainly operating in "discharging" assumptions and within citations of previous results. In this Section, I shall focus on constructive acts, deferring a fuller discussion of deductive steps to Sects. 4.5.1.3–4. To see how the mechanism of validation works in this case, the following tables set out the results referred to in our paradigmatic proposition III.2, namely, III.1, I.post.1, I.5, I.16, and I.19 (*EOO* I, 166.14, 8.7–8, 20.2–3, 42.6–8, and 46.18–19, in the order given in the table, which is also the order in which they are used in III.2):

El.	theme	reprise with variations in III.2
III.1	*τοῦ δοθέντος* **κύκλου τὸ κέντρον** εὑρεῖν	**καὶ** εἰλήφθω τὸ κέντρον *τοῦ ΑΒΓ* κύκλου
post.1	*ἀπὸ παντὸς σημείου ἐπὶ πᾶν σημεῖον εὐθεῖαν γραμμὴν* ἀγαγεῖν	**καὶ** ἐπεζεύχθωσαν *αἱ ΔΑ ΔΒ*
I.5	*τῶν ἰσοσκελῶν τριγώνων αἱ πρὸς τῇ βάσει γωνίαι* ἴσαι ἀλλήλαις εἰσίν	**ἐπεὶ οὖν** ἴση ἐστὶν *ἡ ΔΑ τῇ ΔΒ*, ἴση **ἄρα καὶ** γωνία *ἡ ὑπὸ ΔΑΕ τῇ ὑπὸ ΔΒΕ*
I.16	~~παντὸς~~ **τριγώνου μιᾶς τῶν πλευρῶν** προσεκβληθείσης *ἡ* ἐκτὸς γωνία *ἑκατέρας τῶν ἐντὸς καὶ ἀπεναντίον γωνιῶν* μείζων ἐστίν	**καὶ ἐπεὶ** τριγώνου *τοῦ ΔΑΕ* μία πλευρὰ προσεκβέβληται *ἡ ΑΕΒ*, μείζων **ἄρα** *ἡ ὑπὸ ΔΕΒ* γωνία *τῆς ὑπὸ ΔΑΕ*
I.19	~~παντὸς τριγώνου~~ ὑπὸ τὴν μείζονα γωνίαν ἡ μείζων πλευρὰ ὑποτείνει	ὑπὸ **δὲ** τὴν μείζονα γωνίαν ἡ μείζων πλευρὰ ὑποτείνει

El.	theme	reprise with variations in III.2
III.1	find **the centre of** *a given* **circle**	**and** let the centre of circle *ΑΒΓ* be taken
post.1	to draw *a straight line from any point to any point*	**and** let ‹straight lines› *ΔΑ*, *ΔΒ* be joined
I.5	*the angles at the base of isosceles triangles* are equal to one another	**then since** *ΔΑ* is equal to *ΔΒ*, **therefore** angle *ΔΑΕ* is **also** equal to *ΔΒΕ*
I.16	**one side** of ~~*any*~~ **triangle** being produced, *the external* angle is greater than *each of the internal and opposite angles*	**and since** one side *ΑΕΒ* of a triangle *ΔΑΕ* turns out to be produced, **therefore** angle *ΔΕΒ* is greater than *ΔΑΕ*
I.19	the greater side ~~*of any triangle*~~ extends under the greater angle	**and** the greater side extends under the greater angle

[42] The discussions in this Section and in Sects. 4.5.1.3–4 seem to me to dispel the reservations about Galen's approach expressed in Barnes 2007, 433–438 (cf. Sect. 2.1). A detailed discussion of formulae, of their matrix structure, and of the ways they are combined, can be found in Netz 1999, sect. 4. Netz treats all formulaic expressions on a par (for this reason I cite him here and not in the previous Section). He also frames his discussion in explicit parallel with the formulaic character of Homeric language, following the insight of Aujac 1984 but without endorsing her utterly fantastic claim about this feature proving the oral character of mathematical teaching; this claim has been recently revived, on even flimsier grounds, in Saito 2018 (who does not mention Aujac but of course amply cites Netz). To dissolve these dreams, it is enough to read the analysis of formulae in Dover 1960, 56–65, and his statement "every sustained utterance, colloquial or literary or administrative, is necessarily in some degree formulaic" (60).

The variations dictated by the context are marked as follows:

- (boldface in the right column) coordinants, subordinants, and scope particles are added; the references to I.5 and to I.16 are formulated by paraconditional clauses, which require a richer connective structure than a simple coassumption;
- (italics) the definite descriptions or qualifications that aim at uniquely identifying an object or a class or objects are usually (but not always: exceptions are in boldface in the left column) replaced by lettered designations, or simply eliminated; the latter is always the case when universal quantification occurs in the theme (deleted syntagms);
- (underlined) in the retrieval of constructive acts, the verb forms change from aorist infinitive and participle to passive perfect imperative and indicative, respectively.

Of these five citations, two refer to constructions (a problem and a postulate), three to theorems. The former two citations occur within the construction of III.2, the latter three within its proof. The reference to I.19 is a verbatim quote (cf. Sect. 4.5.4); the sole omission in the quote is the quantified mention of a triangle, useless in the present application. Proposition III.1 is of course referred to by changing the verb εὑρίσκω to λαμβάνω; the structure remains the same, even if the clause is admittedly very short. The reference to the postulate will be discussed in Sect. 4.1. Remarks on the way constructive acts are cited will be offered in a moment.

Let us first focus, however, on the citations of I.5 and I.16, in order to appreciate the variations an enunciation undergoes when it is cited in instantiated form and is deeply embedded in a context. It is a fact that the statement assumed as the template for the citation of any theorem is its enunciation. This is by no means obvious: we shall presently see that the standard template adopted in the citations of a problem is its (instantiated) conclusion.[43] Contrary to the clause that refers to I.19, which is a non-instantiated coassumption, the statements that refer to I.5 and to I.16 are two consecutive "paraconditional" clauses; they make the whole of the "anaphora" of III.2. In any of these two clauses, two references are operative: the antecedent of the paraconditional refers to an immediate consequence of something assumed in the setting-out or in the construction;[44] the paraconditional itself refers to a previous theorem. In the latter kind of reference we are interested now.

As for I.5, the adaptation of this theorem to the context is radical. In the enunciation of I.5, the noun phrase "the angles at the base of isosceles triangles" is transformed into a saturated relation "ΔA is equal to ΔB" (= "isosceles") and into the pair of terms {ΔAE,ΔBE} (= the angles). This pair saturates the relation "* is equal to #" that canonically replaces the nominal groups in the reciprocal formulation typical of enunciations: "*# are equal to one another". Such a replacement obviously results in reducing the number of linguistic items with respect to the template, and aims at giving the clause a shape suited to inferences by transitivity (see Sect. 4.5.1.4). Subsentential templates are in this case a formulaic pole stronger than sentential templates.

As for I.16, the genitive absolute that formulates the geometric constraint in its enunciation is very uncommon (see Sect. 4.1): its use is probably dictated by the choice of phrasing the enunciation as a universal sentence rather than as a conditional, such as for instance *ἐὰν μία τῶν πλευρῶν

[43] This does not occur in the citation of III.1 because this problem does not have a conclusion that restates the enunciation.
[44] First paraconditional: ἡ ΔA is equal to ἡ ΔB because 1) τὸ A and τὸ B are taken on the circumference of a circle of centre τὸ Δ (setting-out), 2) ἡ ΔA and ἡ ΔB have been joined (construction), and 3) of the definition of a circle in I.def.15. Second paraconditional: the first three constructive acts in the construction must be invoked to generate the composite object "triangle τὸ ΔAE with a produced side ἡ AEB". As my annotation to the translation of III.2 shows, *every* supposition is eventually "discharged" within the proof (cf. Sect. 4.1).

τριγώνου προσεκβληθῇ κτλ. “if one of the sides of a triangle be produced etc.” might be. This is in fact the form of enunciation naturally associated with the retrieval of I.16 in the anaphora of III.2. Apart from this variation dictated by the syntax of the anaphora, note the useless persistence, in III.2, of the adjective μία “one” and of the verb προσεκβάλλω—the verb does not refer to an act performed in the setting-out or in the construction, as is instead normally the case for verb forms in the perfect stem—[45]and the linguistic peculiarity of assuming a side of a triangle as the grammatical subject of the antecedent of the paraconditional, while naming the side ἡ AEB, a designation that can only fit the *produced* side of triangle τὸ ΔAE. All of this shows that the formulation of the enunciation of I.16 is strictly adhered to as the template of the citation. The consequent of the paraconditional, instead, is transformed according to the pattern outlined above in the case of I.5.

The very first citation of I.16 in the anaphora of I.17 (*EOO* I, 44.15–17) provides us with complementary pieces of information:[46]

καὶ ἐπεὶ τριγώνου τοῦ ΑΒΓ ἐκτός ἐστι γωνία ἡ ὑπὸ ΑΓΔ, μείζων ἐστὶ τῆς ἐντὸς καὶ ἀπεναντίον τῆς ὑπὸ ΑΒΓ.	And since ΑΓΔ is an external angle of a triangle ΑΒΓ, it is greater than an internal and opposite ‹angle› ΑΒΓ.

The context suggests here to shortcut the constructive step made explicit by προσεκβληθείσης in the enunciation of I.16[47] and to simplify the reference to the external angle: to this effect, the “external angle” is placed in the antecedent of the paraconditional, so that no mention of the produced side is any longer necessary.

Let us return to the way constructive acts are cited. Deferring a fuller discussion to Sect. 4.3, I here point out the matrix structure of the reference. In problems, a general conclusion is frequently absent;[48] the instantiated conclusion contains denotative letters but keeps the indefinite structure of the enunciation (see Sect. 3.3). Let us read the enunciation and the instantiated conclusion of I.2 (*EOO* I, 12.19–20 and 14.13–15) as an example:

πρὸς τῷ δοθέντι σημείῳ τῇ δοθείσῃ εὐθείᾳ ἴσην εὐθεῖαν θέσθαι.	At a given point set a straight line equal to a given straight line.
πρὸς ἄρα τῷ δοθέντι σημείῳ τῷ Α τῇ δοθείσῃ εὐθείᾳ τῇ ΒΓ ἴση εὐθεῖα κεῖται ἡ ΑΛ.	Therefore at a given point, A, a straight line, ΑΛ, turns out to be set equal to a given straight line, ΒΓ.

To check what happens in the applications, it is enough to look at the first application of I.2 in I.3 (*EOO* I, 14.22–23)—the presence of κείσθω (conclusion above) instead of a form of τίθημι (enunciation above) is crucial:

κείσθω πρὸς τῷ Α σημείῳ τῇ Γ εὐθείᾳ ἴση ἡ ΑΒ.	At point A let a ‹straight line›, AB, be set equal to straight line ΓΔ.

[45] This verb only occurs in I.5, 16, and 32; it is peculiar because it has two prefixes (cf. Sect. 4.3).
[46] As we shall see in Sect. 3.3, the preposed adverbial determiner τῆς ἐντὸς καὶ ἀπεναντίον has an indefinite meaning: “an internal and opposite ‹angle›”; there are in fact two such angles in a triangle.
[47] The immediately preceding constructive act in I.17 employs the canonical, one-prefix ἐκβάλλω.
[48] But we read a general conclusion at the end of most problems of Book IV; see Sect 4.3 for detail.

This application shows that the assumed template is the instantiated conclusion;[49] the only structural variation is that the verb form is normally shifted to the first position in the clause; on the fact that all predicates "given" disappear in the application see Sect. 4.3. As for structure, the conclusion of I.2 dictates a three-entry matrix template κείσθω (πρὸς τῷ Α σημείῳ) (τῇ ΓΔ εὐθείᾳ ἴση) (εὐθεῖα ἡ ΑΒ). The first entry of the matrix is usually omitted, as well as one (in I.3) or both (more frequently) of the occurrences of the noun εὐθεῖα.[50] The matrix structure is rigidly adhered to when this construction is cited; as we shall see in Sects. 4.5.1.2–3, this matrix is not a ternary relation (in fact, it is not a relation at all) because it does not interact with deductions.

Let us check again this kind of structural stabilization on the example of the construction licensed by I.23 (*EOO* I, 54.20–22 and 56.17–20), whose enunciation and conclusion are exactly parallel to the enunciation and to the conclusion of I.2, respectively:

πρὸς τῇ δοθείσῃ εὐθείᾳ καὶ τῷ πρὸς αὐτῇ σημείῳ τῇ δοθείσῃ γωνίᾳ εὐθυγράμμῳ ἴσην γωνίαν εὐθύγραμμον συστήσασθαι.	On a given straight line and at a point on it construct a rectilinear angle equal to a given rectilinear angle.
πρὸς ἄρα τῇ δοθείσῃ εὐθείᾳ τῇ ΑΒ καὶ τῷ πρὸς αὐτῇ σημείῳ τῷ Α τῇ δοθείσῃ γωνίᾳ εὐθυγράμμῳ τῇ ὑπὸ ΔΓΕ ἴση γωνία εὐθύγραμμος συνέσταται ἡ ὑπὸ ΖΑΗ.	Therefore on a given straight line, AB, and at a point on it, A, a rectilinear angle, ZAH, turns out to be constructed equal to a given rectilinear angle, ΔΓE.

and whose first application in I.24 (*EOO* I, 58.7–9)

συνεστάτω πρὸς τῇ ΔΕ εὐθείᾳ καὶ τῷ πρὸς αὐτῇ σημείῳ τῷ Δ τῇ ὑπὸ ΒΑΓ γωνίᾳ ἴση ἡ ὑπὸ ΕΔΗ.	On straight line ΔE and at point Δ on it let an ‹angle›, EΔH, be constructed equal to angle BAΓ.

has the matrix structure συνεστάτω (πρὸς τῇ ΔΕ εὐθείᾳ) (καὶ τῷ πρὸς αὐτῇ σημείῳ τῷ Δ) (τῇ ὑπὸ ΒΑΓ γωνίᾳ ἴση) (γωνία ἡ ὑπὸ ΕΔΗ). All subsequent applications of this construction will exhibit the same matrix structure.

A further example comes from the first application of the instantiated conclusion of I.3 in I.5 (*EOO* I, 16.5–7 and 20.12–13), where a clause that is useless in context is eliminated:

δύο ἄρα δοθεισῶν εὐθειῶν ἀνίσων τῶν ΑΒ Γ ἀπὸ τῆς μείζονος τῆς ΑΒ τῇ ἐλάσσονι τῇ Γ ἴση ἀφῄρηται ἡ ΑΕ.	Therefore two unequal straight lines, AB, Γ, being given, from the greater ‹straight line›, AB, a ‹straight line›, AE, turns out to be removed equal to the lesser, Γ.
ἀφῃρήσθω ἀπὸ τῆς μείζονος τῆς ΑΕ τῇ ἐλάσσονι τῇ ΑΖ ἴση ἡ ΑΗ	From the greater ‹straight line›, AE, let a ‹straight line›, AH, be removed equal to the lesser, AZ.

A last example comes from I.31, from whose conclusion derives the matrix ἤχθω (διὰ τοῦ Γ σημείου) (τῇ ΑΒ εὐθείᾳ παράλληλος) (εὐθεῖα ἡ ΓΕ). Let us read in fact the conclusion of I.31 and the first application of this construction in I.32 (*EOO* I, 76.10–12 and 76.24–25):

[49] Among other reasons, an enunciation that contains an *active* directive infinitive cannot provide the template for a constructive act, in which a mathematical object invariably undergoes the construction.
[50] On the indefinite character of the whole citation, see Sect. 3.3.

διὰ τοῦ δοθέντος ἄρα σημείου τοῦ Α τῇ δοθείσῃ εὐθείᾳ τῇ ΒΓ παράλληλος εὐθεῖα γραμμὴ ἦκται ἡ ΕΑΖ.	Therefore through a given point, A, a straight line, EAZ, turns out to be drawn parallel to a given straight line, ΒΓ.
ἤχθω γὰρ διὰ τοῦ Γ σημείου τῇ ΑΒ εὐθείᾳ παράλληλος ἡ ΓΕ.	In fact, through point Γ let a ‹straight line›, ΓΕ, be drawn parallel to straight line AB.

A major feature of all these matrices is that the verb form is in the first position (in a sense, the constructional operator is the frame of the matrix), while the constructed object occupies the last entry of the matrix. The next-to-last entry of the matrix specifies the intended property of this object; the remaining entries prescribe where the object must be located. More generally, this matrix structure allows appreciating the way saliency is given expression in the formulae for constructive acts and their kin. As said, the constructional operator comes first, the constructed object comes last: they occupy the most salient places of the sequence. The entries of the matrix that allow anchoring the construction to pre-existing items of the geometric configuration are inserted between—and in fact nested in—the two saliency poles: objectual anchoring, like adjacency and intersection, comes first; relational anchoring, in the form of characterizing properties like parallelism, comes after. The general tendency of Greek language to place heavy constituents towards the end of the sequence is in principle disregarded. Within each entry of the matrix, denotations precede relational operators like "equal" or "parallel"; each denotation is either neutralized (lettered complex comes first) or indefinite (noun comes first), as we shall see in Sects. 3.2–3.

Before passing to suprasentential validation, I end this section by outlining a beautiful and clear-cut application of argumental validation in Stoic logic; detail will be provided in Sects. 5.1.6 and 5.2.2. Chrysippus recognized five species of atomic argument, which he called ἀναπόδεικτοι (συλλογισμοί) "indemonstrable (syllogisms)". Each species was identified by means of a canonical concrete argument serving as paradigmatic example, by a canonical description, and by a template logical form in which the compounding sentences are replaced by ordinals. A "first indemonstrable" is a form of *modus ponens*; let us read each of the identifying accounts in succession (cf. D.L. VII.76, 80, 76; note the inconsistency in the particle that opens the coassumption):

If it is day, it is light; and [δέ] it is day; therefore it is light.
A first indemonstrable is that ‹compounded› of a conditional and of the antecendent in that conditional ‹as premises›, having the consequent as conclusion.
If the first, the second; but of course [ἀλλὰ μήν] the first; therefore the second.

If the canonical description just read shows that a first indemonstrable is entirely made of items that are codified in Stoic logic (therefore, the definition is generative), the other two formulations serve as templates, and on the same footing. The template with ordinals is probably more perspicuous and formal-looking to modern eyes, but I doubt that the paradigmatic example formulated with stock statements was accorded a lesser logico-epistemic status in origin. In either case, the principle is to replace linguistic items with items homogeneous to them, and equals with equals; the ordinals abstract from the specific features of a concrete statement—which, for instance, can only be either affirmative or negative—but the paradigmatic example makes it immediately clear that the entries of the argumental matrix are statements, a piece of information that must come from outside when the template with ordinals is taken as a reference.

2.4. LARGE-SCALE VALIDATION: ANALYSIS AND SYNTHESIS

Large-scale validation employs the "language of the givens" to provide a unified framework in which deductions, constructive acts, and computations are formalized.[51] This unified framework is put to effect in the proof format of analysis and synthesis,[52] which operates on self-contained mathematical units such as propositions.

2.4.1. *Geometric analysis and synthesis*

The "language of the givens" is a stylistic resource specific to Greek mathematics and one of the major features of the proof format of analysis and synthesis. The basic characteristics of this idiom are best understood if we read the enunciations of some kinds of propositions where it is used.

Let us start with the simplest examples. In the enunciations of problems, an unqualified determiner "given" applies to the geometric objects starting from which the required construction must be performed,[53] as in I.23 (*EOO* I, 54.20–22), which we read again:

πρὸς τῇ δοθείσῃ εὐθείᾳ καὶ τῷ πρὸς αὐτῇ σημείῳ τῇ δοθείσῃ γωνίᾳ εὐθυγράμμῳ ἴσην γωνίαν εὐθύγραμμον συστήσασθαι.

On a given straight line and at a point on it construct a rectilinear angle equal to a given rectilinear angle.

In *Data* 40, it is required to prove that a triangle is "given in form" once its sides are "given in magnitude" (*EOO* VI, 70.2–3)—thus, there exist species of the "givens":[54]

ἐὰν τριγώνου ἑκάστη τῶν γωνιῶν δεδομένη ᾖ τῷ μεγέθει, δέδοται τὸ τρίγωνον τῷ εἴδει.

If each of the angles of a triangle be given in magnitude, the triangle is given in form.

A data-theorem that has a more transparent geometric content (but no species of "given" are introduced) is Pappus, *Coll.* IV.11:[55]

τετράπλευρον τὸ ΑΒΓΔ ὀρθὴν ἔχον τὴν ὑπὸ ΑΒΓ γωνίαν καὶ δοθεῖσαν ἑκάστην τῶν ΑΒ ΒΓ ΓΔ ΔΑ εὐθειῶν. δεῖξαι δοθεῖσαν τὴν ἐπιζευγνύουσαν τὰ Δ Β σημεῖα.

‹Let there be› a quadrilateral, ΑΒΓΔ, having angle ΑΒΓ right and each of the straight lines ΑΒ, ΒΓ, ΓΔ, ΔΑ given. Prove that the ‹straight line› joining points Δ, Β is given.

Less transparent is the enunciation of a locus theorem, for instance one out of the three that Charmandrus added as a preliminary to Apollonius' *Plane loci* (*Coll.* VII.24). This is the transposition of the definition of a circle as a geometric locus (see also Sect. 5.1.1); a point is said "to touch" a line when it is contained in it; the other point is "given in position":

---

[51] The approach of this Section combines Acerbi 2011a (which summarizes Acerbi 2007, 439–523) and 2012b. Most of what we read in Sidoli 2018a and in Sidoli, Isahaya 2018, 4–20, is ill-digested compilation (more embarrassing when it is credited than when it is not) of these papers.

[52] Extant sources on analysis and synthesis are listed in Appendix B.

[53] The point on the straight line is not said to be "given", but it should be: compare the enunciations of I.11–12.

[54] The entire proposition will be read below.

[55] This theorem corresponds to a part of Hero, *Metr.* I.14, which we shall read in Sect. 2.4.2.

ἐὰν εὐθείας τῷ μεγέθει δεδομένης τὸ ἓν πέρας ᾖ δεδομένον, τὸ ἕτερον ἅψεται θέσει δεδομένης περιφερείας κοίλης.	If one extremity of a straight line given in magnitude be given, the other will touch a concave arc given in position.

A *problem* entirely formulated in the idiom of the "givens" is Pappus, *Coll.* VII.294; it introduces the technical clause παρὰ θέσει, where the predicate "given" is understood:

θέσει δεδομένων τῶν ΑΒ ΑΓ ἀγαγεῖν παρὰ θέσει τὴν ΔΕ καὶ ποιεῖν δοθεῖσαν τὴν ΔΕ.	ΑΒ, ΑΓ being given in position, draw ΔΕ parallel ‹to a line given› in position and make ΔΕ given.

Let us finally read an enunciation from Euclid's *Porisms*, in the severely elliptical form transmitted by Pappus, *Coll.* VII.18:

ἐὰν ἀπὸ δύο δεδομένων σημείων πρὸς θέσει δεδομένην εὐθεῖαι κλασθῶσιν ἀποτέμνῃ δὲ μία ἀπὸ θέσει δεδομένης εὐθείας πρὸς τῷ ἐπ' αὐτῆς δεδομένῳ σημείῳ, ἀποτεμεῖ καὶ ἡ ἑτέρα ἀπὸ ἑτέρας λόγον ἔχουσαν δοθέντα.	If straight lines from two given points be inflected on a line given in position and one ‹straight line› cut off ‹a segment› from a straight line given in position up to a given point on it, the other ‹straight line› will also cut off from another ‹straight line given in position a segment› having a given ratio ‹to the first›.

These enunciations make two facts apparent:

- An object can be "given" (*a*) because it is assigned by assumption or (*b*) because it can be obtained from the assigned objects by means of some argument. In case (*b*), the object is proved given. It is not always straightforward to distinguish between the two functions, especially in the case of the Euclidean *Porisms*.[56]
- Several species of "being given" are specified in the *Data*,[57] namely, "in magnitude ", "in position", or "in form", depending on the geometric object to which the predicate is applied and on the point of view from which the object is considered. This point is clarified if we read the archetypal definitions *Data* 1–6 and 11 (*EOO* VI, 2.4–10 and 4.9–11; as explained in the next-to-last note of Sect. 1.5, I take *Data* def. 5–6 to be one single definition):[58]

δεδομένα τῷ μεγέθει λέγεται χωρία τε καὶ γραμμαὶ καὶ γωνίαι οἷς δυνάμεθα ἴσα πορίσασθαι.	Given in magnitude are said regions, lines, and angles for which we can procure equals.
λόγος δεδόσθαι λέγεται ᾧ δυνάμεθα τὸν αὐτὸν πορίσασθαι.	A ratio is said to be given for which we can procure the same.
εὐθύγραμμα σχήματα τῷ εἴδει δεδόσθαι λέγεται ὧν αἵ τε γωνίαι δεδομέναι εἰσὶ κατὰ μίαν καὶ οἱ λόγοι τῶν πλευρῶν πρὸς ἀλλήλας δεδομένοι.	Rectilinear figures are said to be given in form whose angles are given severally and the ratios to one another of whose sides are given.
τῇ θέσει δεδόσθαι λέγονται σημεῖά τε καὶ γραμμαὶ καὶ γωνίαι ἃ τὸν αὐτὸν ἀεὶ τόπον ἐπέχει.	To be given in position are said points, lines, and angles that always hold the same place.

[56] See Hogendijk 1987, 96–102; Acerbi 2007, 733–744.
[57] But no specification is ever found in the *Elements*, where the objects are simply "given". It is *meaningless* to ask what species of "being given" is intended in the *Elements*.
[58] In the applications, the article in determiners such as τῷ μεγέθει "in magnitude" is frequently omitted.

κύκλος τῷ μεγέθει δεδόσθαι λέγεται οὗ δέδοται ἡ ἐκ τοῦ κέντρου τῷ μεγέθει· τῇ θέσει δὲ καὶ τῷ μεγέθει κύκλος δεδόσθαι λέγεται οὗ δέδοται τὸ μὲν κέντρον τῇ θέσει ἡ δὲ ἐκ τοῦ κέντρου τῷ μεγέθει.	A circle is said to be given in magnitude of which the radius is given in magnitude; a circle is said to be given in position and in magnitude of which the centre is given in position and the radius in magnitude.
μέγεθος μεγέθους δοθέντι μεῖζόν ἐστιν ἢ ἐν λόγῳ ὅταν, ἀφαιρεθέντος τοῦ δοθέντος, τὸ λοιπὸν πρὸς τὸ αὐτὸ λόγον ἔχῃ δεδομένον.	A magnitude is greater than in ratio than a magnitude by a given whenever, after the given one is removed, the remainder have a given ratio to the same.

Similar definitions apply to segments of circles, which can be given in magnitude (def. 7) or in position and in magnitude (def. 8). A triangle can be given independently in position, in magnitude, or in form. These definitions allow us to highlight the following points:

(1) The definitions introduce a predicate: the verb form λέγεται "is said" is in fact typical of the definitions of predicates and of relations.[59] The verb πορίζομαι "to procure", "to get" alludes to constructive issues (note the middle voice), whereas the modal connotation δυνάμεθα "we can" underlines the existential import (modal connotation of possibility) and the active role of the mathematician (personal verb form). Finally, the definitions must be referred to all forms of the verb "to give", even if in defs. 1–4 only nominal forms of the perfect stem appear. In the course of the *Data*, however, the aorist participle or finite verb forms are frequently found.

(2) The species of the givens are the "predicative" pendant of some relations, which are of crucial importance both in geometry and in number theory and whose transformations make up the deductive fabric of any proof. These relations are equality (to which "given in magnitude" corresponds), identity ("given" for ratios), similitude ("given in form"), congruence, intended as coincidence by superposition ("given in position"). The mechanism of formation of such predicates, which is explicit in the definitions, is as follows. Each of them is obtained from the corresponding relation by "saturating" one of its entries: in symbols $g_R(*) \equiv Pa{:}R(a,*)$. For instance, to predicate "given in magnitude" of an angle A amounts to claiming "there is an angle that turns out to be made equal to A", where the verb "to make" generically refers to the same constellation of unspecified operations alluded to by the verb πορίζομαι "to procure".[60] A similar mechanism of partial saturation is at work in the notions of "expressibility"[61] and of "right angle". For "right angle" is an ultimate subspecies of angle defined as the angle such that its adjacent is equal to it (I.def.10), and the definition itself gives the mode of "procuring" the two (right) angles involved. As a consequence, to assert that a right angle is given is a paradigmatic application of *Data* def. 1: for this reason, and not because it has a well-defined "size", a right angle is always given.

(3) Ratios deserve a separate definition because they are "given" tout court and, most importantly, because two ratios are said to be "the same", not "equal": the predicate "given" applied to them comes from the saturation of the relation of identity, not of equality.[62]

(4) The definition of "given in form" has its model in VI.def.1: "Similar rectilinear figures are such as have both their angles severally equal and the sides about the equal angles in proportion".

---

[59] In the case in which terms are defined, "to be" or "to call" are normally used (cf. Sect. 1.5); the two verbs are equivalent (this is proved by comparing VII.def.17–18).

[60] This is the operator P in my symbolic transcription; I was once used to identify P with mere existence, but now I am convinced that this interpretation trivializes the issue.

[61] This is the relation of commensurability saturated by a reference straight line, said "expressible"; cf. X.def.1 and 3, and see Sect. 4.2 for detail.

[62] It is a mistake to say, as Proclus and Marinus do (*iE*, 205.13–206.11, and *EOO* VI, 256.12, respectively), that two magnitudes are "given in ratio".

This fact is *prima facie* surprising, since the definition might well have included a direct reference to similitude, in line with what we read in *Data* def. 1: for instance, *"Given in form are said figures for which we can procure similars". The attested definition, which obviously results from combining VI.def.1 and def. 1*, was probably suggested by concerns of deductive economy, since it does nothing but reducing the notion of "given in form" to the predicates "given in magnitude" and "given" for ratios. However, in the applications similitude can directly be used in inferences.

Until now, I have discussed the definitions of the several species of "given". But how is a proof framed that employs the language of the givens? The purest form of application of the idiom as a deductive tool is found, most naturally, in the *Data*: let us read proposition 40 (*EOO* VI, 70.2–23)—references to propositions or to definitions of the *Data* themselves, or of the *Elements*, are added in brackets; the figure simply represents the two equal triangles ΑΒΓ and ΕΔΖ:[63]

ἐὰν τριγώνου ἑκάστη τῶν γωνιῶν δεδομένη ᾖ τῷ μεγέθει, δέδοται τὸ τρίγωνον τῷ εἴδει.	If each of the angles of a triangle be given in magnitude, the triangle is given in form.
τριγώνου γὰρ τοῦ ΑΒΓ ἑκάστη τῶν γωνιῶν δεδομένη ἔστω τῷ μεγέθει.	In fact, let each of the angles of a triangle, ΑΒΓ, be given in magnitude.
λέγω ὅτι δέδοται τὸ ΑΒΓ τρίγωνον τῷ εἴδει.	I claim that triangle ΑΒΓ is given in form.
ἐκκείσθω γὰρ τῇ θέσει καὶ τῷ μεγέθει δεδομένη εὐθεῖα ἡ ΔΕ, καὶ συνεστάτω πρὸς τῇ ΔΕ καὶ τοῖς πρὸς αὐτῇ σημείοις τοῖς Δ Ε τῇ μὲν ὑπὸ ΓΒΑ γωνίᾳ ἴση γωνία εὐθύγραμμος ἡ ὑπὸ ΕΔΖ τῇ δὲ ὑπὸ τῶν ΑΓΒ ἴση ἡ ὑπὸ τῶν ΔΕΖ.	In fact, let a straight line given in position and in magnitude, ΔΕ, be set out, and at points Δ, Ε on it let a rectilinear angle, ΕΔΖ, be constructed equal to angle ΓΒΑ [I.23], and ΔΕΖ equal to ΑΓΒ [I.23].
λοιπὴ ἄρα ἡ ὑπὸ τῶν ΒΑΓ λοιπῇ ἴση τῇ ὑπὸ τῶν ΔΖΕ ἐστιν· δοθεῖσα δὲ ἑκάστη τῶν πρὸς τοῖς Α Β Γ· δοθεῖσα ἄρα καὶ ἑκάστη τῶν πρὸς τοῖς Δ Ε Ζ. ἐπεὶ οὖν πρὸς θέσει δεδομένη εὐθείᾳ τῇ ΔΕ καὶ τῷ πρὸς αὐτῇ σημείῳ δεδομένῳ τῷ Δ εὐθεῖα γραμμὴ ἦκται ἡ ΔΖ δεδομένην ποιοῦσα γωνίαν τὴν πρὸς τῷ Δ, θέσει ἄρα ἐστὶν ἡ ΔΖ. διὰ τὰ αὐτὰ δὴ καὶ ἡ ΕΖ θέσει ἐστίν· δοθὲν ἄρα ἐστὶ τὸ Ζ σημεῖον, ἔστι δὲ καὶ ἑκάτερον τῶν Δ Ε δοθέν· δοθεῖσα ἄρα ἐστὶν ἑκάστη τῶν ΔΖ ΔΕ ΕΖ τῇ θέσει καὶ τῷ μεγέθει· δέδοται ἄρα τὸ ΔΖΕ τρίγωνον τῷ εἴδει· καί ἐστιν ὅμοιον τῷ ΑΒΓ τριγώνῳ· δέδοται ἄρα καὶ τὸ ΑΒΓ τρίγωνον τῷ εἴδει.	Therefore ΒΑΓ as a remainder is equal to ΔΖΕ as a remainder [I.32]; and each of ‹the angles› at Α, Β, Γ is given; therefore each of those at Δ, Ε, Ζ is also given [I.32]. Then since a straight line, ΔΖ, turns out to be drawn making a given angle, namely, the one at Δ, to a straight line given in position, ΔΕ, and at a given point on it, Δ, therefore ΔΖ is ‹given› in position [29]. For the very same ‹reasons› ΕΖ is also in position [29]; therefore point Ζ is given [25], and each of Δ, Ε is also given [27]; therefore each of ΔΖ, ΔΕ, ΕΖ is given in position and in magnitude [26]; therefore triangle ΔΖΕ is given in form [39]; and it is similar to triangle ΑΒΓ [VI.4]; therefore triangle ΑΒΓ is also given in form [def. 1, 3; VI.def.1].

Let us discuss the main features of this proposition.

- The proof runs as follows: a chance straight line is "set out"[64] and two angles are constructed on it that are equal to two of the given angles of the assigned triangle. This amounts to "procuring" an "alias" of the assigned triangle; it is then proved that some of the elements of the "alias" are given: its angles first, then two sides in position, then the intersection of these sides, and

[63] Note the position of the "anaphora", and see Sect. 4.4.
[64] See Sect. 4.2 on this peculiar verb.

finally the remaining two extremities of them. Since their extremities are given, the sides themselves are also given in magnitude. As a consequence, the form of the "alias" triangle is given, and therefore, by similitude and def. 3, also the form of the original triangle is given. The proof may seem uselessly complicated and roundabout, but its form is in fact the only possible one once the fundamental definitions *Data* 1–3 are formulated in the way they are.[65] The creation of the "alias" figure is the analytic counterpart of the "actualization" of the assigned magnitudes by means of the production of duplicates found in synthetic proofs such as that of *El.* I.22 (see Sect. 4.2): in this proposition, the three straight lines with which to construct a triangle are first set out as segments αἱ Α Β Γ, then "reproduced in duplicate" as the sides of the sought for triangle. In I.22 as well as in *Data* 40, in order to construct the "alias", an object must be set out that is fixed by stipulation and is free of constraints induced by the assigned objects (in I.22 a base straight line from which to "cut off" two of the three sides of the triangle, in *Data* 40 the straight line ἡ ΔΕ): the verb ἔκκειμαι "to be set out" is exactly what is needed to fill this function (see again Sect. 4.2). As we shall see in a moment, in the analyses, the same function as the one of ἔκκειμαι is assumed by the initializing verb form γεγονέτω.

- The inferences showing that an object is given are expressed in the language of the givens. It is not permitted to qualify an object as "given" by simply starting from assigned entities, performing constructive acts on them or setting up demonstrative steps in a synthetic language, and finally attaching the label "given" to the object so obtained. As a consequence, it is of primary importance to identify and to collect a series of deductive rules that operate on "givens" and that yield "givens" as output: the *Data* do exactly that, while leaving it undecided what the term "given" must be taken to mean. What is really useful for the working mathematician is to set out, in the form of definitions, zero-grade rules of inference that allow to transfer the predicate "given" from an object to another by equality or identity.[66] This is the primary function of *Data* def. 1–2; these definitions are so cleverly framed that the first theorems of the *Data* do not have the axiomatic character that, for instance, *El.* I.4 has. Propositions with this drawback can be found only from *Data* 25 on, where "given in position" is first introduced.[67]

In the *Data* and in part of the lost "analytic corpus", the language of the givens is also applied in propositions or in contexts that do not require to show that some object or configuration is given.[68] The most celebrated such application is in the so-called "method of analysis and synthesis". Let us read a problem solved by analysis and synthesis: Apollonius, *Con.* II.50 (*AGE* I, 286.26–290.2);[69] the system of boldface, italics and underlining with which I present the text gives prominence to the correspondences between steps of the analysis and steps of the synthesis; references to results in the *Elements*, in the *Data*, and in the *Conica* are bracketed:

[65] Each of the propositions 25, 27, 29 applied in the text of *Data* 40 is nothing but a rewriting of def. 4.

[66] In other words, the definitions in the *Data* do not intend to capture the "essence" of a mathematical entity, but are aimed at formulating operational rules. The same happens for instance with the definition of proportionality in V.def.5 (see Sects. 5.2.2 and 5.3.4 for detail).

[67] In a nutshell, the *Data* is a natural deduction system, an exegetic perspective I shall develop elsewhere. Natural deduction is aptly presented in von Plato 2013.

[68] But recall that the locus theorems and the "porisms" do require this. On locus theorems, see below and Sect. 5.1.1, for Euclid's "porisms", see later in the present Section.

[69] The only notion of the theory of conic sections needed to follow this proof is the so-called "property of the subtangent" to a parabola, which can be formulated in the following way. From a point C of a parabola of vertex A draw a straight line perpendicular to the axis and a straight line tangent to the parabola; both of them fall on the axis of the parabola, say at points B and D, respectively. The segment BD of the axis between the two points of intersection is called "subtangent". The "property of the subtangent" asserts that BD is bisected by the vertex A of the parabola (*Con.* I.35): $DA = AB$.

τῆς δοθείσης κώνου τομῆς ἐφαπτομένην ἀγαγεῖν ἥτις πρὸς τῷ ἄξονι γωνίαν ποιήσει ἐπὶ ταὐτὰ τῇ τομῇ ἴσην τῇ δοθείσῃ ὀξείᾳ γωνίᾳ. ἔστω κώνου τομὴ πρότερον παραβολὴ ἧς ἄξων ὁ ΑΒ. δεῖ δὴ ἀγαγεῖν ἐφαπτομένην τῆς τομῆς ἥτις πρὸς τῷ ΑΒ ἄξονι γωνίαν ποιήσει ἐπὶ τὰ αὐτὰ τῇ τομῇ ἴσην τῇ δοθείσῃ ὀξείᾳ.

Draw a ‹straight line› tangent to a given section of a cone such as will make on the axis, on the same side as the section, an angle equal to a given acute angle. Let there be first as a section of a cone a parabola, whose axis is AB. Thus it is required to draw a ‹straight line› tangent to the section such as will make on axis AB, on the same side as the section, an angle equal to a given acute angle.

γεγονέτω, καὶ ἔστω ἡ ΓΔ· δοθεῖσα ἄρα ἐστὶν ἡ ὑπὸ ΒΔΓ γωνία. **ἤχθω κάθετος ἡ ΒΓ**· ἔστι δὴ καὶ ἡ πρὸς τῷ Β δοθεῖσα· λόγος ἄρα τῆς ΔΒ πρὸς ΒΓ δοθείς· τῆς δὲ ΒΔ πρὸς ΒΑ λόγος ἐστὶ δοθείς· καὶ τῆς ΑΒ ἄρα πρὸς ΒΓ λόγος ἐστὶ δοθείς· καί ἐστι δοθεῖσα ἡ πρὸς τῷ Β γωνία· δοθεῖσα ἄρα καὶ ἡ ὑπὸ ΒΑΓ· καί ἐστι πρὸς θέσει τῇ ΒΑ καὶ δοθέντι τῷ Α· θέσει ἄρα ἡ ΓΑ· θέσει δὲ καὶ ἡ τομή· δοθὲν ἄρα τὸ Γ· καὶ ἐφάπτεται ἡ ΓΔ· θέσει ἄρα ἐστὶν ἡ ΓΔ.

Let it happen to come to be, and let it be ΓΔ; therefore angle ΒΔΓ is given [*Data* def. 1]. **Let a ‹straight line› ΒΓ be drawn perpendicular** [I.12]; then the angle at B is also given [I.def.10 and *Data* def. 1]; therefore the ratio of ΔB to ΒΓ is given [*Data* 40 and def. 3]; *and the ratio of ΒΔ to ΒΑ is given* [prop. of subt.]; therefore the ratio of AB to ΒΓ is also given [*Data* 8]; and the angle at B is given; therefore ΒΑΓ is also given [*Data* 41 and def. 3]; and it is on a ‹straight line› ΒΑ ‹given› in position and at a given point A; therefore ΓΑ is ‹given› in position [*Data* 29]; and the section is also ‹given› in position; therefore Γ is given [*Data* 25]; and ΓΔ is tangent; therefore ΓΔ is ‹given› in position.

συντεθήσεται δὴ τὸ πρόβλημα οὕτως. ἔστω ἡ δοθεῖσα κώνου τομὴ πρότερον παραβολὴ ἧς ἄξων ὁ ΑΒ ἡ δὲ δοθεῖσα γωνία ὀξεῖα ἡ ὑπὸ ΕΖΗ, καὶ εἰλήφθω σημεῖον ἐπὶ τῆς ΕΖ τὸ Ε, **καὶ κάθετος ἤχθω ἡ ΕΗ**, *καὶ τετμήσθω δίχα ἡ ΖΗ τῷ Θ*, καὶ ἐπεζεύχθω ἡ ΘΕ, καὶ τῇ ὑπὸ τῶν ΗΘΕ γωνίᾳ ἴση συνεστάτω ἡ ὑπὸ τῶν ΒΑΓ, **καὶ ἤχθω κάθετος ἡ ΒΓ**, *καὶ τῇ ΒΑ ἴση κείσθω ἡ ΑΔ*, καὶ ἐπεζεύχθω ἡ ΓΔ· ἐφαπτομένη ἄρα ἐστὶν ἡ ΓΔ τῆς τομῆς.

Thus the problem will be synthetized as follows. Let there be first as a section of a cone a parabola, whose axis is AB, and a given acute angle, EZH, and let a point, E, be taken on EZ, **and let a ‹straight line›, EH, be drawn perpendicular** [I.12], *and let ZH be bisected by Θ* [I.10], and let a ‹straight line›, ΘE, be joined [I.post.1], and let an ‹angle›, ΒΑΓ, be constructed equal to angle ΗΘΕ [I.23], **and let a ‹straight line›, ΒΓ, be drawn perpendicular** [I.12], *and let a ‹straight line›, ΑΔ, be set equal to BA* [I.post.3], and let a ‹straight line›, ΓΔ, be joined [I.post.1]; therefore ΓΔ is tangent to the section [*Con.* I.33].

λέγω δὴ ὅτι ἡ ὑπὸ τῶν ΓΔΒ τῇ ὑπὸ τῶν ΕΖΗ ἐστιν ἴση. ἐπεὶ γάρ ἐστιν ὡς ἡ ΖΗ πρὸς ΗΘ οὕτως ἡ ΔΒ πρὸς ΒΑ ἔστι δὲ καὶ ὡς ἡ ΘΗ πρὸς ΗΕ οὕτως ἡ ΑΒ πρὸς ΒΓ, δι' ἴσου ἄρα ἐστὶν ὡς ἡ ΖΗ πρὸς ΗΕ οὕτως ἡ ΔΒ πρὸς τὴν ΒΓ· καί εἰσιν ὀρθαὶ αἱ πρὸς τοῖς Η Β γωνίαι· ἴση ἄρα ἐστὶν ἡ Ζ γωνία τῇ Δ γωνίᾳ.

I now claim that ΓΔΒ is equal to EZH. In fact, since, as ZH is to ΗΘ, so ΔB is to BA, and, as ΘΗ to HE, so AB is also to ΒΓ [VI.4], therefore through an equal, as ZH is to HE, so ΔΒ is to ΒΓ [V.22]; and the angles at H, B are right ‹angles›; therefore angle Z is equal to angle Δ [VI.6].

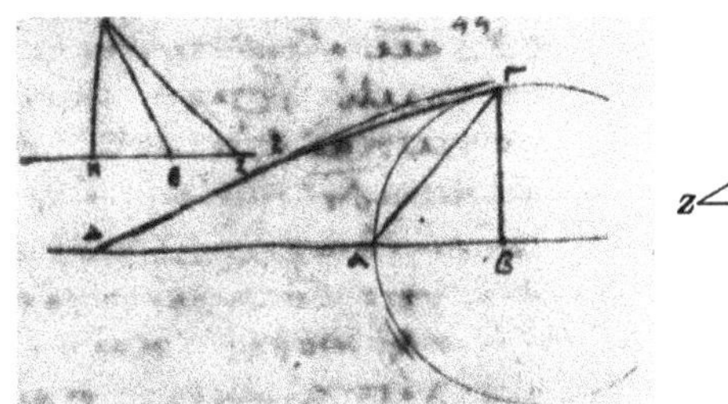

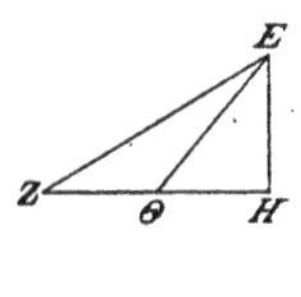

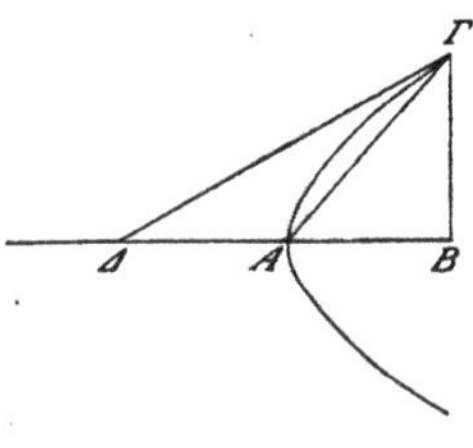

Vat. gr. 206, f. 86r; the conic section is an arc of a circle; ΔA ≠ AB; the two triangles are not similar

Heiberg's diagram for *Con.* II.50, which makes ΔA = AB and the triangles similar

The last statement of the argumental complex I shall call "chain of givens" (second segment of text in the previous page), with which the analysis ends, claims that the predicate "given in position" applies to the same object that at the very beginning was supposed (imperative γεγονέτω "let it happen to come to be") to realize the final configuration of the problem; we might say that this object is "potentially constructible", or maybe we might claim that we have proved its "existence and uniqueness". However, the analysis has a disturbing feature: it opens and closes by asserting that two different states of affairs apply to the *same* object. As a consequence, the synthesis cannot simply consist in an inversion of the analysis. Furthermore, the connection between the chain of givens and the sequence of constructive acts that solve the problem is ineffable. What is certain is that the *Data* does not tell us how to do it. Let us see more closely what happens in this proof.

1) The verb form that initializes the analysis, γεγονέτω "let it happen to come to be", is an invariant stylistic trait with a strong metamathematical connotation. It is a purely stative perfect stem that does not assume that the problem is solved, but only that the final configuration has come to be realized. The problem is in fact solved by performing the construction, not by "displaying" the intended configuration. It is important to stress that assuming the realization of the final configuration does not characterize the analytic method, as is commonly held: no locus theorem begins (and could begin) in this way, as we shall see below.
2) In the present instance,[70] *Data* def. 1 is crucial in making the analytic engine start to work: the angle ἡ ὑπὸ ΒΔΓ is given because, in the final configuration supposed as realized, it is equal to another angle (a given one, but this fact is totally irrelevant); it is not given because the final configuration is "given", which is exactly what the analysis must prove. It is crucial, then, that pairs of "identical" objects are displayed during the proof: this is the reason why the enunciation requires that the tangent "make on the axis an angle *equal to* a given acute angle" and not simply "make on the axis a given acute angle". The inferences subsequent to the first step, which constitute the chain of givens, are transformed and rearranged to make up the synthesis.
3) Some of the constructive acts of the synthesis correspond in the analysis to steps that are included in *deductive* chains formulated in the language of the givens. (Broken underlining and sentences in italics in the text above.) The direction of the deductive progression and the ordering of the sequence of constructive acts is the same. This remark is of the utmost importance: it shows that recent attempts at building a formal system for Euclidean constructive acts were

[70] This qualification is necessary since many analyses start by applying *Data* def. 4; in this case, no statement of equality is needed in the enunciation.

anticipated in antiquity in a most satisfactory way (cf. Sect. 3.3). Here is a correspondence table of the most basic constructive acts that are dressed in deductive clothes in the *Data*:[71]

Definition or theorem of the *Data*	Constructive act
def. 6. A circle is said to be given in position and in magnitude of which the centre is given in position and the radius in magnitude.	I.post.3. Describe a circle with any centre and radius.
26. If the extremities of a straight line be given in position, the line is given in position and in magnitude.	I.post.2. Join any two points by a straight line.
27. If one extremity of a straight line given in position and in magnitude be given, the other will also be given.	I.3. From the greater of two given unequal straight lines remove a straight line equal to the less.
28. If through a given point a straight line be drawn parallel to a straight line given in position, the drawn ‹straight line› is given in position.	I.31. Through a given point draw a straight line parallel to a given straight line.
29. If on a straight line given in position and at a given point on it a straight line is drawn making a given angle, the drawn ‹straight line› is given in position.	I.23. On a given straight line and at a point on it construct an angle equal to a given angle.

4) The truly deductive steps of the synthesis can also be found in the analysis, in (approximately) inverted order and again expressed in the idiom of the givens. (Unbroken underlining above.)
5) The fictitious objects that figure in *Data* def. 1–4 in order to "saturate" a relation is actualized in the synthesis as a "given" of the problem or as an immediate evolution of it, as for instance triangle τὸ ΕΖΘΗ. In its turn, this triangle will be duplicated in the proof as triangle τὸ ΓΔΑΒ, in order to "procure" the sought object.[72] Likewise, logical or geometric objects that figure only once in the analysis can be found in duplicate in the synthesis. (In boldface above is the doubled constructive act of the perpendicular, ἡ ΒΓ or ἡ ΕΗ/ΒΓ, in italics the doubling of the property of the subtangent, transformed into two constructive acts.)
6) What in the analysis is qualified as "given", is found in the synthesis within equalities or identities of ratios: the analysis processes predicates, the synthesis relations. This fact shows that it is meaningless to claim that some chain of inferences in the analysis is the inverse of some chain of inferences in the synthesis. At best, an isomorphism can be established. Yet, the correspondence remains misleading, as propositions in the *Data* and in the *Elements* cannot be related exactly, and, what is worse, those in the *Data* are usually proved by means of the "corresponding" propositions in the *Elements*. For instance, there is an obvious symmetry between the steps applying *Data* 40 in the analysis and VI.4 in the synthesis or *Data* 41 in the analysis and VI.6 in the synthesis, but (*i*) *Data* 40 is proved by means of VI.4 and *Data* 41 by means of VI.6; (*ii*) in order to prove that the relevant ratios and angles are given, *Data* def. 3 must be applied in the analysis to disentangle them from the notion of a figure "given in form", whereas the synthesis may refer directly to VI.4 and VI.6 (where the notion of similar triangles does not appear). It follows that a parallel can be set at best between VI.6 and VI.4. Yet they are not each inverse of the other, but only partial converses. There is more: the inference licensed by V.22 corresponds to what? To *Data* 8, which is proved, with an explicit reference, by means of V.22 itself. This very inference, and not a sort of inverse of it, is used in *Con.* II.50, with the only difference that it is formulated in the language of the givens. In short, the deduction in *Con.* II.50 is a logical labyrinth: it arranges the "isomorphic" theorems of the *Data* and

[71] A complete correspondence table of propositions in the *Elements* (including problems) and in the *Data* can be found in Appendix B.

[72] On doubling objects in Greek mathematics, see Sect. 4.2. None of these triangles is ever mentioned in *Con.* II.50 just read, still they are "there", otherwise *Data* 40 and 41 could not be applied.

propositions of the *Elements* in the same order, thanks to a clever permutation of the angles from which the analysis and the synthesis start and to the fact that VI.4 and VI.6 are partial converses. To appreciate that a true inversion of deductive order has taken place, check the position of the "ratio of ΔB to BΓ" within the two argumental chains.

7) The final part of the analysis gives rise in the synthesis to a sequence of constructive acts: these are the "geometric" steps of the chain of givens. For instance, *Data* 29 corresponds to problem I.23 and both are applied to the same angle, but, on the other hand, what corresponds to *Data* 25 remains unstated.[73] The steps related to the property of the subtangent, even if they are contained in the underlined portion, become two constructive acts: a bisection (within triangle τὸ ZHE) and a doubling (within triangle τὸ ΔΒΓ) of a straight line; the converse of the property of the subtangent is then applied in the form of *Con.* I.33. The only auxiliary[74] constructive act (dropping perpendicular ἡ ΒΓ) is also procured in double in the synthesis, once for each triangle. If it is not immediate to set an isomorphism between the chain of givens and the synthetic proof, it is simply meaningless to claim that a sequence of constructive acts has the same or the opposite direction as a chain of givens, or simply that some linked constructive acts "correspond" to a chain of givens. Two constructive acts cannot be each a consequence of the other; they can at best come one after the other, the former giving rise to a geometric configuration suited to perform the second. In the chains of givens, on the contrary, what is at stake is always a deductive sequence. The only residual correspondence is that, with respect to the actual development of the proof, the chain of givens and the construction have the same direction: see for instance the ordering of the sequence "straight line ἡ ΓΑ (that is, angle ἡ ὑπὸ ΒΑΓ) → straight line ἡ ΓΔ" both at the end of the analysis and in the construction.

Thus, the properly *deductive* steps of a synthesis can be found in the analysis, in an approximately reversed order and expressed in the language of the givens. In this way, a sequence of preconditions is formulated as a "forward" deduction: this is a non-trivial solution to a non-trivial problem.[75] The solution of Greek mathematicians to the problem of giving the cogency of some inferences an adequate linguistic representation was dictated by the perception that lexical and stylistic conventions, when rigidly adhered to, have an actual mathematical import. The rigidity of formulation in a natural language replaces the formalism. This tendency can be viewed as a struggle towards minimizing the "intuitive" component at work in a mathematical proof. As a matter of fact, "intuitions" are required:

- When drawing any conclusion, be it the conclusion of an inference or the conclusion we draw from the whole argument that makes the proof of a proposition. The answer to the second problem, which is the problem of mathematical generality, is provided by the indefinite structure we shall discuss in Sect. 3.3. The first problem is partly answered by the rigidity of the formulaic system and by sentential validation (Sect. 2.3). Some particular deductions, most notably those that involve relations, can be said to conclude in virtue of their form: in a sense, the conclusion shows itself in a suitable notation (see Sect. 4.5.1).

[73] This is for instance a consequence of *Con.* I.17.
[74] That is, a constructive act that is not indicated in the enunciation: see Sect. 4.3 for a typology of constructive acts.
[75] To see where the difference lies, it is enough to compare this solution with the highly contrived solution to the same problem adopted in the so-called theorematic analyses, namely, those preliminary to the synthetic solution to a *theorem*, not to the constructive solution to a problem. See the very end of this Section and Acerbi 2011a, 148–149.

- In the constructions. How is the mathematician to understand which "auxiliary" constructive acts must be performed within the construction? There is no answer, of course. However, tools can be introduced in order to minimize the intuitive component. First of all, the constructive acts are formulated in a rigid matrix structure, which does not interact with deductions but in some sense reproduces their formulation (see Sect. 2.2 and 4.3).[76] This matrix structure has the consequence that only a very limited number of constructive acts may operate on a configuration generated by any assigned constructive act. Second, and most importantly, the chains of givens transform the constructive sequences into deductive sequences, thereby validating them and reducing the issue raised in this item to the issue raised in the following item.
- At every deductively self-contained step. Who gives us a rule to understand which step is to follow a given step? No one, of course, but the number of possible choices can be reduced. For instance, a universe of discourse might be conceived that is so restricted to be decidable: the Stoics did this in their syllogistic, Hero did this in his alternative proofs of II.2–10 (to see this, the reader has to wait until the end of this Section).
- In order to "see" the proof. Or in other terms: we might require that propositions as a whole have to be validated. The chains of givens appear to have this function in the attested analyses. Not only: in late authors, analysis came to be considered a heuristic tool,[77] a technical counterpart of sorts of the Platonic intimation that mathematics must be subordinated to dialectic. This is only a glorious myth (an analysis is as formalized as a synthesis, and this settles the matter of its alleged heuristic import), but it was supported by the idea that there exists only a finite supply of deductive structures, and that after all the *Data* list most of them. Also, it is not difficult to see how this myth took shape and why is it still appealing. The point can be stated as follows: exactly because validation by means of the "givens" is not real metamathematics but remains strictly within the conceptual and stylistic boundaries of the demonstrative code, the validation game ends here: a chain of givens cannot be validated by anything.[78] The only available way to break the deadlock is upgrading analysis from mathematics to some philosophically connotated activity, and heuristic does the job.

The latter point is crucial and deserves a discussion. Our sources attest in fact to a large-scale evolution of the mathematical field, which is best seen by looking at the evolution of the format of locus theorems (see also Sect. 5.1.1). Their enunciation, in a conditional form that attests to their being theorems and not problems, employs the language of the givens to describe the constraints and to identify the line that solves the locus. The proof is framed as an analysis and synthesis. The analysis, of course expressed in the idiom of the givens, consists in identifying as given the line that a point subjected to the assigned constraints comes to ἅψεσθαι "touch"; the nature of the solution, it is to be stressed, is already made explicit in the enunciation. In other terms, if a point satisfies the constraints, then it belongs to a well-determined curve—this corresponds to proving both the existence and the uniqueness of the locus,[79] even if, in the enunciation of a locus theorem, what is sought is formulated as a predication: "a point *A* touches such-and-such a line".

[76] In a strong linguistic sense, I.post.1 has only one single operation of joining a straight line as a correlate; it is a two-entry relation of sorts, the free variables being simply the denotative letters of the extremities. This is what I have called "subsentential validation".

[77] Most celebrated passage Pappus, *Coll.* VII.1, but before him see already Geminus, for whom "analysis is the discovery of a proof", mentioned in Ammonius, *in APr. I*, 5.28, and Galen, *Pecc. dign.*, 54.24–55.2. It goes without saying that this myth is very hard to die. On the issue, see Acerbi 2011b and 2011d.

[78] Only the invention of algebra as a comprehensive metalanguage will allow mathematicians to break these boundaries.

[79] The scanty and late evidence on the issue of existence and uniqueness in the analytic field places emphasis on existence only, probably because uniqueness was taken for granted. This view is corroborated by a passing remark by Pappus about

After Charmandrus' loci (of which we do not have the proofs), let us read the locus theorem in Pappus, *Coll.* IV.78 to see how such a proof worked—Pappus' style is highly elliptical; "to be on" a line is synonymous with "to touch" it:[80]

θέσει εὐθεῖα ἡ ΑΒ, καὶ ἀπὸ δοθέντος σημείου τοῦ Γ προσπιπτέτω τις ἡ ΓΔ, καὶ πρὸς ὀρθὰς τῇ ΑΒ ἡ ΔΕ, ἔστω δὲ λόγος τῆς ΓΔ πρὸς ΔΕ. ὅτι τὸ Ε πρὸς ὑπερβολῇ. ἤχθω διὰ τοῦ Γ τῇ πρὸς ὀρθὰς παράλληλος ἡ ΓΖ· δοθὲν ἄρα τὸ Ζ. καὶ τῇ ΑΒ παράλληλος ἡ ΕΗ, καὶ τῷ τῆς ΓΔ πρὸς ΔΕ λόγῳ ὁ αὐτὸς ἔστω τῆς ΓΖ πρὸς ἑκατέραν τῶν ΖΘ ΖΚ· δοθὲν ἄρα ἑκάτερον τῶν Θ Κ. ἐπεὶ οὖν ἐστιν ὡς τὸ ἀπὸ τῆς ΓΔ πρὸς τὸ ἀπὸ τῆς ΔΕ οὕτως τὸ ἀπὸ τῆς ΓΖ πρὸς τὸ ἀπὸ τῆς ΖΘ, καὶ λοιποῦ ἄρα τοῦ ἀπὸ τῆς ΖΔ – τουτέστιν τοῦ ἀπὸ τῆς ΕΗ – πρὸς λοιπὸν τὸ ὑπὸ τῶν ΚΗΘ λόγος ἐστὶν δοθείς· καὶ ἔστι δοθέντα τὰ Κ Θ· τὸ Ε ἄρα πρὸς ὑπερβολῇ ἐρχομένῃ διὰ τῶν Θ Ε.

‹Let there be› a straight line ‹given› in position, AB, and from a given point, Γ, let some ‹straight line›, ΓΔ, fall ‹on it›, and let a ‹straight line› ΔE be at right ‹angles› with AB, and let the ratio of ΓΔ to ΔE be ‹given›. That E is on a hyperbola. Let a ‹straight line›, ΓZ, be drawn through Γ parallel to the one at right ‹angles› [I.31]; therefore Z is given [*Data* 28, 25]. And ‹let a straight line›, EH, ‹be drawn› parallel to AB [I.31], and let ‹the ratio of› ΓZ to each of ZΘ, ZK be the same as the ratio of ΓΔ to ΔE; therefore each of Θ, K is given [*Data* 27]. Then since, as the ‹square› on ΓΔ is to that on ΔE, so that on ΓZ is to that on ZΘ [*Data* 50], therefore the ratio of that on ZΔ as a remainder—that is that on EH—to the ‹rectangle contained› by KHΘ as a remainder is also given [V.19, *Data* def. 2]; and K, Θ are given; therefore E is on a hyperbola passing through Θ, E [*Con.* I.21].

Par. gr. 2368, f. 103v; there is no hyperbola in the manuscripts

A locus theorem like this does not contain a synthesis; to do the synthesis of a locus amounts to constructing the sought line on the basis of the givens of the theorem and to prove that it satisfies the assigned constraints.[81] Yet, the locus theorems we read in our sources usually contain a synthesis. This is because the format of these propositions underwent an evolution. In a first phase, the

some recent authors who, by applying Euclid's *Porisms*, neglected problems of constructibility, "proving only that what is sought exists without procuring it" (*Coll.* VII.14). Since the *Porisms* contained only analyses, as is obvious from Pappus' account of them, it must be concluded that the authors mentioned by Pappus had it clear that this kind of proposition could be interpreted as an existence proof.

[80] The point is on a hyperbola because of a characteristic property proved in *Con.* I.21 (see the last step). On the notion of "characteristic property", see Sect. 5.1.1.

[81] Since the solution of a locus has to be a previously known line (straight line, circumference, conic section, etc.), many locus theorems are the inverse of known characterizations of such lines, namely, the theorems in which the constraint of the locus is proved to be a property of these lines (in the case of the locus just read, this is *Con.* I.21). Proclus calls "locus theorems" exactly the cluster *El.* I.35–38 (*iE*, 394.11–396.9, and passing mentions at 405.4–6, 412.5–7, and 431.23).

loci were only analyzed. This view is corroborated by the lemmas Pappus proved for the Euclidean *Loci on a surface*, since such lemmas actually are the syntheses of some loci, and by some problems and loci that he very likely extracted from Aristaeus' *Solid loci*, where the syntheses are either absent or are obviously later appendages:[82] this shows that these works contained only analyses. Furthermore, the enunciation itself of a locus theorem requires to prove that a point be on a line "given in position", and the attested analyses, like the analysis just read, end exactly with a proof of this statement: the synthesis performs a construction that is simply not required by the enunciation. In a second phase, the analysis was supplemented by a synthesis where the line that solves the locus is constructed starting from the initial givens. Apollonius is representative of this phase, as he himself asserts when mentioning the locus "on three and four lines".[83] A third phase is attested to only in late texts, such as for instance in Eutocius' "transcription" of the Apollonian construction of a locus already mentioned in Aristotle's *Meteorologica*.[84] This text adheres to a strictly synthetic format: therefore, it must provide a synthetic proof of the uniqueness of the solution (that normally pertains to the analysis), proving that no point that does not belong to the curve satisfies the constraints. The synthetic mode also makes it necessary to change the manner of identification of the solution in the enunciation: one must prove, so Eutocius frames his text, that "it is possible to trace" the line (a formulation mid-way between a theorem and a problem).

It is not surprising that the demonstrative format of loci underwent an evolution, while the formulation of their enunciation was kept fixed: for the enunciation is what makes a locus theorem immediately recognizable as such.

We might wonder why the Greek geometers became dissatisfied with the analysis of the loci. Several phenomena may have contributed to the perception that it was necessary to add a synthesis to a locus theorem, and maybe Apollonius himself is responsible for this reform. Requiring the construction of the solution, which can be performed only in a synthesis, is among the "foundational" concerns that are typical of him.[85] On the other hand, one must be able to complete such syntheses. They normally apply crucial results of the theory of conic sections; a suitable supply of such tools was at the mathematicians' disposal only with Apollonius, as he himself proudly points out in the prefatory letters to *Con.* I and IV.[86] The generality of his approach, in particular the "discovery" of the opposite sections (= hyperbolas with two branches), made the number of tractable (solid) loci considerably larger than in earlier approaches; his complete treatment of the intersections between conics eased the analyses of the determinations of problems (see Sect. 4.2.1).

The same happens in the case of problems of construction: a perfect example is the angle trisection in Pappus, *Coll.* IV.60–66: it is necessary to construct a hyperbola with given asymptotes and that passes through a given point[87] in order to synthetize the problem of trisection, but the analysis does not require that this construction is carried out.[88]

---

[82] See the discussion in Jones 1986, 582–584, 591–595.

[83] Read the passage at *AGE* I, 4.10–17. Arabic sources confirm that Apollonius was used to synthetize the loci: Hogendijk 1986, 206–218. Apparently, scholars have failed to notice that the loci "on one and two lines" (these loci are straight lines or circles, namely, degenerate conic sections) were also solved by Apollonius, as Pappus' summary of Apollonius' *Plane loci* attests: see *Coll.* VII.25.

[84] At *AGE* II, 180.11–184.20; see Vitrac 2002 for a discussion of the locus in the *Meteorologica*, and also Sect. 3.2.1.

[85] See Acerbi 2010b on Apollonius' foundational strategies.

[86] Read these claims at *AGE* I, 4.5–17, and II, 4.5–7, 4.16–17, respectively.

[87] *Con.* II.4, but this is a later addition (we find it proved in Pappus, *Coll.* IV.65–66 and VII.274–275, and again by Eutocius in his commentary on *Sph. cyl.* II.4, at *AOO* III, 176.6–28): Arendt 1913–14 already settled the issue of authenticity of this proposition—this short study was unobtrusively cited by Heiberg in *AOO* II, VIII, and thanks to this cryptomention it was squarely plagiarized, with the addition of an overwhelming scholarly apparatus as usual, in Knorr 1982.

[88] That not every analysis can be inverted was already clear to Aristotle: *SE* 16, 175a26–30, and *APo.* I.12, 78a6–13.

Such an evolution explains a number of phenomena: the perceived "redundancy" of the analysis of a problem once the synthesis is given; the obvious fact that—as we have seen—the mere presence of the synthesis of a locus theorem is incongruous with the form of its own enunciation; and, finally, the disappearance of the entire analytic corpus, and in particular of the segments of it that could not be synthetized, like the *Porisms*:[89] this was mathematics of a past era, almost entirely wiped out by the constructive campaign put forward and pursued with force by Apollonius. Analysis survived as a mathematical fossil in the form of a validating instance, eventually recycling itself in the myth of an heuristic tool. Such a validating instance initially took the form of the residual analyses still attached to perfectly self-contained synthetic arguments as problems or locus theorems (with the modulations explained above). In the hand of Apollonius itself and of later mathematicians, however, analysis as validation became a fashionable stylistic game that gave rise to such amazingly useless elaborations—still perfectly legitimate as fine pieces of (meta)mathematics—as theorematic analyses and as the constellation of arguments I shall study in the next Section.

Before doing this, let me discuss a striking application of theorematic analysis without chains of givens. This is found in Hero's alternative proofs of II.2–10.[90] To understand what happens, let us read first the enunciation of II.8 (*EOO* I, 138.2–7; see also Sect. 4.5.1.3 for the entire Greek text of this proposition, completed by a "symbolic" translation and a diagram):

ἐὰν εὐθεῖα γραμμὴ τμηθῇ ὡς ἔτυχεν, τὸ τετράκις ὑπὸ τῆς ὅλης καὶ ἑνὸς τῶν τμημάτων περιεχόμενον ὀρθογώνιον μετὰ τοῦ ἀπὸ τοῦ λοιποῦ τμήματος τετραγώνου ἴσον ἐστὶ τῷ ἀπό τε τῆς ὅλης καὶ τοῦ εἰρημένου τμήματος ὡς ἀπὸ μιᾶς ἀναγραφέντι τετραγώνῳ.	If a straight line be cut at random, four times the rectangle contained by the whole and by one of the segments with the square on the remaining segment is equal to the square on the whole and on the said segment as if described on one ‹straight line›.

In the *Elements*, the proof of this statement—as well as of all those in the string II.2–10—runs as follows: the whole geometric configuration is first constructed, followed by the proof of the equality stated in the consequent of the above conditional. The proof operates by identification of all the sub-regions that compose the geometric objects in the configuration (squares and rectangles). In this way, the Euclidean proofs of II.2–10 are independent of one another. The Heronian proofs argue in a completely different way. Let us read the first argument of Hero, *El.* II.8*alt* (Tummers 1994, 81.20–82.10); the proof really comprises two analyses; in the Latin translation, the first analysis—which we read—is called *dissolutio*, the second analysis is called *compositio*:

Ponam lineam *ab* quam supra punctum *g* dividam qualitercumque contingat divisio, et adiungam ei lineam *bd* equalem linee *bg*.	I shall set a line *AB* that I shall cut at random at point *G*, and I shall add to it a line *BD* equal to line *BG*.
(1) Cum ergo resolverimus quadratum linee *ad*, resolvetur in probationem figure quarte huius partis. Quod ideo erit quoniam quadratum	(1) Therefore when we shall resolve the square of line *AD*, it will be resolved to the proof of the fourth proposition of this part. Then since it will be that the square

[89] This is borne out, as seen, by Pappus' claim at *Coll.* VII.14, about recent authors who used the *Porisms* to set up mere existence proofs, without "procuring" the sought for object.

[90] At Tummers 1994, 73.25–86.5. Recall that scraps of Hero's commentary on the *Elements* are only preserved, in Arabic and in Arabo-Latin translation, in the analogous commentary of the Persian mathematician an-Nayrīzī (*Anaritius*). For the relation between the Arabic and the Latin text of this commentary, see Brentjes 2001.

factum ex linea *ad* est equale duplo superficiei quam continent due linee *ab bd* cum duobus quadratis factis ex duabus lineis *ab bd* (2) et quia *bd* posita est equalis sectioni *bg*, ergo duplum superficiei que continetur a duabus lineis *ab bg*, cum duobus quadratis factis ex duabus lineis *ab bg* est equale quadrato facto ex linea *ad*. (3) Secundum probationem vero figure septime huius partis erunt duo quadrata facta ex duabus lineis *ab bg* equalia duplo superficiei que continetur a duabus lineis *ab bg* cum quadrato *ag*; (4) cum ergo illud coniungetur, erit quadruplum superficiei que continetur a duabus lineis *ab bg* cum quadrato *ag* equale duplo superficiei que continetur a duabus lineis *ab bg* cum duobus quadratis factis ex lineis *ab bd*; (5) sed iam ostendimus quod ista sunt equalia quadrato facto ex linea *ad*; ergo quadruplum superficiei que continetur a duabus lineis *ab bg*, cum quadrato *ag* est equale quadrato *ad*. Ergo iam resolutum est hoc in figuram quartam prius, post in figuram septimam. Et illud est quod demonstrare voluimus.

made from line *AD* is equal to double the surface that the two lines *AB*, *BD* contain with the two squares made from the two lines *AB*, *BD*, (2) and since *BD* turns out to be set equal to segment *BG*, therefore double the surface that is contained by the two lines *AB*, *BG* with the two squares made from the two lines *AB*, *BG* is equal to the square made from line *AD*. (3) But according to the proof of the seventh proposition of this part, the two squares made from the two lines *AB*, *BG* will be equal to double the surface that is contained by the two lines *AB*, *BG* with the square of *AG*; (4) therefore if we compose that, the quadruple of the surface that is contained by the two lines *AB*, *BG* with the square of *AG* will be equal to double the surface that is contained by the two lines *AB*, *BG* with the two squares made from lines *AB*, *BD*; (5) and we have already proved that these are equal to the square made from line *AD*; therefore the quadruple of the surface that is contained by the two lines *AB*, *BG*, with the square ‹from› *AG* is equal to the square ‹from› *AD*. Therefore this has now been resolved to the fourth proposition first, then to the seventh proposition. And that is what we wanted to prove.

I skip the *compositio*, but the reader will easily reconstruct it from the following symbolic transcription of the entire proof. The enunciation itself is transcribed as follows: if *ab* is cut at *g* and we set $bd = bg$, then $q(ad) = 4r(ab,bg) + q(ag)$.[91]

The *dissolutio* operates on term $q(ad)$ of this equality:

	equality	justified by
(1)	$q(ad) = 2r(ab,bd) + q(ab) + q(bd)$	II.4
(2)	$q(ad) = 2r(ab,bg) + q(ab) + q(bg)$	(1), $bd = bg$
(3)	$q(ab) + q(bg) = 2r(ab,bg) + q(ag)$	II.7
(4)	$2r(ab,bg) + q(ab) + q(bd) = 4r(ab,bg) + q(ag)$	(3) + $2r(ab,bg)$
(5)	$q(ad) = 4r(ab,bg) + q(ag)$	(2), (4)

The *compositio* operates on term $4r(ab,bg) + q(ag)$ of the equality:

	equality	justified by
(0)	$2r(ab,bg) + q(ab) + q(bd) = 4r(ab,bg) + q(ag)$	reference to step (4) of the *dissolutio*
(1)	$2r(ab,bg) + q(ab) + q(bg) = 4r(ab,bg) + q(ag)$	$q(ab) + q(bg) = 2r(ab,bg) + q(ag)$ (II.7) + $2r(ab,bg)$
(2)	$2r(ab,bg) = 2r(ab,bd)$	$bd = bg$
(3)	$2r(ab,bd) + q(ab) + q(bd) = 4r(ab,bg) + q(ag)$	(1), (2)
(4)	$2r(ab,bd) + q(ab) + q(bd) = q(ad)$	II.4
(5)	$4r(ab,bd) + q(ag) = q(ad)$	(3), (4)

[91] The sign $q(a)$ denotes the square on straight line a; the sign $q(a,b)$, the rectangle contained by straight lines a and b.

If we exclude a mathematical trick as the introduction of segment *bd*, which is used in the *Elements* to generate the structure of double gnomon from which the factor "four" in the proof naturally arises, and by Hero to generate a linear configuration suited to applying II.4 and II.7, the above proof is purposely carried out without auxiliary constructions. As a consequence, no trace of language of the givens survives in the analytic part of the proof; this can only happen in a theorematic analysis, for a problem necessarily contains some constructive act. As a consequence, the Heronian proofs are framed as pure reductions, which operate on two levels.

On a first level lie the two geometric objects constructed starting from the objects indicated in the enunciation; it is required to prove that the constructed objects are equal. The mathematical "facts" at issue are some configurations of geometric objects, not the relations between them. In the Heronian proof, one of the configurations is assumed as the starting point [it is called *res nota* in the text; this is $q(ad)$ in the *dissolutio*, $4r(ab,bg) + q(ag)$ in the *compositio*], the other as the end point (*res quesita*); the *res nota* is reduced to the *res quesita* by means of theorems in the same sequence II.2–10 (II.1 serves as a "principle"). Of the two configurations, what is the *res quesita* and what is the *res nota* is arbitrary; the directions of the *dissolutio* and of the *compositio* are determined solely by the order in which the two configurations are mentioned in the enunciation. It is not even true, for instance, that the *dissolutio* operates on the simpler configuration. Thus, the presence of the two terms *dissolutio* and *compositio* (which are certainly translations of ἀνάλυσις and σύνθεσις) is not motivated by the fact that in one of the proofs a figure is "dissolved" and in the other it is "(re)composed", for both operations occur in both proofs.

On a second level, however, the declared goal of the whole analysis is exactly to set up a complete list of the theorems in the sequence II.2–10 to which the theorem at issue is reduced: this is again an analysis, even if on a metamathematical level. In the case of II.8, such theorems (called *figure*) are II.4 and II.7: the items of this micro-list are expressly declared in the course of the proof and at the very end of it: *ergo iam resolutum est hoc in figuram quartam prius, post in figuram septimam*. Of course, the proof is perfectly valid without these metamathematical clauses. Still, their presence cannot simply be explained as a stylistic trait without particular significance or as a commentator's vain endeavour. On the contrary, such clauses show that Hero's approach is eminently metalogical: they intimate that the goal of these alternative proofs is to reduce a proposition of the sequence II.2–10 to some other propositions that precede it in the sequence.

The Heronian proof is thus conceived both as a deployment of the deductive structure in its complete form and as a decomposition of the geometric configuration in its ultimate components.

Hero's achievement finds an ideal field of application in the linear lemmas of Book II, but an interesting parallel can also be drawn with a particular logical doctrine. This is the analysis of those formally valid arguments that in the Stoic tradition got the name of "non-simple indemonstrable syllogisms". Simple indemonstrables are such inferences as it is "immediately clear that they validly conclude, namely, that for them the conclusion is validly deduced from the premises" (Sextus Empiricus, *M* VIII.228). Chrysippus established five basic kinds of indemonstrable syllogism, among which forms of *modus ponens*, *modus tollens*, etc. can be recognized (see Sect. 5.2.2). The arguments made of suitable chains of simple indemonstrables were called "non-simple indemonstrables". The non-simple indemonstrables are also valid arguments, but this becomes apparent only after they are reduced to their simple components. The Stoics called "analysis" this procedure, which was performed by means of rules to manipulate deductions, called θέματα "posits".[92]

[92] On Stoic analysis see Bobzien 1996, Bobzien 1999, 137–151, Bobzien 2019 (whose n. 25 on p. 245, however, needlessly corrects Sextus' text), and Crivelli, forthcoming (with a different reconstruction of the θέματα).

Dedicated theorems sum up several θέματα in a single prescription; the Stoics called their version "dialectical theorem": "when we have the premisses that imply some conclusion, we also have the conclusion that lies potentially in them, even if this is not expressly stated" (*M* VIII.229).

A fundamental feature of the Stoic analysis of syllogisms is that the reduction ends exactly when the simple indemonstrables that compound the syllogism are made explicit. It is not a proof but a reduction method; it works by inserting intermediate elementary inferences and by deploying the complete formulation of an argument; once this is done, the validity of the argument is immediately grasped, insofar as it is a well-formed sequence of valid arguments. The "result" of the reduction, namely, the non-simple indemonstrable to be reduced, is taken for granted from the very outset: the method is analytic. The deductive progression is "backward", but it does not look for preconditions: the operation is metalogical, and does not have a formal structure. Well—and here comes the answer to the question left unanswered at the end of the third point of the last bullet-point list I have set out—both in the Heronian rewriting of II.2–10 and in Stoic analysis, the irreducible inferential structures are known and finite in number (just a few, in fact): they are the preceding theorems in the sequence *El.* II.2–10 and the five simple indemonstrables, respectively. This very limited supply makes the choice of the "subsequent step" easy.

To sum up, the Heronian *compositio* and *dissolutio* are independent, self-contained demonstrative arguments. Any of them is superfluous once the other is carried out, most notably because they comprise the same deductive steps in a different order. Actually, the *compositio* may be said to depend on the *dissolutio* only in a weak sense: the starting point of the *dissolutio* is a relation (dictated by II.7) whose pertinence is in principle established only thanks to the *dissolutio*—but then, we may well wonder what justifies the pertinence of the relation dictated by II.4 at the beginning of the *dissolutio* itself. Apart from these initializations, all deductive steps in the two proofs are simple replacements of equivalent geometric configurations within equalities. Such operations are among the most frequently used tools of Greek mathematics, but here they are used exclusively: the resulting proof has a marked algorithmic connotation.

2.4.2. *Validating algorithms and procedures by the "givens"*

A whole tradition of late Greek mathematicians, like Hero, Ptolemy, and Diophantus, has also read the chains of givens—which are absent in the Heronian proofs just expounded—in an algorithmic perspective only. In these authors, "given" is synonymous with "univocally determined", and hence "findable" or "computable" from the numeric assignments of a problem. In this way, the chains of givens were used to validate the steps that allow to compute one of the magnitudes involved in a "formula" from the others. These computations are processed in the form of an algorithm or of a procedure, whose steps replicate those of the chain of givens. The "formula" itself is proved by means of a standard geometric proof in synthetic format: this happens both in the case of "Hero's formula" read in Sect. 1.3 and in the case of Diophantus' definition of a polygonal number read in Sect. 1.2. As a consequence, within the framework of validation, the Greek mathematicians came to establish a hierarchy among the three main codes, apparently on the basis of a perceived different degree of "argumentative power". Let us see how the mechanism just outlined worked in detail.

Validating a calculation by a chain of givens does not require so many tools. These are some definitions and theorems of the *Data*, which correspond to operations on the numeric values assigned to the given magnitudes, as we see in the table set out in the next page. In the right column, the signs *a* and *b* denote magnitudes or relations that have an explicitly expressed numeric value;

$q(a)$ is the square on straight line a or the square of its numeric value; the arrow separates the input from the output of the operation.[93]

Definition or theorem of the *Data*	Operation
def. 1. Given in magnitude are said regions, lines, and angles for which we can procure equals. def. 2. A ratio is said to be given for which we can procure the same.	$a, a = b \rightarrow b$
1. The ratio to one another of given magnitudes is given.	$a, b \rightarrow a{:}b$
2. If a given magnitude have a given ratio to some other magnitude, the other is also given in magnitude.	$a, a{:}b \rightarrow b$
3. If as many given magnitudes as we please be compounded, the magnitude compounded of them will also be given.	$a, b, \ldots, c \rightarrow a + b + \ldots + c$
4. If a given magnitude be removed from a given magnitude, the remainder will be given.	$a, b \rightarrow a - b$
5. If a magnitude have a given ratio to some part of itself, it will also have a given ratio to the remainder.	$a{:}b \rightarrow a{:}(a - b)$
6. If two magnitudes having a given ratio to one another be compounded, the whole will also have a given ratio to each of them.	$a{:}b \rightarrow (a + b){:}a, (a + b){:}b$
7. If a given magnitude be divided in a given ratio, each of the segments is given.	$a + b, a{:}b \rightarrow a, b$
8. Items that have a given ratio to the same will also have a given ratio to one another.	$a{:}b, c{:}b \rightarrow a{:}c$
22. If two magnitudes have a given ratio to some magnitude, both together will also have a given ratio to the same.	$a{:}c, b{:}c \rightarrow (a + b){:}c$
52. If on a straight line given in magnitude a form given in form be described, the described ‹form› is given in magnitude.	$a \rightarrow q(a)$
55. If a region be given in form and in magnitude, its sides will also be given in magnitude.	$q(a) \rightarrow a$
57. If a given ‹region› be applied to a given ‹straight line› in a given angle, the width of the application is given.	$ab, a \rightarrow b$
85. If two straight lines contain a given region in a given angle, and they be given together, each of them will also be given.	$ab, a + b \rightarrow a, b$

The "by division" operation on ratios, $a{:}b \rightarrow (a - b){:}\ b$, requires combining two theorems of the *Data*, namely, propositions 5 and 8.

Other theorems validate such geometric constructive acts as have an immediate operational import in the geometric metrological field,[94] most notably in the divisions of surfaces in Hero, *Metr.* III. In Hero's *Metrica* these are taken as operations and not as constructive acts: this is confirmed by the fact that they are formulated in the aorist imperative: see for instance τοσούτου ἀπόλαβε τὴν ΑΕ, καὶ ἐπίζευξον τὴν ΔΕ "of that much cut off AE, and join ΔE" in *Metr.* III.2 or, in the same proposition, the clause in suppositional mode with an aorist subjunctive ὥστε ἐὰν ἀπολάβωμεν τὴν ΑΔ μονάδων ια δ^{ον} καὶ παράλληλον ἀγάγωμεν τὴν ΔΕ, ἔσται τὸ προκείμενον "so that if we cut off AΔ of 11 ¼ units and draw a parallel ΔE, what has been proposed will be the case".[95] The fact that the first operation is sometimes formulated by means of a perfect imperative (ἀπειλήφθω, for instance in *Metr.* III.5–6) is probably due to the stylistic inertia of the model of geometric constructions. Note, finally, that the calculation of the area of a triangle of assigned sides $a, b, c \rightarrow tr(a,b,c)$—whose synthetic validation we read in *Metr.* I.8—has the combination of *Data* 39 and 52 as an immediate analytic counterpart: if the three sides of a triangle are given in magnitude, the triangle is also given in magnitude.

[93] The operation is sometimes weaker than the *Data*-theorem. Therefore, such operations are validated a fortiori.
[94] See the table under point 3 of the numbered list in Sect. 2.4.1.
[95] Acerbi, Vitrac 2014, 312.20–21 and 312.10–11, respectively.

The crucial point of this approach is that a calculation, very much as a geometric synthesis, is considered accessible only after the analysis has shown that the sought object is univocally determined by the assignments of the proposition: in metrical contexts, "given" means "given in magnitude" and has the numeric determination of that "magnitude" as a "synthetic" counterpart. This guarantees that the magnitude can actually be "procured", by means of either a calculation or a constructive act. This is the reason why the analytic format is identical in a metrical and in a geometric proposition. Let us read three examples.

The aim of Hero's *Metrica* is to set demonstrative grounds to the procedures of calculation of areas (Book I) and volumes (Book II) of some basic geometric figures, and to problems of division of plane regions (Book III): he offers rigorous geometric proofs that validate algorithms. The geometric proofs are sometimes in synthetic form, but they are most often in analytic form. That the sought geometric magnitude can be determined is proved in strictly geometric terms by means of an analysis, the magnitude itself being the final product of a chain of givens. In the synthesis, the calculation of the same magnitude is performed, either as a description of a procedure further supported by numeric examples, or by an algorithm.[96] Both the procedure and the algorithm are exactly parallel to the chain of givens.

Let us read the chain of givens and the validated algorithm of *Metr.* I.14 (Acerbi, Vitrac 2014, 180.8–182.6) set in parallel;[97] it is required to calculate the area of a quadrilateral τὸ ΑΒΓΔ having a right angle at τὸ Γ and no two sides parallel. The sides are given: ἡ AB is of 13 units, ἡ ΒΓ of 10 units, ἡ ΓΔ of 20 units, ἡ ΔΑ of 17 units. The algorithm divides the quadrilateral in two triangles, computes their areas and adds them. The area of the right-angled triangle τὸ ΒΓΔ is readily computed; to find the area of triangle τὸ ΑΒΔ, the hypothenuse ἡ ΒΔ of triangle τὸ ΒΓΔ must first be computed, and then, by means of *El.* II.13, the height ἡ AE of triangle τὸ ΑΒΔ. Note the clauses that open the two extracts: Hero sets out to prove that the area of the quadrilateral is given and calls the whole argument an "analysis"; the associated calculation is called "synthesis".

δεῖξαι αὐτοῦ τὸ ἐμβαδὸν δοθέν.	Prove that its area is given.	And as a consequence of the analysis it will be synthetized as follows.	συντεθήσεται δὲ ἀκολούθως τῇ ἀναλύσει οὕτως.
ἐπεὶ ἑκατέρα τῶν ΒΓ ΓΔ δοθεῖσά ἐστιν καὶ ὀρθὴ ἡ πρὸς τῷ Γ, δοθὲν ἄρα ἐστὶ τὸ ΒΓΔ τρίγωνον.	Since each of ΒΓ, ΓΔ is given and the ‹angle› at Γ is right, therefore triangle ΒΓΔ is given.	The 10 by the 20: it yields 200; and half of these: it yields 100.	τὰ ι ἐπὶ τὰ κ· γίγνεται σ· καὶ τούτων τὸ ἥμισυ· γίγνεται ρ.
καὶ ἔτι τὸ ἀπὸ τῆς ΒΔ ἔσται δοθέν	And further, the ‹square› on ΒΔ will be given	And again, the 10 by themselves: it yields 100. And the 20 by themselves: it yields 400; compose: it yields 500.	καὶ πάλιν τὰ ι ἐφ' ἑαυτά· γίγνεται ρ. καὶ τὰ κ ἐφ' ἑαυτά· γίγνεται υ· σύνθες· γίγνεται φ.
– ἔστι γὰρ μονάδων φ –·	—for it is of 500 units—;		
ἀλλὰ καὶ τὸ ἀπὸ τῆς ΑΒ δοθέν·	but that on AB is also given;	And the 13 by themselves: it yields 169;	καὶ τὰ ιγ ἐφ' ἑαυτά· γίγνεται ρξθ·
δοθέντα ἄρα ἐστὶ τὰ ἀπὸ τῶν ΑΒ ΒΔ·	therefore those on AB, ΒΔ are also given;	these with the 500: it yields 669;	ταῦτα μετὰ τῶν φ· γίγνεται χξθ·
καὶ ἔστι μείζονα τοῦ ἀπὸ τῆς ΑΔ· ὀξεῖα ἄρα ἐστὶν ἡ ὑπὸ ΑΒΔ·	and they are greater than that on ΑΔ; therefore angle ΑΒΔ is acute;		

[96] Hero expressly states this at the very end of *Metr.* I.6, Acerbi, Vitrac 2014, 160.16–17.
[97] All propositions of the *Metrica* are rearranged in this way ibid., 373–409.

τὰ ἄρα ἀπὸ τῶν ΑΒ ΒΔ τοῦ ἀπὸ τῆς ΑΔ μείζονά ἐστιν τῷ δὶς ὑπὸ τῶν ΔΒ ΒΕ·	therefore those on AB, BΔ are greater than that on AΔ by twice the ‹rectangle contained› by ΔB, BE;		
δοθὲν ἄρα ἐστὶν τὸ δὶς ὑπὸ τῶν ΔΒ ΒΕ·	therefore twice that by ΔB, BE is given;	subtract the 17 by themselves: 380 as a remainder;	ἄφελε τὰ ιζ ἐφ᾽ ἑαυτά· λοιπαὶ τπ·
ὥστε καὶ τὸ ἅπαξ ὑπὸ τῶν ΔΒ ΒΕ δοθέν ἐστι·	so that once that by ΔB, BE is also given;	half of these: it yields 190;	τούτων τὸ ἥμισυ· γίγνεται ρϙ·
καὶ ἔστι πλευρὰ τοῦ ἀπὸ τῆς ΒΔ ἐπὶ τὸ ἀπὸ ΒΕ·	and it is a side of that on BΔ by that on BE;		
δοθὲν ἄρα καὶ τὸ ἀπὸ ΔΒ ἐπὶ τὸ ἀπὸ ΒΕ·	therefore that on BΔ by that on BE is also given;	these by themselves: it yields 36100;	ταῦτα ἐφ᾽ ἑαυτά· γίγνεται μ$^{\gamma}$,ςρ·
καὶ ἔστι δοθὲν τὸ ἀπὸ ΒΔ· δοθὲν ἄρα καὶ τὸ ἀπὸ ΒΕ·	and that on BΔ is given; therefore that on BE is also given;	these ‹divided› by the 500: it yields 72 $^1/_5$;	ταῦτα παρὰ τὸν φ· γίγνεται οβ ε΄·
ἀλλὰ καὶ τὸ ἀπὸ τῆς ΕΑ	but also that on EA	subtract these from the 169: they yield 96 $^1/_2$ $^1/_5$ $^1/_{10}$ as a remainder;	ἄφελε ταῦτα ἀπὸ τῶν ρξθ· γίγνονται λοιπαὶ ϙς ∠΄ ε΄ ι΄·
ἐπὶ τὸ ἀπὸ ΒΔ·	by that on BΔ;	these by the 500: it yields 48400;	ταῦτα ἐπὶ τὸν φ· γίγνεται μ$^{\delta}$,ηυ·
καὶ ἔστιν αὐτοῦ πλευρὰ τὸ ὑπὸ ΒΔ ΑΕ·	and a side of it is that by BΔ, AE;		
δοθὲν ἄρα καὶ τὸ ὑπὸ ΒΔ ΑΕ·	therefore that by BΔ, AE is also given;	a side of these: it yields 220;	τούτων πλευρά· γίγνεται σκ·
καὶ ἔστι διπλάσιον τοῦ ΑΒΔ τριγώνου·	and it is double of triangle ABΔ;		
δοθὲν ἄρα καὶ τὸ ΑΒΔ τρίγωνον·	therefore triangle ABΔ is also given;	half of these: it yields 110. Of that much will be the area of ABΔ;	τούτων τὸ ἥμισυ· γίγνεται ρι. τοσούτου ἔσται τοῦ ΑΒΔ τὸ ἐμβαδόν·
ἀλλὰ καὶ τὸ ΒΓΔ·	but also BΓΔ;	but ‹the area› of BΓΔ is also of 100 units;	ἀλλὰ καὶ τοῦ ΒΓΔ μονάδων ρ·
ὥστε καὶ ὅλον τὸ ΑΒΓΔ τετράπλευρον δοθὲν ἔσται.	so that the quadrilateral ABΓΔ as a whole will also be given.	therefore the area of quadrilateral ABΓΔ will be of 210.	τοῦ ἄρα ΑΒΓΔ τετραπλεύρου τὸ ἐμβαδὸν ἔσται σι.

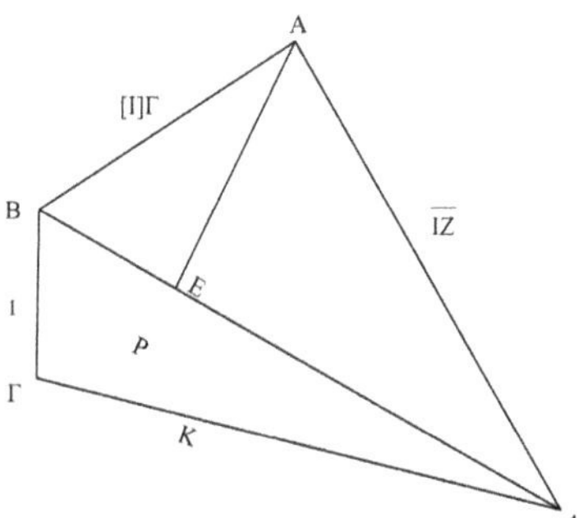

A faithful reproduction of the diagram in Seragl. G.İ.1, f. 75r, the sole manuscript witness of Hero's *Metrica*; the letters attached to the sides indicate the assumed numerical values

Note, once again and according to the discussion in Sect. 1, the striking difference between the two registers: an impersonal register in the proof, a fictitious interlocutory register in the algorithm. Even more striking is the difference between the passive impersonal imperatives that figure in the geometric construction that precedes the chain of givens (I have not transcribed this construction) and the active second-person imperatives of the algorithm. The personal or impersonal connotation must be regarded as a stylistic trait in both cases.

We find the same format in a purely geometric problem: in *Metr.* III.4 Hero deals with a problem of division of a figure. This is solved by a canonical analysis, after which we might expect to find a geometric synthesis, but Hero directly calculates the positions, with respect to the vertices of the figure, of the points at which the sides must be cut in order to have the required division performed. The analysis shows that such points are univocally determined by the assignments of the problem, and the calculation of their position can be carried out without further ado.

The two parts of *Metr.* I.14 read in parallel above nicely show the sentence-by-sentence correspondence between an algorithm and the chain of givens that validates it. However, the match between the two kinds of mathematical argument is seldom so exact. In the *Metrica*, for instance, we find algorithms in which sequences of operations are performed that differ from those explicitly indicated in the validating chain of givens.[98] More generally, as we shall see at the end of this section when reading a text by Ptolemy, the chains of givens dictate the form of the associated operations in simple cases only: see already the first two steps of *Metr.* I.14 read above. The origin of such a divarication lies in the fact that the chains of givens were originally intended to back up in deductive form the constructive acts and the deductive steps of a geometric proposition, and not to validate a computation. An extreme but very significant example of this phenomenon are the propositions *Data* 87–88. They prove in a few lines that, in a circle given in magnitude, a chord that extends under a given arc is given, and vice versa. In the *Almagest*, the operation that corresponds to these two data-theorems *is* using the chord tables in *Alm.* I.11 (interpolation included), for it is possible to determine by proof the value of a chord associated with a given arc only in particular cases—and even in these cases, only as the outcome of a long sequence of propositions that Ptolemy formulates as data-theorems (this is *Alm.* I.10).[99]

Let us now see how validation works in the case of a procedure; to do this, let us return to the paradigmatic example of a procedure in Diophantus' *On polygonal numbers*. Recall the relation that defines a polygonal number P in terms of its side s and of the multiplicity of its angles v:

$$8P(v-2)+(v-4)^2=[2+(v-2)(2s-1)]^2$$

As seen in Sect. 1.2, Diophantus explains, by means of a procedure that operates with this relation, how to find, once v is fixed, a polygonal number P whose side s is given, and vice versa. Before doing this, however, he sets out a chain of givens that validates the direct procedure:[100]

[98] See for instance *Metr.* II.13, III.8, and III.22.

[99] The chord associated with some specific arcs is computed, and then theorems are proved that correspond to our addition, subtraction, and bisection formulae for sines. The entire argument allows computing approximately the values of the chords associated with integer or half-integer values of arcs. Of course, the chord table can be used both ways; for this reason, using it is the algorithmic counterpart of both *Data* 87 and 88.

[100] The text is at Acerbi 2011e, 197.5–16; the inverse procedure is liquidated with a potential proof (ibid. 197.14–15): "and similarly, a polygonal being given, we shall find its side ΗΘ". The correspondence between the signs in the above relation and Diophantus' lettered designations in the text we shall read in a moment is as follows (this is already quite contrived, isn't it?): ΗΘ = s, ΗΜ = 1, ΝΚ = 2, and therefore ΘΜ = $s-1$ and ΗΘΜ = ΗΘ + ΘΜ = $2s-1$; again, one sets Νξ = $(v-2)(2s-1)$, hence Κξ = ΝΚ + Νξ = $2+(v-2)(2s-1)$; ΚΒ = $v-2$, hence ΝΒ = ΚΒ − ΝΚ = $v-4$.

<u>ἔχοντες</u> γὰρ πλευρὰν δοθεῖσαν τινὸς πολυγώνου τὸν ΗΘ <u>ἔχοντες</u> δὲ καὶ τὸ πλῆθος αὐτοῦ τῶν γωνιῶν ἔχομεν καὶ τὴν ΚΒ δοθέντων· *ὥστε* καὶ τὸν ὑπὸ συναμφοτέρου τοῦ ΗΘ ΘΜ καὶ τοῦ ΚΒ ἕξομεν δοθέντα, ὅς ἐστιν ἴσος τῷ Νξ· *ὥστε* ἕξομεν καὶ τὸν Κξ δοθέντα – ἐπείπερ δυάς ἐστιν ὁ ΝΚ –· *ὥστε* καὶ τὸν ἀπὸ τοῦ Κξ ἕξομεν δοθέντα· καὶ ἀπὸ τούτου <u>ἀφελόντες</u> τὸν ἀπὸ τοῦ ΝΒ τετράγωνον ὄντα δοθέντα ἕξομεν καὶ τὸν λοιπὸν δοθέντα, ὅς ἐστιν τοῦ ζητουμένου πολυγώνου πολλαπλασίων κατὰ τὸν ὀκταπλάσιον τοῦ ΚΒ· ὥστε εὑρετός ἐστιν ὁ ζητούμενος πολύγωνος.

In fact, <u>having</u> a side of some polygonal ΗΘ given and also <u>having</u> the multiplicity of its sides we also have ΚΒ among the givens; *so that* we shall also have the ‹rectangle contained› by ΗΘ, ΘΜ, both together, and by ΚΒ given, which is equal to Νξ; *so that* we shall also have Κξ given—since ΝΚ is a dyad—*so that* we shall also have ‹the square› on Κξ given; and <u>subtracting</u> from this the square on ΝΒ that is given we shall also have the remaining ‹number› given, which is multiple of the sought polygonal according to the octuple of ΚΒ; so that the sought polygonal is findable.

As a matter of fact, validation is here operated by a hybrid between a procedure (note the three conjoined participles, underlined) and a proof (note the three ὥστε "so that", italicized).[101] The predicate "given" is transferred from the assigned objects to the objects whose determination is required: it is remarkable the transition, in the very last step, from δοθείς "given" to the verbal adjective εὑρετός "findable",[102] which corresponds to the finite verb form εὑρήσομεν "we shall find" in the procedure. Let us read the chain of givens and the procedure set in parallel; the signs [...] and the boldface denote corresponding dislocated steps.

ἔχοντες γὰρ πλευρὰν δοθεῖσαν τινὸς πολυγώνου τὸν ΗΘ	λαβόντες γὰρ τὴν πλευρὰν τοῦ πολυγώνου
ἔχοντες δὲ καὶ τὸ πλῆθος αὐτοῦ τῶν γωνιῶν ἔχομεν καὶ τὴν ΚΒ δοθέντων·	
ὥστε καὶ τὸν ὑπὸ	**πολλαπλασιάσαντες ἐπὶ**
συναμφοτέρου τοῦ ΗΘ ΘΜ	ἀεὶ διπλασιάσαντες ἀφελοῦμεν μονάδα, καὶ τὸν λοιπὸν
καὶ τοῦ ΚΒ	**[...]** τὸν δυάδι ἐλάσσονα τοῦ πλήθους τῶν γωνιῶν
ἕξομεν δοθέντα, ὅς ἐστιν ἴσος τῷ Νξ·	τῷ γενομένῳ
ὥστε ἕξομεν καὶ τὸν Κξ δοθέντα – ἐπείπερ δυάς ἐστιν ὁ ΝΚ	προσθήσομεν ἀεὶ δυάδα,
–· ὥστε καὶ τὸν ἀπὸ τοῦ Κξ ἕξομεν δοθέντα·	καὶ λαβόντες τὸν ἀπὸ τοῦ γενομένου τετράγωνον
καὶ ἀπὸ τούτου ἀφελόντες τὸν ἀπὸ τοῦ ΝΒ τετράγωνον ὄντα δοθέντα	ἀφελοῦμεν ἀπ' αὐτοῦ τὸν ἀπὸ τοῦ τετράδι ἐλάσσονος τοῦ πλήθους τῶν γωνιῶν,
ὅς ἐστιν τοῦ ζητουμένου πολυγώνου πολλαπλασίων κατὰ τὸν ὀκταπλάσιον τοῦ ΚΒ·	καὶ τὸν λοιπὸν μερίσαντες εἰς τὸν ὀκταπλασίονα τοῦ δυάδι ἐλάσσονος τοῦ πλήθους τῶν γωνιῶν
ὥστε εὑρετός ἐστιν ὁ ζητούμενος πολύγωνος.	εὑρήσομεν τὸν ζητούμενον πολύγωνον.

In fact, having a side of some polygonal ΗΘ given	In fact, taking the side of the polygonal
and also having the multiplicity of its sides we also have ΚΒ among the givens;	
so that we shall also have the ‹rectangle contained› by	**multiplying [...] by**
ΗΘ, ΘΜ, both together,	always doubling we shall subtract a unit,
and by ΚΒ	and **[...]** the remainder **[...]** the ‹number› less by a dyad than the multiplicity of the angles
given, which is equal to Νξ;	to the result
so that we shall also have Κξ given—since ΝΚ is a dyad—;	we shall always add a dyad
so that we shall also have ‹the square› on Κξ given;	and taking the square on the result

[101] The last ὥστε "so that" (not italicized) has a metadiscursive function.
[102] Thus, Diophantus treats these predicates as strict synonyms.

and by subtracting from this the square on NB that is given we shall also have the remaining ‹number› given,	we shall subtract from it the ‹square› on the ‹number› less by a tetrad than the multiplicity of the angles,
which is multiple of the sought polygonal according to the octuple of KB;	and dividing the remainder by the octuple of the ‹number› less by a dyad than the multiplicity of the angles
so that the sought polygonal is findable.	we shall find the sought polygonal.

A fact that is striking, and obviously so: the definite descriptions that refer to the objects in the procedure make it cumbersome but definitely more transparent than the lettered designations in the validating chain of givens. Reading the validating argument without a list of the correspondences between the signs in the formula and the lettered designations is virtually impossible.

Other features make the chain of givens more opaque than the associated procedure. The former systematically anticipates the results of the operations; these results are, in part, explicitly mentioned by means of designations, in part, hidden in the lettered designations and thereby deferred to the reader's memory. This happens when the lettered designations refer to identifications made a few lines before—like those of ὁ KB and of ὁ NB—[103]or within proposition 4 of the treatise, as in the case of ὁ ΘM, which is the number less by a unit than the side of the polygonal. In this way, most of the operations actually performed in the procedure are left understood in the chain of givens. On the other hand, some of the arithmetic objects that correspond to results attained in the procedure are not declared to be "given". The chain of givens also lacks two steps that secure the fact that ὁ ΘM is given once ὁ HΘ is (one must subtract a unit) and that the sum ὁ HΘM is given if its addends are. Moreover, this very step is formulated differently, since ὁ HΘM ($= 2s - 1$) is obtained in the chain of givens as the sum of ὁ HΘ ($= s$) and of ὁ ΘM ($= s - 1$), whereas the procedure doubles the side of the polygonal and subtracts a unit from the result.

These remarks and the comparative table above show that the two arguments cannot match exactly, if for no other reason because the syntactic structures employed in them are not isomorphic. In turn, this shows that the use of the demonstrative code in validations is not forced by logical constraints: on the contrary, its use creates irreducible tensions between the validated procedure and the validating argument.

Such tensions eventually drove the validation model to disintegration, as the following example shows. At *Alm.* III.5 (*POO* I.1, 241.1–242.13), Ptolemy sets out to compute, within the eccentric model and assuming that the arc the Sun traverses on the eccentric circle is given, the arc of the equation of the solar anomaly and the arc the Sun traverses on the ecliptic:[104]

ἔστω δὴ πρῶτον μὲν ὁμόκεντρος τῷ ζῳδιακῷ κύκλος ὁ ΑΒΓ περὶ κέντρον τὸ Δ ὁ δ' ἔκκεντρος ὁ ΕΖΗ περὶ κέντρον τὸ Θ ἡ δὲ δι' ἀμφοτέρων τῶν κέντρων καὶ τοῦ Ε ἀπογείου διάμετρος ἡ ΕΑΘΔΗ, καὶ ἀπολῃφθείσης τῆς ΕΖ περιφερείας ἐπεζεύχθωσαν ἥ τε ΖΔ καὶ ἡ ΖΘ. δεδόσθω δὲ πρῶτον ἡ ΕΖ περιφέρεια μοιρῶν οὖσα λόγου ἕνεκεν λ, καὶ ἐκβληθείσης τῆς ΖΘ κάθετος ἐπ' αὐτὴν ἤχθω ἀπὸ τοῦ Δ ἡ ΔΚ.	Thus let there be first a circle homocentric to the ecliptic, ΑΒΓ, about centre Δ, the eccentre ΕΖΗ about centre Θ, the diameter ΕΑΘΔΗ through both centres and the apogee Ε, and arc ΕΖ being cut off let ‹straight lines›, ΖΔ and ΖΘ, be drawn. Let first arc ΕΖ be given of, say, 30 parts, and ΖΘ being produced let a ‹straight line›, ΔΚ, be drawn from Δ perpendicular to it.

[103] The identification of these designations is provided in the clause "multiplied by the octuple of the ‹number› less by a dyad—that is of the excess: it will be by the octuple of KB—and taking in addition the ‹square› on the ‹number› less by a tetrad—that is that on NB" at Acerbi 2011e, 196.31–197.1.

[104] The three arcs at issue are so related that, any of them being appointed given, the other two are proved to be given; see also below. Note the five-letter name of the diameter.

ἐπεὶ τοίνυν ἡ ΕΖ περιφέρεια ὑπόκειται μοιρῶν λ, καὶ ἡ ὑπὸ ΕΘΖ ἄρα γωνία – τουτέστιν ἡ ὑπὸ ΔΘΚ – οἵων μέν εἰσιν αἱ δ ὀρθαὶ τξ τοιούτων ἐστὶν λ, οἵων δὲ αἱ δύο ὀρθαὶ τξ τοιούτων ξ· καὶ ἡ μὲν ἐπὶ τῆς ΔΚ ἄρα περιφέρεια τοιούτων ἐστὶν ξ οἵων ὁ περὶ τὸ ΔΘΚ ὀρθογώνιον κύκλος τξ, ἡ δὲ ἐπὶ τῆς ΚΘ τῶν λοιπῶν εἰς τὸ ἡμικύκλιον ρκ· καὶ αἱ ὑπ᾽ αὐτὰς ἄρα εὐθεῖαι ἔσονται ἡ μὲν ΔΚ τοιούτων ξ οἵων ἐστὶν ἡ ΔΘ ὑποτείνουσα ρκ, ἡ δὲ ΚΘ τῶν αὐτῶν ργ νε΄· ὥστε καὶ οἵων ἐστὶν ἡ μὲν ΔΘ εὐθεῖα β λ΄ ἡ δὲ ΖΘ ἐκ τοῦ κέντρου ξ, τοιούτων καὶ ἡ μὲν ΔΚ ἔσται α ιε΄ ἡ δὲ ΘΚ τῶν αὐτῶν β ι΄ ἡ δὲ ΚΘΖ ὅλη ξβ ι΄. καὶ ἐπεὶ τὰ ἀπ᾽ αὐτῶν συντεθέντα ποιεῖ τὸ ἀπὸ τῆς ΖΔ, ἔσται καὶ ἡ ΖΔ ὑποτείνουσα τοιούτων ξβ ια΄ ἔγγιστα· καὶ οἵων ἄρα ἐστὶν ἡ ΖΔ ρκ τοιούτων ἔσται καὶ ἡ μὲν ΔΚ εὐθεῖα β κε΄, ἡ δ᾽ ἐπ᾽ αὐτῆς περιφέρεια τοιούτων β ιη΄ οἵων ἐστὶν ὁ περὶ τὸ ΖΔΚ ὀρθογώνιον κύκλος τξ· ὥστε καὶ ἡ ὑπὸ ΔΖΚ γωνία, οἵων μέν εἰσιν αἱ δύο ὀρθαὶ τξ τοιούτων ἐστὶν β ιη΄, οἵων δὲ αἱ δ ὀρθαὶ τξ τοιούτων α θ΄. τοσούτων ἄρα ἐστὶν τὸ παρὰ τὴν ἀνωμαλίαν τότε διάφορον· τῶν δ᾽ αὐτῶν ἦν ἡ ὑπὸ ΕΘΖ γωνία λ· καὶ λοιπὴ ἄρα ἡ ὑπὸ ΑΔΒ γωνία – τουτέστιν ἡ ΑΒ τοῦ ζῳδιακοῦ περιφέρεια – μοιρῶν ἐστιν κη να΄.

Now then, since arc EZ has been supposed of 30 parts, therefore angle ΕΘΖ—that is ΔΘΚ—is also 30 where 4 right ‹angles› are 360 and 60 where two right ‹angles› are 360; therefore the arc on ΔK is also 60 where the circle about the right-angled ‹triangle› ΔΘΚ is 360, and the ‹arc› on ΚΘ is 120, those remaining to a semicircle; therefore the chords under them will also be, ΔΚ of 60 where hypotenuse ΔΘ is 120, and ΚΘ 103;55 of the same; so that ΔK will also be 1;15 where chord ΔΘ is 2;30 and radius ΖΘ 60, and ΘΚ 2;10 of the same, and ΚΘΖ as a whole 62;10. And since the ‹squares› on them compounded make that on ΖΔ, the hypotenuse ΖΔ will also be about 62;11 of such; therefore chord ΔK will also be 2;25 where ΖΔ is 120, and the arc on it 2;18 where the circle about the right-angled ‹triangle› ΖΔΚ is 360; so that angle ΔΖΚ is also 2;18 where 2 right ‹angles› are 360 and 1;9 where 4 right ‹angles› are 360. Therefore of that much is the equation of anomaly at issue; and angle ΕΘΖ was 30 of the same; therefore angle ΑΔΒ as a remainder—that is arc AB of the ecliptic—is also of 28;51 parts.

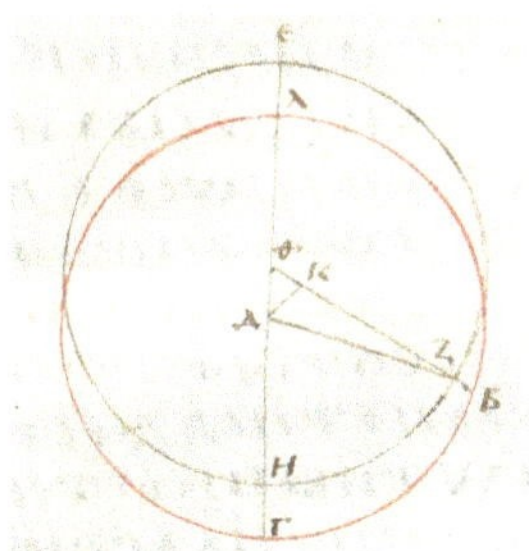

Par. gr. 2389, f. 82v, with a disturbing "bedheaded" triangle

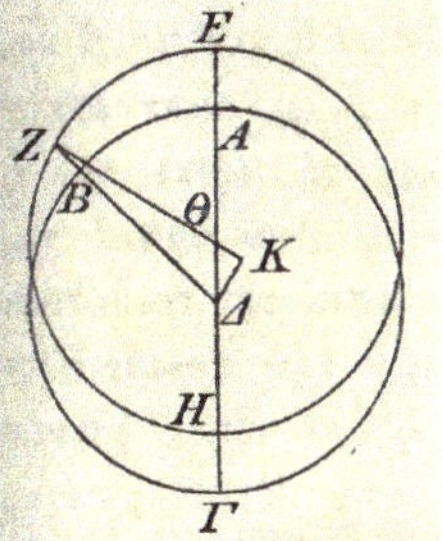

Heiberg's diagram for *Alm.* III.5, *POO* I.1, 241.1–242.13, displays a "combed" triangle

A computation-proof[105] as this is incomplete because it is not validated by a superior argumentative instance. This problem was addressed by commentators like Pappus and Theon. For instance, at *Alm.* V.5 Ptolemy calculates the direction of the mean apogee of the Moon. He only partly describes the related diagram, directly proceeding to perform the calculation with the actual numeric data. Pappus writes in his commentary: "we shall analyze the 5th theorem of the *Composition* in this way" (*iA*, 35.21–22). The analysis Pappus provides is a chain of givens exactly parallel to

105 The demonstrative style is non-canonical: the setting-out and the construction are not compartmented; the scope particle τοίνυν "now then" is (almost) never used in the *Elements*. The presence of numbers is a minor trait in this perspective.

Ptolemy's calculations, does not include a clarification of the related diagram, and lasts 36 lines in Rome's edition. When Pappus comes to show that intermediate quantities of some interest are given, he stops for a while and writes: "if the numbers are inserted one proves …", and accordingly calculates the numeric value of the quantity arrived at—these computational steps exactly reproduce the steps we read in the *Almagest*.

In the present instance, however, Ptolemy himself implicitly provides a validating argument. For he sets out two chains of givens that show how to determine any two of the three arcs that figure in the above computation-proof if the third is given (*Alm*. III.5, *POO* I.1, 242.14–243.15)—the case just treated is of course excluded; in this way, Ptolemy treats all possible cases:[106]

ὅτι δέ, κἂν ἄλλη τις τῶν γωνιῶν δοθῇ, καὶ αἱ λοιπαὶ δοθήσονται, φανερὸν αὐτόθεν ἔσται καθέτου ἀχθείσης ἐπὶ τῆς αὐτῆς καταγραφῆς ἀπὸ τοῦ Θ ἐπὶ τὴν ΖΔ τῆς ΘΛ.

ἐάν τε γὰρ τὴν ΑΒ τοῦ ζῳδιακοῦ περιφέρειαν ὑποθώμεθα δεδομένην – τουτέστιν τὴν ὑπὸ ΘΔΛ γωνίαν – διὰ τοῦτο ἔσται καὶ ὁ τῆς ΔΘ πρὸς ΘΛ λόγος δεδομένος· δεδομένου δὲ καὶ τοῦ τῆς ΔΘ πρὸς ΘΖ δοθήσεται καὶ ὁ τῆς ΘΖ πρὸς ΘΛ· διὰ τοῦτο δὲ ἕξομεν δεδομένας τήν τε ὑπὸ ΘΖΛ γωνίαν – τουτέστιν τὸ παρὰ τὴν ἀνωμαλίαν διάφορον – καὶ τὴν ὑπὸ ΕΘΖ – τουτέστιν τὴν ΕΖ τοῦ ἐκκέντρου περιφέρειαν.

ἐάν τε τὸ παρὰ τὴν ἀνωμαλίαν διάφορον ὑποθώμεθα δεδομένον – τουτέστιν τὴν ὑπὸ ΘΖΔ γωνίαν – ἀνάπαλιν τὰ αὐτὰ συμβήσεται, δεδομένου μὲν διὰ τοῦτο τοῦ τῆς ΘΖ πρὸς ΘΛ λόγου δεδομένου δὲ ἐξ ἀρχῆς καὶ τοῦ τῆς ΘΖ πρὸς ΘΔ, ὥστε δεδόσθαι μὲν καὶ τὸν τῆς ΔΘ πρὸς ΘΛ λόγον δεδόσθαι δὲ διὰ τοῦτο καὶ τὴν ὑπὸ ΘΔΛ γωνίαν – τουτέστιν τὴν ΑΒ τοῦ ζῳδιακοῦ περιφέρειαν – καὶ τὴν ὑπὸ ΕΘΖ – τουτέστιν τὴν ΕΖ τοῦ ἐκκέντρου περιφέρειαν.

That, if some other angle is also given, the remaining ones will also be given, will be immediately manifest once, in the same figure, a ‹straight line›, ΘΛ, is drawn from Θ perpendicular to ΖΔ.

In fact, if we suppose arc AB of the ecliptic—that is angle ΘΔΛ—given, for this reason the ratio of ΔΘ to ΘΛ will also be given; and the ‹ratio› of ΔΘ to ΘZ being also given that of ΘZ to ΘΛ will also be given; and for this reason we shall also have both angle ΘZΛ—that is the equation of the anomaly—and EΘZ—that is arc EZ of the eccentre—given.

If we suppose the equation of the anomaly—that is angle ΘZΔ—given, the same will be concluded inversely, the ratio of ΘZ to ΘΛ being for this reason given and the ‹ratio› of ΘZ to ΘΔ being also originally given, so as to be the ratio of ΔΘ to ΘΛ also given and so as to be angle ΘΔΛ—that is arc AB of the ecliptic—and EΘZ—that is arc EZ of the eccentre—also given for this reason.

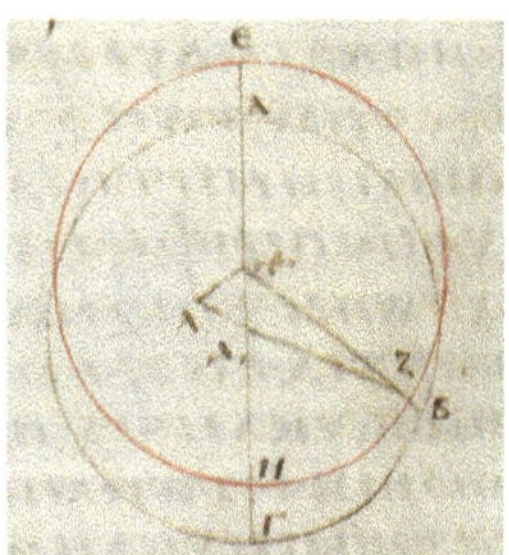

Par. gr. 2389, f. 82v

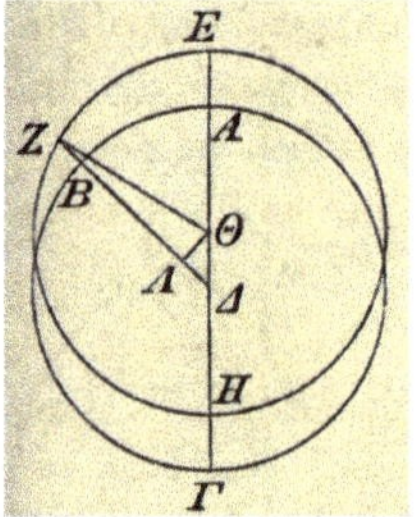

Heiberg's diagram for *Alm*. III.5, *POO* I.1, 242.14–243.15

[106] This argument also displays a non-canonical demonstrative code: note the personal verb forms, the genitive absolute, the metadiscursive expression ἀνάπαλιν τὰ αὐτὰ συμβήσεται "the same will be concluded inversely", the construct ὥστε + infinitive. Each of the two arguments might be used, *mutatis mutandis*, to validate the computation-proof read above.

If we compare any of these two arguments with the computation-proof above, we note the following anomalies in the correspondence between the computation and the chain of givens:

i. some operations "disappear": this happens to the determination of the chord associated with a given arc and vice versa (as said, they are validated by *Data* 87 and 88) and to the conversions of units of measurement in the transitions from a circle to another circle and from angles at the centre to angles at the circumference; these operations are formulated in a cumbersome way and give rise to paradoxical phenomena (see just below);

ii. some operations are indicated only implicitly: the reader must have had in mind *Data* 8, 40, and 43,[107] but the deductions in which these theorems are used tell nothing as to how to compute the magnitudes that are deduced to be "given" (this is the operation we would call "solution of a right-angled triangle").

The following table sets out a possible correspondence between a part of the computation-proof read above and a suitable modification of the first chain of givens in Ptolemy's argument read in the previous page. The "missing" steps I have supplied are in italics; the sign […] and the boldface identify corresponding dislocated steps:

ἐπεὶ τοίνυν ἡ ΕΖ περιφέρεια ὑπόκειται μοιρῶν λ´, καὶ ἡ ὑπὸ ΕΘΖ ἄρα γωνία	ἐὰν γὰρ τὴν ΕΖ τοῦ ἐκκέντρου περιφέρειαν ὑποθώμεθα δεδομένην, τουτέστιν τὴν ὑπὸ ΕΘΖ γωνίαν
– τουτέστιν ἡ ὑπὸ ΔΘΚ – οἵων μέν εἰσιν αἱ δ ὀρθαὶ τξ τοιούτων ἐστὶν λ, οἵων δὲ αἱ δύο ὀρθαὶ τξ τοιούτων ξ·	*– τουτέστιν τὴν ὑπὸ ΔΘΚ γωνίαν,*
καὶ ἡ μὲν ἐπὶ τῆς ΔΚ ἄρα περιφέρεια τοιούτων ἐστὶν ξ οἵων ὁ περὶ τὸ ΔΘΚ ὀρθογώνιον κύκλος τξ, ἡ δὲ ἐπὶ τῆς ΚΘ τῶν λοιπῶν εἰς τὸ ἡμικύκλιον ρκ·	*τουτέστιν τὴν ἐπὶ τὴν ΔΚ περιφέρειαν καὶ τὴν ἐπὶ τὴν ΚΘ –*
καὶ αἱ ὑπ᾽ αὐτὰς ἄρα εὐθεῖαι ἔσονται ἡ μὲν ΔΚ τοιούτων ξ οἵων ἐστὶν ἡ ΔΘ ὑποτείνουσα ρκ, ἡ δὲ ΚΘ τῶν αὐτῶν ργ νε´·	*δεδομένης τῆς ΔΘ ὑποτεινούσης δοθήσεται καὶ ἡ ΔΚ καὶ ἡ ΚΘ·*
	διὰ τοῦτο ἔσται καὶ ὁ τῆς ΔΘ πρὸς ΔΚ λόγος δεδομένος·
ὥστε καὶ οἵων ἐστὶν ἡ μὲν ΔΘ εὐθεῖα β λ´ ἡ δὲ ΖΘ ἐκ τοῦ κέντρου ξ,	δεδομένου δὲ καὶ τοῦ τῆς ΔΘ πρὸς ΘΖ
	δοθήσεται καὶ ὁ τῆς ΘΖ πρὸς ΔΚ·
τοιούτων καὶ ἡ μὲν ΔΚ ἔσται α ιε´ ἡ δὲ ΘΚ τῶν αὐτῶν β ι´	*δεδομένης δὲ καὶ τῆς ΘΖ δοθήσεται καὶ ἡ ΔΚ καὶ ἡ ΘΚ·*
ἡ δὲ ΚΘΖ ὅλη ξβ ι´.	*ὥστε καὶ ὅλη ἡ ΚΘΖ δοθήσεται·*
καὶ ἐπεὶ τὰ ἀπ᾽ αὐτῶν συντεθέντα ποιεῖ τὸ ἀπὸ τῆς ΖΔ, ἔσται καὶ ἡ ΖΔ ὑποτείνουσα τοιούτων ξβ ια´ ἔγγιστα·	*ὥστε καὶ ἡ ΖΔ ὑποτεινούση δοθήσεται·*
καὶ οἵων ἄρα ἐστὶν ἡ ΖΔ ρκ τοιούτων ἔσται καὶ ἡ μὲν ΔΚ εὐθεῖα β κε´, ἡ δ᾽ ἐπ᾽ αὐτῆς περιφέρεια τοιούτων β ιη´ οἵων ἐστὶν ὁ περὶ τὸ ΖΔΚ ὀρθογώνιον κύκλος τξ·	*δεδομένου δὲ τοῦ τῆς ΔΖ πρὸς ΔΚ λόγου ἕξομεν δεδομένην καὶ τὴν ΔΚ εὐθεῖαν καὶ τὴν ἐπ᾽ αὐτῆς περιφέρειαν·*
ὥστε καὶ ἡ ὑπὸ ΔΖΚ γωνία, οἵων μέν εἰσιν αἱ δύο ὀρθαὶ τξ τοιούτων ἐστὶν β ιη´, οἵων δὲ αἱ δ ὀρθαὶ τξ τοιούτων α θ´·	διὰ τοῦτο δὲ ἕξομεν δεδομένας τήν τε ὑπὸ ΔΖΚ γωνίαν,
τοσούτων ἄρα ἐστὶν τὸ παρὰ τὴν ἀνωμαλίαν τότε διάφορον·	– τουτέστιν τὸ παρὰ τὴν ἀνωμαλίαν διάφορον –
τῶν δ᾽ αὐτῶν ἦν ἡ ὑπὸ ΕΘΖ γωνία λ· καὶ λοιπὴ ἄρα ἡ ὑπὸ ΑΔΒ γωνία, **[…]** μοιρῶν ἐστιν κη να´.	καὶ τὴν ὑπὸ ΑΔΒ
τουτέστιν ἡ ΑΒ τοῦ ζῳδιακοῦ περιφέρεια,	– τουτέστιν τὴν ΑΒ τοῦ ζῳδιακοῦ περιφέρειαν.

[107] The enunciation of *Data* 8 is translated in the table set out at the beginning of this Section; *Data* 40 and 43 state that a triangle is given in form if the angles are given in magnitude (40), or if an angle is right and the sides that contain one of the acute angles are in a given ratio (43).

Now then, since arc EZ has been supposed of 30 parts, therefore angle EΘZ	In fact, if we suppose arc EZ of the ecliptic—that is angle ΘΔΛ—given
—that is ΔΘK—is also 30 where 4 right ‹angles› are 360 and 60 where two right ‹angles› are 360;	*—that is angle ΔΘK,*
therefore the arc on ΔK is also 60 where the circle about the right-angled ‹triangle› ΔΘK is 360, and the ‹arc› on KΘ is 120 of those remaining to a semicircle;	*that is the arc on ‹chord› ΔK and that on ‹chord› KΘ—*
therefore the chords under them will also be, ΔK of 60 where hypotenuse ΔΘ is 120, and KΘ 103;55 of the same;	*the hypotenuse ΔΘ being given ΔK and KΘ will also be given;*
	for this reason the ratio of ΔΘ to ΔK will also be given;
so that **[...]** where chord ΔΘ is 2;30	and the ‹ratio› of ΔΘ to ΘZ being also given
	that of ΘZ to ΔK will also be given;
and radius ZΘ 60, **ΔK will also be 1;15** and ΘK 2;10 of the same,	*and ΘZ being also given ΔK and ΘK will also be given;*
and KΘZ as a whole 62;10.	*so that ‹chord› KΘZ as a whole will also be given;*
And since the ‹squares› on them compounded make that on ZΔ, the hypotenuse ZΔ will also be about 62;11 of such;	*so that hypotenuse ZΔ will also be given;*
therefore chord ΔK will also be 2;25 where ZΔ is 120, and the arc on it 2;18 where the circle about the right-angled ‹triangle› ZΔK is 360;	*and the ratio of ΔZ to ΔK being given we shall also have chord ΔK given and the arc on it;*
so that angle ΔZK is also 2;18 where 2 right ‹angles› are 360 and 1;9 where 4 right ‹angles› are 360.	and for this reason we shall also have both angle ΔZK
Therefore of that much is the equation of anomaly at issue;	—that is the equation of the anomaly—
and angle EΘZ was 30 of the same; therefore angle AΔB as a remainder—**[...]**—is also of 28;51 parts.	and AΔB
that is arc AB of the ecliptic	—that is arc AB of the eccentre—given.

Thus, what in the chain of givens is an application of the operation of composition of ratios (*Data* 8) and an implicit reference to three theorems of the *Data*, in the corresponding computation-proof becomes an intricate solution of right-angled triangles of which we know the angles (and for this reason the triangles are "given in form") and one side: in the validating argument, thus, the intricacies of the calculation are hidden behind the fact—which is recognized only implicitly—that the triangles to be solved are given in form and hence *Data* 40 and 43 can be applied—such triangles are not even mentioned.

Even more bewildering is the fact that Ptolemy computes more than once the same quantity. This happens because some straight lines are determined by referring to triangles whose linear elements are expressed in different units of measurement. The norm in such computations is to set the diameter of the reference circle equal to 120 parts, but several such circles are used in the course of this computation: first the circle whose diameter is ἡ ΔΘ, then the circle whose radius is ἡ ZΘ, finally the circle whose diameter is ἡ ZΔ. Thus, Ptolemy computes more than once the length of straight lines ἡ ΔK (60 parts in the first norm, 1;15 in the second, and 2;25 in the third), ἡ ΘK (123;55 parts in the first norm, 2;10 in the second), ἡ ΔΘ (assumed as 120 parts in the first norm and then calculated as 2;30 in the second), and ἡ ZΔ (calculated as 62;11 parts in the second norm and assumed as 120 in the third). For this reason, in the chains of givens I have supplied in the right column, at least one of the straight lines must be proved "given" twice: this is ἡ ΔK, which opens and closes with two different values (!) the portion of the algorithm that computes the rectilinear elements. All of this could not be allowed in a well-formed chain of givens, and shows that the isomorphism between the two argumentative structures cannot but be imperfect—for this reason Ptolemy's chain of givens is concise and "reticent".

Summing up, the language of the givens is applied to the determination of any mathematical object, either by a constructive approach, or by an abstract approach (for instance by means of chains of equalities), or by computation. The same language is a large-scale attempt at validating, under the unifying logical form of a "predicate logic" whose rules of inference are set out in the *Data*, the following argumentative patterns:

- *Deductions*, mainly in the form of manipulations of relations reduced by saturation to subspecies of the predicate "given".
- *Constructions*, for which a deductive ordering replaces a non-deductive sequence of prescriptions and data-theorems validate constructive acts.[108]
- *Computations*, for which a deductive ordering replaces the algorithmic chain and data-theorems validate operations on numbers.

The language of the givens also formulates several kinds of enunciation, thereby contributing in a decisive way to creating stable domains of research: this is a metalinguistic function that induces a partition of the mathematical propositions that is transverse to the canonical partition between theorems and problems. Such concerns are strictly related to ontological issues, in this case the demarcation between existence and constructibility. We must also recognize that the unifying character of this language is also its main weakness, insofar as it conflates under the same lexical range a galaxy of demonstrative and enunciative practices that the ancient tradition tried to keep separated. The tensions displayed by the *Almagest* texts just read attest to this.

Validation is strictly linked with a literary phenomenon abundantly practised by late authors: the contamination of genres. If adhering to a specific stylistic code or contaminating them is a choice left to the author, it remains, as seen at the end of Sect. 1.2, that it is not the case that each code is suited to express any self-contained piece of mathematical discourse. The very phenomenon of validation shows that the mathematical contents embedded in the three main stylistic codes were perceived as different, and in fact as incompatible: the procedural code provides the only possible formulation of iterative procedures; the operations performed in an algorithm cannot be framed in every instance in the language of the givens, that is, as a sub-idiolect of the demonstrative code. Thus, if validation reaffirms the prominence of the demonstrative code, it nevertheless shows that the procedural and the algorithmic code were granted the same dignity—exactly because they were validated by, not simply eliminated in favour of, proofs.

Finally, we might argue that, after all, the directive connotation of procedures and algorithms suggests to set a parallel between these two codes and the only specific parts of a proposition where such a connotation is displayed: the setting-out and the construction. In the same way, as we have seen in Sect. 1, the discursive arrow is realized in these codes by linguistic items other than connectors: repeated designations and (in procedures only) syntactic subordination. Moreover, sequences of constructive acts are validated, albeit in an imperfect way, by those chains of givens that have the same function for procedures and algorithms. It remains to see how validation contributes to solving the problem of mathematical generality.

[108] A chain of givens to which a construction corresponds without any accompanying (synthetic) proof can be found in Hero, *De speculis* 16 (*HOO* II, 352.3–356.10 = [22] in Jones 2001, 163–164).

3. THE PROBLEM OF MATHEMATICAL GENERALITY

The questions I shall answer in this Section are: how can a proof like the one in our paradigmatic proposition III.2 be a general proof? How can referring to an object such as circle ὁ ΑΒΓ—apparently a particular object taken as a generic specimen of a whole class—secure generality?[1] The traditional answer—erected to a *vulgata* that has never been questioned—dates back at least to Proclus (*iE*, 207.4–25) and runs as follows:

τό γε μὴν συμπέρασμα διπλοῦν εἰώθασι ποιεῖσθαί τινα τρόπον· καὶ γὰρ ὡς ἐπὶ τοῦ δεδομένου δείξαντες καὶ ὡς καθόλου συνάγουσιν ἀνατρέχοντες ἀπὸ τοῦ μερικοῦ συμπεράσματος ἐπὶ τὸ καθόλου. διότι γὰρ οὐ προσχρῶνται τῇ ἰδιότητι τῶν ὑποκειμένων ἀλλὰ πρὸ ὀμμάτων ποιούμενοι τὸ δεδομένον γράφουσι τὴν γωνίαν ἢ τὴν εὐθεῖαν, ταὐτὸν ἡγοῦνται τὸ ἐπὶ ταύτης συναγόμενον καὶ ἐπὶ τοῦ ὁμοίου συμπεπεράνθαι παντός. μεταβαίνουσι μὲν οὖν ἐπὶ τὸ καθόλου ἵνα μὴ μερικὸν ὑπολάβωμεν εἶναι τὸ συμπέρασμα· εὐλόγως δὲ μεταβαίνουσιν, ἐπειδὴ τοῖς ἐκτεθεῖσιν, οὐχ ᾗ ταῦτά ἐστιν, ἀλλ' ᾗ τοῖς ἄλλοις ὅμοια, χρῶνται πρὸς τὴν ἀπόδειξιν. οὐ γὰρ ᾗ τοσήδε ἐστὶν ἡ ἐκκειμένη γωνία, ταύτῃ τὴν διχοτομίαν ποιοῦμαι, ἀλλ' ᾗ μόνον εὐθύγραμμος, ἔστι δὲ τὸ μὲν τοσόνδε τῆς ἐκκειμένης ἴδιον τὸ δὲ εὐθύγραμμον πασῶν τῶν εὐθυγράμμων κοινόν. ἔστω γὰρ ἡ δεδομένη ἡ ὀρθή. εἰ μὲν οὖν τῇ ἀποδείξει τὴν ὀρθότητα παρελάμβανον, οὐκ ἠδυνάμην ἐπὶ πᾶν τὸ εἶδος τῆς εὐθυγράμμου μεταβαίνειν· εἰ δὲ τὸ μὲν ὀρθὸν αὐτῆς οὐ προσποιοῦμαι τὸ δὲ εὐθύγραμμον σκοπῶ μόνον, ὁμοίως ὁ λόγος ἐφαρμόσει καὶ παρὰ ταῖς εὐθυγράμμοις γωνίαις.

In truth, ‹geometers› are accustomed to draw what is in a way a double conclusion: and in fact, after proving about a given, they also deduce in general, returning from the particular to the general conclusion. For, because they do not make use of the specific properties of the subjects but trace *that* angle or *that* straight line in order to place what is given before our eyes, they hold that what is deduced about *this* turns out to be also concluded for every similar case. Then, they pass to the general in order that we may not suppose that the conclusion is particular—and they rightly pass, since in the proof they really make use of what has been set out, not qua such-and-such, but qua similar to the others. For it is not qua of such-and-such a size that I bisect the angle set out, but only qua rectilinear, and "of such-and-such a size" is a specific property of the ‹angle› set out, while "rectilinear" is a common feature of all rectilinear ‹angles›. For let the given ‹angle› be the right ‹angle›. Then, if I used its rightness in my proof, I should not be able to pass to every species of rectilinear ‹angle›; but if I make no use of its being right and consider only its being rectilinear, the argument will also similarly apply to rectilinear angles.

All ingredients of the traditional answers to the above questions figure in Proclus' text: the primary role assigned to the diagram and the confusion between mathematical objects and graphic entities, the setting-out taken to introduce particular objects, the need to supply a generalizing step after the proof, the (entirely appropriate) emphasis on generalizing over the features of an object that are not mentioned. It is not difficult to read, in Proclus' text, echoes of passages like Plato, *Rsp.* VI, 510D–E, but Proclus surely adds the claim that every geometric proposition has a general conclusion of his own, a contention that is already falsified by the *Elements* (see Sect. 4.1).

Moreover, Proclus conflates two problems that must be kept distinct. What we are interested in is the status of mathematical objects like the "circle ΑΒΓ" set out in the ἔκθεσις of our paradigmatic proposition III.2 (Sect. 1): to this problem, which has ontological, denotative, syntactic, and

[1] The approach expounded in this Section was already argued in Acerbi 2011c. An updated version is Acerbi 2020a, which I use, with enrichments, small changes, corrections, and permutation of the material.

F. Acerbi, *The Logical Syntax of Greek Mathematics*, Sources and Studies in the History of Mathematics and Physical Sciences,
https://doi.org/10.1007/978-3-030-76959-8_3

diagrammatic facets, has simply been given a wrong solution so far. A totally different problem is how to identify the largest class of objects a specific proof applies to. This is a linguistic and stylistic problem; on a theoretical level, it was already solved, in a way perfectly adequate to Greek mathematical practice, by Aristotle in *APo*. I.4–5: generalization as far as the largest possible class is automatic on the features of an object that are not used—and hence should not be mentioned—in a proof. Proclus alludes to this theory too in the passage above.

The problem of the generality of the geometric objects involved in a proof entered the modern philosophical debate, under the name of "Locke-Berkeley problem", as the archetypal example of the issue of the ontological status of generic objects.[2] The gist of the standard solution to this problem was put forward by B. Russell in the following terms—note the presence of the deictic "this one", exactly as in Proclus' account (italics in the original):[3]

The distinction [*scil. between* all *and* any] is roughly the same as that between the general and particular enunciation in Euclid. The general enunciation tells us something about (say) all triangles, while the particular enunciation takes one triangle and asserts the same thing of this one triangle. But the triangle taken is *any* triangle, not some one special triangle; and thus, although, throughout the proof, only one triangle is dealt with, yet the proof retains its generality. If we say "Let *ABC* be a triangle, then the sides *AB* and *AC* are together greater than the side *BC*", we are saying something about *one* triangle, not about *all* triangles; but the one triangle concerned is absolutely ambiguous, and our statement consequently is also absolutely ambiguous. We do not affirm any one definite proposition, but an undetermined one of all the propositions resulting from supposing *ABC* to be this or that triangle.

One might well wonder what features of the objects set out in Russell's instantiated enunciation just read and in a Greek ἔκθεσις make them particular (albeit generic) objects. Maybe their being denoted by letters? Or their being represented in a specific, concrete diagram? Of course, such contentions are untenable, and in fact utterly naive.[4] It remains that even showing that they are naive amounts to elaborating an argument in favour of the generality of the objects involved in the proof. I shall take a radically different route, instead: showing that the modern debate is grounded on a standard translation of a Greek mathematical proposition that simply falsifies syntactic elements crucially relevant to the issue.[5] It is no surprise, then, that such a translated proposition does

---

[2] Among scholars studying the Greek mathematical and philosophical tradition, it is enough to mention Heath 1949, 219–220; Mueller 1981, 11–14; Smith 1982, 123–125; Hussey 1991, 126–127; and, in dealing with denotative letters, Barnes 2007, 347–354. Among "professional philosophers" of classical ages, think of Berkeley 1734, *Introduction*, §§ 12–16, reacting to Locke's theory of generic objects (see in particular Locke 1700, IV.vii, § 9); and also Kant 1787, 744; De Morgan 1860, sect. 72, Russell (to be read just below).

[3] Russell 1908, 227. Russell apparently introduces quantification as a device to enhance generality, but he is also unfaithful to Greek practice in so doing: as we shall see in Sect. 5.1, only about 30 propositions out of 465 in the *Elements* include a quantifier.

[4] See the criticism of Berkeley's position in Husserl 1984, II ("Die ideale Einheit der Spezies und die neueren Abstraktionstheorien"), in particular §§ 20 and 30.

[5] By "standard translation" I mean the one canonized in Heath's translation of the *Elements*. A very interesting solution to the problem of restoring the alleged defect of generality was proposed, along with the technical resource of semantic *tableaux*, by E.W. Beth (see for instance Beth 1953/54 and 1956/57). He thereby prepared for Hintikka's revision of the Kantian theory of mathematics. Hintikka's strategic aim was to reaffirm the ampliativity of mathematical reasoning (see Hintikka 1973, sects. VI–IX, and 1977; Hintikka, Remes 1974, sects. IV–V). To Kant and Hintikka, geometric proofs are synthetic or ampliative (that is, they increase information) because new individuals are introduced in them, both in the setting-out and in the construction. Diagram-based solutions to the problem of generality of a Greek geometric proposition have been set forth for instance by Manders 2008a (written in 1995 and amply circulating since then), and, in the wake of it, in Mumma 2010 (very stimulating from the technical point of view) and Panza 2012. I shall not discuss these approaches, nor shall I discuss the one argued for in Netz 1999, sect. 6, according to which "Greek generality is the repeatability of necessity" (270), a slogan I have been unsuccessfully trying to attach a meaning since 22 years.

present a defect of generality, so that additional arguments, such as the one outlined by Proclus and Russell, must be provided in order to restore full generality.

The syntactic elements falsified by the standard translation are both contained in the liminal clause of Russell's instantiated enunciation, namely, "Let *ABC* be a triangle"; the corresponding Greek clause opens the setting-out. As I shall show in Sect. 3.1, this translation is incorrect on two counts. First, the value of the verb "to be" in the setting-out is not copulative but "presential", a value akin, but not identical, to "existential". Second, and consequently, the denotative letters are not the grammatical subject of the clause that opens the setting-out but the *name* of a non-particular triangle. Thus, if Russell's liminal clause has to be a faithful example of an Euclidean setting-out, "Let *ABC* be a triangle" should read "Let there be a triangle, <called> *ABC*", where *ABC* is as perfectly fitting a name for a triangle as "(1)" or "William Thurston" could be.

The last remark suggests that a fully-fledged solution to the generality problem in a Greek mathematical proposition can only be provided (as I shall do in Sects. 3.2 and 3.3) after carefully assessing the use and meaning of the denotative letters, like the complex *ABC* Russell attaches to his triangle (ancient pragmatics was surveyed in Sect. 2.2). The second reason for the incorrectness of the standard translation lies in fact in not recognizing that the denotative letters are "letter-labels", namely, itemizers in a list (Sect. 3.2.1). This reading is corroborated by the above-mentioned syntactic function of the denotative letters, by their use in very early mathematical-style arguments, and finally by epigraphic evidence, which I bring to bear for the first time on the issue. This kind of evidence will also provide a satisfactory explanation of such archaic forms of lettered designations as they are for instance found in Aristotelian texts, thereby removing—a crucial step in my perspective—any historical ground to the seemingly obvious link between lettered designations and diagrammatic representation of a geometric configuration established by considering the just-mentioned archaic forms of lettered designation locative and not simply labelling.

As a consequence of all of this, I shall outline how indefiniteness innervates the whole structure of a Greek mathematical proposition, and draw some philosophically-oriented conclusions: the bottom line of my discussion will be that, on the basis of the textual and stylistic evidence, Greek mathematics is not committed to any ontological stance on the nature of mathematical objects (Sect. 3.3). However, this does not mean that the Greek mathematicians did not take any such stance; this can in fact be detected by assessing a different kind of evidence. To this end, I shall describe how the Greek geometers generated the building blocks of their mathematical universe, namely, the curves; how they rationalized their own practice; how they handled geometric objects by cutting and pasting them. Archimedes, who is uncommonly outspoken in these matters and who wrote the *Method*, a key treatise in this perspective, will play a prominent role in this outline (Sect. 3.4). In the final Section (3.5), I shall explain a peculiar phenomenon that occurs in geometric diagrams as they are transmitted in medieval manuscripts, and which I have briefly described in Sect. 1.1, namely, the fact that these diagrams systematically, and conspicuously, display symmetry properties that are not required by the enunciation of the proposition they are associated with.

A strategic aim of this Section, especially because a fair amount of literature on the issue has accumulated in recent years, is to reconstruct ancient mathematical pragmatics in presence of a diagram. It will be clear to any reader, I hope, that I believe that the quite remarkable, and amply argumented, interpretive insight that a "Greek" diagram has a deductive import is not only false but also straightforwardly falsified by any conceivable evidence. A long—too long and rhetorically overloaded—footnote will present a part of this evidence, along with a part of a secondary literature that is likewise too abundant, to the reader.

3.1. The presential value of the verb "to be" in the setting-out

We need to anticipate a bit on Sect. 4.2. The ἔκθεσις "setting-out" of a mathematical proposition derives from an enunciation in conditional form as follows: the antecedent of the conditional is transformed into a supposition with the verb in the imperative; this is the ἔκθεσις. The consequent of the conditional is transformed into a sentence introduced by λέγω ὅτι "I claim that"; as seen in Sect. 1.1, Proclus calls it διορισμός "determination". The declarative enunciations of theorems and the directive enunciations of problems undergo similar transformations. From the logical point of view, the dismemberment of an enunciation into two independent clauses singles out the assumptions of a proposition for use as premises in the subsequent proof. The verb forms and the structure of an enunciation and of the associated setting-out are strictly related, with two systematic variants: the introduction of the denotative letters and the transformation of all verb moods into the imperative, most frequently in the perfect stem.

There is a further feature of the setting-out that is highly relevant to our purposes: the objects mentioned in the enunciation are frequently introduced,[6] in the setting-out, by a liminal imperative ἔστω of the verb "to be". Let us read again the enunciation and the setting-out + determination of our paradigmatic proposition III.2 (*EOO* I, 168.17–23):

ἐὰν κύκλου ἐπὶ τῆς περιφερείας ληφθῇ δύο τυχόντα σημεῖα, ἡ ἐπὶ τὰ σημεῖα ἐπιζευγνυμένη εὐθεῖα ἐντὸς πεσεῖται τοῦ κύκλου.	If on the circumference of a circle two random points be taken, the straight line joined at the points[7] will fall within the circle.
<u>ἔστω κύκλος ὁ ΑΒΓ</u>, καὶ ἐπὶ τῆς περιφερείας αὐτοῦ εἰλήφθω δύο τυχόντα σημεῖα τὰ Α Β.	<u>Let there be a circle, ΑΒΓ</u>, and let two random points, A, B, be taken on its circumference.
λέγω ὅτι ἡ ἀπὸ τοῦ Α ἐπὶ τὸ Β ἐπιζευγνυμένη εὐθεῖα ἐντὸς πεσεῖται τοῦ κύκλου.	I claim that the straight line joined from A to B will fall within the circle.

Let me state tersely the point implicitly made in my translation: the verb "to be" that introduces a setting-out does not have a copulative value, but a "presential" value.[8] It is a way to make the objects at issue in a specific proposition "present", ready for use within the proof.

---

[6] See Sect. 4.2 for the range of variability (there are no "exceptions", but different kinds of enunciation).

[7] The polarity definite / indefinite of this expression is neutralized (cf. Sect. 1.1): the meaning is definite because there is only one straight line joining two assigned points (I.post.1) and not because the noun phrase ἡ ἐπὶ τὰ σημεῖα ἐπιζευγνυμένη εὐθεῖα is articular (the article is in this case forced by the presence of the nested prepositional determiner ἐπὶ τὰ σημεῖα of the attributive participle ἐπιζευγνυμένη): see Sect. 3.3. The order modifier-noun in this designation is dictated by saliency (see Sect. 2.2).

[8] M. Federspiel (1992, 15–17, and 1995, but see already all translations in Toomer 1984, and Mendell 1986, 271–273; Federspiel 2010 is just parasitic of the Italian redaction of the present book), employs "existential" to designate the value complementary to "copulative", but the set of such complementary values cannot be reduced to a singleton (still, the relevant opposition in our perspective is copulative / non copulative). On the values of the verb "to be", see Kahn 1973, in particular sect. VI, and the fundamental Ruijgh 1979 and Ruijgh 1984. The great Dutch linguist introduces the term "presential" for a value of "to be" identical to that employed in a setting-out, of which however he was unaware: as a matter of fact, the setting-out of mathematical propositions provides by far the largest and earliest supply of such a syntactic construction in the entire Greek literary corpus: compare Ruijgh's difficulties in finding examples in early literary texts. More generally, on this kind of construction (also called "presentative" or 'thetic'), see Bailey 2009. Apart from Acerbi 2011, Federspiel's remark has been stressed in Netz 1999, 43–44—who did not draw any conclusion from it—and by Lattmann, e.g. 2018, 117–118 and 122, who moves within an a priori interpretive framework, so that any specific feature of a Greek mathematical proposition can be bent to confirm it. Further detail will be found in Sect. 4.2; for existential "to be" see also Sect. 5.1.3.

In a strong sense, the verb "to be" that introduces a setting-out initializes an itemized list, in this way: "Here are: (*a*) a circle, (*b*) two random points on it, etc.".[9] Other verbs might have been used, but all possible candidates ("to be given", "to be conceived", "to be taken", "to be set out", "to be supposed") have other, specific functions within a mathematical proposition, and are infrequently found in a setting-out.[10] Thus, the presence of liminal ἔστω is a stylistic trait introducing and thereby marking off the setting-out: the syntactically marked thetic construction with ἔστω subsumes the function elsewhere performed by a liminal scope particle or by the expressions λέγω ὅτι "I claim that" and δεῖ δή "thus it is required", which canonically mark off other specific parts of a mathematical proposition (see Sects. 1.1 and 4.2): different linguistic items may perform the same syntactic function, in this instance a structuring function.[11]

That the value of "to be" initializing a setting-out is "presential" is formally proved by the existence of enunciations in conditional form like those of *Data* 17–19 or of II.1 and V.22 (*EOO* I, 118.10–11, 118.15–16, and II, 60.18–26, setting-outs included). In them, very short antecedents only feature a grammatical subject and a form of "to be" as the predicate, which in this syntactic configuration can only have a presential value. Still, the setting-out associated with such enunciations is formulated exactly as any other is:

ἐὰν ὦσι δύο εὐθεῖαι, τμηθῇ δὲ ἡ ἑτέρα αὐτῶν εἰς ὁσαδηποτοῦν τμήματα [...]	If there be two straight lines, and one of them be cut in as many segments as we please [...]
ἔστωσαν δύο εὐθεῖαι αἱ Α ΒΓ, καὶ τετμήσθω ἡ ΒΓ ὡς ἔτυχεν κατὰ τὰ Δ Ε σημεῖα.	Let there be two straight lines, A, BΓ, and let BΓ be cut at random at points Δ, E.[12]
ἐὰν ᾖ ὁποσαοῦν μεγέθη καὶ ἄλλα αὐτοῖς ἴσα τὸ πλῆθος σύνδυο λαμβανόμενα καὶ ἐν τῷ αὐτῷ λόγῳ, καὶ δι' ἴσου ἐν τῷ αὐτῷ λόγῳ ἔσται.	If there be as many magnitudes as we please and others equal to them in multiplicity taken two and two together and in a same ratio, they will also be in a same ratio through an equal.
ἔστω ὁποσαοῦν μεγέθη τὰ Α Β Γ καὶ ἄλλα αὐτοῖς ἴσα τὸ πλῆθος τὰ Δ Ε Ζ σύνδυο λαμβανόμενα ἐν τῷ αὐτῷ λόγῳ, ὡς μὲν τὸ Α πρὸς τὸ Β οὕτως τὸ Δ πρὸς τὸ Ε, ὡς δὲ τὸ Β πρὸς τὸ Γ οὕτως τὸ Ε πρὸς τὸ Ζ.	Let there be as many magnitudes as we please, A, B, Γ, and others equal to them in multiplicity, Δ, E, Z, taken two and two together in a same ratio, as A is to B, so Δ is to E, and, as B is to Γ, so E is to Z.

But if this is true, then, in the setting-out of III.2, the *indefinite* expression κύκλος ὁ ΑΒΓ "a circle, ABΓ" *as a whole* is the grammatical subject of the clause ἔστω κύκλος ὁ ΑΒΓ: the standard translation "let ABΓ be a circle" is erroneous—and crucially so, because the contention that particular objects are introduced in a proof rests on assuming that a bare lettered designation is the subject of any supposition in the setting-out. If the lettered designations are not the grammatical subject of the suppositional clauses in the setting-out, one can legitimately wonder what they are.

[9] Languages such as French or Italian even allow the elimination of the verb form: liminal ἔστω can very appropriately be translated "voilà" or "ecco". Purely nominal sentences as the one just presented are not unknown in the technical corpus: see [Aristotle], *Pr.* XV.9–10, to be discussed in Sect. 3.2.1, where a final remark will complete our picture.

[10] For these verbs, see Sects. 2.4 ("to be given"), 4.3 ("to be taken"), and 4.2 (the others). A quick survey of presential verbs in Greek mathematics is also in Federspiel 2010, 112–115.

[11] As seen in Sect. 1.1, the scope particles in question are γάρ ("in fact") introducing the construction, οὖν ("then") opening the proof, ἄρα ("therefore") introducing the conclusion; in this function, they have a suprasentential scope.

[12] The first clause here translated constitutes an exception to the rule for translating δέ and καί which I have set out at the beginning of Sect. 1.1, for here δέ coordinates two subclauses of the antecedent of a conditional, καί, two independent suppositional clauses. I might have deleted the comma in the enunciation of II.1; I have kept it because the two subclauses are governed by different verbs.

3.2. The function of the denotative letters

In the setting-out, the logical and grammatical subject κύκλος "*a* circle" is modified by a lettered designation that follows the noun and is preceded by an article: the modifier is a noun and is *appositive*, and provides the intended circle with its *name*. The article that precedes the letters has two functions. The first is distinguishing between objects designated by identical strings of letters, because the gender of the article is the same as that of the noun modified by the string of letters: ὁ ΑΒΓ is a circle but τὸ ΑΒΓ is a triangle—for instance inscribed in circle ὁ ΑΒΓ (see Sect. 2.2). The second function is to produce a linguistic item suited to be a noun, which must have a declension: the case of the noun can only be deduced from the case of the article. Stated otherwise, the simple presence of the article makes ὁ ΑΒΓ a noun: there is no need to postulate that the lexical item κύκλος "circle" is understood (even if the referent is indeed some circle). The same is true for any other lettered designation, even for those that include linguistic items other than denotative letters and an article, as in the case of the name of an angle, ἡ ὑπὸ ΑΒΓ (again Sect. 2.2): we need not supply a lexical item γωνία "angle", even if the syntagm does refer to an angle.

We are thus led to conclude—a fact that is by no means obvious—that the meaning of the setting-out is as indefinite as the meaning of the enunciation, for all mathematical objects are in both of them designated by indefinite nouns, of the same kind as "a circle".[13] The lettered complexes give these designations a name—and these designations are linguistic objects.[14] Only by the intermediation of the primary designations by means of indefinite nouns can the lettered complexes be taken to refer, in the course of the proof, to the associated mathematical objects. In sum, the letters do not have an ostensive, or indexical function,[15] but an anaphoric function to other linguistic units: they act within discourse itself.[16] Thus, the particularization allegedly induced by the lettered expressions, as well as the alleged reference to the diagram, are not just apparent, but simply non-existent, since they are the consequence of a wrong interpretation of the syntax of the setting-out.

There is more to the issue. Let us read again the setting-out of III.2 (*EOO* I, 168.20–21 and 168.24–170.2), together with the subsequent construction:

ἔστω κύκλος ὁ ΑΒΓ, καὶ ἐπὶ τῆς περιφερείας αὐτοῦ εἰλήφθω δύο τυχόντα σημεῖα τὰ Α Β.	Let there be a circle, ΑΒΓ, and let two random points, Α, Β, be taken on its circumference.
μὴ γάρ, ἀλλ' εἰ δυνατόν, πιπτέτω ἐκτὸς ὡς ἡ ΑΕΒ, καὶ εἰλήφθω τὸ κέντρον τοῦ ΑΒΓ κύκλου καὶ ἔστω τὸ Δ, καὶ ἐπεζεύχθωσαν αἱ ΔΑ ΔΒ, καὶ διήχθω ἡ ΔΖΕ.	In fact not, but if possible, let it fall outside as ΑΕΒ, and let the centre of circle ΑΒΓ be taken and let it be Δ, and let ‹straight lines›, ΔΑ, ΔΒ, be joined, and let a ‹straight line›, ΔΖΕ, be drawn through.

[13] This crucial remark is made for the first time in Federspiel 1995. The way the article is used in ancient Greek is in several respects different from the way it is used in most Western languages. Thus, it frequently happens in Greek that articular expressions are nevertheless indefinite (see Sect. 3.3 for more detail). However, the absence of an article certainly points to the indefiniteness of the referent.

[14] Part of the problem lies in the fact that we are used to associate a name only to a particular.

[15] For the indexical function, see Peirce 1931–35 II, 305. For a Peircean approach to Greek diagrams, see the quite idiosyncratic Netz 1999, sect. 1, and the more orthodox Lattmann 2018 and 2019, sect. 3. In this approach, the Locke-Berkeley problem is automatically solved (but not at all satisfactorily: see below); Netz did not realize this, Lattmann did. In this Peircean footnote, I cannot resist citing Stjernfelt 2007, who has at least had the decency not to try to explain what is Greek mathematics.

[16] On the other hand, mathematical letters cannot convey generality, as they instead do in Aristotelian syllogistic (so already Alexander, *in APr.*, 53.28–54.2, who aptly uses the adjective δεικτικά "indexical" to denote their function), and as ordinals do in the Stoic "modes" and in some very peculiar mathematical propositions (see Sects. 5.1.6 and 5.2.2).

Get a close look at the underlined syntagms above:[17] a pair of points is named τὰ A B and not τὸ A καὶ τὸ B, and the same for the two straight lines αἱ ΔA AB. Thus, pluralities of objects are named as a whole in the plural. The presence of one article in the plural shows that the string of letters A B is appositive of the indefinite noun δύο τυχόντα σημεῖα "two random points". There is no obstruction, of course, to assigning each point its name, and exactly this will be done in the immediately subsequent determination. But on the other hand, such designations in the plural show that the strings of letters provide nothing but names of linguistic entities; they name bizarre lexical complexes that correspond to no objects in Greek geometry: there is no single entity like "points A and B". The fact that we currently write the denotative letters without preposing them an article neutralizes this feature in translation.

In principle, the complexes of letters serving as names are arbitrary. Read again the entire proof of III.2: the letter Γ is only mentioned within the denomination of circle ὁ ABΓ, so that it is not legitimate to assume that it denotes a point. It is only part of the name of the circle (as we have seen in Sect. 1.1, August-Heiberg even forgot to insert the letter in their diagram). This remark is further corroborated if we read again the construction of III.2 (*EOO* I, 168.24–170.2):[18]

μὴ γάρ, ἀλλ' εἰ δυνατόν, πιπτέτω ἐκτὸς ὡς ἡ AEB, καὶ εἰλήφθω τὸ κέντρον τοῦ ABΓ κύκλου, καὶ ἔστω τὸ Δ, καὶ ἐπεζεύχθωσαν αἱ ΔA ΔB, καὶ διήχθω ἡ ΔZE.	In fact not, but if possible, let it fall outside as AEB, and let the centre of circle ABΓ be taken and let it be Δ, and let ‹straight lines›, ΔA, ΔB, be joined, and let a ‹straight line›, ΔZE, be drawn through.

Consider the letter E in the denomination of straight line ἡ AEB. If E were to designated a point *in this syntagm*,[19] the subsequent constructive act of drawing straight line ἡ ΔZE would require the imperative ἐπεζεύχθω "let it be joined"—exactly as we read in the immediately preceding constructive act and exactly as is required in any application of I.post.1, namely, whenever both extremities to be joined by a straight line are mentioned—not διήχθω "let it be drawn through", for this verb is used exclusively of straight lines one of whose extremities is unconstrained (cf. Sect. 4.3). From this it follows that, not only the letter E, but also the letter Z is simply a part of the name of straight line ἡ ΔZE. What happens is in fact that only in the middle of the proof of proposition III.2 will the designation τὸ Z come to be assigned to a point, that is, to the intersection of straight line ἡ ΔZE and of circle ὁ ABΓ.

A striking example of the practice of including letters within designations before identifying them as points can be found the construction of the cube in XIII.15 (*EOO* IV, 300.19–302.5):[20]

---

[17] On the fact that an indefinite noun εὐθεῖαι "straight lines" has to be understood before the complex αἱ ΔA ΔB we shall return in the following Section.

[18] The discussion in Netz 1999, 19–25, misses the crucial point, thereby making the presence of allegedly "unspecified" or "underspecified" denotative letters the main reason for the diagram being indispensable in the process of fixing the reference of the letters.

[19] That τὸ E is the name of a point will become apparent only when triangle ὁ ΔAE is first named in the proof. A triangle is designated by naming its vertices: since a vertex is a point, there is no need expressly to assign the name τὸ E to a point. The same holds for the extremities of straight line ἡ AEB.

[20] The name of the cube has two letters, the name of its base square has four. Another example: in the setting-outs of IV.4, 11, and 15, the names of the circles in which suitable regular polygons have to be inscribed (a square, a pentagon, and a hexagon, respectively) comprise four, five, and six letters, respectively, but any such circle is of course the sole object named in the setting-out.

καὶ ἐκκείσθω τετράγωνον τὸ ΕΖΗΘ ἴσην ἔχον τὴν πλευρὰν τῇ ΔΒ, καὶ ἀπὸ τῶν Ε Ζ Η Θ τῷ τοῦ ΕΖΗΘ τετραγώνου ἐπιπέδῳ πρὸς ὀρθὰς ἤχθωσαν αἱ ΕΚ ΖΛ ΗΜ ΘΝ, καὶ ἀφῃρήσθω ἀπὸ ἑκάστης τῶν ΕΚ ΖΛ ΗΜ ΘΝ μιᾷ τῶν ΕΖ ΖΗ ΗΘ ΘΕ ἴση ἑκάστη τῶν ΕΚ ΖΛ ΗΜ ΘΝ, καὶ ἐπεζεύχθωσαν αἱ ΚΛ ΛΜ ΜΝ ΝΚ· κύβος ἄρα συνέσταται ὁ ΖΝ ὑπὸ ἓξ τετραγώνων ἴσων περιεχόμενος.	And let a square, ΕΖΗΘ, be set out having the side equal to ΔΒ, and from Ε, Ζ, Η, Θ let straight lines, ΕΚ, ΖΛ, ΗΜ, ΘΝ, be drawn at right ‹angles› with the plane of square ΕΖΗΘ, and from each of ΕΚ, ΖΛ, ΗΜ, ΘΝ let each of ΕΚ, ΖΛ, ΗΜ, ΘΝ be removed equal to one of ΕΖ, ΖΗ, ΗΘ, ΘΕ, and let ‹straight lines› ΚΛ, ΛΜ, ΜΝ, ΝΚ be joined; therefore a cube ΖΝ turns out to be constructed contained[21] by six equal squares.

In the underlined segments of text, the designation αἱ ΕΚ ΖΛ ΗΜ ΘΝ of a quartet of straight lines recurs three times. In the first two occurrences, the letters Κ, Λ, Μ, Ν are just part of the name of the straight lines. In the third occurrence, these letters finally come to designate points: these are the extremities of the straight lines, equal to assigned straight lines, cut off from the straight lines bearing the same name.

Of course, requirements of transparency motivate assigning the lettered designations in such a way that the geometric configuration is represented by them in the simplest way. The simplest way is to use a system of designation that is compositional, and the only way to make geometric designation compositional is to assign the "atomic" designations to points: thus, two adjacent straight lines ἡ ΑΒ and ἡ ΒΓ, if compounded and by simple elimination of the name of the common extremity, give rise to straight line ἡ ΑΓ—this is the prevailing practice, and such a practice has nothing to do with the alleged requirement of indicating the conspicuous points in a diagram. Accordingly, a complex figure comes to be designated by means of points on it that univocally determine the figure. Still, I stress again that this happens in an incidental way: in principle, a lettered designation is just a graphic device suited to assign a name.

But finally, why use lettered designations at all? The reason is obvious: conventional designations allow effective and unambiguous reference, and the only resources Greek mathematicians availed themselves of to the end of forming such conventional designations were internal to the universe of discourse as represented by alphabetic signs.[22] Granted, a purely conventional system like "first straight line", "second straight line" etc., may in principle work, if one is accustomed to it; one might even eliminate names by using definite descriptions to designate the mathematical objects in a proof: after all, such objects are always very limited in number.[23] As we have seen in Sect. 1.2, this is the solution adopted in the procedural code.

Well, let us read such a denotative monster, showing what a construction (not even a whole proof!) would look like if only definite descriptions of all the objects mentioned were used (the diagrams are also attached). I know of just one example in the technical corpus, and this is Ptolemy, *Alm.* IV.6 (*POO* I.1, 306.7–307.18), where he sets out to provide one single proof for two,

[21] Note again the strongly marked discontinuous constructions ἐκκείσθω [...] ἔχον and συνέσταται [...] περιεχόμενος (at the beginning and at the end of the quote, respectively). As seen in Sect. 1.1, ἔκκειμαι "to be set out" is a standard presential verb; it has a stative value and must be translated as a perfect stem; see Sect. 4.2 for detail.

[22] Recall that technical texts employed the alphabetic numeral system. Paragraphemes like bars and apices are used in Greek manuscripts, but they do not perform the same function as the current graphic resources that make up the bulk of the sign-supply of any symbolic language.

[23] When letters are used, a complete alphabetic sequence is exhausted in almost no Greek mathematical proposition (as for the *Elements*, *episēma* are used in XI.31, 34, XII.17, XIII.16; our texts also tend to avoid using the letter *iota*). I assume that 24 items are very limited in number. Using ordinals amounts to writing ἡ α′ εὐθεῖα, ἡ β′ εὐθεῖα, that is, to a (non-compositional) form of lettered designation.

essentially different, geometric configurations—still, Ptolemy does not resist assigning denotative letters and referring to one configuration only:[24]

ὅτι μὲν οὖν οὐ δυνατὸν ἐπὶ τῆς ΒΑΓ περιφερείας τὸ περιγειότατον εἶναι τοῦ ἐπικύκλου, φανερὸν ἐκ τοῦ ἀφαιρετικήν τε αὐτὴν ὑπάρχειν καὶ ἐλάσσονα ἡμικυκλίου τῆς μεγίστης κινήσεως κατὰ τὸ περίγειον ὑποκειμένης. ἐπεὶ δὲ πάντως ἐπὶ τῆς ΒΕΓ, εἰλήφθω τὸ κέντρον τοῦ τε διὰ μέσων τῶν ζῳδίων κύκλου καὶ τοῦ φέροντος τὸ κέντρον τοῦ ἐπικύκλου καὶ ἔστω τὸ Δ, καὶ ἐπεζεύχθωσαν ἀπ' αὐτοῦ ἐπὶ τὰ τῶν γ ἐκλείψεων σημεῖα εὐθεῖαι αἱ ΔΑ ΔΕΒ ΔΓ.

Now then, that the perigee of the epicycle cannot lie on arc ΒΑΓ, is manifest because this arc has a subtractive effect and is less than a semicircle, while the greatest speed occurs at the perigee. And since ‹the perigee lies› in every instance on ‹arc› ΓΕΒ, let the centre of the ecliptic be taken, which is also the centre of the deferent, and let it be Δ, and let straight lines ΔΑ, ΔΕΒ, ΔΓ be joined from it to the points ‹representing the positions of the Moon at› the 3 eclipses.

καθόλου τοίνυν, ἵνα καὶ πρὸς τὰς ὁμοίας δείξεις εὐεπίβολον τὴν μεταγωγὴν τοῦ θεωρήματος ποιώμεθα, ἐάν τε διὰ τῆς κατ' ἐπίκυκλον ὑποθέσεως αὐτὰς ὡς νῦν δεικνύωμεν ἐάν τε διὰ τῆς κατ' ἐκκεντρότητα τοῦ Δ κέντρου τότε ἐντὸς λαμβανομένου,

Now then, in order to make the sequence of the proof readily transferable for computations of this kind, whether we use the epicyclic model (as now) for our demonstration, or the eccentric model, in which case centre Δ is taken inside the circle, ‹we give the following› generally ‹applicable description›,

μία μὲν τῶν ἐπιζευγνυμένων τριῶν εὐθειῶν ἐκβαλλέσθω ἐπὶ τὴν ἀντικειμένην περιφέρειαν, *ὡς ἐνθάδε τὴν ΔΕΒ αὐτόθεν ἔχομεν διεκβεβλημένην ἐπὶ τὸ Ε σημεῖον ἀπὸ τοῦ Β τῆς δευτέρας ἐκλείψεως*, τὰ δὲ λοιπὰ δύο σημεῖα τῶν ἐκλείψεων ἐπιζευγνύτω εὐθεῖα, *ὡς ἐνθάδε ἡ ΑΓ*, καὶ ἀπὸ τῆς γενομένης τομῆς ὑπὸ τῆς ἐκβεβλημένης, *οἷον τοῦ Ε*, ἐπιζευγνύσθωσαν μὲν ἐπὶ τὰ λοιπὰ δύο σημεῖα εὐθεῖαι, *ὡς ἐνθάδε αἱ ΕΑ ΕΓ*, κάθετοι δὲ ἀγέσθωσαν ἐπὶ τὰς ἀπὸ τῶν λοιπῶν δύο σημείων ἐπὶ τὸ τοῦ ζῳδιακοῦ κέντρον ἐπιζευγνυμένας εὐθείας, *ἐπὶ μὲν τὴν ΑΔ ἡ ΕΖ ἐπὶ δὲ τὴν ΓΔ ἡ ΕΗ*, καὶ ἔτι ἀπὸ τοῦ ἑτέρου τῶν εἰρημένων δύο σημείων, *ὡς ἐνθάδε ἀπὸ τοῦ Γ*, κάθετος ἀγέσθω ἐπὶ τὴν ἀπὸ τοῦ ἑτέρου αὐτῶν, *οἷον τοῦ Α*, ἐπὶ τὴν γενομένην ὑπὸ τῆς διεκβολῆς περισσὴν τομήν, *οἷον τὸ Ε*, ἐπιζευχθεῖσαν εὐθεῖαν, *ὡς ἐνθάδε ἐπὶ τὴν ΑΕ ἡ ΓΘ*.

let one of the three straight lines joined [*scil.* ΔΑ, ΔΒ, ΔΓ] be produced as far as the opposite circumference (*here we already have ΔΕΒ produced through from Β of the second eclipse as far as point Ε*), and let a straight line join the remaining points of the eclipses (*as here ΑΓ*), and from the intersection resulting from the produced ‹straight line› (*namely, Ε*) let straight lines be joined as far as the remaining two points (*as here ΕΑ, ΕΓ*), and ‹from the same point› let ‹straight lines› be drawn perpendicular to the straight line joined from the remaining two points as far as the centre of the ecliptic (*ΕΖ to ΑΔ and ΕΗ to ΓΔ*), and further from one of the said two points (*as here from Γ*) let a ‹straight line› be drawn perpendicular to the straight line (*as here ΓΘ to ΑΕ*) joined from the other of them (*namely, Α*) as far as the extra intersection ‹with the circumference› (*namely, Ε*) resulting from ‹the first straight line, namely, ΔΒ,› being produced through.

ὁπόθεν γὰρ ἂν χρησώμεθα τῇ τῆς καταγραφῆς ἀγωγῇ, τοὺς αὐτοὺς εὑρήσομεν ἐκβαίνοντας λόγους διὰ τῶν τῆς δείξεως ἀριθμῶν τῆς ἐκλογῆς πρὸς τὸ εὔχρηστον μόνον καταλειπομένης.

As a matter of fact, whichever point we start drawing the figure from, we shall find that the same ratios result from the numbers used in the demonstration, our choice being guided merely by convenience.

[24] Toomer's translation, modified. The definite descriptions are underlined; the complements of designation deemed necessary by Ptolemy are in italics. The geometric configurations intended by Ptolemy are represented by the diagrams that follow the text in the next page.

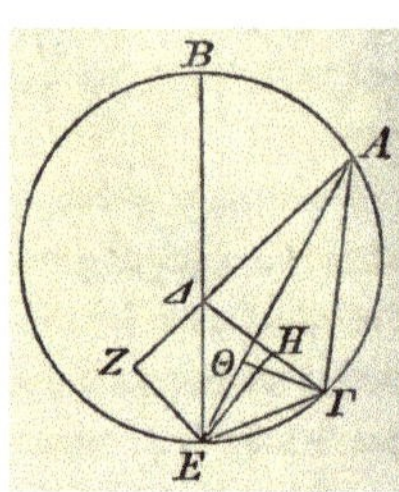

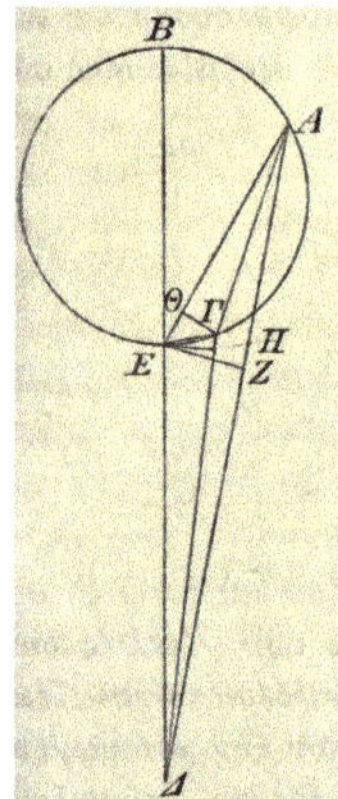

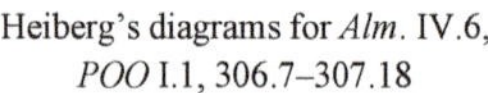
Heiberg's diagrams for *Alm*. IV.6, *POO* I.1, 306.7–307.18

3.2.1. *Denotative letters as "letter-labels"*

I have outlined the function of the denotative letters in the setting-out, and, more generally, in a mathematical proposition. This interpretation conflicts with the standard view, which runs as follows. The setting-out is the description of the diagram that represents the initial configuration of a proposition; the letters are signposts identifying mathematical objects in the diagram that are relevant to the construction and to the proof. In this way, the diagram becomes an integral part of the proof.[25] Such a view, admittedly quite "natural" and which I shall call "iconic", has been erected to

[25] The seminal paper arguing for this interpretation is Manders 2008a; see also Manders 2008b; this view is endorsed in Netz 1999 (who does not mention Manders). I refrain from citing the legions of epigones. This view, as a historically sound interpretation of Greek mathematical practice, is straightforwardly refuted by the massive presence of diagrams in Books V (proportion theory) and VII–IX (number theory), where they not only are deductively sterile, but systematically misrepresent the configuration at issue: in our most ancient manuscripts, the segments that represent the numbers in a given propositions are all of equal length—thus, the presence of diagrams, besides making the proof easier to follow, is just a stylistic trait like our ending a theorem with "QED". Note, however, that the bewildering feature of equal-length segments might well be the result of oversymmetrization (see Sect. 3.5) induced by repeated acts of copying. The curious belief that Greek diagrams are not simply a visual aid but have a full deductive import is entertained by many scholars. They apparently do not realize that any such diagram (better: the geometric configuration it represents) can easily—and fully—be reconstructed from the text of the proposition (any editor of Greek texts since Eutocius does this; this is the role of the setting-out and of the construction, after all: how on earth can one think that the information that point τὸ Z in I.16 lies in the right half-plane pops up from the diagram?) but the converse is obviously false; nor that a diagram taken in isolation is as dumb as a stone (take the diagram in Pappus' alternative proof of I.5 reported by Proclus in *iE*, 249.20–250.12, and try to reconstruct the enunciation and the proof); nor that exactly those geometric items are introduced in the construction as are strictly required to complete the subsequent proof, so that the latter determines the former, and not the converse; nor that the text does not refer to the diagram but describes the geometric configuration (explicit references in Apollonius' *Conica*—see II.27, IV.11, 27–29, 55—belong beyond any possible doubt to the metadiscursive layers introduced by Eutocius' recension); nor that the alleged gaps in ancient geometric proofs are a matter of axiomatic incompleteness (usually involving issues of existence, as in the case of the infamous intersection of the two circles in the geometric configuration of I.1—by the way, these circles are closed curves each of which has one point internal and one external to the other—this property is an immediate consequence of the givens of the problem and of the definition of a circle—so that they necessarily intersect: shall we add this "obvious" continuity postulate to the *Elements*?), not of some information being necessarily drawn from the diagram (so Manders 2008a and e.g. Giaquinto 2011, their strategic goal being to show that the proof of I.1 is not "flawed": this approach is marred en bloc by their wrong conception of a "Greek line" as made of points): Zeno of Sidon's criticisms of geometry exactly point to this drawback (Proclus, *iE*, 214.15–218.11). After all, a totally wrong diagram attached to a valid proof does not turn it into an invalid one. More generally, in the approach to the issue of the status of diagrams in Greek mathematics, there appears to be a remarkable confusion between, on the one side, (*i*) the practice of the working mathematician (heuristic included) and (*ii*) the presentation of mathematical results in formal treatises; on the other, (*a*) a geometric configuration, (*b*) its intended graphic representation, (*c*) the perception of such a representation, and (*d*) the intentionality associated with the representation as a consequence of such a perception. This said, I urge the reader to approach Giaquinto's fascinating studies, as for instance Giaquinto 2007, 2008a and 2008b (partly coinciding with chapters of the first item), and 2011.

a *vulgata*,[26] as is shown for instance by its being accepted as a matter of course in fields of study of ancient thought other than history of mathematics.[27]

The iconic interpretation is allegedly corroborated by a historical reconstruction of the origin of the designation by means of letters. There are two main factors that make such a reconstruction far from easy: the complexity of the textual record and the variety of the designations. As for the textual record, it mainly comprises strictly mathematical texts (like for instance the treatises of Archimedes, Euclid, and Apollonius), but also pieces of mathematics embedded in philosophical writings, which make the totality of the earliest documents: Eudemus' account of Hippocrates' quadrature of lunes preserved as an excerpt in Simplicius' commentary on Aristotle's *Physics*; two long passages in Aristotle's *Meteorologica*; many, very short, mathematical-style arguments scattered in the entire Aristotelian corpus. Another kind of evidence comes from the presence of schematic letters in Aristotelian syllogistic, in particular in *APr.* I: strictly speaking, these are not mathematical-style arguments, but it is a safe and universally accepted assumption that the mathematical model guided Aristotle in his use of letters.

As for the variety of the lettered designations, it is at least threefold, according to the nature of the syntagm that precedes the complexes of letters. First, the model described in detail in Sect. 2.2, adopted in the overwhelming majority of the mathematical corpus and best represented by the practice in the *Elements*: article + prepositional syntagm[28] + letters, as in τὸ Α (point), ἡ ΑΒ (straight line), or ἡ ὑπὸ ΑΒΓ (angle). Second, a model adopted, with a very specific function, in early mathematical texts and in Aristotelian syllogistic: article + relative syntagm[29] + letters, as in ἡ ἐφ' ᾗ ΑΒ ('the ‹straight line› by which AB'). In Aristotelian syllogistic, a major function of this model is to present concrete terms serving as counterexamples to syllogismhood of suitable pairs of premises, as in ἔτι καὶ ἐκ τῶν ὅρων φανερὸν ὅτι οὐκ ἔσται τὸ συμπέρασμα ἀναγκαῖον, οἷον εἰ τὸ μὲν Α εἴη κίνησις, τὸ δὲ Β ζῷον, ἐφ' ᾧ δὲ τὸ Γ ἄνθρωπος.[30] Finally, there is a "citation model", for instance adopted by Aristotle himself to denote the terms in a syllogism: neuter article + letters, as in τὸ Α ("A") just read. This model must be carefully distinguished from the first model applied to designate a point or an undifferentiated σχῆμα ("figure"; neuter in Greek). Here, the article performs the function of the abstract citation operator and must be translated, as I have just done, by putting the letters between quotation marks.[31] Historically, the first model is almost certainly an

[26] The *vulgata* dates back at least to the Proclean passage we have read at the very beginning of Sect. 3. All discussions cited in the footnotes that immediately follow Proclus' quote endorse the iconic view as a matter of course.

[27] See for instance Barnes 2006 and 2007, 347–354.

[28] The prepositional syntagm is not mandatory: see below.

[29] This syntagm is made of a preposition and of a relative pronoun. We shall see below that, contrary to what is commonly held, the relative syntagm is not locative.

[30] *APr.* I.9, 30a28–30. This example shows that this model is not applied exclusively. Only in modal syllogistic are the concrete terms of the counterexamples assigned letters; in assertoric syllogistic, the counterexamples are just set out by naming the triads of terms. Assigning letters to concrete terms is not a generality mistake, for setting out the terms by their names may also carry metalogical connotations of generality: the specific terms set out simply suggest how to select a generic counterexample: see Lear 1980, 64. Another feature pertaining to lettering suggests that assertoric syllogistic is a more refined (or maybe simply edited) elaboration than modal syllogistic: the former carefully picks up different letters for the terms featuring in syllogistic pairs of premises in the different figures (A, B, Γ for the first figure; M, N, Ξ for the second; Π, P, Σ for the third), the latter only uses the letters A, B, Γ.

[31] Thus, the article does not serve the sole goal of making the complex of letters a noun, which must show a case-structure in a declensional language as ancient Greek (this sole function is pointed out for instance in Malink 2008, 524). The "citation model" is also at work when a premise is called τὸ ΑΒ (at e.g. *APr.* I.4, 26a29 and 26a39; I.9, 30a23–24; I.11, 31a26–39; but cf. also the type ἡ ΑΒ πρότασις at I.2, 25a14, and I.6, 28b35). We need not (and in fact must not) think that a noun is understood in syllogistic lettered designations: the only natural candidate, namely, ὅρος ("term"), is masculine (see already Barnes 2007, 347). Supposing instead that the neuter στοιχεῖον ("letter") is understood is just philosophically silly. If the presence of the "citation model" may be doubted in the case of syllogistic (after all, it is just a

evolution and a regularization of the other two. The difficult point is the meaning of the second model. More generally, one of the problems to be settled is to explain the fact that different models happen to be used within one and the same mathematical argument, and by Aristotle within one and the same logical or mathematical-style argument (as in the above clause).

Let us start with the mathematical arguments embedded in the philosophical writings mentioned above. The strings of letters are there preceded by prima facie locative expressions, such as for instance in ἡ ἐφ' ᾗ AB "the ‹straight line› by which AB", or in τὸ ἐφ' ᾧ τὸ A "the ‹point› by the ‹letter› A".[32] The locative expressions are currently assumed to stress in an obvious way the iconic character of the designations by means of letters. Since such complex syntagms are typical of an archaic approach to the lettered diagrams—so it is argued on the basis of the kind of text where they mainly appear (as said, their presence is massive in the Aristotelian corpus)—a historical evolution of the lettering conventions can be outlined, starting from an initial stage where complex expressions were employed and ending with the canonical assignments:[33]

τὸ ἐφ' οὗ A → *τὸ σημεῖον ἐφ' οὗ A → *τὸ σημεῖον τὸ ἐφ' οὗ A → *τὸ ἐφ' οὗ A σημεῖον →
→ τὸ A σημεῖον → *τὸ σημεῖον τὸ A → τὸ A

Such a reconstruction implicitly rests on some assumptions: (1) that there was such an evolutionary line linking formula τὸ ἐφ' οὗ A to formula τὸ A;[34] (2) that the "complex designations"—so I shall name them—are restricted to archaic writings; (3) that the locative expressions figuring in the complex designations do have an iconic connotation.

I think that the whole picture is wrong: among other drawbacks, it entails identifying references to the geometric configuration and references to the diagram representing it. I propose here a different picture, whose qualifying point is that it brings epigraphic evidence, never taken into account in this context, to bear on the issue. Before doing this, a historiographic survey is necessary.

The presence of complex designations as an exegetic marker was first brought to scholarly attention in order to disentangle the Eudemian text on Hippocrates' quadrature of lunes from Simplicius' additions.[35] It was first applied as a criterion, namely, as a necessary and sufficient condition: complex designations = archaism = Eudemus; standard designations = Simplicius.[36] Strong reservations were immediately raised against the general validity of the criterion. H. Diels rightly asserted that nothing could be inferred from the absence of complex formulations, since "breviore ratione signandi interdum usum [est] Eudemum"[37]; F. Rudio rejected outright the criterion. P. Tannery, who had also applied it in his study of the mathematical passage on the form of the rainbow in Aristotle, *Mete.* III.5, pointed out that complex designations can also be found in later authors

matter of thinking of an appropriate neuter noun to be understood), it is certainly at work in some mathematical arguments in the Aristotelian corpus: see below.

[32] Both genitive and dative are attested, see below. I translate here as an iconodule would do—ἐπί means here locative "by", not "upon". My own translation of the second designation would be "a point identified by 'A'". An account of the standard interpretation of the complex syntagms as locative expression is for instance in Netz 1999, 44–49.

[33] See Federspiel 1992, 21–22.

[34] This is irrelevant to my present purposes, but none of the steps marked by * is attested in any mathematical argument.

[35] Of course, we cannot hope to reconstruct Hippocrates' text: in the words of Tannery reported by Diels at Simplicius, *in Ph.*, XXIX, "Hippocratis verba ex Eudemeis enucleari posse vanum est somnium" (such a dream has been had in Netz 2004). In the analogous proofs Simplicius extracts from Alexander the complex designations are absent. The relevant passages are at Simplicius, *in Ph.*, 55.25–57.24 (Alexander) and 61.1–68.32 (Eudemus).

[36] See Tannery 1883 and Allman 1889, 72 n. 45.

[37] At *in Ph.*, XXIX; cf. also Becker 1936b. The subsequent references are to Rudio 1902; Tannery 1886, 1883, and 1902, respectively; Vitrac 2002.

(he mentioned Philo of Byzantium), and that their survival was entirely in the hands of the copyists. Finally, B. Vitrac has shown that the presence of complex designations can only be used as a supporting criterion, showing that, besides being typical of mathematical passages in the Aristotelian corpus, complex designations also occur in classical authors as Archimedes and Apollonius,[38] and to a larger extent in writers of applied mathematics as Philo of Byzantium.[39]

The main features of the complex designations can be listed on the basis of the texts where they were first detected and studied: Simplicius excerpting Eudemus on the quadrature of lunes and the two mathematical passages in Aristotle's *Meteorologica*; there are 83 complex designations in them.[40] Of course, the manuscript tradition of these passages attests to variant readings (mainly concerning the gender of the relative pronoun), but these can safely be ascribed to personal initiatives of the copyists and, what is more, they are in any case statistically irrelevant: if their number is low, they are not significant, if it is high, they tend to cancel out. Recall also that the same passages contain plenty of canonical designations. The main features of the complex designations can be summarized as follows:

1) The relative pronoun in the "standard" locative syntagm can indifferently be in the dative (ἐφ' ᾧ) or in the genitive (ἐφ' οὗ), with a strong preference for the dative (these are 57 items out of 70 with the relative syntagm).
2) The gender of the relative pronoun may not be the same as the gender of the object designated by the complex expression, even if such exceptions are sporadic.[41]
3) An article is sometimes interposed between the locative expression and the letters, but in the majority of cases it is absent (47 items out of 70, most notably in all 40 occurrences in Simplicius). This article is always in the gender neuter. When more than one letter is involved, this article can be either in the plural (*Mete*. III.5, 376a7) or in the singular (377a5).
4) In the Aristotelian texts there is also a form with a different locative syntagm (1 item: οὗ τὸ H at *Mete*. III.5, 375b30) and a bewildering form of the kind ἡ τὸ AB, with two singular articles preceding a multi-lettered complex (12 occurrences; 6 of them are at 373a7–13, 6 at 376a15–376b1; cf. 377a3–5 for the same phenomenon occurring with "standard" complex expressions).
5) Note also the forms ἐν ᾧ ἡ (τὸ) A, where τὸ is added in some manuscripts, and ἐν ᾧ τὸ HKM at *Mete*. III.5, 375b31–32.

[38] Beware of "fake" complex designations in mathematical writings. These are: (1) locative phrases with the relative in the accusative, meaning "on the same / opposite side as", as in Archimedes, *Fluit*. II.8, *AOO* II, 374.11; *Con. sph*. 19, *AOO* I, 338.5–6, 8–11; (2) descriptions of the design of a technological device, as in Philo of Byzantium, *Belopoeika*, 128–132, 144–148: the iconic reference is explicit but what is done is definitely not assigning letters in a mathematical proposition; (3) locative phrases whose meaning is that a point lies on a line, as in Apollonius, *Con*. III.13, IV.8, 13, 17, 20–21, and Archimedes, *Spir*. 12, *AOO* II, 46.27.

[39] But they are absent in *Sph. cyl*., a "treatise" that we very likely read in a later recension, since it shows no traces of Doric dialect (with the only exception of the pronoun τῆνος = ἐκεῖνος). (I use quotation marks because what Archimedes writes in the prefatory epistle of *Sph. cyl*. II makes it clear that he did not conceive this writing as a "second book" to be attached to *Sph. cyl*. I.) For a list of the occurrences in mathematical texts, see Vitrac 2002, 252–254. A very interesting typology is constituted by indefinite series of items all designated by the same letter: τὸ ἐφ' ὧν τὰ A is a string AAAAA, where the number of As is left undetermined. (The complete phrase is πλῆθος ἀριθμῶν τὸ ἐφ' ὧν τὰ A "a collection of numbers noted by As"; the construction is *ad sensum*.) The presence of the locative expression is here a stylistic trait, introduced in order to avoid ambiguities. There are 25 such occurrences in Pappus, *Coll*. II.2–6, in the description of a writing by Apollonius: the locative phrases were almost surely in Apollonius, too. In this special instance, there cannot be any iconic connotations: the listed items are numbers; therefore, the letters designate nothing but themselves.

[40] These are Simplicius, *in Ph*., 64.7–68.32 (40 occurrences, all with the relative syntagm), Aristotle, *Mete*. III.3, 373a4–19 (7 occurrences, 1 of which with the relative syntagm), and *Mete*. III.5, 375b9–377a28 (36 occurrences, 29 of which with the relative syntagm).

[41] See, elsewhere in the Aristotelian corpus, *Ph*. VI.4, 235a19–20; VI.7, 237b35; VII.5, 250a2; *Cael*. I.5, 272b27–28.

Apparently, then, the locative phrases admit of a straightforward iconic interpretation. This interpretation is seemingly corroborated by the second feature just listed, which suggests that the relative pronoun refers in fact to the object designated by the complex expression (the exceptions are bewildering but they are ruled out insofar as they are sporadic). The presence of the article τό before the letters is usually taken to stand for γράμμα or στοιχεῖον "letter". In the iconic interpretation, the article τό (sing.) before two letters is quite surprising. Even more surprising are the non-canonical non-locative designations of the kind ἡ τὸ AB.

The complex designations occurring in the rest of the Aristotelian corpus offer a large sample of occurrences; this constitutes in fact the most ancient set of testimonies.[42] In some cases, the just-mentioned surprise turns into bewilderment, and in fact paves the way to a different view. Let us see in detail what happens in [Aristotle], *Pr.* XV.9–10, 912a34–b10:

9 διὰ τί τὰς σκιὰς ποιεῖ ὁ ἥλιος ἀνίσχων καὶ δύνων μακρὰς αἰρόμενος δὲ ἐλάττους ἐπὶ τῆς μεσημβρίας δ' ἐλαχίστας; ἢ ὅτι ἀνίσχων τὸ μὲν πρῶτον παράλληλον ποιήσει τὴν σκιὰν τῇ γῇ καὶ ἄπειρον ὡς ἄνισον ὑπερτείνει, ἔπειτα μακρὰν ἀεὶ δ' ἐλάττω διὰ τὸ ἀεὶ τὴν ἀπὸ τοῦ ἀνωτέρου σημείου εὐθεῖαν ἐντὸς πίπτειν. γνώμων, τὸ ΑΒ, ἥλιος, οὗ τὸ Γ καὶ οὗ τὸ Δ· ἡ δὲ ἀπὸ τοῦ Γ ἀκτὶς, ἐφ' ἧς τὸ ΓΖ, ἐξωτέρω ἔσται τῆς ΔΕ. ἔστι δὲ σκιὰ ἡ μὲν ΒΕ ἀνωτέρω ὄντος τοῦ ἡλίου ἡ δὲ ΒΖ κατωτάτω ἐλαχίστη δὲ ὅσῳ ἀνωτάτω, ἢ καὶ ὑπὲρ τῆς κεφαλῆς.	**9** Why does the Sun make long shadows as it rises and sets, shorter when it is high, and shortest of all at midday? It is because, as it rises, it will at first make a shadow parallel to the earth and casts it unbounded far away, and then will make a long one, which, however, is ever less because a straight line from a higher point falls within ‹that from a lower one›. Gnomon, "AB", Sun, where Γ and where Δ; the ray from Γ, by which "ΓΖ", will be outside ΔΕ. And there is shadow BE when the Sun is higher, and BZ when it is lower, and it will be shortest when the Sun is at its highest and over our head.
10 διὰ τί αἱ ἀπὸ τῆς σελήνης σκιαὶ μείζους τῶν ἀπὸ τοῦ ἡλίου, ὅταν ἀπὸ τῆς αὐτῆς ὦσι καθέτου; ἢ διότι ἀνώτερος ὁ ἥλιος τῆς σελήνης· ἀνάγκη οὖν ἐντὸς πίπτειν τὴν ἀπὸ τοῦ ἀνωτέρω ἀκτῖνα. γνώμων, ἐφ' ᾧ ΑΔ, σελήνη, Β, ἥλιος, Γ. ἡ μὲν οὖν ἀπὸ τῆς σελήνης ἀκτὶς ΒΖ· ὥστε ἔσται σκιὰ ἡ τὸ ΔΖ· ἡ δὲ ἀπὸ τοῦ ἡλίου ἡ τὸ ΓΕ· ὥστε ἔσται σκιὰ ἐξ ἀνάγκης ἥττων – ἔσται γὰρ τὸ ΔΕ.	**10** Why are the shadows from the Moon longer than those from the Sun, whenever both come directly from the same place? It is because the Sun is higher than the Moon; then the ray from a higher object necessarily fall inside. Gnomon, by which ΑΔ, Moon, B, Sun, Γ. Then, the ray from the Moon is BZ; so that the shadow will be "ΔΖ"; and the ray from the Sun is "ΓΕ"; so that its shadow will of necessity be less—for it will be "ΔΕ".

Thus, besides standard complex designations, we find again the rare form οὗ τὸ A, some non-iconic occurrences like ἡ AB, three lettered complexes without the article, and three instances like ἡ τὸ AB. As the two designations γνώμων τὸ AB and ἐφ' ἧς τὸ ΓΖ show (two letters with article in

[42] This is because the three texts listed above have been heavily reworked by later revisers. This is obvious in the case of the Eudemian passage in Simplicius; as for the *Meteorologica* passages, see Vitrac 2002. Reworking can only result in regularization, which in our case does not mean elimination of the complex designations, but a mimetic attitude tending to increase their number, with alignment on the prevailing species. Letters designating syllogistic terms in *APr.* are schematic letters—and in fact complex designations never appear in *APr.* I.2–22, unless, as seen, concrete terms, mainly serving as counterexamples, are assigned their "name" (at I.9, 30a30; I.11, 31b5, 31b28; I.15, 34a7–8 [a very peculiar assignment!], 34b33–34, 34b39; I.17, 37b4; I.19, 38a31–32, 38a42, 38b20). Complex designations in presenting concrete terms of syllogistic deductions also abound in e.g. *APr.* I.31 and I.33–38.

the singular), the neuter article in forms like ἡ τὸ ΓΕ stands for the abstract citation operator. Accordingly, I have translated the article with quotation marks.

These two texts also make a general phenomenon obvious: complex designations in (pseudo)Aristotelian and mathematical texts are mostly used when a mathematical object is first introduced.[43] But the central fact is that neither clause in which the letters are introduced in *Pr.* XV.9–10 contains any verb form: these clauses are complex noun phrases, and in fact, as seen in Sect. 3.2, they correspond to providing an itemized list, in this way "(*a*) a circle, (*b*) two points on it, etc."—what we gather from the previous discussion is the following, fundamental clue: the itemizers *a* and *b* are used to designate the items in the subsequent proof. The letters introduced in mathematical arguments are *letter-labels*, deprived of any iconic connotation.

This seemingly counter-intuitive interpretation is corroborated in a decisive way by epigraphic evidence. This is constituted by the use of letter-labels in Greek inscriptions, mainly coming from 4th-century Athens. I refrain from citing the several inscriptions involved: I am wholly dependent on M.N. Tod's studies; checking his data would be vain labour.[44] The nature of the evidence is always the same: lists of objects made for inventory purposes; the listed material was mainly stored in temples. The earliest recorded inscription dates to 371 BCE, while a widespread use of letter-labels is attested from 343 BCE onwards. The inscriptions described by Tod come from Athens and Delos; in Delos, the period from 314 BCE onwards is covered. These dates exactly fit the dates of such actors in our story as Aristotle and Eudemus.

The letters are used as labels to identify the items in the inventory list. The nature of the listed objects makes it impossible that the letters were materially marked on or near them. Therefore, the letter-labels were used as metalinguistic markers, very much like the headings "a" to "d" in the following typological list of para-locative (that is: labelling) expressions found in inscriptions:

a) ἵνα τὸ Α: the first attested form; the relevant inscriptions date from 371 to 343 BCE;[45]
b) ἐφ' οὗ τὸ Α: the standard form, first attested in an Athenian inscription of 320 BCE;
c) οἱ τὸ Α: an alternative system, attested in a group of inscriptions about 334–331 BCE;
d) οὗ τὸ Α: a single occurrence in Delos, 267 BCE.

Other features of this system are pointed out by Tod:

1) The letter-labels are expressed by letters (like Α), not by letter-names (like ἄλφα).
2) The letters are real alphabetic letters, not ordinals or cardinals: lists are found that contain both letter-labels and ordinal adjectives spelled in full words.[46]
3) The letters are always preceded by the neuter article τό or τά; as explained above, it is not said that the word στοιχεῖον "letter" must be understood.
4) The phrases containing the letters are normally noun phrases. Where a verb form is attested, the perfect παρασεσήμανται "turns out to be marked beside" is employed for para-locative expressions, the participle ἔχοντες "carrying" for expressions like οἱ τὸ Α. According to what has

[43] This is shown in Vitrac 2002, 248–255, after a detailed analysis of all Aristotelian arguments in mathematical style and of the occurrences in strictly mathematical texts. Such a first occurrence mainly takes place in suppositions, that is, in noun phrases (as in *Pr.* XV.9–10 read above) or in clauses with the verb form in the imperative.
[44] The evidence is presented in Tod 1954.
[45] The particle ἵνα also has the value of adverb of place.
[46] The Athenians adopted the acrophonic numeral system as far as the end of the 2nd century BCE. The best introduction to the Greek numeral systems are Tod 1911–12, 1913, 1926–27, 1936–37, 1950.

been said above, the subject of any of these verbs must be the linguistic unit describing the object(s) falling under the item identified by the letter-label.

5) All sorts of variants of form (b) are recorded: the relative pronoun can be in the dative or in the genitive, all genders alike, showing no correlation with the gender of the object referred to, in the singular or in the plural depending on whether one or more objects are described by the inventory item.

The evidence coming from letter-labels in Greek inscriptions corroborates my interpretation of the lettered designations in Greek mathematical texts as letter-labels for several reasons.

- Such an evidence comes from texts earlier than any Greek mathematical text, possibly—but not necessarily—rewritten by later authors, as is the case with the Eudemian fragment.
- It deploys a range of para-locative expressions that almost exactly coincide with those featuring in the complex designations of mathematical objects.
- It naturally gives rise to the phenomenon of imperfect correlation between the gender of the relative in the labelling locative phrase and the gender of the labelled item that bewilders the supporter of the iconic interpretation of complex designations.
- It assigns letters to objects in a way that perfectly fits the way letters are introduced in a setting-out, as we have seen in the previous Section.[47]
- The ordering principle of letter-labels is, and must be, strictly alphabetic. The letters of a mathematical proposition are sometimes introduced in non-alphabetic order, and some texts display hints of an acrophonic system of designations.[48] Still, introducing the letters in their alphabetic order is by far the most widespread—and canonical—practice.

It goes without saying that we only have access to stages of the evolving convention for lettered designations. The earliest of them display labelling expressions and multi-lettered complexes, and could be taken to be a non-initial stage, in which the process of standardization towards non-complex designations is still in progress. This is not necessarily the case, however. The mathematical convention may well have combined from the very outset labelling syntagms such as ἐφ' οὗ and a lettering practice, as the one canonized in the *Elements* and surveyed in Sect. 2.2, which fits better the requirement of identifying complex objects like geometric figures and not simply structureless inventory items. The key point of my interpretation is that it severs the seemingly obvious link between lettered designations and diagrammatic representation of a geometric configuration established by considering syntagms like ἐφ' οὗ locative and not simply labelling.

Once this link is severed, both as a reliable historical connection and as a support to the thesis of the deductive import of a diagram, and taking into account our acquisitions about the "presential value" of the verb "to be" in the setting-out (Sect. 3.1) and of the syntactic function of the lettered syntagms in the specific part of a proposition in which they are introduced (Sect. 3.2), the way is paved to a full understanding of the "Greek" solution to the Locke-Berkeley problem.

[47] Of course, the interpretation of the previous Section does not depend on, while being corroborated by, the epigraphic evidence just adduced.

[48] A striking example is Euclid, *Optica* both redactions, in which significant points such as the eye or the centre of circles and spheres are designated by using letter K (from κέντρον). Even more striking are the just-discussed mathematical passages in *Mete.* III.3, 373a4–19, and III.5, 375b9–377a28. In the former, the letters are assigned in nearly alphabetic order and range without gaps from A to Z. In the latter, they are not, and moreover the centre of a circle is designated by using letter K, the Sun by using letter H (from ἥλιος), the pole of a circle on a sphere by using letter Π (from πόλος).

3.3. The indefinite structure

We have repeatedly seen that, in the setting-out, the complex article + letters follows the associated noun syntagm, as in ἔστω κύκλος ὁ ΑΒΓ "let there be a circle, ΑΒΓ": the lettered syntagm is appositive, and provides the intended circle with its name. Yet, in the determination and in any further occurrence (with specific exceptions, as we shall see in a moment), the denotative letters change their position. They now precede the noun syntagm, as in ἡ ΑΒ εὐθεῖα. In translation we may supply an article "the straight line AB", since the object referred to has already been mentioned: this is warranted by the well-known, strongly marked anaphoric value of the article in Greek language.[49] Still, the Greek noun syntagm ἡ ΑΒ εὐθεῖα need not be considered definite (and even if we consider it definite, let me repeat, we are entitled to do so just insofar as it is anaphoric of the primary reference),[50] for a phenomenon of neutralization takes place of the polar opposition definite / indefinite. This amounts to the following. Since the complex of letters AB in the syntagm ἡ ΑΒ εὐθεῖα requires an article, two readings are possible:[51]

- The prenominal modifier ἡ ΑΒ is attributive, the meaning is definite: the article is required in Greek both because it determines the letters and because the noun phrase is definite: "the straight line AB", or, better, "straight line AB".[52]
- The prenominal modifier ἡ ΑΒ is appositive, the meaning is indefinite: the article is required but it only determines the string of letters: "*a* straight line, ‹namely, the one called› AB".[53]

This state of affairs is summarized by saying that in the articular noun phrase ἡ ΑΒ εὐθεῖα the opposition definite / indefinite is neutralized. Neutralization is also at work when, as happens in the overwhelming majority of case, the noun εὐθεῖα is omitted, only the name ἡ ΑΒ being used. In the sole complete noun phrase that is certainly definite, ἡ ΑΒ is postnominal attributive, thus: ἡ εὐθεῖα ἡ ΑΒ—which in its pure form[54] is *never* used in the Greek mathematical corpus.

The phenomenon of neutralization allows us to recognize articular expressions as indefinite, and even more so if the presence of the article is forced by grammar. This happens when the noun

[49] See already Apollonius Dyscolus, who makes the anaphoric value of the article the *leitmotiv* of his exposition in *Synt.* I.25, 43–44, 48, 78, 87, 97–98, 133–135, 144, II.9. See also *Synt.* I.111, in *GG* II.2, 94.7–17, for an attributive participle whose σύνταξις "construction" was, according to the Stoics, ἀοριστώδης "indefinite". An overview of scholarship on the Greek article can be found in Peters 2014, sect. 1; an illuminating discussion is in Bakker 2009a, sect. 5.

[50] Referents that are univocally identified in a given context can be designated by an articular expression even in their first occurrence (this is the so-called "associative anaphoric use" of the article): τοῦ δοθέντος κύκλου τὸ κέντρον εὑρεῖν "find the centre of a given circle" (III.1, *EOO* I, 166.14), for the centre of a given circle is unique—still, it is crucial that the circle is mentioned before its centre. Recall (cf. Sect. 2.2, and Sect. 5.1.5 for further detail) that the articular phrase τοῦ δοθέντος κύκλου is indefinite; the article τοῦ marks the prepositive determiner δοθέντος for saliency.

[51] For the articulation of noun phrases, see again the illuminating discussion in Bakker 2009a, sect. 6, and in particular 220 n. 10 for aXN noun phrases (a = article; X = modifier; N = noun).

[52] Contrary to other scholarly languages, such as French or Italian, English has the remarkable resource of allowing seemingly neutralized expressions like "straight line AB". Since English, contrary to ancient Greek, does have an indefinite article, the meaning of a non-articular expression like this is in fact definite.

[53] The modifier is prepositive because of its higher saliency with respect to the head noun "straight line", a piece of information that has been provided earlier in the proposition and that in principle does not univocally identify the referent.

[54] This means that the noun is not accompanied by a further modifier, as for instance in the formula συνεστάτω πρὸς τῇ ΔΕ εὐθείᾳ καὶ τῷ πρὸς αὐτῇ σημείῳ τῷ Δ τῇ ὑπὸ ΒΑΓ γωνίᾳ ἴση ἡ ὑπὸ ΕΔΗ "on straight line ΔΕ and at point Δ on it let an ‹angle›, ΕΔΗ, be constructed equal to angle ΒΑΓ" (I.24, *EOO* I, 58.7–9; it is the first application of I.23): the first article τῷ in the underlined syntagm is forced by the presence of the prepositional determiner πρὸς αὐτῇ. Recall that aXaN noun phrases simply do not exist in Greek.

phrase includes complex modifiers, like entire prepositional expressions. Let us read the enunciation and the setting-out of I.29 (*EOO* I, 70.20–72.1) as an example:[55]

ἡ εἰς τὰς παραλλήλους εὐθείας εὐθεῖα ἐμπίπτουσα τάς τε ἐναλλὰξ γωνίας ἴσας ἀλλήλαις ποιεῖ καὶ τὴν ἐκτὸς τῇ ἐντὸς καὶ ἀπεναντίον ἴσην καὶ τὰς ἐντὸς καὶ ἐπὶ τὰ αὐτὰ μέρη δυσὶν ὀρθαῖς ἴσας.	A straight line falling on parallel straight lines makes both the alternate angles equal to one another and an external one equal to the internal and opposite one and those internal and on the same side equal to two right ‹angles›.
εἰς γὰρ παραλλήλους εὐθείας τὰς ΑΒ ΓΔ εὐθεῖα ἐμπιπτέτω ἡ ΕΖ.	In fact, let a straight line, EZ, fall on parallel straight lines, AB, ΓΔ.

The setting-out shows that the noun phrase underlined in the enunciation is fully indefinite. The first article, ἡ, in this noun phrase is required because the noun εὐθεῖα is modified by a prepositive participial expression that includes a further prepositional modifier, namely, ἡ εἰς τὰς παραλλήλους εὐθείας [...] ἐμπίπτουσα.[56] In the setting-out, where the participle is canonically transformed into an imperative, the article disappears and the designation ἡ ΕΖ follows the entire noun phrase. The second article τάς, in εἰς τὰς παραλλήλους εὐθείας, enhances the saliency of the prepositive modifier: the intended straight line falls on straight lines that are first and foremost parallel straight lines. In the setting-out, this supplement of saliency is eliminated. To conclude: both the enunciation and the setting-out use indefinite noun phrases.

Thus, expressions like ἡ ΑΒ εὐθεῖα, exclusive of the setting-out and of the construction, and in fact quite rare, may be considered to be definite not in virtue of their form, but because they refer to objects that have already been introduced in the universe of discourse. In other terms: mathematical species introduced by means of indefinite noun phrases may be considered to carry a definite reference starting from their second occurrence in a specific part of a proposition.[57]

But, if the designation of a mathematical species *may be considered* to carry a definite reference starting from its second occurrence, then it *may also be considered* not to carry it: neutralization and the practice of omitting nouns like εὐθεῖα or σημεῖον in constructive acts [see point (*iii*) below] and in the proof—which almost uniquely contains short designations of the form ἡ ΑΒ—make a proposition totally blind to the polarity definite / indefinite. We are thus led to conclude that a Greek mathematical proposition is *potentially* carried out by using only indefinite noun phrases.

In order to see more clearly that a Greek mathematical proposition is potentially carried out by using only indefinite noun phrases, let me first list the specific parts of a proposition whose indefinite character is explicit:

(*i*) Enunciation. The denotative letters are absent. The geometric species are designated in their first occurrence by indefinite noun phrases; from the second occurrence on, these expressions take on an article, with purely anaphoric value.[58]

---

[55] Note again the plural article in παραλλήλους εὐθείας τὰς ΑΒ ΓΔ.

[56] The participle follows the head noun, probably because the whole modifier is too heavy to be entirely prepositive; a polyptoton εὐθείας εὐθεῖα also results.

[57] The modal connotation in my statement is necessary once denotative letters are present, again by the phenomenon of neutralization; see just below.

[58] There are obvious exceptions: as seen above, the designation of the centre of a circle is articular from its very first occurrence since the centre of a circle is uniquely identified once the circle is provided.

(*ii*) Setting-out. Denotative letters are introduced as appositives of the same indefinite noun phrases featuring in the enunciation; from the second occurrence on, these expressions may be considered to take on an article, with purely anaphoric value.

(*iii*) Construction. This is the specific part in which the phenomenon of preterition of nouns occurs systematically. For mathematical objects are newly introduced in the construction, in order to complete the initial configuration in a way that fits the subsequent proof (cf. Sect. 4.3). These objects are designated by indefinite noun phrases in their first occurrence, accompanied again by letters in apposition. Now, after a few primary occurrences of a specific constructive act, such indefinite noun phrases are systematically omitted. To appreciate the phenomenon, let us first read I.2 (*EOO* I, 12.24–14.3), where some basic constructive acts are formulated fully:

ἐπεζεύχθω γὰρ ἀπὸ τοῦ Α σημείου ἐπὶ τὸ Β σημεῖον εὐθεῖα ἡ ΑΒ, καὶ συνεστάτω ἐπ' αὐτῆς τρίγωνον ἰσόπλευρον τὸ ΔΑΒ, καὶ ἐκβεβλήσθωσαν ἐπ' εὐθείας ταῖς ΔΑ ΔΒ εὐθεῖαι αἱ ΑΕ ΒΖ, καὶ κέντρῳ μὲν τῷ Β διαστήματι δὲ τῷ ΒΓ κύκλος γεγράφθω ὁ ΓΗΘ.	In fact, from point A to point B let a straight line, AB, be joined, and let an equilateral triangle, ΔAB, be constructed on it, and let straight lines, AE, BZ, be produced in a straight line with ΔA, ΔB, and with centre B and radius BΓ let a circle, ΓΗΘ, be described.

As I have explained in Sect. 2.2 and shall better argue in Sect. 4.3, the formulation of every constructive act rigidly conforms to a template deriving, by a series of standard linguistic transformations, from a primary formulation; the templates of the constructive acts in I.2 are, in the order, provided by I.post.1, I.1,[59] I.post.2, and I.post.3. In these templates, as well as in the construction of I.2, each newly constructed geometric object is explicitly named by means of an indefinite noun phrase (underlined above). In I.2, these noun phrases are further accompanied by appositive letters. Thus, *a* straight line is constructed whose name is ἡ AB, etc. In subsequent propositions, as for instance in the construction of our paradigmatic III.2, the constructive acts are systematically provided in abbreviated form: what is omitted are for instance some determinations implicit in the lettered designations,[60] and, most importantly, the indefinite noun phrases that designate the newly constructed geometric objects. Since every constructive act introduces a new object afresh in each proposition, even if the act is each time the same (for instance, joining two points according to I.post.1), all constructive acts introduce generic geometric species. As a consequence, we must integrate their designations between angular brackets in translation.

It may even happen, as in the case of the constructive act "through a given point draw a straight line parallel to a given straight line", that the noun serving as the grammatical subject is understood in all the applications. To see this, let us read the primary formulation of this constructive act in the conclusion of I.31 and the first application in I.32 (*EOO* I, 76.10–12 and 76.24–25):

διὰ τοῦ δοθέντος ἄρα σημείου τοῦ Α τῇ δοθείσῃ εὐθείᾳ τῇ ΒΓ παράλληλος εὐθεῖα γραμμὴ ἦκται ἡ ΕΑΖ.	Therefore through a given point, A, a straight line, EAZ, turns out to be drawn parallel to a given straight line, BΓ.
ἤχθω γὰρ διὰ τοῦ Γ σημείου τῇ ΑΒ εὐθείᾳ παράλληλος ἡ ΓΕ.	In fact, through point Γ let a ‹straight line›, ΓE, be drawn parallel to straight line AB.

[59] Recall that, in the case of I.1, the template is the instantiated conclusion.
[60] For instance, the mention of the extremities of any straight line joined according to I.post.1 is omitted.

Subsequent applications will also lack the nouns σημείου and εὐθείᾳ.[61]

(*iv*) Citations of previous results, either instantiated or not instantiated. They constitute a key component of the proof, since they feed in, in the form of coassumptions, deductive material extraneous to the proposition (see Sects. 4.5.4–5), and accentuate its formulaic character. Such citations keep the indefinite structure of the template constituted by the enunciation of the cited theorem, while adapting it to the geometric configuration at issue (see Sect. 2.3).

(*v*) General conclusion. When it is present (this mainly happens in theorems, see Sect. 4.1), it is identical to the enunciation. In problems, the general conclusion is frequently absent; the instantiated conclusion contains denotative letters but explicitly restores the indefinite structure (see Sect. 4.3). We check this by reading again the enunciation and the instantiated conclusion of proposition I.2 (*EOO* I, 12.19–20 and 14.13–15):

πρὸς τῷ δοθέντι σημείῳ τῇ δοθείσῃ εὐθείᾳ ἴσην <u>εὐθεῖαν</u> θέσθαι.	At a given point set <u>a straight</u> line equal to a given straight line.
πρὸς ἄρα τῷ δοθέντι σημείῳ τῷ Α τῇ δοθείσῃ εὐθείᾳ τῇ ΒΓ ἴση <u>εὐθεῖα</u> κεῖται ἡ ΑΛ.	Therefore at a given point, Α, <u>a straight line</u>, ΑΛ, turns out to be set equal to a given straight line, ΒΓ.

Let us now see the segments of a proposition in which indefiniteness is implicit or neutralized:

(*vi*) Anaphora. This is the sequence of statements that open the proof and whose formulation as sentences are introduced by ἐπεί "since" (see Sects. 4.4 and 5.3.2). The anaphora may contain either (*a*) indefinite expressions or (*b*) bare lettered designations or (*c*) neutralized expressions (all three cases are found in III.2). The fact that the anaphora directly refers to what has been supposed in the setting-out or in the construction allows us to read case (*b*) as resulting from ellipsis of an indefinite noun phrase and case (*c*) as an indefinite expression.

(*vii*) Proof. It is almost always neutralized, and mainly resorts to bare lettered designations; if the related objects have already been introduced, the designations *may be considered* to carry a definite reference. Exceptions are the indefinite citations of previous results [point (*iv*) above] and the mentions, very frequently without an article, of specific nouns like πλευρά "side", βάσις "base", γωνία "angle", and λόγος "ratio".[62]

To summarize: in the course of a proposition, the following path is followed by any designation of a mathematical object:

εὐθεῖα (indefinite noun: enunciation) → εὐθεῖα ἡ ΑΒ (indefinite noun with a name: setting-out) → ἡ ΑΒ εὐθεῖα (neutralized syntagm: proof—very infrequent) → ἡ ΑΒ (pure name: proof—by far the most frequent) → εὐθεῖα (ἡ ΑΒ) (indefinite noun, with a name in the case of problems: conclusion).

[61] These nouns necessarily refer to mathematical objects that have already been mentioned (in fact, they are even "given"), so that the neutralized expressions διὰ τοῦ Γ σημείου and τῇ ΑΒ εὐθείᾳ do have a definite referent. On the fact that the forms of the participle δοθείς "given" carry an article even in their first occurrence, see above and Sect. 5.1.5.

[62] On this bewildering feature of Greek mathematical style see Federspiel 2008a, 325–330, who suggests that it is a fossil of an archaic linguistic format. If Federspiel is right, this fact confirms my interpretation. However, one must recall that several Greek nouns behave non-referentially more often than expected: Bakker 2009a passim, mentions βασιλεύς, γνώμη, γῆ, θάλασσα, ἥλιος, ὄνομα.

Such a path is dictated by the necessity of setting up an effective reference to the mathematical objects introduced in the enunciation and in the construction.

Shall we feel entitled to squeeze some philosophically-oriented conclusion out of this linguistic analysis? If so, we may be tempted to submit that linguistic indefiniteness entails that the referents of the noun phrases are also generic: and these referents are the mathematical objects. Thus, a "Greek" mathematical proof does not refer to generic objects insofar as they are fictitiously particular. The carriers of the reference to mathematical objects are in fact the indefinite noun phrases; the lettered complexes—whose compositional structure ensures a radical denotative economy—operate instead within the universe of discourse, for they are the names of the indefinite noun phrases. One might object that the indefinite structure is after all a surface grammar, a Greek proof really dealing with individuals. To this I reply that we are not entitled to presuppose anything "deeper" than linguistic evidence. The denotative letters, thus, are not the "free variables" of the proof. The compositional structure of the lettered complexes, which is the *only* means by which we have access to the intended geometric configuration (the diagram is not such a means), make the configuration itself a graph—that is, a discrete network of lines.[63]

We may proceed a step further in this argument and observe that, if the geometric configuration is a graph and if the proof is carried out using indefinite noun phrases, quantification becomes irrelevant both within a proof and, in the metadiscourse, as a way to secure generality by means of the standard UE-IE transition from $\forall xF(x)$ to $F(a)$ and vice versa, with a a generic representative of the class on which the quantifier operates.[64] A "Greek" theorem requires to, and very effectively does, prove $F(a)$, that is all.

But then, if the issue of the indefinite character of the linguistic items featuring in a Greek proof can now be said to be settled once and for all, much more arduous is to draw positive conclusions about the ontological status of the objects such items refer to. Granted, we may now claim that, contrary to Neo-Kantian beliefs, particular objects have finally set behind the horizon of Greek geometry, but we are not thereby forced to accept that Greek geometry deals with Kit Fine-style arbitrary objects: in this view, the indefinite expressions figuring in the enunciation and in the setting-out pick up any arbitrary object that satisfies the indefinite description to be associated, possibly by definition, with that expression.[65] This might well be acceptable, but an alternative

[63] Ontologically, this amounts to saying that Greek geometric lines, as mathematical objects, are not made of the points on them (cf. Sect. 3.4 and 5.1.1). In a semiotic approach, diagrams can only be graphs: see Netz 1999, sect. 1, and Lattmann 2018, 114–115.

[64] Just see Lemmon 1990, 104–109, who, in providing his Euclidean example at 106–107, expressly relies on the standard translation. Of course, the UE-IE transition is the transition from the enunciation to the proof and vice versa; the presence of a "conclusion" is immaterial. Quantification is irrelevant *within* the proof because only a finite (and very limited) number of objects is there present. This fact is the gist of the Tarskian proof that a suitable formalization (most importantly, it does not take constructions into account) of elementary geometry is decidable: see Tarski 1951 and 1959; Moler and Suppes 1968 extended Tarski's system to cover the Euclidean constructive postulates. The fact that the Euclidean diagram is a graph also lies behind the recent wealth of successful axiomatizations of Euclidean-style constructive acts. See for instance Mäenpää, von Plato 1990; Mäenpää 1997; Pambuccian 2008; Beeson 2010; Gulwani, Korthikanti, Tiwari 2011; read also Gardies 2003. An informal approach along similar lines is developed in Sidoli 2018b, which also puts forward fanciful reasons for distinguishing between "problem-constructions" and "proof-constructions" (the non-fanciful part of this paper is pure compilation of the Italian version of the present book and of a couple of studies cited above). The technical papers just mentioned (to which Avigad, Dean, Mumma 2009 can be added) claim to formalize the act of "inspecting the diagram" in order to spell out "precisely what inferences can be 'read off'" from it (ibid., 701), while as a matter of fact they simply formalize (a suitable extension of) the Euclidean postulates.

[65] It is in this way (or better, as "representative objects") that Hussey 1991 appears to make Aristotle construe the mathematical objects of Greek geometry—on the basis, however, of the *vulgata* translation of the setting-out (cf. 126). See also Mueller 1970. My reference in the text is to Fine 1985; see also the very effective Fine 1983. The mathematical objects resulting from the semiotic approach to Greek diagrams inevitably are generic objects (see e.g. Lattmann 2018, 122–123)—actually, they are simply an avatar of Platonic forms. However, the scholars advocating such an approach do not

point of view is suggested by the fact that a proof works with second-order linguistic objects used as abbreviations of designations, namely, within the universe of discourse. This, as well as the care in maximizing the indefinite character of the entire proposition, suggests to see an ontological (de-)commitment of the kind underlying the modern approach in terms of substitutional quantification as partly underlying the most distinctive style of Greek mathematics, too.[66]

This said, clues about the ontological stance of the Greek mathematicians in the matter of mathematical objects can be extracted from other pieces of evidence. Before doing this, however, I shall briefly present evidence about the categorial status of ultimative and non-ultimative geometric species according to philosophical sources. As usual, the starting point of the whole chain is Aristotle.[67] He held that any kind of figure, like a triangle, a square, or a circle, belongs to the category of quality, together with other attributes of mathematical objects like εὐθύτης "straightness" and καμπυλότης "curvature". Neoplatonic exegesis, in particular Proclus and Simplicius, adopted a standard syncretistic approach, holding that, for instance, triangles partake of quantity (because they are said to be equal or unequal) and of quality (because of their shape, in virtue of which they are said to be similar or dissimilar).

Angles are a tricky geometric species: Aristotle's position is not clear, but the Peripatetic tradition, and in particular Eudemus, took them to be qualities. The definition of an angle in I.def.8 makes it a subspecies of κλίσις "inclination" and hence a relation, even if angles can obviously be said to be equal or unequal and can be divided, and hence is a magnitude, and even if an angle is a shape of sorts, thereby partaking of quality. Of course, any of these categorizations can be shown to be inadequate; the solution of the exegetes was the same as above: an angle is a combination of these three categories. According to Proclus, Apollonius defined an angle as a "contraction [συναγωγή] of a surface or of a solid into one single point under an inflected [κεκλασμένη] line or surface", hence possibly a relation, but just afterwards Proclus himself asserts that Apollonius, as well as Proclus' teacher Plutarch, held that an angle is "the first interval under the point" (hence a quantity), a definition that has a manifest Epicurean ring. This dangerous feature is immediately neutralized by Proclus, who first develops his opponents' argument: "for there must be, he [scil. Plutarch] says, some first interval under the inflection of the containing lines or surfaces", and then counterargues as follows: "yet since the interval under the point is certainly continuous, it is impossible to take the first one, for every interval is indefinitely divisible. Add that, even if we were able to determine in some way the first one and we draw a straight line through that, a triangle results, and not a single angle".

Lines are not figures. In the next Section, we shall extract some information on their status by reviewing the several ways they are generated in our sources. The rest of the Section will present some amazing Archimedean material.

seem to realize that these objects, unless they are conceived e.g. *à la* Fine, immediately run into the nest of contradictions first pointed out by Berkeley.

[66] The second occurrence of "underlying" must be taken *cum grano salis*: here we are just moving within an interpretive framework, after all. The best account of substitutional quantification is still Kripke 1976; see also Hand 2007. The context dependent quantifiers approach advocated in King 1991 also allows disposing of arbitrary objects, and bears some resemblance with ancient mathematical practice as here reconstructed.

[67] See the overview in Wilck 2020, 380–381, who also sets forth stylistic reasons for believing that the underlying metaphysics of the *Elements* takes many geometric species as substances, and Acerbi 2010b, 161–162 for Apollonius' conception of an angle, and 160 for his clarifications of the conceptions of a line and of a surface. Aristotle's texts on the categorial status of geometric species are *Cat.* 8, 10a11–15, and consequently *Metaph.* Δ.14, 1020a35. The main source for the ancient conceptions of an angle (which I have partly summarized) is the long doxographic survey in Proclus, *iE*, 121.12–126.6 (the quoted passages are at 123.16–17, 125.15–16, and 125.18–24, in this order). Simplicius' categorization of a triangle, whose motivations coincide with Proclus' at *iE*, 123.24–124.2, is explained at *in Cat.*, 153.3–5.

3.4. Ontological commitment

We may distinguish three approaches to the ontological problem: a naive metaphysical stance, a well-informed stance, and an operational stance. The first and the third stance manifest themselves by opinions or acts whose unquestionable metaphysical import is not presented as such by the relevant actors; these actors may be unknowledgeable about the detail or even the existence of the metaphysics they implicitly adhere to: we may be Platonists without knowing who was Plato. It is a fact that the naive metaphysics of most contemporary mathematicians is outright realism: the mathematical objects exist and mathematical activity consists in discovering them and their properties. Some of these mathematicians, however, are well-informed realists.[68]

The species and subspecies of ontological stance in the matter of mathematical objects have exploded in number with the development of mathematical logic and analytic philosophy. Greek antiquity contributed already to this proliferation. It is simply impossible to try to summarize the main Greek ontological doctrines about mathematical objects[69]—but, in a nutshell: early Pythagoreanism held that numbers are the essence of everything; arithmetic thereby gets an epistemic status higher than geometry. Plato maintained that the *mathemata* are immutable, actually existing entities, and that their status is intermediate between the forms and the sensible world. Plato's pupil Speusippus denied existence to forms but assigned the *mathemata* a form of existence separated from physical reality. Aristotle held that the mathematical objects do not exist separately from the physical objects: mathematical objects somehow coincide with physical objects, some features of which are projected out by "removal" (ἀφαίρεσις). The Stoics probably granted mathematical objects the same (weak) existential status as they accorded to the four basic incorporeals: time, place, void, and propositional contents. We must not forget that Aristotle is the sole author who offers a sustained discussion of his own view. He is even the main or unique source for some of the other views just mentioned.[70] I shall now turn to Greek mathematics and discuss two examples of operational metaphysical stance and one example of naive metaphysical stance.

My first example of operational ontology is implicit in the way Greek mathematicians defined lines. For lines are the real objects Greek mathematics deals with: the way in which lines are defined demonstrates its appreciation of the mode of existence of mathematical entities.[71]

The Greek geometric corpus does not contain a general, independent characterization of a geometric line as a mathematical genus on which some ordering principle is imposed; even the designation "curved line" (καμπύλη γραμμή) fell out of usage after Apollonius. Each single curve was defined or identified as a mathematical object on the basis of a variety of techniques:

[68] Such is the position of the outstanding contemporary mathematician and Fields Medalist A. Connes, expressed for instance in Changeux, Connes 1989. As customary with him, K. Gödel formulated his Platonist views in a radical and concise fashion: see Gödel 1944, 137–138, and Gödel 1983, 483–485. For an overview of the modern conceptions of mathematical objects, see Shapiro 2005.

[69] It is also simply impossible to provide a satisfactory overview of the scholarship on the issue. Here is a title for each stance, also or mainly for the bibliographical record they contain: early Pythagoreans: Burkert 1972; Plato: Pritchard 1995; Speusippus: Tarán 1981 (to be checked against Mueller 1986); Aristotle: Pettigrew 2009 (to be compared with Mueller 1970); Chrysippus: Robertson 2004.

[70] As for Plato, see the divided line passage at *Rsp.* 509D–511E; his statement at *Rsp.* 527A–B that the operational language of the mathematicians is "most ridiculous yet necessary"; *Euthd.* 290B–C, implying that the *mathemata* have an objectual existence; and finally *Phlb.* 55D–57E. Aristotle expounds his and others' conceptions in *Metaph.* A and M–N. He mainly deals with the issue of the existence of numeric entities (but geometry is specifically treated in M.1–3), for which the ontological problem is obviously more urgent, and which is discussed in all competing doctrines. Aristotle's conception of mathematical objects is strictly intertwined with his very refined theory of the continuum.

[71] See Acerbi 2010a, 13–19, on the ancient approaches to the problem of defining curves; I use here part of the discussion I have presented in that article.

1) Constructions by intersection. These normally amount to cutting a surface with a plane. In this way were defined the conic sections (*Con.* I.11–13) and the toric sections.[72] More generally, surfaces, in their turn usually generated by rotation of plane figures [conic, cylindrical, toric surfaces: this is mode of generation (3) below], can be intersected. This is the case of the curves implicit in Archytas' method for solving the problem of duplication of a cube.[73]
2) Pointwise geometric constructions: a construction by intersection of suitable lines yields isolated points that lie on the curve; the curve is then approximated by joining these points by line segments or by arcs of other known lines. Pointwise constructions are for instance employed by Diocles to generate the curve (which in modern times came quite improperly to be called "cissoid") by means of which he solves the problem of duplication of a cube,[74] and to identify the parabola as the curve that enjoys the focus-directrix property.[75]
3) "Mechanical" constructions, in which some geometric object, for instance a straight line, is allowed to move, the curve being generated by the motion of a suitable point on it. Of such a kind are the definitions of most higher curves, such as: (*i*) all kinds of spiral and of helix,[76] (*ii*) Nicomedes' conchoid,[77] (*iii*) the quadratrix,[78] (*iv*) very likely other special curves of which we only have the name.[79] Of such a kind are also the definitions of the solids of revolution in *El.* XI.def.14, 18, 21 (sphere, cone, and cylinder, respectively). Some sources suggest that attempts were made to replace the "mechanical" generations with other methods perceived as more geometric. Pappus, *Coll.* IV.51–52, mentions such attempts with approval, showing that the quadratrix can be obtained from the plectoid surface (a screw) generated by the cylindrical helix, after intersection with a suitable plane and subsequent orthogonal projection on the base plane.
4) Setting forward a property that univocally identifies the curve as a locus (like the definition of a circle at *El.* I.def.15); in ancient technical jargon, this is a σύμπτωμα "characteristic property" (see Sects. 2.4.1 and 5.1 for detail).[80]
5) There is only one example of a σύμπτωμα identifying a whole class of lines that did not receive, as a class, a generative definition: these are homeomeric lines, defined as follows "A line is

[72] Proclus, *iE*, 111.23–112.15. These curves were studied by some Perseus.
[73] Eutocius, on the authority of Eudemus, in *AOO* III, 84.12–88.2. On Archytas' construction, see most recently Masià 2016; Menn 2015 is just a heap of groundless conjectures.
[74] Excerpt in Eutocius at *AOO* III, 66.8–70.5, Arabic translation at Toomer 1976, 97–113 = *CG*, 131–141. All constructions of curves here mentioned are thoroughly discussed in Knorr 1986, sects. 4–6; see also Tannery 1883–84.
[75] Only in Arabic translation, at Toomer 1976, 63–71 = *CG*, 112–116.
[76] The plane spiral is defined and studied—after a suggestion by Conon, if we are to believe Pappus, *Coll.* IV.30—in Archimedes' *Spir.*; an abridged exposition is in Pappus, *Coll.* IV.31–38; the spherical helix is described ibid., IV.53–55. The Archimedean spiral is generated by a point that translates uniformly on a uniformly rotating ray.
[77] Generation and application of the curve to solve the problem of duplication of a cube in Eutocius, *AOO* III, 98.1–106.24, and in Pappus, *Coll.* IV.39–44.
[78] Generation, criticisms thereon by Sporus, and application of the curve to solve the problem of squaring the circle in Pappus, *Coll.* IV.45–50; at IV.45 he claims that the curve was used by Dinostratus, Nicomedes, and "some other moderns". The quadratrix is obtained as the intersection between two adjacent sides of a square set in uniform motion, one translating parallel to itself, the other rotating around the extremity that initially is not in common with the first side. The two motions are synchronized: at the end of the process the two sides are to coincide. Sporus notes first that synchronizing the component motions entails rectifying the circumference, the very problem the quadratrix was set out to solve. Second, he remarks, the final extremity of the quadratrix, whose position is what is really needed to solve the same problem, cannot be obtained by intersection, since at that stage of motion the two sides coincide. This extremity can only be obtained by producing the curve (by continuity, we should say), which again entails supposing the problem solved.
[79] See for instance Pappus, *Coll.* IV.58 (who mentions Philo of Tyana and Demetrius of Alexandria's *Linear Investigations*), and Simplicius, *in Phys.*, 60.10–16 = *in Cat.*, 192.18–24, quoting Iamblichus.
[80] The curve may lie in a plane ("plane" loci, as in Apollonius' *Plane Loci*, described in Pappus, *Coll.* VII.21–26, with lemmas thereon at VII.185–192; "solid" loci, as in Aristaeus' *Solid loci*; and "linear" loci) or on a surface ("loci on a surface", as in Euclid's eponymous treatise, cf. Pappus, *Coll.* IV.51–52, VII.3, 312–318). The main source for ancient loci is Pappus' *Collectio*; see Appendix B.

homeomeric which has every part coincident with every part equal to it, whenever the parts have the concavity on the same side".[81]

This quick survey shows that the ontological stance underlying the conception of such central mathematical objects as lines cannot be naive realism: a line token may exist as a boundary (a section),[82] as a subset of a surface singled out by a characteristic property, as a subset of the general class of lines (!) singled out by a characteristic property, as something generated by the flux of a point.[83] Finally, a line may exist because we are able to find isolated points on it by intersecting suitable lines. As only a finite number of points can be generated in this way, one might well wonder what is the ontological status of such curves. Recall in this connection that, as we have noted at the end of Sect. 3.3, and as is confirmed by the Greek conception of a geometric locus (cf. Sect. 5.1.1), a line is not composed of the points on it.[84]

Imposing an ordering principle on the variegated population of lines entails classifying the corpus of problems in which these lines enter as necessary mathematical tools. In its turn, this entails fixing the rules of the game, namely, setting limitations on the very techniques of solution to problems. According to the standard scheme we find in ancient sources, in fact, problems were classified as "plane" if their solution involved only straight line and circle, "solid" if their solution involved conic sections, "linear" if any other curve was needed.[85] It was obvious to ancient geometers that any plane problem could also be solved by solid methods etc., and quite clear to them that any single problem falls essentially into a single category, if employing a minimal set of mathematical tools is required—even if no proof is given of this.[86]

The only Greek mathematician who gave voice to his naive and operational metaphysics is Archimedes.[87] His naive metaphysics about mathematical objects is mainly expounded in the prefatory epistle of *Sph. cyl.* I.[88] Archimedes takes occasion when he presents the results contained in this treatise to Dositheus, his addressee. For unknown reasons, the philosophical core of the epistle amounts to a long dialectical entrenchment in which Archimedes declares that the main results of *Sph. cyl.* I first just "occurred" (ὑποπεσόντων) to him, and only afterwards he put himself to work on their proofs. Archimedes then lists these results, and points out the simplicity of the involved properties, which manifest itself in the συμμετρία "simple numeric relations" occurring between the figures he studied. Now, these properties "already subsisted by nature" (τῇ φύσει προυπῆρχεν) in the said figures, still they "passed unnoticed by previous mathematicians" (ἠγνο-εῖτο δὲ ὑπὸ τῶν πρὸ ἡμῶν). The importance of these results is stressed by an analogy with some of Eudoxus' major achievements.[89] However, an assessment of the validity and importance of the results is deferred to the mathematical community (τοῖς περὶ τὰ μαθήματα ἀναστρεφομένοις).

---

[81] See Acerbi 2010a, especially 7 for the (reconstructed) definition.

[82] Cf. I.def.6, which, however, is not a definition but a characterization that coordinates the notions of line and of surface (cf. Sect 1.5). For mathematical objects as boundaries in Chrysippus, see Robertson 2004.

[83] Some of these definitions were compiled or debated in antiquarian or philosophical sources (see Dye, Vitrac 2009). A line is a flowing point in Theon of Smyrna, *Exp.*, 83.22–23, on the authority of Eratosthenes; same definition in Sextus, *M* III.19, with refutation at III.29–36 (cf. III.71–76); Sextus again has a line defined as the boundary of a surface at *M* III.20, refuted at III.60–64. A definition of line as the flux of a point is also compiled in *Def.* 1 and 2, at *HOO* IV, 14.21–23 and 16.2–3, respectively; the latter is followed by a clarification that originates in Apollonius (Acerbi 2010b, 160).

[84] We shall also see in Sect. 5.1.2 that the Greek geometers resisted taking random points in a plane.

[85] See Pappus, *Coll.* III.20–21 and IV.57–59; a similar classification of loci is expounded at *Coll.* VII.22; again, a similar classification of *neuseis* is at VII.27.

[86] See *Coll.* III.21 and IV.59; the normative character of Pappus' prescription is likely to originate with Apollonius.

[87] But Apollonius carried out extensive research on foundational topics: Acerbi 2010b again. I use here Acerbi 2013c.

[88] The passages I summarize are at *AOO* I, 2.6–8, 2.19–4.5, 4.13–21.

[89] These achievements we now read as *El.* XII.10 and XII.7por.

These statements obviously point to a realist position as to the ontological status of mathematical objects: the properties of such objects are inherent to them; a mathematician perceives these properties by means of some form of intuition: Archimedes always employs the verb νοέω "to conceive" to this effect. On the contrary, the proof techniques undergo a process of evaluation among peers, thereby amounting to a shared practice.[90]

The prefatory epistle of the *Method* adds a further dimension to Archimedes' reflections.[91] This celebrated treatise contains a detailed description, in terms of worked-out examples, of the mechanical "procedure" (τρόπος) that enabled Archimedes to discover some of his most outstanding results. In the preface to the *Method*, Archimedes sharply demarcates heuristic and proof, by asserting that these results (θεωρήματα) first appeared to him by mechanical means, but then he proved them geometrically because establishing them by means of his procedure lies outside the boundaries of probative techniques.[92]

Can we make realist Archimedes a strict Platonist? We cannot, because his astounding operational metaphysics, again to be found in the *Method*, forbids us to do this.

In the *Method*, Archimedes employs three different argumentative techniques. Two of them, the celebrated procedure by means of a virtual balance and an infinitary method in *Meth.* 14, are deemed "mechanical" by Archimedes himself, and contrasted with his geometric proofs. The third technique is an application, in *Meth.* 15, of the method of exhaustion.[93] We are only interested in the mechanical procedure by means of a virtual balance. Its argumental structure can naturally be divided into self-contained deductive units. To show this, I take *Meth.* 4 as an example; this proposition argues that a segment of paraboloid is $\frac{3}{2}$ of the cone inscribed in it.[94]

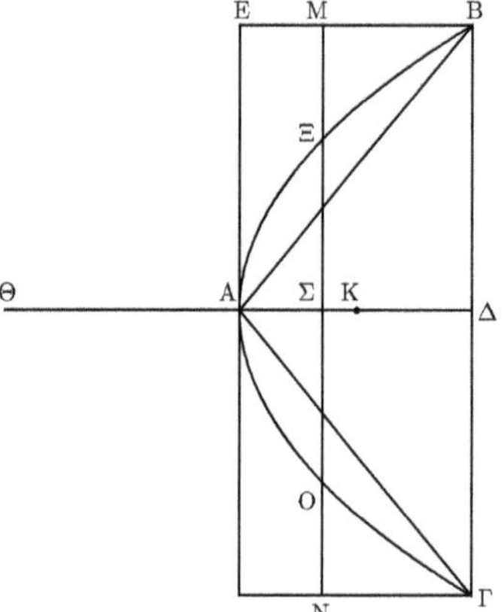

(1) *Construction.* Let there be a segment of paraboloid cut by a plane giving a segment of parabola ΑΒΓ of vertex Α, axis ΑΔ, and base ΒΓ as a section. Associated virtual balance: one harm of the beam ΑΔ coinciding with the paraboloid's axis, pivot in vertex Α, other harm ΘΑ equal to the paraboloid's axis. Cone ΑΒΓ inscribed in the paraboloid, cylinder ΒΕ{Ζ}Γ circumscribed to it. The sections to

[90] Archimedes refers again to Eudoxus and to the mathematical community when he has to advocate in favour of his own adopting the principle now known as "Archimedes' lemma". He does this, in slightly different terms, in the prefatory epistles of *Sph. cyl.* I, *Quadr.*, and *Spir.* See in particular the historical outline Archimedes sketches in the preface to *Quadr.*, at *AOO* II, 262.13–264.26.

[91] Read in particular *AOO* II, 428.26–430.1.

[92] This paraphrases τινα τῶν πρότερόν μοι φανέντων μηχανικῶς ὕστερον γεωμετρικῶς ἀπεδείχθη διὰ τὸ χωρὶς ἀποδείξεως εἶναι τὴν διὰ τούτου τοῦ τρόπου θεωρίαν. But the remark at the end of *Meth.* 2 (*AOO* II, 446.4–15) suggests that the mechanical procedure lies somehow mid-way between heuristic and proof: τούτου τεθεωρημένου, διότι πᾶσα σφαῖρα τετραπλασία ἐστὶ τοῦ κώνου τοῦ βάσιν μὲν ἔχοντος τὸν μέγιστον κύκλον, ὕψος δὲ ἴσον τῇ ἐκ τοῦ κέντρου τῆς σφαίρας, ἡ ἔννοια ἐγένετο ὅτι πάσης σφαίρας ἡ ἐπιφάνεια τετραπλασία ἐστὶ τοῦ μεγίστου κύκλου τῶν ἐν τῇ σφαίρᾳ· ὑπόληψις γὰρ ἦν καὶ διότι πᾶς κύκλος ἴσος ἐστὶ τριγώνῳ τῷ βάσιν μὲν ἔχοντι τὴν τοῦ κύκλου περιφέρειαν, ὕψος δὲ ἴσον τῇ ἐκ τοῦ κέντρου τοῦ κύκλου, καὶ διότι πᾶσα σφαῖρα ἴση ἐστὶ κώνῳ τῷ βάσιν μὲν ἔχοντι τὴν ἐπιφάνειαν τῆς σφαίρας, ὕψος δὲ ἴσον τῇ ἐκ τοῦ κέντρου τῆς σφαίρας. Thus, heuristic here reduces to seeing the right analogy between a circle and a sphere. *Meth.* 2 is rigorously proved in *Sph. cyl.* I.34 and porism. The result that came to Archimedes's mind once he had established the one in *Meth.* 2 is rigorously proved in *Sph. cyl.* I.33.

[93] As a consequence, in *Meth.* 12–15, three different arguments are mobilized to establish one and the same result (the quadrature of the so-called "cylindrical nail"). Only the method of exhaustion is a rigorous technique, according to ancient and modern standards.

[94] The Greek text is at *AOO* II, 454.9–458.18.

be weighed are parallel to the base circle of the paraboloid. A generic section is: in the cylinder, a circle of diameter MN; in the segment of paraboloid, a circle of diameter ΞO. Both diameters lie on a straight line MN that cuts the axis of the paraboloid at Σ, which is the barycentre of the circles.

(2) *Deduction of the fundamental proportion between sections.* Because of the basic property of the parabola,[95] and since ΔA = AΘ, one has ΘA : AΣ :: q(MΣ) : q(ΣΞ). Passing from radii to diameters and by *El.* XII.2, one gets ΘA : AΣ :: c(MN) : c(ΞO).

(3) *Equilibrium between remaining and transferred sections.* By the fundamental law of equilibrium applied to plane objects (they equilibrate with one another at distances from the pivot that are inversely proportional to their surface contents), the last proportion in the previous step entails that the circle of diameter MN, while remaining in its own place, equilibrates, with respect to point A, with the circle of diameter ΞO, transferred (the operation is expressed by forms of μεταφέρω) on the other harm of the beam of the balance at Θ, so as to be Θ its barycentre.[96]

(4) *Generalization to all sections.* By means of a standard "potential proof" (see Sect. 4.5.2), the results attained in units (2) and (3) are generalized to any section of the cylinder and of the paraboloid cut off by one and the same plane.

(5) *Recomposition of the figures and their equilibrium.* The figures are recomposed starting from their circular sections. The recomposed cylinder, any of whose sections remains in its own place, coincides with the cylinder set out at the beginning. The segment of paraboloid is not reconstructed around its new barycentre Θ, but it is transferred there "disarticulated": the transferred circles that originally compose the paraboloid are superposed to one another in such a way that the barycentre of each of them—and hence of their aggregate—coincides with point Θ. Since the corresponding circular sections of the cylinder and of the paraboloid equilibrate with one another in the said positions, and since the paraboloid and the cylinder are composed of these circular sections, therefore the (disarticulated) paraboloid and the cylinder also equilibrate with one another.

(6) *Fundamental proportion between the figures and deduction of the required property.* By the fundamental law of equilibrium applied to solid objects, and since the barycentre of the cylinder is the middle point K of AΔ and the barycentre of the disarticulated paraboloid is Θ, equilibrium entails ΘA : AK :: (cylinder) : (segment of paraboloid). But ΘA = 2AK; therefore the circumscribed cylinder, which is three times the inscribed cone, is twice the segment of paraboloid; therefore the segment of paraboloid is $\frac{3}{2}$ of the cone.

The way the sections are operated upon in the Archimedean procedure just exemplified entails a stance as to their nature. Let us see in detail how this can be argued.

(a) *The sections.* The weighing procedure operates on sections of the intended figures that are not homogeneous to the figures of which they are sections: these are plane sections of solids and linear sections of plane regions. Archimedes expressly asserts that the intended figures are "compounded" (verbs συνίστημι or σύγκειμαι) of their sections (*Meth.* 1 and 14). Thus, to Archimedes a geometric entity such as a solid or a plane region is made of lower-dimensional entities.

[95] We read it at Apollonius, *Con.* I.20. The signs q(AB) and c(AB) denote the square on straight line ἡ AB and the circle having the same diameter, respectively. Theorems of *El.* V must also be invoked to make Archimedes' argument run, but they are basic stuff when proportions are manipulated.

[96] Step (3) is followed by a deduction ending with a proportion that repeats the one obtained in step (2), no supplementary information being fed in or deduced. This useless paragraph figures in *Meth.* 3–6 and could possibly not be added by a reviser; its presence is a major deductive flaw of Archimedes' treatise.

(b) *The "disarticulation"*. One of the figures decomposed into sections remains in its own place together with all of its sections (this amounts to a virtual decomposition, discharged in the course of the argument), the other figure is transferred on the other harm of the beam of the balance. The tricky point is the way in which such a transfer is carried out: as seen above, the segment of paraboloid of *Meth.* 4 is not reconstructed as such around its own new barycentre Θ, but it is transferred there disarticulated; the transferred circles that originally compose the paraboloid are superposed to one another in such a way that the barycentre of each of them—and hence of their aggregate—coincides with Θ. Thus, what equilibrates with the figure that remains on the other harm of the beam of the balance is not a geometric figure,[97] nor a fortiori the original figure decomposed, transferred, and recomposed, but a mathematical artefact generated by the procedure and necessary to its being conclusive. From the mathematical point of view, the trick of disarticulation performs the function of neutralizing the problem of determining the barycentre of the transferred figure. This step is inevitable if the transferred figure has to be reconstructed in its original form, but it gives rise to a circular argument. In fact, the Archimedean procedure can indifferently be applied to determine the content of a figure or its barycentre, but any of these results depends on the other.[98] For the purposes of the argument, thus, the transferred figure is concentrated in its own barycentre.

Archimedes is pretty explicit in his formulation of the procedure, and in fact plays dirty with disarticulation. He employs verbs such as "to complete", "to fill up" (συμπληρόω) when he refers to the recomposition both of the figures that remain in their own place and of the figures transferred in disarticulated form. If these verbs, insofar as they are applied to the figures that remain in their own place, may well be taken as mere linguistic stipulations, the same lexicon applied to the transferred figures is at best misleading. Tertium non datur: either the lexicon is appropriate to the figures that remain in their own place, or it is appropriate to the transferred figures. Add to this that Archimedes asserts that the transferred figures are "transferred and placed on the balance" as such, not as aggregates of their sections.

(c) *Proliferation of figures*. The disarticulation allows Archimedes to operate freely on his figures: the transferred figures may be parts of the figures that remain in their own place; some figures proliferate before being transferred; in two instances, one of them remains in its own place but—oh miracle!—it is also transferred (*Meth.* 6 and 9). This happens because Archimedes is forced to graft the mechanical part of his procedure on the fundamental proportion between sections [see step (2) above]. If, as happens in *Meth.* 6 and 9, he gets $\Theta A : A\Sigma :: [c(MN) + c(\Xi O)] : c(\Xi O)$ instead of just $\Theta A : A\Sigma :: c(MN) : c(\Xi O)$, the only strategy Archimedes can adopt is duplicating the circle $c(\Xi O)$, leaving one of them in its own place and transferring the other disarticulated on the other harm of the beam of the balance.

Interpenetration of figures may occur without connection with disarticulation. In *Meth.* 9, two cylinders M and N are placed on the same barycentre and their sum MN is again called "cylinder". Of course, no diagram can represent this configuration; the two cylinders are depicted one by the side of the other both in the palimpsest, f. 41v, and by Heiberg.

[97] In *Meth.* 4, this would coincide with a circle equal to the base of the paraboloid, and hence of the cylinder; as a consequence, a cylinder would equilibrate with its own base. Of course, all of this reminds us of Democritus' dilemma of the cone and of Chrysippus' solution thereto; see again Robertson 2004. See also the argument in Sextus, *M* III.66–70.

[98] The pairs *Meth.* 4/5 and 7/9 (content and barycentre of a segment of a paraboloid and of a sphere, respectively) show that only small modifications are required to transform one proof into the other. Archimedes chose the ordering in which the former result is required to establish the latter.

(d) *What kind of conception?* Thus, we cannot make realist Archimedes a strict Platonist simply because, in the entire Greek mathematics, there are no geometric objects less Platonic than those Archimedes operates upon in the *Method*. For we must sharply differentiate the destructuring Archimedean actions and the operational practice, eminently constructive and cut-and-paste, that characterizes mainstream Greek geometry. In fact, the operations on objects that are not codified in the first three postulates of the *Elements* mainly consist in translations of objects and in cutting-and-pasting. Basic operations like the translation of a line segment and of an angle (I.2–3 and I.23, respectively) belong to the first family. The second family includes the fundamental theorems on the equivalence of plane rectilinear regions and the problems of parabolic application of areas (I.35–45), as well as the combinatorics of regions underlying all theorems of Book II.[99] The cutting-and-pasting technique operates on regions and on subregions of them that are dimensionally homogeneous to one another and to the original regions. The regions are usually provided as partitioned in subregions, which in their turn come to be by means of basic operations of sectioning. In this game, the lower-dimensional sections (usually straight lines cutting rectangles) are actualized only as boundaries of homogeneous regions, and never figure in the proof. Some operations in the *Method* can be included in this category of harmless manipulations. Among these are for instance removing two cones from the equality "three identical cones = two cones identical to one another and to the former three + two spheres", so that any of the cones is equal to the two spheres (*Meth.* 2); or the argument in *Meth.* 6, according to which, from the two "sides" of an equilibrium relation (this is not an equality!) between a hemisphere and a cone remaining in their own places and the same cone transferred on the other extremity Θ of the beam of the balance [as in point (5) above], the remaining cone and $^3/_8$ of the transferred cone (which in their turn equilibrate with one another) are removed, so as to get a new equilibrium between the hemisphere and the residual $^5/_8$ of the transferred cone. Such a practice does not affect the nature of the manipulated objects (thereby conflicting with a Platonic view of them), but pertains to the operational approach of the mathematician:[100] it is a stylistic, not an ontological stance.

But after all, maybe we should resist the temptation of assigning too modern a conception to Archimedes. It is a plain fact, on the one hand, that he sees his solid and plane objects as respectively composed of plane and rectilinear sections. It is also true, on the other hand, that lines constitute an insurmountable inner bound, insofar as a "Greek" line is not composed of the points on it.[101] Thus, maybe we are not entitled to think that the Archimedean manipulations are grounded on an insight into the "point content" of a regular geometric region such that every such region can produce—without "emptying" itself—an arbitrary number of replicas of itself, or such that the point content of any region is invariant under transformations as those occurring in disarticulation. Still, it is difficult not to think of the Banach-Tarski paradox when reading the *Method*:[102] give me a sphere to stand, and I shall create the world …

---

[99] Hero advocated an alternative approach to these theorems, as we have seen in Sect. 2.4.1.

[100] As we have seen, even Plato regarded this approach as "necessary".

[101] This was noted at the end of Sect. 3.3, and it is clearly borne out by the Greek conception of a geometric locus: see also Sect. 5.1.1.

[102] The Banach-Tarski paradox is in fact a theorem: given any two bounded sets A and B in three-dimensional space, each having nonempty interior—stated more simply and less cryptically: take any two spheres—one can partition A into *finitely many* disjoint parts and rearrange them by rigid motions to form B. Of course, the chunks of the decomposition can only be *very* weird point sets; actually, the Axiom of Choice is almost necessary to get them. On this theorem, see the comprehensive monograph Tomkowicz, Wagon 2016.

3.5. Oversymmetrized Diagrams

The diagrams in Greek manuscripts are usually more symmetric than the geometric configurations they are intended to represent; this phenomenon is usually called "overspecification",[103] but it should more properly be termed "oversymmetrization". Thus, in medieval manuscripts, triangles intended to be generic are almost always represented as isosceles, points taken at random on a segment actually bisect it, parallelograms are represented as rectangles. As seen, in the case of our paradigmatic proposition III.2, all manuscripts make straight line ἡ ΔΖΕ bisect both arc ἡ ΑΒ and the straight line joined from τὸ Α to τὸ Β (which is unnecessarily represented as an arc of a circumference), place letter Γ exactly opposite to Ζ, and make the whole diagram left-right symmetric. Again, the main diagram of manuscript **B** even makes radii ἡ ΔΑ and ἡ ΔΒ belong to the same diameter.[104] This phenomenon calls for an explanation insofar as we are used to think that oversymmetrized diagrams may in some sense harm the generality of a proposition—and in fact all diagrams of Heiberg's edition of the *Elements* are tacitly redrawn in order to have them de-oversymmetrized. At least two explanations can be envisaged.

First, the diagrams were oversymmetrized in the original treatises, maybe as a way to make, by contrasting an oversymmetrized figure with the real configuration it is intended to represent, the general character of a mathematical proposition more manifest, or maybe because this was the standard graphic code.[105] I must say that I now find the first possibility quite implausible: I cannot see any reason why Archimedes would like to have his diagrams oversymmetrized. On the contrary, in Greek mathematical *texts* we find an opposite phenomenon of "underspecification": geometric objects may happen to be designated by expressions that make them more generic than they are; a case in point is Archimedes' denomination παραλληλόγραμμον (ὀρθογώνιον *Stom.*) for a square in the *Stomachion* and in the *Method.*[106] As for symmetric diagrams constituting a primary graphic code, this is not an explanation, but simply a way of stating the evidence.

I regard a second explanation as more plausible. The goal of medieval copyists was to reproduce as faithfully as possible the diagrams they found in their models. Diagrams in different manuscripts are often so stunningly identical as to suggest that some copyists contrived conformal reproductions of the figures by employing suitable devices or simple tricks (such as superposing the sheets of parchment before a source of light). Now, the successive steps of the manuscript transmission of a figure can quite obviously be modelled as the evolution of a dynamical system: what evolves, by discrete time-steps, is the form of the diagram, which any act of copying modifies in a more or less appreciable way. Now, if external constraints are not imposed, the forms of a diagram that enjoy

---

[103] This phenomenon was well-known well before it became a fashionable research theme (for instance, see the clear statement in Jones 1986, 76). For the manuscript evidence see Saito (2006) and, for a more general assessment, Saito, Sidoli (2012). The explanation I propose was first argued in Acerbi 2017, 244–246, which I use here.

[104] This manuscript sometimes contains several replicas of the diagram associated with a proposition (see the image in Sect. 4.5.1.3). One of these is the "original" diagram; it is located, as usual, in an indentation of the text. The replicas are found in the margins. At least one of these replicas, sometimes accompanied by the inscription "this is the most accurate", is a de-oversymmetrized version of the original diagram. In the case of III.2, de-oversymmetrization amounts to make ἡ ΑΔΒ a line inflected at τὸ Δ (f. 44v, external margin, upper half of the page). All these diagrams in **B** (copied in 888) are drawn by the main copyist, who reproduced those in his model; thus, the campaign of regularization of their symmetry properties took place in late antiquity.

[105] I have suggested the first proposal in Acerbi 2007, 296. Saito and Sidoli (2012, 143) favour the idea of a stabilized graphic code: "[f]or us, an irregular triangle is somehow a more satisfying representation of 'any' triangle, whereas for the ancient and medieval mathematical scholars an arbitrary triangle might be just as well, if not better, depicted by a regular triangle". Of course, the medieval mathematical scholars may only perpetuate this graphic code, for early manuscripts were intended for conservation and entrusted to professional copyists, who simply reproduced their models.

[106] See *AOO* II, 418.5, 426.11.24, 428.2–3.

additional symmetries work as points of stable equilibrium in such an evolution.[107] An external constraint may be, for instance, a mismatch between the diagram and the indentation reserved to it, making it necessary to deform the diagram (figures are usually added after the text was written).[108] What makes oversymmetrized diagrams work as points of stable equilibrium is the obvious perceptional and psychological mechanism that makes a limiting case to be perceptually (and hence graphically) more significant than a less limiting one: an isosceles triangle versus a scalene triangle, a diameter versus a generic chord, the middle point of a segment versus a generic point on it. Add to this that, from a graphic point of view, the notion "isosceles triangle" is not as sharply defined as the mathematical notion is: quasi-isosceles yet scalene triangles are simply perceived as isosceles. After all, *we ourselves* are victims of the same psychological mechanism when we deem the diagrams in medieval manuscripts oversymmetrized—for of course a diagram cannot display *exact* properties of symmetry.

Now, copyists do not get a look at the construction-part of a proposition in order to draw their diagrams: they simply reproduce pre-existing figures, and, if they have to do their job accurately without resorting to conformal copying by mechanical tricks, they must ask themselves some questions about the structure of the drawing they have before their eyes: basic—such as which points must be assigned the denotative letters, which lines intersect, etc.—and less basic issues must be addressed just by looking at the model diagram. The less basic issues include deciding whether a chord drawn near to the centre of a circle really is a diameter or not, whether two nearly equal segments are equal or not, whether an angle is a right angle or not, or, apparently a very difficult task, which angle of a right-angled triangle is the right angle. As no one will transform a triangle perceived as isosceles into a decidedly scalene triangle, convergence towards limiting cases is the only alternative to stationary evolution. In this way, elements of a diagram intended to be drawn in a generic position tend to drift, during the process of copying, towards a limiting position, which thereby works as an "attractor" in the sense of the theory of dynamical systems: a chord near the centre of a circle will converge towards a diameter. If mistakes or external factors do not intervene, such limiting positions, being perceived as such, will keep stable under the subsequent acts of copying. Experience shows that just a few steps in the process of copying are enough to make a diagram converge towards an oversymmetrized form.

We all know that transferring information involves its degradation: order is replaced by disorder; meaning, by meaningless white noise and thermal death: in short, entropy increases. Philologists are wont to represent this phenomenon as corruption of a text in the course of manuscript transmission—and they use medical, not thermodynamic, metaphors.

An increase of entropy is found in texts and in tables—in the latter, the way this happens is particularly insidious, since the rules of meaning that govern tables are opaque to any copyist. Ptolemy shows himself well aware of this when he states, in *Alm.* I.10, that he has exposed in detail a long series of theorems so as to allow us to recalculate the values in the table of chords if we suspect a copying error: ἐὰν δισταγμῷ γενώμεθα γραφικῆς ἁμαρτίας (*POO* I.1, 47.14–15). As for diagrams, the phenomenon I have just described—and the explanation of it which I have just expounded—highlight a fundamental difference in the manuscript transmission of scientific treatises: texts get

[107] That this is the case when geometric patterns are to be reproduced by memory was proved experimentally within the framework of Gestaltpsychologie: see for instance Perkins 1932, and references therein. Even if, contrary to what happens in these experiments, the period during which the copyist is exposed to the original pattern (a text or a diagram) can be arbitrarily long, we must not forget that the act of copying is first and foremost a process that involves memory.

[108] A case in point is the diagram of I.47, which is usually compressed in the vertical direction, so that the squares on the two legs are represented as rhombi.

increasingly blurred by copying mistakes, yet diagrams get increasingly symmetrized: in the transmission of diagrams, *graphic* entropy decreases—still, it is open to debate whether an oversymmetrized diagram makes information increase or decrease.

Issues of symmetry are also prominent in the layout of the non-textual parts of mathematical treatises. Let us get a quick look at what happens with diagrams and tables.[109]

In majuscule and early minuscule manuscripts, the text was arranged on two or more columns. Locating a diagram in this configuration is simple: just open a full-column window in the text. With the introduction of minuscule script, a transition took place from two-column to full-page layout. Where to place the diagrams in a codex in which writing occupies the entire page? There are four possibilities: in the outer margin of each page, in the inner margin of each page, on the right of the text, on the left. Unlike the two-column layout, there is no natural option. The most frequent choice (always on the right) can be explained by the fact that we write from left to right: the opening margin is more important than the closing margin and should thereby be reserved for writing. In this case, the symmetry of the layout of the open codex is subordinate to the repetition of a pattern that is stable and therefore easier to reproduce.

As for tables, a well-defined graphic convention for their layout was soon established, as attested to by the oldest manuscripts of the *Almagest* and of the *Handy Tables*: the tables are highly symmetric and the material is organized hierarchically. In the *Almagest*, the standard number of rows in a table is 45; this calibration is fixed already with the table of chords, the first in the treatise. Ptolemy gives as a reason that 45 is an exact divisor of 360, the number of tabulated values (διὰ τὸ σύμμετρον "because of congruence" at *POO* I.1, 47.3). As for other divisors of 360, 60 lines were apparently too many for a standard papyrus, 40 do not fit other tables set out later in the treatise, 30 lines were too few for the size of the *Almagest*. This choice of layout has substantial consequences: the cycles of mean solar motion are set to 18 years διὰ τὸ φανησόμενον σύμμετρον τῆς κανονογραφίας "for a symmetric presentation of the layout of the tables" (*Alm*. III.1, at *POO* I.1, 209.14–15). In this way, the values of the mean motion for cycles (of course, there are 45 cycles), for simple years from 1 to 18, for months (in groups of 30 days, from 30 to 360), for days from 1 to 30, for hours from 1 to 24, are displayed in a congruent set of 45-line tables (*Alm*. III.2): 45 cycles of 18 years (+ 1 title) = 18 years within a cycle + 24 hours (+ 2 titles) = 12 months + 30 days (+ 2 titles). In short: the tables of the *Almagest* are conceived to fill the papyrus column.

After the transition from the papyrus roll to the codex, a beautiful example of adaptation of the format of tables to the needs of a fully symmetric presentation is offered by the table of the mean and anomalistic motions in longitude of the five planets (*Alm*. IX.4) in Vat. gr. 1594, ff. 178r–185r. For each planet, a sequence of three tables has the same structure as that just described for the mean motion of the Sun: 45 cycles of 18 years; 18 years within a cycle + 24 hours; 12 months + 30 days. In each of the three tables, the numerical argument, which is common to the two sets of values, occupies the first column, the values of the mean motion are located in the next seven columns (for the position is calculated up to the sixth sexagesimal order), the values of the anomalistic motion, in the next seven columns. Well, these columns are followed by a column, regularly divided into cells, which is totally empty: in this way, the table is symmetrically divided into two sub-tables of eight columns each. The organization of the titles of the table accentuates the symmetric aspect; the only discrepancy lies in the fact that the empty column is not entirely delimited by double line separators, as is the case for the column of the arguments at the opposite end of the table.

---

[109] See Acerbi 2020d for detail.

4. THE DEDUCTIVE MACHINE

This part of the book will present the main features of the suprasentential logical system induced by the traditional division of a mathematical proposition into specific parts: enunciation (Sect. 4.1), setting-out (4.2), construction (4.3), and proof (4.5). The Section on the setting-out, which includes a subsection on the so-called "determination" (4.2.1), also contains a detailed discussion of seemingly similar notions developed in Stoic and in Aristotelian logic; in particular, I shall show that none of the kinds of Aristotelian ἔκθεσις fits the mathematical notion. As for the proof, after a Section on its liminal portion, called "anaphora" (4.4), I shall focus on the logic of relations (4.5.1): I shall define what a relation must be taken to mean in a Greek mathematical text; I shall also pay specific attention to word order (4.5.1.2 and 4.5.1.3), an issue that will be investigated using mild statistical methods. Before doing this, however, I shall discuss the relevant texts by Aristotle, Galen, and Alexander on relations and relational syllogisms (4.5.1.1). Interactions between relations and the other components of the "deductive machine" (4.5.1.4) will complete this Section. The rest of this part will present specific features of the deductive system that constitutes a proof: metamathematical markers (4.5.2), postposed arguments (4.5.3), instantiated and non-instantiated citations of theorems (4.5.4), assumptions and coassumptions (4.5.5).

4.1. Enunciation and conclusion

The enunciation is the statement that opens a mathematical proposition, which is closed by the conclusion. Enunciation and conclusion go together: they are nearly identical and are characterized by the absence of denotative letters. The *Elements* exhibits three species of enunciation:

1) Conditional statement.[1] This is the most frequent form.
2) Declarative statement, possibly quantified. The property that would figure in the antecedent of a conditional statement expressing the same mathematical content is normally predicated as an attributive participle of the grammatical subject of the sentence that formulates the declarative statement. The non-conditional form seems to be preferred whenever, if the same enunciation were formulated as a conditional, its antecedent and its consequent would have the same grammatical subject.[2] The non-conditional form is also preferred when a property is predicated of a fairly general class of objects—[3]or of a not-so-general class, still identified by a specific name.[4]
3) Directive clause with the verb in the aorist infinitive. This is the canonical form of enunciation for problems, namely, propositions[5] that require to construct or to find some object. The aorist stem abstracts from any aspectual or temporal connotation: the intended constructive act is a punctual operation. The infinitive may be interpreted either as directive or as being governed by an understood verb form with directive connotation, like an imperative or δεῖ "it is required".

[1] See Sect. 5.3.1 for a logico-grammatical analysis of this non-simple statement.
[2] This claim presupposes that there is a normal form of the conditional statement an enunciation in the form of a declarative statement can be transformed into. Strictly speaking, this is false, but the shared practice and the "mathematical content" play here a decisive role. The rule I have formulated is not strictly adhered to, though: exceptions in one sense are I.14–15 and 41 (the first and the third of them contain a repeated grammatical subject), in the other sense is XI.14.
[3] Like "all triangles", as in I.16–20.
[4] In the first part of the enunciation of I.5 "The angles at the base of isosceles triangles are equal to one another", the alternative in conditional form "If a triangle have two equal sides, the angles at its base are equal to one another" is nullified by the existence of the genus name "isosceles triangle".
[5] See Sect. 4.3 for the enunciation of problems in number theory, and more generally for the structure of a problem.

F. Acerbi, *The Logical Syntax of Greek Mathematics*, Sources and Studies in the History of Mathematics and Physical Sciences,
https://doi.org/10.1007/978-3-030-76959-8_4

The logical subject of this enunciation is "someone" required to perform the construction of the object or to find it. The same grammatical structure is found in the first three postulates of Book I, governed by ἠτήσθω "let it be required". A marginal subspecies of this form are the propositions enunciated with a modal connotation of possibility, as IX.18–19 and XI.22 are.

The following table sets out the distribution of the three species of enunciation in the *Elements*:[6]

Book	# prop.	conditional enunciation	non-conditional enunciation	problems
I	48	4, 6, 8, 13–15, 21, 24–28, 41, 48	5, 7, 16–20, 29–30, 32–40, 43, 47	1–3, 9–12, 22–23, 31, 42, 44–46
II	14	1–10	12–13	11, 14
III	37	2–9, 11–13, 18–19, 26–27,[7] 32, 35–37	10, 14–16, 20–24, 28–29, 31	1, 17, 25, 30, 33–34
IV	16	/	/	1–16
V	25	1–6, 12–14, 16–25	7–11, 15	/
VI	33	2–3, 5–8, 16–17, 22, 26, 32–33	1, 4, 14–15, 19–21, 23–24, 27, 31	9–13, 18, 25, 28–30
VII	39	1, 5–19, 23–28, 30, 35, 37–38	4, 20–22, 29, 31–32	2–3, 33–34, 36, 39
VIII	27	1, 3, 6–10, 13–17, 20–25	5, 11–12, 18–19, 26–27	2, 4
IX	36	1–17, 21–31, 33–36	20, 32	18–19
X	115	1–2, 6, 8, 11, 13–18, 20, 36–41, 54–59, 73–78, 91–96, 114	5, 7, 9, 12, 19, 21–26, 42–47, 60–72, 79–84, 97–113, 115	3–4, 10, 27–35, 48–53, 85–90
XI	39	2–8, 10, 15–20, 22, 24–25, 28, 35–39	1–2,[8] 9, 13–14, 21, 29–34	11–12, 23, 26–27
XII	18	4, 13	1–3, 5–12, 14–15, 18	16–17
XIII	18	1–12	/	13–18 (*bis* each)
tot.	465	224	146(7)	95(101)

The numbers for each Book are as follows:

	I	II	III	IV	V	VI	VII	VIII	IX	X	XI	XII	XIII	tot.
# prop.	48	14	37	16	25	33	39	27	36	115	39	18	18	465
conditional	14	10	19	/	19	12	26	18	32	37	23	2	12	224
non conditional	20	2	12	/	6	11	7	7	2	54	11(2)	14	/	146(7)
problem	14	2	6	16	/	10	6	2	2	24	5	2	6(12)	95(101)

Let us read an example for each species of enunciation: these are, in the order given above, VII.18, III.21, and IV.13 (*EOO* II, 224.22–25, and I, 220.16–17 and 306.21–22):

ἐὰν δύο ἀριθμοὶ ἀριθμόν τινα πολλαπλασιάσαντες ποιῶσί τινας, οἱ γενόμενοι ἐξ αὐτῶν τὸν αὐτὸν ἕξουσι λόγον τοῖς πολλαπλασιάσασιν.	If two numbers multiplying some number make some ‹numbers›, those resulting from them will have the same ratio as those that multiply.
ἐν κύκλῳ αἱ ἐν τῷ αὐτῷ τμήματι γωνίαι ἴσαι ἀλλήλαις εἰσίν.	In a circle, the angles in a same segment are equal to one another.
εἰς τὸ δοθὲν πεντάγωνον ὅ ἐστιν ἰσόπλευρόν τε καὶ ἰσογώνιον κύκλον ἐγγράψαι.	In a given pentagon that is both equilateral and equiangular, inscribe a circle.[9]

[6] In the *Data*, 91 enunciations out of 94 are in conditional form; exceptions are propositions 1, 8, 47.
[7] The difficulty represented by the enunciations of III.13, 26–27, and VI.33 will be discussed in Sect. 5.3.5.
[8] XI.2 has a double enunciation; one is in conditional form, the other is not.
[9] Note the long designation, prepositive because strongly marked for saliency; I have kept this feature in my translation.

The constraints required in an enunciation are seldom formulated as a genitive absolute; this happens more frequently in problems than in theorems. An interesting example is found in Book X (cf. also VII.1, read in Sect. 1.2)—let us read the enunciation of X.1 (*EOO* III, 4.5–10), where a genitive absolute introduces a whole conditional,

δύο μεγεθῶν ἀνίσων ἐκκειμένων, ἐὰν ἀπὸ τοῦ μείζονος ἀφαιρεθῇ μεῖζον ἢ τὸ ἥμισυ καὶ τοῦ καταλειπομένου μεῖζον ἢ τὸ ἥμισυ καὶ τοῦτο ἀεὶ γίγνηται, λειφθήσεταί τι μέγεθος ὃ ἔσται ἔλασσον τοῦ ἐκκειμένου ἐλάσσονος μεγέθους.

Two unequal magnitudes being set out, if from the greater a ‹magnitude› greater than the half be removed and from the remainder one greater than the half and this come about continually, some magnitude will have remained that will be less than the lesser magnitude set out.

whereas an analogous constraint in X.2 (*EOO* III, 6.12–15; underlined below) is embodied as a determiner in a genitive absolute that formulates in a compact way a constraint similar to the one required in the first conjunct in the antecedent of X.1 (in italics below):[10]

ἐὰν δύο μεγεθῶν ἀνίσων *ἀνθυφαιρουμένου ἀεὶ τοῦ ἐλάσσονος ἀπὸ τοῦ μείζονος* τὸ καταλειπόμενον μηδέποτε καταμετρῇ τὸ πρὸ ἑαυτοῦ, ἀσύμμετρα ἔσται τὰ μεγέθη.

If, *from the greater* of two unequal magnitudes *the lesser being continually removed in turn*, the remainder never measure out the ‹magnitude› before itself, the magnitudes will be incommensurable.

The two problems X.3–4 canonically replace the participle ἐκκειμένων "set out" of X.1 with δοθέντων "given", as it must be in the enunciation of a problem. A genitive absolute as a means to introduce a given object is more frequent in the enunciation of number-theoretical problems,[11] but a geometric example can be found in III.25 (*EOO* I, 226.17–18):[12]

κύκλου τμήματος δοθέντος προσαναγράψαι τὸν κύκλον οὗπέρ ἐστι τμῆμα.

A segment of a circle being given, describe out the circle of which it is a segment.

From Book X again, let us read a non-conditional enunciation whose assumption is formulated by a genitive absolute. This occurs in theorem X.71 (*EOO* III, 212.16–19)—the prepositional constructs that designate the species of irrationals do not have an article, as they are the nominal complements of the copula (cf. Sect. 5.1.5):

ῥητοῦ καὶ μέσου συντιθεμένου τέσσαρες ἄλογοι γίγνονται ἤτοι ἐκ δύο ὀνομάτων ἢ ἐκ δύο μέσων πρώτη ἢ μείζων ἢ ῥητὸν καὶ μέσον δυναμένη.

An expressible and a medial ‹region› being compounded, there result four irrationals, either a binomial or a first bimedial or a major or a ‹straight line› worth an expressible and a medial ‹region›.

Mixed cases of species (2) and (3) in the typology above are the enunciations of the constructions of the regular polyhedra in problems XIII.13–17. Let us read XIII.17 (*EOO* IV, 316.9–12):

---

[10] A marginal ἐκκειμένων in the first hand of manuscript **P** tries to align the text of X.2 with that of X.1.

[11] Enunciations in which a genitive absolute introduces the given objects are III.25, VII.2, 3, 33–34, 36, VIII.4, IX.18–19, X.3–4, XII.17 (the last with "to be").

[12] In the Euclidean corpus, the verb προσαναγράφω "to describe out" occurs only in proposition III.25 (the double prefix is very rare, too).

δωδεκάεδρον συστήσασθαι καὶ σφαίρᾳ περιλαβεῖν ᾗ καὶ τὰ προειρημένα σχήματα, καὶ δεῖξαι ὅτι ἡ τοῦ δωδεκαέδρου πλευρὰ ἄλογός ἐστιν ἡ καλουμένη ἀποτομή.

Construct a dodecahedron and comprehend ‹it› with a sphere with which the said figures too, and prove that the side of the dodecahedron is an irrational ‹line›, the so-called apotome.

What is required is, first, to construct a dodecahedron, then, to circumscribe a given sphere to it, finally, to prove a property of the dodecahedron itself. The directive infinitive of the first two assignments is canonical for an enunciation of species (3). The third assignment must be ranged among the enunciations of species (2), adapted for stylistic reasons to the species-(3) format of the other two assignments. To do this, it is necessary to include the verb form δεῖξαι "to prove" in the enunciation, which is thereby loaded with metamathematical connotations. The same stylistic register is found in the first retrieval of the enunciation, in the form of the partial "determination" (see Sect. 4.2.1) that follows the first construction of XIII.17 (*EOO* IV, 322.14–16):

δεῖ δὴ αὐτὸ καὶ σφαίρᾳ περιλαβεῖν τῇ δοθείσῃ καὶ δεῖξαι ὅτι ἡ τοῦ δωδεκαέδρου πλευρὰ ἄλογός ἐστιν ἡ καλουμένη ἀποτομή.

Thus it is required to comprehend ‹it› with a given sphere, and to prove that the side of the dodecahedron is an irrational ‹line›, the so-called apotome.

The metadiscursive character of the verb form δεῖξαι is confirmed by the fact that it is omitted in the "determination" that precedes the part of the proof of XIII.17 (*EOO* IV, 326.1–2) in which the predicated property is finally proved:

λέγω δὴ ὅτι ἡ τοῦ δωδεκαέδρου πλευρὰ ἄλογός ἐστιν ἡ καλουμένη ἀποτομή.

I now claim that the side of the dodecahedron is an irrational ‹line›, the so-called apotome.

Within a proof, enunciations of species (1)—conditional clauses—are frequently cited in non-instantiated form and are employed as a premise of a *modus ponens*; enunciations of species (2)—declarative statements—are normally cited in instantiated form. As said, there are no examples of non-instantiated citations of constructive acts; such citations conform instead to the matrix templates sanctioned in the (instantiated) conclusions of the associated problems (see Sect. 2.2): as a consequence, citations of enunciations of species (3) cannot be found within a proof.

We might wonder why to provide all propositions of the *Elements* and of the *Data* with a general enunciation.[13] As we have seen in Sect. 3.3, in fact, instantiated suppositions and proofs already secure the highest degree of generality—and after all, the Greek mathematical corpus abounds with propositions directly enunciated in instantiated form.[14] Here the difference between the generality of a statement and of a proposition comes into play.[15] If we want to apply a mathematical result of general validity in a general context, what counts is the general formulation of its statement, not its

[13] At the end of this Section, we shall see that the general conclusions are exclusive to the *Elements*; moreover, they are present only in a minority of propositions.

[14] Archimedes (199 propositions in the whole corpus): 22 occurrences of instantiated enunciations and all within theorems, in *Sph. cyl.* I.23, 28, 36, 39, 41 (they all have the same structure and establish similar results), *Quadr.* 4–16 (all and only the theorems related to the "mechanical" quadrature; *nota bene*: these are theorems, not problems), *Meth.* 1, 13–15. Apollonius (*Con.* I–IV comprise 226 propositions): 24 occurrences of instantiated enunciations, in *Con.* I.53, 57 (these are two problems), III.8–10, 25–26, IV.2–5, 7–8, 10–14, 16–17, 19, 21–23, 28–29; all these are additional cases of previous propositions. Almost every geometric proposition in the *Almagest* is directly formulated in instantiated form.

[15] Also, a general enunciation is in principle better suited to encompass within one single statement all cases a theorem may branch into.

general validity insofar as this is secured by a general proof. But a statement formulated in general terms cannot contain denotative letters—if for no other reasons, just to forestall the objection that, by adapting the denotative letters to the host proposition, we are not using *the same* enunciation. In this perspective, there is a difference between using a theorem and using a problem in a subsequent proposition, since, as said, the template of any citation of a problem is its (instantiated) conclusion.

In Archimedes, there is still a fair balancing between conditional (72 occurrences) and non-conditional (98)[16] enunciations, to which 27 problems must be added. The conditional form becomes pervasive in later authors; one of the reasons may be that, in less elementary treatises, the specific assumptions on which a theorem rests tend to multiply. However, the overload of designations may introduce tensions in the stylistic code. Let us read an extreme example in *Con.* I.11 (*AGE* I, 36.27–38.14), where the parabola is introduced.[17] This is the shortest enunciation among those in the group I.11–13; it contains the definition of the curve (cf. Sect. 1.5)—I have separated the antecedent of the conditional from the consequent by interposing a vertical slash "|", and underlined some relevant syntagms, to be discussed below:

ἐὰν κῶνος ἐπιπέδῳ τμηθῇ διὰ τοῦ ἄξονος τμηθῇ δὲ καὶ ἑτέρῳ ἐπιπέδῳ τέμνοντι τὴν βάσιν τοῦ κώνου κατ' εὐθεῖαν πρὸς ὀρθὰς οὖσαν τῇ βάσει τοῦ διὰ τοῦ ἄξονος τριγώνου ἔτι δὲ ἡ διάμετρος τῆς τομῆς παράλληλος ᾖ μιᾷ πλευρᾷ τοῦ διὰ τοῦ ἄξονος τριγώνου, | ἥτις ἂν ἀπὸ τῆς τομῆς τοῦ κώνου παράλληλος ἀχθῇ τῇ κοινῇ τομῇ τοῦ τέμνοντος ἐπιπέδου καὶ τῆς βάσεως τοῦ κώνου μέχρι τῆς διαμέτρου τῆς τομῆς δυνήσεται τὸ περιεχόμενον ὑπό τε τῆς ἀπολαμβανομένης ὑπ' αὐτῆς ἀπὸ τῆς διαμέτρου πρὸς τῇ κορυφῇ τῆς τομῆς καὶ ἄλλης τινὸς εὐθείας, ἣ λόγον ἔχει πρὸς τὴν μεταξὺ τῆς τοῦ κώνου γωνίας καὶ τῆς κορυφῆς τῆς τομῆς ὃν τὸ τετράγωνον τὸ ἀπὸ τῆς βάσεως τοῦ διὰ τοῦ ἄξονος τριγώνου πρὸς τὸ περιεχόμενον ὑπὸ τῶν λοιπῶν τοῦ τριγώνου δύο πλευρῶν. καλείσθω δὲ ἡ τοιαύτη τομὴ παραβολή.

If a cone be cut with a plane through the axis and be also cut with another plane cutting the base of the cone in a straight line that is at right ‹angles› with the base of the triangle through the axis and further the diameter of the section be parallel to one side of the triangle through the axis, | any ‹straight line› which be drawn from the section of the cone parallel to the common section of the cutting plane and of the base of the cone as far as the diameter of the cone will be worth the ‹rectangle› contained both by the ‹straight line› cut off by it from the diameter as far as the vertex of the cone and by some other straight line, which has a ratio to the ‹straight line› between the angle of the cone and the vertex of the cone that ‹has› the square on the base of the triangle through the axis to the ‹rectangle› contained by the remaining two sides of the triangle. And let such a section be called parabola.

In this enunciation, syntactic tools are displayed that are finely suited to express the relations of coordination and subordination between the assumptions and their consequences.[18] The antecedent is in weakly conjunctive form, thanks to the three-place correlative … δέ … ἔτι δέ … "… and … and further …" (cf. Sect. 5.3.5); the second conjoined clause is further determined by two nested participial constructs. The relative clause with generalizing connotation introduced by ἥτις ἄν "any" and that opens the consequent was probably devised to avoid a nested conditional.

[16] Possibly add *Meth.* 9 and 11, whose enunciations cannot be reconstructed with certitude but were quite likely non-conditional statements.

[17] In the Greek Apollonius, there are 176 conditional enunciations, 32 in declarative form, and 18 problems. Recall that, in the Greek *Conica*, several propositions that carry a number are nothing but further cases of previous propositions. It is not clear whether this is an original feature or the result of Eutocius' recension.

[18] An approach more in line with ancient logical doctrines will be adopted by Pappus, as we shall see in Sect. 5.3.4.

The description of the rectangle in the principal clause of the consequent is expressed by the canonical correlative τε ... καί ... "both ... and ..."; the second conjoined clause is further determined by two nested relative clauses. It may be that Apollonius, a sophisticated stylist as his prefatory epistles show, introduced a balancing between the nested determiners in the antecedent and in the consequent, with the further touch of varying their form.[19]

To the requirement of blending fairly articulated complexes of assumptions in a compact way, as is the case in the enunciation just read, we may add the fact that the conditional form is more rigid and therefore came to be perceived as more canonical. Thus, in the 3 Books of Theodosius' *Sphaerica* there are 44 conditional enunciations, 9 non-conditional statements, and 7 problems. All enunciations that Pappus and Theon of Alexandria concoct for Euclid's *Porisms* and for Ptolemy's *Almagest* are in conditional form.[20]

Finally, the conditional form sharply differentiates, on mere grounds of syntax, that which is assumed from that which is to be proved. Let us compare the enunciations of propositions I.18 and I.19 (*EOO* I, 46.2–3 and 46.18–19) as an example:[21]

παντὸς τριγώνου ἡ μείζων πλευρὰ τὴν μείζονα γωνίαν ὑποτείνει.	The greater side of every triangle extends under the greater angle.
παντὸς τριγώνου ὑπὸ τὴν μείζονα γωνίαν ἡ μείζων πλευρὰ ὑποτείνει.	Under the greater angle of every triangle extends the greater side.

The two Greek sentences differ in word order (as they do in English) and in the (pleonastic) presence of the preposition ὑπό "under". However, the two theorems are one the inverse of the other: their enunciations definitely do not help in clarifying in which of the two propositions the inequality between the angles is assumed and in which it is proved. Another example is provided by the pair III.28 (first part) and 29 (*EOO* I, 234.19–21 and 238.2–3):

ἐν τοῖς ἴσοις κύκλοις αἱ ἴσαι εὐθεῖαι ἴσας περιφερείας ἀφαιροῦσι.	In equal circles, equal straight lines remove equal arcs.
ἐν τοῖς ἴσοις κύκλοις τὰς ἴσας περιφερείας ἴσαι εὐθεῖαι ὑποτείνουσιν.	In equal circles, under equal arcs extend equal straight lines.

Here the ambiguity cannot be eliminated by the difference of the verb form: ἀφαιρεῖν "to remove" in this sense is never defined and is used only in this proposition and in III.30 (whithin an instantiated citation of III.28), in a context in which we would expect to read ἀπολαμβάνω "to cut off"; moreover, ὑποτείνω "to extend under" is not defined if referred to arcs and their chords and is here used for the first time in Book III. Granted, in III.28–29 (but not in I.18–19 above) the presence of the article discriminates between the assumption and that which is to be proved, but which of them should receive the article? A look at the four proofs shows that saliency is used as a demarcating criterion: assumptions come first. So the trick is just getting accustomed to a shared practice. Better use conditional enunciations, though.

[19] The two relative clauses in the consequent are the normal form of such determiners.

[20] See for instance *Coll.* VII.16 and *in Alm. I.13*, respectively.

[21] In I.19, all modern translators turn to the passive a verb that is always intransitive: the accusative τὴν μείζονα γωνίαν in I.18 is in fact governed by the preverbal ὑπό. As a consequence, the preposition ὑπό in I.19 simply repeats the preverbal and cannot be treated as if it introduced an agent (which should be in the genitive, by the way).

The conditional form of an enunciation is also connected with a major logical problem: in what sense the proof of a proposition validates its enunciation?[22] I write "enunciation" because the general conclusion of a theorem, when it is present (see below), is in fact nearly identical to the enunciation, the only difference being an added scope particle ἄρα "therefore". In the general conclusion of a problem, instead, the verb is also changed to a perfect stem. To check this, let us read the general conclusions associated with two of the three enunciations presented at the beginning of this Section, namely, III.21 and IV.13 (*EOO* I, 222.5–6 and 310.19–21):

ἐν κύκλῳ ἄρα αἱ ἐν τῷ αὐτῷ τμήματι γωνίαι ἴσαι ἀλλήλαις εἰσίν.	Therefore, in a circle, the angles in a same segment are equal to one another.
εἰς ἄρα τὸ δοθὲν πεντάγωνον ὅ ἐστιν ἰσόπλευρόν τε καὶ ἰσογώνιον κύκλος ἐγγέγραπται.	Therefore in a given pentagon that is both equilateral and equiangular a circle turns out to be inscribed.

Thus, validating the last step in a proof (or the conclusion) coincides with validating the enunciation of the theorem that includes the proof. Two principles may be invoked to justify this crucial metadeductive step. The first is the principle of the corresponding conditional, or detachment theorem: it was explicitly formulated in Stoic logic to test the validity of an argument (Sextus, *P* II.137, and cf. *P* II.249 and *M* VIII.415–417):

τῶν δὲ λόγων οἱ μέν εἰσι συνακτικοὶ οἱ δὲ ἀσύνακτοι· συνακτικοὶ μέν, ὅταν τὸ συνημμένον τὸ ἀρχόμενον μὲν ἀπὸ τοῦ διὰ τῶν τοῦ λόγου λημμάτων συμπεπλεγμένου λῆγον δὲ εἰς τὴν ἐπιφορὰν αὐτοῦ ὑγιὲς ᾖ.	Of arguments, some are deductive, others non-deductive; they are deductive, whenever the conditional beginning with the conjunction of the as-sumptions of the argument and ending with its conclusion be sound.

A formulation of the inverse rule can be read in D.L. VII.77, where it is stated that such arguments are non-deductive for which "the contradictory of the conclusion is not in conflict with the conjunction of the premises", so that the corresponding Chrysippean conditional is not valid. The two formulations entail that the validity of a Chrysippean conditional was assumed to be a necessary and sufficient condition for the validity of the corresponding argument.[23]

The principle of the corresponding conditional is a semantic metatheorem, and must not be confused with the deduction theorem, which is a rule to manipulate deductions. The deduction theorem states that, if from a premise together with other premises a conclusion follows, then from the additional premises the conditional follows that has the singled out premise as antecedent and the said conclusion as the consequent. The assumption that gets transferred in the antecedent of the conditional is said to be "discharged" or "closed". There are no ancient formulations of this rule, which is in fact the rule that underlies the mathematical practice (cf. Sect. 2.3).

As just said, a theorem validates its own enunciation, not its own conclusion.[24] I have transcribed above the general conclusions of III.21 and IV.13 simply because VII.18 does not have one, and ends with the instantiated statement "therefore, as A is to B, so Δ is to E" (*EOO* II, 226.9–10). Still, in Proclus' account of the specific parts of a proposition, a general conclusion is said to be necessary

[22] This issue is not the same as the issue discussed in Sect. 3.
[23] This is already implicit in the Sextan use of ὅταν "whenever". However, in Stoic syllogistic the test is not a criterion, for there are sound conditionals whose corresponding argument is not deductive: Bobzien 2019, 261 n. 57.
[24] To repeat: in problems, what is validated as a template is the (instantiated) conclusion, as we have seen in Sect. 2.2.

and always present. Proclus takes occasion for this claim when commenting on I.1, which the entire manuscript tradition hands down without a conclusion, and within a commentary on Book I, in which only two-thirds of the propositions do have a conclusion.[25] A quick check shows that Proclus' blind commonplace has no grounds: the *Elements* is the only treatise in the Greek mathematical corpus in which general conclusions are present. Even in the case of the *Elements*, the indirect tradition supports the view that most general conclusions must be later additions. The Arabo-Latin translations exhibit in fact far fewer general conclusions (66 out of 466 propositions in the richer case, which is 14.16%) than the Greek tradition (200 out of 471, which is 42.46%). Within the Greek tradition, theorems give the ratio 176/370 = 47.57%, problems 22/101 = 21.78% (recall that XIII.13–18 require to perform two constructions).

The distribution of the general conclusions by Books is set out in the following table. The second row gives the number of propositions in each Book, the third, fourth, fifth, and sixth row give the number of attested general conclusions in the Greek tradition and its percentage with respect to the total number of propositions, and the number of attested general conclusions in Gerard of Cremona's and Adelard's Arabo-Latin translations, respectively.[26]

	I	II	III	IV	V	VI	VII	VIII	IX	X	XI	XII	XIII
# props.	48	14	37	16	25	33	39	27	36	115	39	18	18
Greek	32	12	31	10	25	22	3	1	0	20	26	7	11
%	66.67	85.71	83.78	62.5	100	66.67	7.69	3.70	0.00	17.39	66.67	38.89	61.11
Gerard	30	12	7	1	0	0	4	0	1	7	5	0	0
Adelard	5	0	5	3	0	0	3	0	/	(7)	4	0	2

Note the striking data as regards Books V–VI, and more generally the clear-cut difference between the direct and the indirect tradition for Books IV–VI. Referring to the discussion in Sect. 1.1, and as far as the Greek tradition is concerned, the general conclusion is an obvious negative stylistic marker of the arithmetic Books. Comparison with the indirect tradition, and especially with Gerard, suggests that the number of general conclusions in the Greek *Elements* has gradually increased, starting from a good approximation of zero, as the cumulative outcome of early campaigns of regularization, which for reasons we are unable to perceive did not affect the arithmetic Books. Of particular interest are the general conclusions of problems, most notably their fairly systematic presence in Book IV; see Section 4.3 for detail.

---

[25] Heiberg absurdly adds a conclusion to his critical text of I.1, even if within brackets: *EOO* I, 12.16–17. In Book I, no problem has a general conclusion, and 2 theorems lack it. Proclus' statement is at *iE*, 203.17–18. The QED formula ὅπερ ἔδει δεῖξαι (ποιῆσαι) "which it was really required to prove (do)" that closes a theorem or a problem is certainly a spurious feature of the *Elements*—and in fact it is absent in the rest of the Euclidean corpus. The only exception is the last (!) theorem (94) of the *Data*. In the manuscripts of the *Elements*, the entire phrase is not infrequently abbreviated by a small circle partly enclosed on the right by a semicircle (obviously an abbreviation of ὅπερ); the sign also ends, in the Archimedean palimpsest, *Meth*. 3 and 4, at ff. 63v, col. 2, line 30, and 44v, col. 1, line 36 (Heiberg writes οι at *AOO* II, 454.7). Other occurrences of a QED formula in Archimedes are at *Sph. cyl*. II.2, *Aequil*. II.4, 8, 9, *AOO* I, 178.29, II, 174.2, 190.21, 202.13, respectively (ὅπερ ἔδει δεῖξαι); *Aequil*. II.6, *AOO* II, 184.30 (ἔδει δὲ τοῦτο δεῖξαι); *Con. sph*. 8, 9, 28, 30, *Fluit*. I.6, *AOO* I, 296.8, 300.6, 408.25, 428.22, II, 332.19 (φανερὸν / δῆλον (οὖν ἐστιν) ὃ ἔδει δεῖξαι); *Meth*. 2, *AOO* II, 446.2 (ὅπερ ἔδει δειχθῆναι); *Sph. cyl*. I.2, *AOO* I, 16.7 (ὅπερ προέκειτο εὑρεῖν); *Con. sph*. 3, *AOO* I, 274.27–28 (δῆλον οὖν ἐστι τὸ προτεθέν). As is clear, Archimedes displays a fair variability of QED clauses; note in particular the third item. As for general conclusions, they can be found only in *Sph. cyl*. I.2, *AOO* I, 12.19–20, in the obviously metadiscursive clause εὑρημέναι εἰσὶν ἄρα δύο εὐθεῖαι ἄνισοι ποιοῦσαι τὸ εἰρημένον ἐπίταγμα, and in *Sph. cyl*. I.30 and partly II.8 (I.39 has a general conclusion but does not have an enunciation: this is Archimedes, baby); recall, however, that some Archimedean enunciations are extremely long, and that some strings of propositions, as for instance *Quadr*. 1–16, do not have a general enunciation (the first three propositions are only enunciated).

[26] In Adelard, the entire Book IX and the first 35 propositions of Book X are missing.

4.2. SUPPOSITIONS AND "SETTING-OUT"

The instantiated supposition, or ἔκθεσις "setting-out" according to Proclus' denomination, is the second specific part of a proposition. The setting-out derives from an enunciation in conditional form as follows:[27] the antecedent of the conditional is transformed into a conjoined sequence of suppositions with the verb in the imperative; this is the ἔκθεσις. The consequent of the conditional is transformed into a statement formulated by a sentence that is introduced by λέγω ὅτι "I claim that";[28] Proclus (and no one before him) calls it διορισμός "determination".[29] From the logical point of view, the dismemberment of an enunciation into two independent clauses singles out the assumptions of a proposition for use as premises in the subsequent proof. The proof contains in fact "paraconditional" clauses (see Sect. 5.3.2), whose antecedents eventually pick up all assumptions—or some immediate consequences of them. The suppositions posited in the setting-out are thus "discharged" the one after the other.

The verb forms and the structure of an enunciation and of the associated setting-out are strictly related, with two systematic variants. First, denotative letters are introduced in the setting-out; normally such letters do not replace the indefinite designations found in the enunciation, but only supplement them as appositions: they are the *names* of the mathematical objects at issue, as I have shown in Sect. 3.2. Second, in the setting-out, the mood of the verb is invariably an imperative; this means that the setting-out is in the "suppositional mode" (see below). The setting-outs of VII.18, III.21, and IV.13 (*EOO* II, 226.1–3, and I, 220.18–21 and 306.23–25), whose enunciations we have read at the beginning of the previous Section, are as follows:

δύο γὰρ ἀριθμοὶ οἱ Α Β ἀριθμόν τινα τὸν Γ πολλαπλασιάσαντες τοὺς Δ Ε ποιείτωσαν. λέγω ὅτι ἐστὶν ὡς ὁ Α πρὸς τὸν Β οὕτως ὁ Δ πρὸς τὸν Ε.	In fact, let two numbers, Α, Β, multiplying some number, Γ, make Δ, Ε. I claim that, as Α is to Β, so Δ is to Ε.
ἔστω κύκλος ὁ ΑΒΓΔ, καὶ ἐν τῷ αὐτῷ τμήματι τῷ ΒΑΕΔ γωνίαι ἔστωσαν αἱ ὑπὸ ΒΑΔ ΒΕΔ. λέγω ὅτι αἱ ὑπὸ ΒΑΔ ΒΕΔ γωνίαι ἴσαι ἀλλήλαις εἰσίν.	Let there be a circle, ΑΒΓΔ, and let there be angles, ΒΑΔ, ΒΕΔ, in a same segment, ΒΑΕΔ. I claim that angles ΒΑΔ, ΒΕΔ are equal to one another.
ἔστω τὸ δοθὲν πεντάγωνον ἰσόπλευρόν τε καὶ ἰσογώνιον τὸ ΑΒΓΔΕ. δεῖ δὴ εἰς τὸ ΑΒΓΔΕ πεντάγωνον κύκλον ἐγγράψαι.	Let there be a given pentagon both equilateral and equiangular, ΑΒΓΔΕ. Thus it is required to inscribe a circle in pentagon ΑΒΓΔΕ.

It is a mistake to think that the distribution of assumptions and *demonstranda* between setting-out and determination follows rigid rules. Conflicting examples can even be found when the enunciation is in conditional form. Let us check this claim on the basis of a beautiful example, which involves the entire enunciation of III.11 (*EOO* I, 194.19–196.3) and the second conditional in the enunciation of VI.2 (*EOO* II, 78.1–4 and 78.23–26):

[27] The declarative enunciations of theorems and the directive enunciations of problems undergo similar, but more elastic, transformations, as several subsequent examples will show.

[28] The standard translation "I say that" is less appropriate: after all, everyone read it as "I claim that". Note the strong metalinguistic connotation of the first person singular.

[29] For the sake of simplicity, in this Section I shall sometimes present the determination as if it were a part of the setting-out. The determination will be specifically discussed in Sect. 4.2.1.

enunciation	setting-out
ἐὰν δύο κύκλοι ἐφάπτωνται ἀλλήλων ἐντὸς καὶ ληφθῇ αὐτῶν τὰ κέντρα, ἡ ἐπὶ τὰ κέντρα αὐτῶν ἐπιζευγνυμένη εὐθεῖα καὶ ἐκβαλλομένη ἐπὶ τὴν συναφὴν πεσεῖται τῶν κύκλων.	δύο γὰρ κύκλοι οἱ ΑΒΓ ΑΔΕ ἐφαπτέσθωσαν ἀλλήλων ἐντὸς κατὰ τὸ Α σημεῖον, καὶ εἰλήφθω τοῦ μὲν ΑΒΓ κύκλου κέντρον τὸ Ζ τοῦ δὲ ΑΔΕ τὸ Η. λέγω ὅτι ἡ ἀπὸ τοῦ Η ἐπὶ τὸ Ζ ἐπιζευγνυμένη εὐθεῖα ἐκβαλλομένη ἐπὶ τὸ Α πεσεῖται.
ἐὰν αἱ τοῦ τριγώνου πλευραὶ ἀνάλογον τμηθῶσιν, ἡ ἐπὶ τὰς τομὰς ἐπιζευγνυμένη εὐθεῖα παρὰ τὴν λοιπὴν ἔσται τοῦ τριγώνου πλευράν.	ἀλλὰ δὴ αἱ τοῦ ΑΒΓ τριγώνου πλευραὶ αἱ ΑΒ ΑΓ ἀνάλογον τετμήσθωσαν, ὡς ἡ ΒΔ πρὸς τὴν ΔΑ οὕτως ἡ ΓΕ πρὸς τὴν ΕΑ, καὶ ἐπεζεύχθω ἡ ΔΕ. λέγω ὅτι παράλληλός ἐστιν ἡ ΔΕ τῇ ΒΓ.

enunciation	setting-out
If two circles be internally tangent to one another and their centres be taken, the straight line joined at theirs centres and produced will fall on the point of tangency of the circles.	In fact, let two circles, ΑΒΓ, ΑΔΕ, be internally tangent to one another at point A, and let the centre Z of circle ΑΒΓ be taken and the ‹centre› H of ΑΔΕ. I claim that the straight line joined from H to Z, once produced, will fall on A.
If the sides of a triangle be cut in proportion, the straight line joined at the sections will be parallel to the remaining side of the triangle.	But now, let sides AB, ΑΓ of a triangle, ΑΒΓ, be cut in proportion, as ΒΔ is to ΔΑ, so ΓΕ is to ΕΑ, and let a ‹straight line›, ΔΕ, be joined. I claim that ΔΕ is parallel to ΒΓ.

In both conditional enunciations, the grammatical subject of the consequent is a noun qualified by a participial expression with attributive value, further determined by a prepositional complement: the two participial expressions are identical both as to their form and as to their syntactic function; they identify a specific straight line by means of the (in)definite description ἡ ἐπὶ *** ἐπιζευγνυμένη εὐθεῖα "a straight line joined at ***".[30] Now, the setting-out of III.11 reproduces the structure of the enunciation, so that the intended straight line is introduced in the determination,[31] whereas in VI.2 the straight line is already posited in the setting-out, by means of a further conjunct in suppositional form (= a constructive act). As a consequence, in the determination of VI.2 the designation of the straight line is replaced by the name ἡ ΔΕ that is assigned to it in the setting-out.

If the enunciation is divided into cases or into conjoined sub-enunciations, a "partial" setting-out may be placed after the first part of the proof. In this case, the partial setting-out and the partial determination that follows it are introduced by suitable coordinants, as in I.26 (*EOO* I, 64.13–18):

ἀλλὰ δὴ πάλιν ἔστωσαν αἱ ὑπὸ τὰς ἴσας γωνίας πλευραὶ ὑποτείνουσαι ἴσαι, ὡς ἡ ΑΒ τῇ ΔΕ. λέγω πάλιν ὅτι καὶ αἱ λοιπαὶ πλευραὶ ταῖς λοιπαῖς πλευραῖς ἴσαι ἔσονται, ἡ μὲν ΑΓ τῇ ΔΖ ἡ δὲ ΒΓ τῇ ΕΖ καὶ ἔτι ἡ λοιπὴ γωνία ἡ ὑπὸ ΒΑΓ τῇ λοιπῇ γωνίᾳ τῇ ὑπὸ ΕΔΖ ἴση ἐστίν.

But now, again, let the sides extending under the equal angles be equal, as AB is to ΔΕ. I claim again that the remaining sides will also be equal to the remaining sides, ΑΓ to ΔΖ and ΒΓ to ΑΓ, and further the remaining angle ΒΑΓ is equal to the remaining angle ΕΔΖ.

[30] As we have seen in Sect. 3.2, the expression is indefinite: the article ἡ is forced by grammar, and moreover it does not by itself make a noun phrase of the type aXN definite.
[31] The denomination ἡ ἀπὸ τοῦ Η ἐπὶ τὸ Ζ ἐπιζευγνυμένη εὐθεῖα will be kept throughout the proof. The reason is the same as the reason seen in the case of our paradigmatic proposition III.2 (cf. Sect. 1.1).

The setting-out may be absent when a theorem proves a negative result and its enunciation is in non-conditional form. Thus, after the enunciation, the canonical conditional clause that initializes a reduction to the impossible is directly found (see Sect. 5.2.1); this clause normally contains an instantiation of the "impossible" configuration (underlined below; it is not a "real" construction), and is in its turn immediately followed by the "real" construction,[32] as in X.26 (*EOO* III, 74.8–13)—other examples are in propositions I.7, III.10, 13, 23, XI.1, 13:

μέσον μέσου οὐχ ὑπερέχει ῥητῷ.	A medial does not exceed a medial by an expressible.
εἰ γὰρ δυνατόν, μέσον τὸ ΑΒ μέσου τοῦ ΑΓ ὑπερεχέτω ῥητῷ τῷ ΔΒ, καὶ *ἐκκείσθω* ῥητὴ ἡ ΕΖ, καὶ τῷ ΑΒ ἴσον παρὰ τὴν ΕΖ παραβεβλήσθω παραλληλόγραμμον ὀρθογώνιον τὸ ΖΘ πλάτος ποιοῦν τὴν ΕΘ, τῷ δὲ ΑΓ ἴσον ἀφῃρήσθω τὸ ΖΗ.	In fact, if possible, let a medial, AB, exceed a medial, ΑΓ, by an expressible, ΔΒ, and *let* an expressible, EZ, *be set out*, and let a rectangular parallelogram, ΖΘ, equal to AB be applied to EZ making a width ΕΘ, and let a ‹region›, ZH, be removed equal to ΑΓ.

The setting-out is also absent in problems for which no givens are provided,[33] as in IV.10, X.27–35, 48–53, 85–90,[34] XIII.13–18 (cf. Sect. 4.3). This entails, as we shall see in a moment, that the construction must be opened by the verb form ἐκκείσθω "let it be set out". This verb marks the introduction of a new object, yet not a given one, on which to carry out the construction: this is exactly the role of the "expressible line" in X.26 just read (the verb form ἐκκείσθω is in italics).

As we have seen in Sect. 3.1, the imperative of "to be" that introduces many setting-outs has a "presential" value. A liminal ἔστω in a setting-out may, however, have a copulative value if this was already the case in the enunciation;[35] the first occurrence of such a value for liminal "to be" in the *Elements* is in I.30 (*EOO* I, 74.4–7), of which we also read the enunciation:

αἱ τῇ αὐτῇ εὐθείᾳ παράλληλοι καὶ ἀλλήλαις εἰσὶ παράλληλοι.	‹Straight lines› parallel to a same straight line are also parallel to one another.
ἔστω ἑκατέρα τῶν ΑΒ ΓΔ τῇ ΕΖ παράλληλος. λέγω ὅτι καὶ ἡ ΑΒ τῇ ΓΔ ἐστι παράλληλος.	Let each of two ‹straight lines›, AB, ΓΔ, be parallel to a ‹straight line›, EZ. I claim that AB is also parallel to ΓΔ.

The verb "to be" has a locative value in specific theorems; let us read the enunciation and the setting-out + determination of XII.5 (*EOO* IV, 164.16–23) as an example:[36]

αἱ ὑπὸ τὸ αὐτὸ ὕψος οὖσαι πυραμίδες καὶ τριγώνους ἔχουσαι βάσεις πρὸς ἀλλήλας εἰσὶν ὡς αἱ βάσεις.	Pyramids that are under a same height and have triangular bases are to one another as the bases.
ἔστωσαν ὑπὸ τὸ αὐτὸ ὕψος πυραμίδες ὧν βάσεις μὲν τὰ ΑΒΓ ΔΕΖ τρίγωνα κορυφαὶ δὲ τὰ Η Θ σημεῖα. λέγω ὅτι ἐστὶν ὡς ἡ ΑΒΓ βάσις πρὸς τὴν ΔΕΖ βάσιν οὕτως ἡ ΑΒΓΗ πυραμὶς πρὸς τὴν ΔΕΖΘ πυραμίδα.	Let pyramids be under a same height whose bases are triangles, ΑΒΓ, ΔΕΖ, and vertices points Η, Θ. I claim that, as base ΑΒΓ is to base ΔΕΖ, so pyramid ΑΒΓΗ is to pyramid ΔΕΖΘ.

---

[32] The instantiation of the "impossible" configuration and the construction are in the suppositional mode and are conjoined by a καί "and", but only the former belongs to the consequent of the conditional introduced by εἰ "if".
[33] Read also Proclus, *iE*, 204.5–13, on this.
[34] All constructions of Book X contain the verb form εὑρεῖν "find" (see Sect. 4.3 and Appendix A).
[35] A non-liminal ἔστω with a copulative value we shall read later, in proposition XI.10.
[36] Note the adjective τριγώνους "triangular" (fem.) in the enunciation.

Of some interest are the divided forms ἔστω + participle:[37] they emphasize the presential import of the verb form that figures in the enunciation. Let us read two examples, in the following extracts from I.36 and X.42 (*EOO* I, 86.12–17, and III, 120.21–24):[38]

τὰ παραλληλόγραμμα τὰ ἐπὶ ἴσων βάσεων ὄντα καὶ ἐν ταῖς αὐταῖς παραλλήλοις ἴσα ἀλλήλοις ἐστίν.	Parallelograms that are[39] on equal bases and in the same parallels are equal to one another.
ἔστω παραλληλόγραμμα τὰ ΑΒΓΔ ΕΖΗΘ ἐπὶ ἴσων βάσεων ὄντα τῶν ΒΓ ΖΗ καὶ ἐν ταῖς αὐταῖς παραλλήλοις ταῖς ΑΘ ΒΗ.	Let there be parallelograms, ΑΒΓΔ, ΕΖΗΘ, that are on equal bases, ΒΓ, ΖΗ, and in the same parallels, ΑΘ, ΒΗ.
ἡ ἐκ δύο ὀνομάτων κατὰ ἓν μόνον σημεῖον διαιρεῖται εἰς τὰ ὀνόματα.	A binomial can be divided[40] in the names at one point only.
ἔστω ἐκ δύο ὀνομάτων ἡ ΑΒ διῃρημένη εἰς τὰ ὀνόματα κατὰ τὸ Γ.	Let there be a binomial, AB, that turns out to be divided into the names at Γ.

The setting-out may not be introduced by a form of "to be". The verb in the imperative in the first clause of the setting-out may in fact be the same as the verb used in the enunciation. The choice between the two possibilities is a matter of style: the two formulations can be easily transformed into one another, for instance by using a circumstantial participle or a conjoined clause. Let us check this in the setting-outs of XIII.3–4 (*EOO* IV, 254.21–23 and 256.26–27), whose conditional enunciations have identical antecedents:[41]

εὐθεῖα γάρ τις ἡ ΑΒ ἄκρον καὶ μέσον λόγον τετμήσθω κατὰ τὸ Γ σημεῖον, καὶ ἔστω μεῖζον τμῆμα τὸ ΑΓ.	In fact, let some straight line, AB, be cut in extreme and mean ratio at point Γ, and let ΑΓ be the greater segment.
ἔστω εὐθεῖα ἡ ΑΒ, καὶ τετμήσθω ἄκρον καὶ μέσον λόγον κατὰ τὸ Γ, καὶ ἔστω μεῖζον τμῆμα τὸ ΑΓ.	Let there be a straight line, AB, and let it be cut in extreme and mean ratio at Γ, and let ΑΓ be the greater segment.

The presence of a liminal imperative of "to be" in a setting-out is connected with the structure of the enunciation: if this is not a conditional, the setting-out is normally introduced by ἔστω presenting the object that is the logical subject of the enunciation; if the enunciation is in conditional form, the verb of the setting-out is normally the same as that of the antecedent of the enunciation and is placed in the same position in both clauses. Since the verb form of the antecedent of a conditional enunciation almost never immediately follows the subordinant ἐάν "if", liminal verb forms of the setting-out other than ἔστω are exceedingly rare. Such occurrences may simply be stylistic choices or may be forced by the interference with formulaic expressions, as in the first part of the enunciation of XII.10 (*EOO* IV, 186.11–14)—here, it is noteworthy that no denotative letters are assigned to the cylinder or to the cone:

[37] A quick survey of such constructions in mathematical texts is in Federspiel 2010, 104–107; see also Sect. 2.2.
[38] In the first example, the locative value of "to be" in the enunciation is attached to the participle ὄντα, not to ἔστω; the second example is representative of the "hexad" X.42–47 (see below for this notion).
[39] The participle ὄντα marks the attributive syntagm in which it is embedded as a transitory state of the parallelograms, and at the same time supports a very heavy modifier with a "crutch". Heaviness also justifies the aNaX noun phrase.
[40] On the potential connotation of this present stem see Sect. 5.2.
[41] The antecedent is in both cases ἐὰν εὐθεῖα γραμμὴ ἄκρον καὶ μέσον λόγον τμηθῇ "if a straight line be cut in extreme and mean ratio". In both setting-outs, note the copulative value of the second occurrence of the verb "to be", for in a straight line cut in extreme and mean ratio the greater segment is univocally determined.

πᾶς κῶνος κυλίνδρου τρίτον μέρος ἐστὶ τοῦ τὴν αὐτὴν βάσιν ἔχοντος αὐτῷ καὶ ὕψος ἴσον.
ἐχέτω γὰρ κῶνος κυλίνδρῳ βάσιν τε τὴν αὐτὴν τὸν ΑΒΓΔ κύκλον καὶ ὕψος ἴσον.

Every cone is a third part of a cylinder having the same base as it and equal height.
In fact, let a cone have both as base the same as a cylinder, namely, a circle ΑΒΓΔ, and equal height.

Setting-outs governed by a verb different from the main verb of a conditional enunciation are also infrequent. This happens if the verb in the enunciation specifies an action undergone by a mathematical object, as in our paradigmatic proposition III.2, or in the first conjunct of the antecedent of the enunciation of III.9 (*EOO* I, 190.12–18),[42]

ἐὰν κύκλου ληφθῇ τι σημεῖον ἐντὸς ἀπὸ δὲ τοῦ σημείου πρὸς τὸν κύκλον προσπίπτωσι πλείους ἢ δύο ἴσαι εὐθεῖαι, τὸ ληφθὲν σημεῖον κέντρον ἐστὶ τοῦ κύκλου.
ἔστω κύκλος ὁ ΑΒΓ ἐντὸς δὲ αὐτοῦ σημεῖον τὸ Δ, καὶ ἀπὸ τοῦ Δ πρὸς τὸν ΑΒΓ κύκλον προσπιπτέτωσαν πλείους ἢ δύο ἴσαι εὐθεῖαι αἱ ΔΑ ΔΒ ΔΓ.

If some point be taken inside a circle and from the point more than two equal straight lines fall on the circle, the taken point is the centre of the circle.
Let there be a circle, ΑΒΓ, and a point inside it, Δ, and from Δ let more than two equal straight lines, ΔΑ, ΔΒ, ΔΓ, fall on circle ΑΒΓ.

or in some theorems of Book I that deal with triangles, of which we read the SSS criterion of congruence in I.8 (*EOO* I, 26.13–22):

ἐὰν δύο τρίγωνα τὰς δύο πλευρὰς ταῖς δύο πλευραῖς ἴσας ἔχῃ ἑκατέραν ἑκατέρᾳ ἔχῃ δὲ καὶ τὴν βάσιν τῇ βάσει ἴσην, καὶ τὴν γωνίαν τῇ γωνίᾳ ἴσην ἕξει τὴν ὑπὸ τῶν ἴσων εὐθειῶν περιεχομένην.
ἔστω δύο τρίγωνα τὰ ΑΒΓ ΔΕΖ τὰς δύο πλευρὰς τὰς ΑΒ ΑΓ ταῖς δύο πλευραῖς ταῖς ΔΕ ΔΖ ἴσας ἔχοντα ἑκατέραν ἑκατέρᾳ, τὴν μὲν ΑΒ τῇ ΔΕ τὴν δὲ ΑΓ τῇ ΔΖ, ἐχέτω δὲ καὶ βάσιν τὴν ΒΓ βάσει τῇ ΕΖ ἴσην.

If two triangles have two sides equal to two sides, respectively, and also have the base equal to the base, they will also have the angle contained by the equal straight lines equal to the angle.
Let there be two triangles, ΑΒΓ, ΔΕΖ, having two sides, ΑΒ, ΑΓ, equal to two sides, ΔΕ, ΔΖ, respectively, ΑΒ to ΔΕ and ΑΓ to ΔΖ, and let them also have base ΒΓ equal to base ΕΖ.

The antecedent of the conditional of proposition I.8 contains two identical forms of the verb ἔχω "to have". The first is transformed into an attributive participle governed by a presential ἔστω, the second follows the rule of transformation into an imperative.

Non-liminal imperatives of the verb "to be" are found in a setting-out if "to be" is the main verb of the enunciation and if it has a copulative value there; conversely, if "to be" has a presential value in the enunciation, ἔστω is regularly in a liminal position in the setting-out. Let us read the enunciation and the setting-out of XI.10 and of V.22 (*EOO* IV, 30.2–7, and II, 60.18–26) as examples:

ἐὰν δύο εὐθεῖαι ἁπτόμεναι ἀλλήλων παρὰ δύο εὐθείας ἁπτομένας ἀλλήλων ὦσι μὴ ἐν τῷ αὐτῷ ἐπιπέδῳ, ἴσας γωνίας περιέξουσιν.
δύο γὰρ εὐθεῖαι αἱ ΑΒ ΒΓ ἁπτόμεναι ἀλλήλων παρὰ δύο εὐθείας τὰς ΔΕ ΕΖ ἁπτομένας ἀλλήλων ἔστωσαν μὴ ἐν τῷ αὐτῷ ἐπιπέδῳ.

If two straight lines touching one another be parallel to two straight lines touching one another not in a same plane, they will contain equal angles.
In fact, let two straight lines touching one another, ΑΒ, ΒΓ, be parallel to two straight lines touching one another not in a same plane, ΔΕ, ΕΖ.

[42] Note the construction ἀπὸ κοινοῦ of ἔστω.

ἐὰν ᾖ ὁποσαοῦν μεγέθη καὶ ἄλλα αὐτοῖς ἴσα τὸ πλῆθος σύνδυο λαμβανόμενα καὶ ἐν τῷ αὐτῷ λόγῳ, καὶ δι' ἴσου ἐν τῷ αὐτῷ λόγῳ ἔσται.
ἔστω ὁποσαοῦν μεγέθη τὰ Α Β Γ καὶ ἄλλα αὐτοῖς ἴσα τὸ πλῆθος τὰ Δ Ε Ζ σύνδυο λαμβανόμενα ἐν τῷ αὐτῷ λόγῳ, ὡς μὲν τὸ Α πρὸς τὸ Β οὕτως τὸ Δ πρὸς τὸ Ε, ὡς δὲ τὸ Β πρὸς τὸ Γ οὕτως τὸ Ε πρὸς τὸ Ζ.

If there be as many magnitudes as we please and others equal to them in multiplicity taken two and two together and in a same ratio, they will also be in a same ratio through an equal.
Let there be as many magnitudes as we please, A, B, Γ, and others equal to them in multiplicity, Δ, E, Z, taken two and two together in a same ratio, as A is to B, so Δ is to E, and, as B is to Γ, so E is to Z.

As noted in Sect. 3.1, examples such as V.22 just read formally prove that the liminal ἔστω in a setting-out has a presential value.

A major stylistic trait is the presence of γάρ at the beginning of all and only those setting-outs as are introduced by a verb different from "to be" or from one of its presential substitutes (see below). In this function, the particle γάρ does not carry any explicative value: it is simply a scope particle. This is formally proved by the fact that the standard form of setting-out does not contain any γάρ, nor any linguistic item that may convey an explicative connotation. The exceptions to the rule of alternance liminal ἔστω / liminal γάρ in the main text of the *Elements* are sparse, and all with liminal ἔστω followed by γάρ.[43] No exceptions can be found in the *Data* or in Apollonius.[44] Such exceptions are nothing but stylistic variants,[45] as is confirmed by the similar setting-outs + determinations of propositions I.30 and XI.9 (*EOO* I, 74.6–7, and IV, 28.7–9), both featuring a liminal copulative "to be"—well, these setting-out are not only similar: if we eliminate the participial modifier (italicized below) and due allowance being made for the γάρ, they are strictly identical:

ἔστω ἑκατέρα τῶν ΑΒ ΓΔ τῇ ΕΖ παράλληλος. λέγω ὅτι καὶ ἡ ΑΒ τῇ ΓΔ ἐστι παράλληλος.	Let each of two ‹straight lines›, AB, ΓΔ, be parallel to a ‹straight line›, EZ. I claim that AB is also parallel to ΓΔ.
ἔστω γὰρ ἑκατέρα τῶν ΑΒ ΓΔ τῇ ΕΖ παράλληλος *μὴ οὖσαι αὐτῇ ἐν τῷ αὐτῷ ἐπιπέδῳ*. λέγω ὅτι παράλληλός ἐστιν ἡ ΑΒ τῇ ΓΔ.	In fact, let each of two ‹straight lines›, AB, ΓΔ, *which are not in a same plane as it*, be parallel to a ‹straight line›, EZ. I claim that AB is parallel to ΓΔ.

The alternance of liminal verb form and of liminal scope particle (where "liminal" has a specific meaning in both cases) confirms that liminal ἔστω is a stylistic marker of the setting-out qua specific part of a mathematical proposition (cf. Sect. 3.1).[46]

[43] These are I.18, 20, III.24, V.11, 15, 19; VI.21, VII.20, X.80, 105, XI.9, XII.9, 14. They are distributed among the several values of "to be" as follows: presential: I.20, VII.20, X.80, 105; copulative: V.11, 15, 19, VI.21, XI.9; locative: III.24, XII.14; discontinuous participial clause: I.18, XII.9. It is noteworthy that V.11, VI.21, and XI.9 prove the transitivity of identity of ratios, of similitude, and of parallelism in space, respectively.

[44] Archimedes exhibits several occurrences, in *Sph. cyl.* I.4, 27, 28*, 30, 32*, 33, 34, 38, 40, 41, 43, II.3por; *Con. sph.* 4, 5, 21, 26, 27*, 29; *Spir.* 19; *Quadr.* 17, 18, 19, 20, 22, 24; *Fluit.* I.4 (* = non-liminal setting-out).

[45] The exceptions date back to early campaigns of revision: only for X.105 Heiberg's critical apparatus reports a variant reading: the Theonine manuscripts omit γάρ; see *EOO* III, 336.16 *app.*

[46] Federspiel 2010, 109–112, has surmised that the compartmentation liminal ἔστω / liminal γάρ in the setting-out is the result of an evolution of the format of a proposition; originally, this was directly enunciated in instantiated form, such an enunciation being opened by a presential ἔστω. The presence of γάρ, even in the function of a scope particle, is obviously impossible in this syntactic configuration.

I now offer a discussion of the use and the meaning of the term ἔκθεσις and its kin in mathematical texts. This will lead us to outline the structure of Book X.

The noun ἔκθεσις is the *nomen actionis* of ἐκτίθημι "to set out"; in English, it is very appropriately translated by a gerund + adverb: "setting-out". In mathematical contexts, this verb introduces a mathematical entity that is not assumed in the enunciation of a proposition but that is put to use in a proof even if its presence cannot be anchored to supposed objects by means of a construction: such an entity is "fixed" or "appointed" but it is not "given"; it is *set out* "there" for us.[47] In the Euclidean corpus, the noun ἔκθεσις is found only twice, with an obvious metadiscursive connotation, at the very beginning of the alternative proofs of propositions 12 and 14 of *Phaenomena* redaction **b**.[48] In the same corpus, instead, the verb exhibits 177 occurrences, 174 of which as forms of the presential verb ἔκκειμαι "to be set out",[49] which frequently serves as the passive voice of (ἐκ)τίθημι and that must always be taken as a *praesens pro perfecto* (cf. Sect. 1.1). None of these verb forms is found in a setting-out: they all occur in the construction.

Book X alone consumes 121 of the 159 occurrences of forms of ἔκκειμαι in the *Elements*. The reason is that the theory of irrational lines presupposes that a reference line, called ῥητή "expressible", be set out. This assumption is necessary since no straight line can be irrational per se, but only with respect to a line assumed as a standard.[50] Such an assumption cannot be made once and for all at the very beginning of Book X because the expressible line suitable for any specific proof varies from proposition to proposition: thus, an expressible must be "set out" afresh at the beginning of every proposition.[51] The standard form of the assumption is ἐκκείσθω ῥητὴ ἡ AB "let an expressible, AB, be set out"; use is made of ἔκκειμαι and not of εἰμί since the expressible line almost never figures among the objects mentioned in the enunciation of a proposition: in particular, an expressible is not "given". For this reason, the ἐκκείσθω-suppositions are not included in the setting-out, but in the construction.[52] Of course, when an expressible line or region is mentioned in the enunciation, it is regularly "presented" in the setting-out. Let us read the beginning of X.108 (*EOO* III, 342.10–18) as an example (*another* expressible is set out in the construction):

ἀπὸ ῥητοῦ μέσου ἀφαιρουμένου ἡ τὸ λοιπὸν χωρίον δυναμένη μία δύο ἀλόγων γίνεται ἤτοι ἀποτομὴ ἢ ἐλάσσων.	From *an expressible ‹region›* a medial ‹region› being removed, the ‹straight line› worth the remaining region yields one of two irrationals, either an apotome or a minor.
ἀπὸ γὰρ *ῥητοῦ* τοῦ ΒΓ μέσον ἀφῃρήσθω τὸ ΒΔ. λέγω ὅτι ἡ τὸ λοιπὸν δυναμένη τὸ ΕΓ μία δύο ἀλόγων γίνεται ἤτοι ἀποτομὴ ἢ ἐλάσσων.	In fact, from *an expressible ‹region›*, ΒΓ, let a medial ‹region›, ΒΔ, be removed. I claim that the ‹straight line› worth the remaining ‹region› ΕΓ yields one of two irrationals, either an apotome or a minor.
ἐκκείσθω γὰρ ῥητὴ ἡ ΖΗ, καὶ τῷ μὲν ΒΓ ἴσον παρὰ τὴν ΖΗ παραβεβλήσθω ὀρθογώνιον παραλληλόγραμμον τὸ ΗΘ, τῷ δὲ ΔΒ ἴσον ἀφῃρήσθω τὸ ΗΚ.	In fact, let an expressible, ΖΗ, be set out, and let a rectangular parallelogram, ΗΘ, equal to ΒΓ be applied to ΖΗ, and let a ‹region›, ΗΚ, be removed equal to ΔΒ.

[47] The first occurrence in the *Elements* is at the beginning of the construction of I.22: ἐκκείσθω τις εὐθεῖα ἡ ΔΕ "let some straight line ΔΕ be set out" (*EOO* I, 52.26). We shall read the entire proposition below.
[48] At *EOO* VIII, 116.4 and 122.14. There are two recensions of Euclid's *Phaenomena*, called **a** and **b**.
[49] The 3 occurrences of forms of ἐκτίθημι are in IX.36, XIII.18, and *alt*X.23.
[50] Such a dialectic is formulated, in the scholarly material that accompanies the *Elements*, under the traditional opposition between "nature" and "convention"; see for instance *sch*. X.1 in *EOO* V, 414.1–16.
[51] In X.27–35, several interrelated expressible lines must be supposed.
[52] Likewise, a unit is "set out" in VIII.9 and IX.32.

However, an imperative ἐκκείσθω "let it be set out" instead of ἔστω is sometimes used in the setting-out, a remarkable exception to the standard practice of the *Elements*. As an example, let us read the enunciation and the setting-out of X.60 (*EOO* III, 180.26–182.5):[53]

τὸ ἀπὸ τῆς ἐκ δύο ὀνομάτων παρὰ ῥητὴν παραβαλλόμενον πλάτος ποιεῖ τὴν ἐκ δύο ὀνομάτων πρώτην. ἔστω ἐκ δύο ὀνομάτων ἡ ΑΒ διῃρημένη εἰς τὰ ὀνόματα κατὰ τὸ Γ ὥστε τὸ μεῖζον ὄνομα εἶναι τὸ ΑΓ, καὶ ἐκκείσθω ῥητὴ ἡ ΔΕ.

The ‹square› on a binomial applied to an expressible makes a first binomial width. Let there be a binomial, AB, that turns out to be divided into the names at Γ so as to be ΑΓ the greater name, and let an expressible ΔΕ be set out.

The range of use of ἔκκειμαι "to be set out" is wide, even within Book X, and deserves a detailed discussion. Let us start with some raw data. The following table sets out the distribution, in the main text of the *Elements*, of the verbs ἐκτίθημι / ἔκκειμαι, keyed to the verb form and to the kind of mathematical object set out:

	geometric object	number	magnit.	expressible
ἐκκείσθω	I.22, IV.10–11, VI.23, XI.36, XIII.13, 13/14, 14–16, 18	VIII.9, IX.32		X.23, 25–26, 29–30, 38, 41, 41/42, 44, 47, 48–53, 60–61, 64, 71–72, 75, 78, 81, 84, 85–90, 108–109, 111, 114–115
ἐκκείσθωσαν	VI.12, XI.23/24, XII.13, XIII.17	IX.36, X.10, 28/29, 29–30, 48–53, 85–90		X.27–28, 31–35
ἐκκειμεν–	XII.11	VII.1*, IX.36	X.1*, XII.2	X.18/19, def.II, 49, 52–54, 60, 63, 65–66, 71–72 X.85–103, def.III, 108–113
ἐκτίθημι	XIII.18*	IX.36*		

The occurrences of participial forms of ἔκκειμαι "to be set out" in an enunciation (marked by an asterisk above) are particularly interesting.[54] We shall read the occurrence in XIII.18 later. As for the others, VII.1, IX.36, and X.1 are theorems of a particular kind: numbers or magnitudes are set out and acted upon by an iterative procedure with a termination condition. In VII.1, the procedure is the Euclidean algorithm of reciprocal subtractions applied to numbers, the termination condition is attaining a unit (see Sect. 1.2). In X.1, the procedure consists in continually removing more than half of a set out magnitude, the termination condition is attaining the threshold represented by another magnitude set out.[55] Therefore, in these propositions ἔκκειμαι serves as a synonym of δίδομαι "to be given"; the stylistic demarcation is strong, and is probably induced by the fact that VII.1 and X.1 are theorems, not problems. In fact, when the above procedures are applied within problems, as in VII.2–3 and X.3–4, respectively, the assigned magnitudes are regularly qualified by the predicate "given". In IX.36, which proves the celebrated sufficient condition for a number to be perfect, the procedure consists in progressively setting out a geometric progression of ratio 2, the termination condition is attained when the sum of the progression is a prime number.[56] Let us now read the setting-out of proposition IX.36 (*EOO* II, 408.7–16):

[53] On the use of the article in these texts, see Sect. 5.1.5.

[54] A further occurrence in X.2 is certainly spurious.

[55] X.1 is characterized by an ἔστω in the setting-out that replaces the ἐκκείσθω we would have expected. This confirms that the verb "to be" has a presential value in this context.

[56] In number theory, objects presented by ἐκκείσθω are usually imposed more rigid constraints that geometric magnitudes: in IX.36, a concrete geometric progression is in fact at issue.

ἀπὸ γὰρ μονάδος ἐκκείσθωσαν ὁσοιδηποτοῦν ἀριθμοὶ ἐν τῇ διπλασίονι ἀναλογίᾳ ἕως οὗ ὁ σύμπας συντεθεὶς πρῶτος γένηται, οἱ Α Β Γ Δ, καὶ τῷ σύμπαντι ἴσος ἔστω ὁ Ε, καὶ ὁ Ε τὸν Δ πολλαπλασιάσας τὸν ΖΗ ποιείτω.	In fact, starting from a unit let as many number as we please in double proportion be set out until the sum total become prime, A, B, Γ, Δ, and let E be equal to the sum total, and let E multiplying Δ make ZH.

Other numeric assignments can be found in problems X.10, 28/29, 29–30, 48–53, and 85–90.[57] Let us read the assignment in proposition X.48 (*EOO* III, 136.22–138.1)—of course, the expressible line is also set out at the end:[58]

ἐκκείσθωσαν δύο ἀριθμοὶ οἱ ΑΓ ΓΒ ὥστε τὸν συγκείμενον ἐξ αὐτῶν τὸν ΑΒ πρὸς μὲν τὸν ΒΓ λόγον ἔχειν ὃν τετράγωνος ἀριθμὸς πρὸς τετράγωνον ἀριθμόν, πρὸς δὲ τὸν ΓΑ λόγον μὴ ἔχειν ὃν τετράγωνος ἀριθμὸς πρὸς τετράγωνον ἀριθμόν, καὶ ἐκκείσθω τις ῥητὴ ἡ Δ.	Let two numbers, AΓ, ΓB, be set out so as to have the ‹number› compounded of them, AB, to BΓ a ratio that a square number ‹has› to a square number, and so as not to have to ΓA a ratio that a square number ‹has› to a square number, and let some expressible, Δ, be set out.

In the Neopythagorean tradition, represented for instance by the writings of Nicomachus and of Iamblichus,[59] an ἐκκείσθω-clause came to be the canonical formulation for introducing concrete numeric sequences.

An explanation of the structure of Book X is necessary to understand the distribution of forms of ἔκκειμαι in this Book. After a series of preliminary propositions and the construction of the first irrational line—called "medial"—six irrational lines are constructed (and in fact, defined)[60] by adding two expressible or medial straight lines related by suitable relations of (in)commensurability (X.36–41). An analogous operation allows constructing six irrational lines by removal (X.73–78): the two basic straight lines are not added but one of them is removed from the other. The two sets of propositions correlate as follows—the names of the irrational lines are also indicated:

addition		removal	
X.36	binomial	X.73	apotome
X.37	first bimedial	X.74	first apotome of a medial
X.38	second bimedial	X.75	second apotome of a medial
X.39	major	X.76	minor
X.40	‹straight line› that is worth an expressible and a medial ‹region›	X.77	‹straight line› that produces with an expressible ‹region› a medial whole
X.41	‹straight line› that is worth two medial ‹regions›	X.78	‹straight line› that produces with a medial ‹region› a medial whole

The names of the last four irrational lines expressly refer to the method of their construction; this shows that they were introduced for systematic purposes. The remaining propositions of Book X prove that the six species of irrational lines defined by addition and the six species defined by removal are univocally determined and identify disjoint classes of irrationals. Five propositions for

[57] In X.29–30 and 90 the verb form is "factored out" between the expressible and the numeric assignment.
[58] In the corresponding proposition X.85, the numeric assignment is interchanged with the geometric setting-out.
[59] The nominal form ἔκθεσις and the verbs ἐκτίθημι and ἔκκειμαι exhibit nearly the same number of occurrences in Nicomachus' *Introductio arithmetica* and in Iamblichus' rewriting of it, with a slight prevalence of the noun.
[60] Each of the twelve irrational lines is in fact defined in the very same proposition in which it is constructed.

each species are proved to this effect, including the construction of six subspecies of the binomial and of six subspecies of the apotome. Since the proofs are repeated for each of the twelve species of irrationals constructed so far, the size of Book X is easy to explain: 1 proposition in which the species is introduced + 5 needed for the "classification", which, multiplied by 12, gives 72 propositions. Actually, we only find 70 propositions: X.67 and X.104, whose proofs are strictly parallel, dispose of two species (the two bimedials and the two apotomes of a medial) in a single theorem. Now, the theory starts at X.36, which gives 105 propositions as a sum; add 5 theorems that make explicit the link with a constraint on suitable regions constructed from the species (X.71–72 and X.108–110) and the theorem where it is proved that an apotome is not identical to a binomial (X.111). This is the entirety of Book X, except for the four final results, which all scholars deem spurious. Add to this a series of lemmas and porisms, coming from several layers of interpolations.

The following table sets out the structure of the "core" 70 propositions of Book X; they are naturally arranged in groups of six, called "hexads":

	addition	removal
construction of the species of irrationals	36–41	73–78
uniqueness of the partition into straight lines	42–47	79–84
definition of the six subspecies	defII	defIII
construction of the six subspecies	48–53	85–90
first relation between the six species and the six subspecies	54–59	91–96
second relation between the six species and the six subspecies	60–65	97–102
closure of the species under commensurability	66–70	103–107
identification by means of regions	71–72	108–110

Now, the following tables set out the distribution of the verb forms ἐκκείσθω and ἐκκειμεν– in the hexads of Book X: these verb forms mark how and when an expressible is introduced, and within the proof of which theorems it is used, respectively.[61] In two of the instances in which the expressible is introduced by ἔστω, it is subsequently referred to by a participle ἐκκειμένη "set out" (fem.). Let us see first the table of the imperative:

I	II	III	IV	V	VI	identif.
/ / 38 / / 41	/ / 44 / / 47	48–53	/	60 61 / / 64 /	/	71 72
/ / 75 / / 78	/ / 81 / / 84	85–90	/	/ / / / / /	/	108 109 /

then the table of the participle:

I	II	III	IV	V	VI	identif.
/	/	/ 49 / / 52 53	54 / / / / /	60 / / 63 / 65	66 / / / / /	71 72
/	/	85–90	91–96	97–102	103 / / / / /	108 109 110

If we exclude the fifth hexad, the occurrences in the imperative display a remarkable symmetry. The sporadic (but parallel) occurrences in the first two hexads are explained by the fact that only two of the six irrational lines are generated by a plane construction, in which a suitable region is applied to a set out expressible. The theorems of the fourth and of the fifth hexad mention the expressible in the enunciation. The setting-outs of the fourth hexad keep the same verb as the enunciation: therefore, there are no presential verbs and the expressible is directly designated by letters;

[61] This usually happens in an instantiated citation of X.defII or X.defIII (we shall read X.defII in Sect. 5.3.1).

the same happens with the two expressibles in the enunciations of X.36 and 73. In the fifth hexad, the enunciation is transformed into a setting-out with liminal ἔστω, and the subsequent list also includes the expressible line; only in three propositions (X.60, 61, and 64) is this line further presented by ἐκκείσθω. Among the "identifications" that close Book X, the entire construction is omitted in X.110. All occurrences of participial forms are induced by the primary occurrences in X.defII and X.defIII (*quater* each, in the definitions of the first four subspecies); in these definitions, the expressible is presented within a genitive absolute ὑποκειμένης ῥητῆς "an expressible being supposed". The hexads of the irrationals generated by removal display the participle in all possible instances, those generated by addition only occasionally do, even if in two cases out of three the participle is contained in the first proposition of the hexad.

Outside the theory of irrational lines, the *geometric* entities presented or qualified by forms of ἔκκειμαι are in a strong sense left to the mathematician's choice, no further constraints being imposed. As a matter of fact, this verb is employed to appoint such objects in situations of complete indetermination, namely, when the mentioned objects are neither introduced in the enunciation nor constrained as to their construction (see I.22, to be read entirely just below).[62]

In Apollonius, forms of ἔκκειμαι can only be found in the problems *Con.* II.51 (*bis*), 52, 53.[63] These are 4 occurrences in the imperative within constructions; they present arbitrary auxiliary objects (a straight line and two circles), totally disconnected from the givens of the problem.

Archimedes makes a peculiar use of ἔκκειμαι; the occurrences are set out in the following table—with one exception, all of them are in *Sph. cyl.*:

	setting-out	construction	conclusion
ἐκκείσθω	I.5, 16, 18–20, 24, *Con. sph.* 21	I.15, 18, 25	
ἐκκείσθωσαν		I.16, 19, 20, 24, 34, II.3	
ἐκκειμένῳ			I.26

The occurrences in the constructions fit the typology just expounded for the *Elements*; in *Sph. cyl.* I.34, ἔκκειμαι is synonymous with ἔστωσαν εἰλημμέναι "let them turn out to be taken". In all but one setting-out in the above list,[64] instead, the presence of ἔκκειμαι is explained by the peculiarity of the results to be proved, of a kind that is not found in the *Elements* or in Apollonius. To see this, let us read the enunciation and the setting-out of *Sph. cyl.* I.18 (*AOO* I, 76.18–78.2):

παντὶ ῥόμβῳ ἐξ ἰσοσκελῶν κώνων συγκειμένῳ ἴσος ἐστὶ κῶνος ὁ βάσιν μὲν ἔχων ἴσην τῇ ἐπιφανείᾳ τοῦ ἑτέρου κώνου τῶν περιεχόντων τὸν ῥόμβον ὕψος δὲ ἴσον τῇ ἀπὸ τῆς κορυφῆς τοῦ ἑτέρου κώνου καθέτῳ ἀγομένῃ ἐπὶ μίαν πλευρὰν τοῦ ἑτέρου κώνου.

To every rhombus compounded of isosceles cones is equal a cone having base equal to the surface of one cone among those containing the rhombus and height equal to the ‹straight line› drawn from the vertex of the other cone perpendicular to one side of the one cone.

[62] These degrees of freedom always and only obtain in constructions, namely, those in I.22, IV.10, VI.12, 23, X.29, 48, 50, 52–53, 62, XI.23/24 XII.13 (straight line), IV.11 (triangle), XI.36 (solid angle), XIII.11, 13–16, 18 (diameter serving as an expressible), XIII.18 (edges of regular polyhedra), XIII.13, 16 (circle), XIII.14–15 (square), XIII.17 (two faces of a cube), to which we may add *Data* 6 (magnitude), 24 (straight line given tout court), 39 (straight line given in position), 42 (in magnitude), 40–41, 43, 55, 80 (in magnitude and in position).

[63] At *AGE* I, 300.4, 300.27, 308.1, and 312.18, respectively. The double occurrence in II.51 is forced by the format by analysis and synthesis. The straight line set out in II.51 is incongruously called δεδομένη "given"; only in the first occurrence in this proposition is the arbitrariness of the straight line emphasized by τις "some" (see Sect. 5.1.3 for τις).

[64] In the setting-out of *Sph. cyl.* I.5, the verb is synonymous with δεδόσθω or with ἔστω + participle δοθείς.

ἔστω ῥόμβος ἐξ ἰσοσκελῶν κώνων συγκείμενος ὁ ΑΒΓΔ οὗ βάσις ὁ περὶ διάμετρον τὴν ΒΓ κύκλος ὕψος δὲ τὸ ΑΔ, ἐκκείσθω δέ τις ἕτερος ὁ ΗΘΚ τὴν μὲν βάσιν ἔχων τῇ ἐπιφανείᾳ τοῦ ΑΒΓ κώνου ἴσην τὸ δὲ ὕψος ἴσον τῇ ἀπὸ τοῦ Δ σημείου καθέτῳ ἐπὶ τὴν ΑΒ ἢ τὴν ἐπ' εὐθείας αὐτῇ ἠγμένῃ, ἔστω δὲ ἡ ΔΖ, τὸ δὲ ὕψος τοῦ ΘΗΚ κώνου ἔστω τὸ ΘΛ· ἴσον δή ἐστιν τὸ ΘΛ τῇ ΔΖ. λέγω ὅτι ἴσος ἐστὶν ὁ κῶνος τῷ ῥόμβῳ.

Let there be a rhombus compounded of isosceles cones, ΑΒΓΔ, whose base is the circle around ΒΓ and height ΑΔ,[65] and let some other ‹cone›, ΗΘΚ, be set out having base equal to the surface of cone ΑΒΓ and height equal to the ‹straight line› drawn from point Δ perpendicular to ΑΒ or to the ‹straight line› drawn in a straight line with it, and let it be ΔΖ, and let the height of cone ΘΗΚ be ΘΛ; thus ΘΛ is equal to ΔΖ. I claim that the cone is equal to the rhombus.

It is required to prove that two geometric objects are equal: one of them is a given conical rhombus, the other is a cone whose dimensional parameters (its base and its height) are specified by their being equal to suitable dimensional parameters of the conical rhombus. The second cone is "actualized" at the end of an argument that *proves* that it is equal to a *third* cone equal to the given conical rhombus; the third cone is regularly presented by ἐκκείσθω *at the beginning of the construction.* Everyone sees that the "second cone" is a fictitious entity, introduced in the enunciation in order to fill an empty place in a chain of equalities. And in fact, it is presented *in the setting-out* by ἐκκείσθω: it is not an object that figures in the proof, but an anticipation of the final result: it is the (in)definite description of a complex entity generated as the "right-hand side" of an equality.

This is not a further sign of the alleged Archimedean heterodoxy. On the contrary, this is a move that is deeply rooted in a surprising conceptual framework of Greek geometry, namely, what I call the "actualization" of specific magnitudes by "production" of their duplicates, in particular within the "language of the givens".[66] In this categorial framework, the use of forms of ἔκκειμαι is exactly what we should expect to find. Let us see how "actualization" works by reading Theodosius, *Sph.* I.18 (32.6–24), in fact a theorem of plane geometry:

τοῦ δοθέντος ἐν σφαίρᾳ κύκλου τὴν διάμετρον ἐκθέσθαι.

Set out the diameter of a given circle in a sphere.

ἔστω ὁ δοθεὶς ἐν σφαίρᾳ κύκλος ὁ ΑΒΓ. δεῖ δὴ τοῦ ΑΒΓ κύκλου τὴν διάμετρον ἐκθέσθαι.

Let there be a given circle in a sphere, ΑΒΓ. Thus it is required to set out the diameter of circle ΑΒΓ.

εἰλήφθω ἐπὶ τῆς τοῦ κύκλου περιφερείας τυχόντα σημεῖα τὰ Α Β Γ, καὶ ἐκ τριῶν εὐθειῶν τρίγωνον συνεστάτω τὸ ΔΕΖ ὥστε ἴσην εἶναι τὴν μὲν ΔΕ τῇ ἀπὸ τοῦ Α ἐπὶ τὸ Β τὴν δὲ ΔΖ τῇ ἀπὸ τοῦ Α ἐπὶ τὸ Γ τὴν δὲ ΕΖ τῇ ἀπὸ τοῦ Β ἐπὶ τὸ Γ, καὶ ἀπὸ μὲν τοῦ Ε σημείου τῇ ΕΔ πρὸς ὀρθὰς ἤχθω ἡ ΕΗ ἀπὸ δὲ τοῦ Ζ τῇ ΔΖ πρὸς ὀρθὰς ἤχθω ἡ ΖΗ, καὶ ἐπεζεύχθω ἡ ΔΗ· ἤχθω δὴ διάμετρος τοῦ ΑΒΓ κύκλου ἡ ΑΘ, καὶ ἐπεζεύχθωσαν αἱ ΑΒ ΒΓ ΓΑ ΓΘ.

Let random points, Α, Β, Γ, be taken on the circumference of the circle, and from three straight lines let a triangle, ΔΕΖ, be constructed so as to be ΔΕ equal to the ‹straight line› from Α to Β and ΔΖ to the ‹straight line› from Α to Γ and ΕΖ to the ‹straight line› from Β to Γ, and from point Ε let a ‹straight line›, ΕΗ, be drawn at right ‹angles› with ΕΔ, and from Ζ let a ‹straight line›, ΖΗ, be drawn at right ‹angles› with ΔΖ, and let a ‹straight line›, ΔΗ, be joined; thus let a diameter, ΑΘ, of circle ΑΒΓ be drawn, and let ‹straight lines›, ΑΒ, ΒΓ, ΓΑ, ΓΘ, be joined.

ἐπεὶ οὖν δύο αἱ ΑΒ ΒΓ δύο ταῖς ΔΕ ΕΖ ἴσαι εἰσὶν ἑκατέρα ἑκατέρᾳ καὶ βάσις ἡ ΑΓ βάσει τῇ ΔΖ ἴση

Then since two ‹sides›, ΑΒ, ΒΓ, are equal to two ‹sides›, ΔΕ, ΕΖ, respectively, and base ΑΓ is equal to

[65] Note that the setting-out does not end here. The construction of this proposition follows the quoted passage!
[66] In this case, the "production" is secured by forms of πορίζομαι "to procure". See Sect. 2.4.1 and Acerbi 2011a on this.

ἐστί, γωνία ἄρα ἡ ὑπὸ ΑΒΓ γωνίᾳ τῇ ὑπὸ ΔΕΖ ἴση ἐστίν· ἀλλ' ἡ μὲν ὑπὸ ΑΒΓ τῇ ὑπὸ ΑΘΓ ἴση ἐστίν ἡ δὲ ὑπὸ ΔΕΖ τῇ ὑπὸ ΔΗΖ ἴση ἐστί· καὶ ἡ ὑπὸ ΑΘΓ ἄρα τῇ ὑπὸ ΔΗΖ ἐστιν ἴση· ἀλλὰ καὶ ὀρθὴ ἡ ὑπὸ ΑΓΘ ὀρθῇ τῇ ὑπὸ ΔΖΗ ἴση ἐστί· καί ἐστιν ἡ ΑΓ τῇ ΔΖ ἴση· καὶ ἡ ΑΘ ἄρα τῇ ΔΗ ἴση ἐστί· καί ἐστιν ἡ ΑΘ διάμετρος τοῦ κύκλου· ἡ ΔΗ ἄρα ἴση ἐστὶ τῇ διαμέτρῳ τοῦ κύκλου.

base ΔΖ, therefore angle ΑΒΓ is equal to angle ΔΕΖ; but ΑΒΓ is equal to ΑΘΓ and ΔΕΖ is equal to ΔΗΖ; therefore ΑΘΓ is also equal to ΔΗΖ; but a right ‹angle›, ΑΓΘ, is also equal to a right ‹angle›, ΔΖΗ; and ΑΓ is equal to ΔΖ; therefore ΑΘ is also equal to ΔΗ; and ΑΘ is a diameter of the circle; therefore ΔΗ is equal to the diameter of the circle.

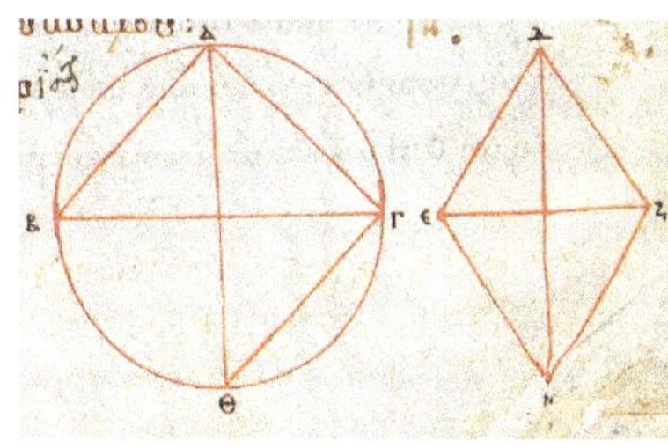

Vat. gr. 204, f. 7v, obviously oversymmetrized

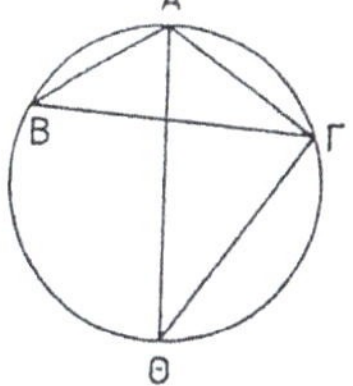

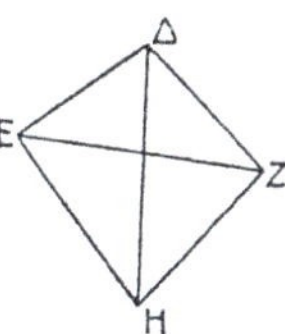

Heiberg's diagram for *Sph.* I.18

Thus, the mathematician who is asked to "set out" an object—here, the diameter of a given circle on a sphere—does not simply draw a diameter ἡ ΑΘ and says "here it is",[67] but duplicates the geometric configuration and calls "set out" the element of the "alias" that is equal to ἡ ΑΘ.

To better appreciate the issue, let us read the archetypal occurrence of "duplication" in *El.* I.22 (*EOO* I, 52.15–54.18)—this very proposition is used by Theodosius at the beginning of the construction of *Sph.* I.18, in order to produce the "alias" of the triangle generated by taking three random points on the circumference of the given circle:[68]

ἐκ τριῶν εὐθειῶν αἵ εἰσιν ἴσαι τρισὶ ταῖς δοθείσαις εὐθείαις τρίγωνον συστήσασθαι. δεῖ δὴ τὰς δύο τῆς λοιπῆς μείζονας εἶναι πάντῃ μεταλαμβανομένας.

Construct a triangle from three straight lines that are equal to three given ‹straight lines›. Thus it is required that two, however permuted, are greater than the remaining one.

ἔστωσαν αἱ δοθεῖσαι τρεῖς εὐθεῖαι αἱ Α Β Γ, ὧν αἱ δύο τῆς λοιπῆς μείζονες ἔστωσαν πάντῃ μεταλαμβανόμεναι, αἱ μὲν Α Β τῆς Γ αἱ δὲ Α Γ τῆς Β καὶ ἔτι αἱ Β Γ τῆς Α. δεῖ δὴ ἐκ τῶν ἴσων ταῖς Α Β Γ τρίγωνον συστήσασθαι.

Let there be three given straight lines, Α, Β, Γ, of which let two, however permuted, be greater than the remaining one, Α, Β of Γ and Α, Γ of Β and further Β, Γ of Α. Thus it is required to construct a triangle from the ‹straight lines› equal to Α, Β, Γ.

ἐκκείσθω τις εὐθεῖα ἡ ΔΕ πεπερασμένη μὲν κατὰ τὸ Δ ἄπειρος δὲ κατὰ τὸ Ε, καὶ κείσθω τῇ μὲν Α ἴση ἡ ΔΖ τῇ δὲ Β ἴση ἡ ΖΗ τῇ δὲ Γ ἴση ἡ ΗΘ, καὶ κέντρῳ μὲν τῷ Ζ διαστήματι δὲ τῷ ΖΔ κύκλος

Let some straight line, ΔΕ, be set out bounded at Δ and unbounded at Ε, and let a ‹straight line›, ΔΖ, be set equal to Α and ΖΗ equal to Β and ΗΘ equal to Γ, and let a circle, ΔΚΛ, be described with centre Ζ and

[67] Exactly this is done at the end of the construction. There is no difficulty in doing this with the *Elements* tool-box, for it is enough to find the centre of the circle (*El.* III.1).

[68] The idea behind the proof of I.22 is the same as that behind the proof of I.1, but the difference in formulation is striking. On the meaning of the participial phrase πάντῃ μεταλαμβανόμεναι "however permuted" see Federspiel 2006c. The standard translation "taken together in any manner" is just wrong.

γεγράφθω ὁ ΔΚΛ. πάλιν κέντρῳ μὲν τῷ Η διαστήματι δὲ τῷ ΗΘ κύκλος γεγράφθω ὁ ΚΛΘ, καὶ ἐπεζεύχθωσαν αἱ ΚΖ ΚΗ.

radius ΖΔ. Again, let a circle, ΚΛΘ, be described with centre Η and radius ΗΘ, and let ‹straight lines›, ΚΖ, ΚΗ, be joined.

λέγω ὅτι ἐκ τριῶν εὐθειῶν τῶν ἴσων ταῖς Α Β Γ τρίγωνον συνέσταται τὸ ΚΖΗ.

I claim that a triangle ΚΖΗ turns out to be constructed from three straight lines equal to Α, Β, Γ.

ἐπεὶ γὰρ τὸ Ζ σημεῖον κέντρον ἐστὶ τοῦ ΔΚΛ κύκλου, ἴση ἐστὶν ἡ ΖΔ τῇ ΖΚ· ἀλλὰ ἡ ΖΔ τῇ Α ἐστιν ἴση· καὶ ἡ ΚΖ ἄρα τῇ Α ἐστιν ἴση. πάλιν ἐπεὶ τὸ Η σημεῖον κέντρον ἐστὶ τοῦ ΛΚΘ κύκλου, ἴση ἐστὶν ἡ ΗΘ τῇ ΗΚ· ἀλλὰ ἡ ΗΘ τῇ Γ ἐστιν ἴση· καὶ ἡ ΚΗ ἄρα τῇ Γ ἐστιν ἴση· ἐστὶ δὲ καὶ ἡ ΖΗ τῇ Β ἴση· αἱ τρεῖς ἄρα εὐθεῖαι αἱ ΚΖ ΖΗ ΗΚ τρισὶ ταῖς Α Β Γ ἴσαι εἰσίν.

In fact, since point Ζ is the centre of circle ΔΚΛ, ΖΔ is equal to ΖΚ; but ΖΔ is equal to Α; therefore ΚΖ is also equal to Α. Again, since point Η is the centre of circle ΛΚΘ, ΗΘ is equal to ΗΚ; but ΗΘ is equal to Γ; therefore ΚΗ is also equal to Γ; and ΖΗ is also equal to Β; therefore the three straight lines ΚΖ, ΖΗ, ΗΚ are equal to the three Α, Β, Γ.

ἐκ τριῶν ἄρα εὐθειῶν τῶν ΚΖ ΖΗ ΗΚ αἵ εἰσιν ἴσαι τρισὶ ταῖς δοθείσαις εὐθείαις ταῖς Α Β Γ τρίγωνον συνέσταται τὸ ΚΖΗ, ὅπερ ἔδει ποιῆσαι.

Therefore a triangle, ΚΖΗ, tuns out to be constructed from three straight lines, ΚΖ, ΖΗ, ΗΚ, that are equal to three given straight lines, Α, Β, Γ, which it was really required to do.

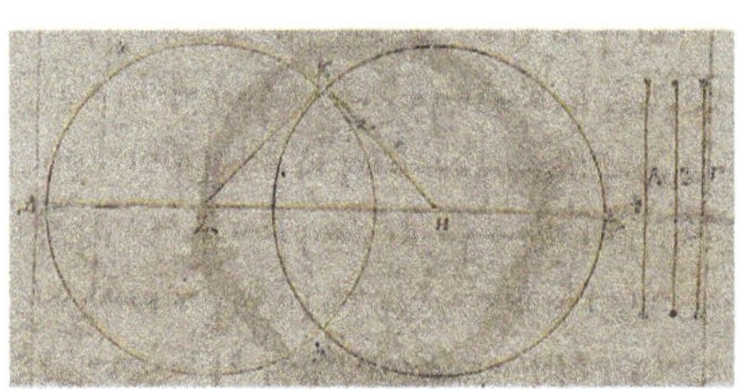

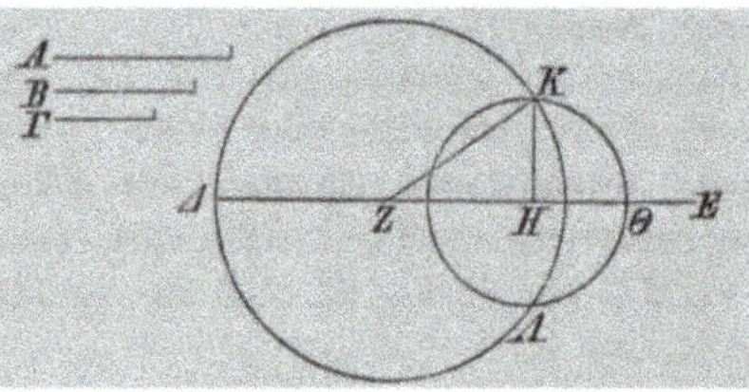

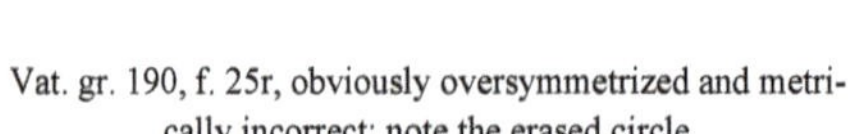

Vat. gr. 190, f. 25r, obviously oversymmetrized and metrically incorrect; note the erased circle

Heiberg's diagram for *El.* I.22, metrically correct and de-oversymmetrized

We are only proposed the "alias" in XIII.18 (*EOO* IV, 328.6–9), where it is assigned to set out and to compare the edges of the regular polyhedra:[69]

τὰς πλευρὰς τῶν πέντε σχημάτων ἐκθέσθαι καὶ συγκρῖναι πρὸς ἀλλήλας.

Set out and compare to one another the sides of the five figures.

ἐκκείσθω ἡ τῆς δοθείσης σφαίρας διάμετρος ἡ ΑΒ.

Let a diameter of a given sphere, ΑΒ, be set out.

In order to compare the edges of the five regular polyhedra, a reference line must be assumed serving as a standard. The assumed standard is a diameter of the invariant sphere assumed as reference in the constructions of the polyhedra themselves in XIII.13–17.[70] This diameter is preliminarily "set out" in the construction of XIII.18 (which does not have a setting-out). In their turn, the edges of the regular polyhedra cannot be "constructed", but only "set out", because the enunciation does not exhibit given objects—it does not even exhibit the sphere, which is given only because it is invariantly attached to the polyhedra: the edges to be compared can only be generated, as parts

[69] The earliest mention of the five regular polyhedra as τῶν πέντε σχημάτων τῶν Πλάτωνος καλουμένων "the so-called five figures of Plato" is in Hero, *Metr.* II.15, at Acerbi, Vitrac 2014, 292.7.

[70] This can be done on the lines of Theodosius, *Sph.* I.19, whose proof is similar to the proof of I.18 just read.

of an auxiliary geometric configuration, as items equal to the original edges of the polyhedra that are actually inscribed in the given invariant sphere.

To summarize the above discussion, whenever a geometer arbitrarily "sets out" an object in order to root some specific constructive act, we must suppose that this amounts to produce (πορίσασθαι), as is required in the *Data*, a copy of the object that is totally or partly conformal (that is, only specific features such as its form are "copied"; see Sect. 2.4.1): a copy of an expressible line, a copy of a unit, or even a copy of the particular isosceles triangle, whose construction is licensed by IV.10, used in IV.11 in order to inscribe a pentagon in a circle. Such a triangle "set out" in IV.11 (*EOO* I, 298.18–300.1) is in its turn duplicated "in form" according to IV.2 (the citation is underlined); the pentagon will finally be constructed by means of the duplicate triangle:

ἐκκείσθω τρίγωνον ἰσοσκελὲς τὸ ΖΗΘ διπλασίονα ἔχον ἑκατέραν τῶν πρὸς τοῖς Η Θ γωνιῶν τῆς πρὸς τῷ Ζ, καὶ ἐγγεγράφθω εἰς τὸν ΑΒΓΔΕ κύκλον τῷ ΖΗΘ τριγώνῳ ἰσογώνιον τρίγωνον τὸ ΑΓΔ ὥστε τῇ μὲν πρὸς τῷ Ζ γωνίᾳ ἴσην εἶναι τὴν ὑπὸ ΓΑΔ ἑκατέραν δὲ τῶν πρὸς τοῖς Η Θ ἴσην ἑκατέρᾳ τῶν ὑπὸ ΑΓΔ ΓΔΑ.	Let an isosceles triangle, ΖΗΘ, be set out having each of the angles at Η, Θ double of that at Ζ, and let a triangle, ΑΓΔ, equiangular to triangle ΖΗΘ be inscribed in circle ΑΒΓΔΕ so as to be ΓΑΔ equal to the angle at Ζ and those at Η, Θ equal to ΑΓΔ, ΓΔΑ, respectively.

A function similar to the function of ἔκκειμαι have the sparse constructions introduced by presential νενοήσθω "let it be conceived", like those in IV.12 and XI.12 (*EOO* I, 302.10–16, and IV, 34.22–25)[71]—note the absence of γάρ: this shows that νενοήσθω has exactly the same function, *within a construction*, as presential ἔστω in the setting-out:

νενοήσθω τοῦ ἐγγεγραμμένου πενταγώνου τῶν γωνιῶν σημεῖα τὰ Α Β Γ Δ Ε ὥστε ἴσας εἶναι τὰς ΑΒ ΒΓ ΓΔ ΔΕ ΕΑ περιφερείας, καὶ διὰ τῶν Α Β Γ Δ Ε ἤχθωσαν τοῦ κύκλου ἐφαπτόμεναι αἱ ΗΘ ΘΚ ΚΛ ΛΜ ΜΗ, καὶ εἰλήφθω τοῦ ΑΒΓΔΕ κύκλου κέντρον τὸ Ζ, καὶ ἐπεζεύχθωσαν αἱ ΖΒ ΖΚ ΖΓ ΖΛ ΖΔ.	Let points of the angles of the inscribed pentagon, Α, Β, Γ, Δ, Ε, be conceived so as to be the arcs ΑΒ, ΒΓ, ΓΔ, ΔΕ, ΕΑ equal; and through Α, Β, Γ, Δ, Ε let ‹straight lines›, ΗΘ, ΘΚ, ΚΛ, ΛΜ, ΜΗ, be drawn tangent to the circle, and let there be taken centre of circle ΑΒΓΔΕ, Ζ, and let ‹straight lines›, ΖΒ, ΖΚ, ΖΓ, ΖΛ, ΖΔ, be joined.
νενοήσθω τι σημεῖον μετέωρον τὸ Β, καὶ ἀπὸ τοῦ Β ἐπὶ τὸ ὑποκείμενον ἐπίπεδον κάθετος ἤχθω ἡ ΒΓ, καὶ διὰ τοῦ Α σημείου τῇ ΒΓ παράλληλος ἤχθω ἡ ΑΔ.	Let some elevated point, Β, be conceived, and from Β let a ‹straight line›, ΒΓ, be drawn perpendicular to the underlying plane, and through point Α let a ‹straight line›, ΑΔ, be drawn parallel to ΒΓ.

In Book XII, the verb νοέω is used to present solids; in XII.17 and 18, this verb is found in the setting-out and presents spheres. Apollonius employs this verb in a similar way in the constructions in *Con.* I.52, 54, 56,[72] where it is required to "conceive" a cone whose vertex and base circle are given (the cone is thereby univocally determined). As is to be expected since νοέω specifically refers to setting out solids, Archimedes uses this verb frequently and in a wide range of geometric situations: it is a stylistic trait of his prose.[73]

[71] Other occurrences in the main text of the *Elements* are in XII.4/5, 13 (νοείσθω), 14–15, 17 (*ter*, verb ἐννοέω), 18 (*bis*).
[72] At *AGE* I, 160.18, 168.14, and 178.12, respectively; only the form νοείσθω is present.
[73] There are 66 occurrences in the Archimedean corpus.

I now turn to a discussion of the philosophical notions connected with the operation of setting out and with suppositions. As is often the case, evidence coming from Aristotelian doctrines is richer than evidence pertaining to Stoic logic, which, moreover, we mainly read through the filter of late Aristotelian commentators.

In Aristotelian syllogistic, the term ἔκθεσις and the associated verb have two main meanings.[74] The first meaning identifies a kind of argument that proves the syllogistic validity of E-conversion (*APr.* I.2, 25a14–19), of the assertoric schemes *Darapti* (I.6, 28a17–26), *Datisi* (I.6, 28b14–15), and *Bocardo* (I.6, 28b20–21), and of the modal schemes NNN-*Baroco* and NNN-*Bocardo* (I.8, 30a6–14). In this kind of argument, an auxiliary syllogism is constructed by setting out an item among those contained in the extension of a suitable term that figures in one of the original premises. This reading is confirmed by the fact that Aristotle repeatedly refers to the item set out by means of the indefinite expression τι τῶν … "some of …".[75] The procedure *resembles* what is nowadays called "existential exemplification" of a generic individual of a class of terms:[76] this operation exactly amounts to a θέσις "setting" of the individual ἐκ "out" of the class.[77]

The second meaning of ἔκθεσις is connected with expressions like ἔκθεσις τῶν ὅρων "setting out of the terms".[78] The phrase designates either the procedure that allows extracting from an informal argument the three terms necessary to formulate it in a suitable syllogistic scheme or, in the first and in the last occurrence in Aristotle, the exemplification of a syllogistic scheme by means of concrete terms, in order to show, by means of a counterexample, that some premises cannot syllogize a conclusion. Neither procedure has anything to do with the denotative letters included in a syllogism, nor has it any connection with known mathematical practices.

A further, cryptic series of Aristotelian texts deserve a more detailed discussion, insofar as they allow clarifying the issue of the meaning of ἔκθεσις in mathematical contexts. In *APr.* I.41, 49b33–50a4, and in a handful of other passages, Aristotle makes the point that incidental or material features of the mathematical objects set out in a diagram by the geometers do not lead them to assert something false. Let us read the text:[79]

οὐ δεῖ δ' οἴεσθαι παρὰ τὸ <u>ἐκτίθεσθαί</u> τι συμβαίνειν ἄτοπον· οὐδὲν γὰρ προσχρώμεθα τῷ τόδε τι εἶναι, ἀλλ' ὥσπερ ὁ γεωμέτρης τὴν ποδιαίαν καὶ εὐθεῖαν τήνδε καὶ ἀπλατῆ εἶναι λέγει οὐκ οὔσας, ἀλλ' οὐχ οὕτως χρῆται ὡς ἐκ τούτων συλλογιζόμενος.

One should not think that any absurdity results from <u>setting</u> something <u>out</u>: for we make no use of the fact that it is a certain "this"; instead, it is like a geometer who calls this a one foot long ‹line›, this a straight line, and says that it is breadthless, though they are not, but does not use these features as though he were syllogizing from them.

[74] The verb is expressly used at *APr.* I.6, 28a23, and I.8, 30a9–12. The discussion in Smith 1982 can be dismissed insofar as grounded on a wrong conception of what happens in a mathematical setting-out.

[75] For instance at *APr.* I.2, 25a16, and I.6, 28a24; for a discussion of the textual evidence see Malink 2008.

[76] Alexander (*in APr.*, 99.19–100.26) and other commentators after him interpreted this procedure as the identification of a singular term by means of perception, a meaning that is not found in the Aristotelian text. The categorial status of the term set out (whether it is a singular term or a categorial term) is not immediately made clear by Aristotle, as "some of" such-and-such terms is not necessarily an individual and as Aristotle is quite explicit about the fact that the item set out is a categorial term only when he expounds *modal* syllogistic (*APr.* I.8, 30a10). Interpreters like Łukasiewicz 1957, 59–67, and Patzig 1968, 156–168, hold that the instantiated term is a subclass of one of the terms contained in the premises, and not a representative individual. The point on the issue has recently been made in Malink 2013, 67–80.

[77] A syllogistic entirely grounded on this kind of ἔκθεσις was first presented in Smith 1983, and subsequently amended and completed in Joray 2014.

[78] It can be found at *APr.* I.10, 30b31; I.34, 48a1–a8 and 48a25; I.35, 48a29; I.39, 49b6; II.4, 57a35.

[79] I use here Acerbi 2008a. The other passages are in *Metaph.* I.1, 1052b31–33; M.3, 1078a17–21; N.2, 1089a21–26. In the second of these texts, the geometer "draws in the sand and calls 'one foot long' a ‹straight line› that is not one foot long". By the way, one of these passages must be the source of what we read in Berkeley 1734, *Introduction*, § 12.

In the other passages, the reference to the line one foot long remains, but the geometer is engaged in the activity of ὑποτίθεσθαι "supposing", as in *APo.* I.10, 76b39–77a3:[80]

οὐδ' ὁ γεωμέτρης ψευδῆ ὑποτίθεται, ὥσπερ τινὲς ἔφασαν, λέγοντες ὡς οὐ δεῖ τῷ ψεύδει χρῆσθαι, τὸν δὲ γεωμέτρην ψεύδεσθαι λέγοντα ποδιαίαν τὴν οὐ ποδιαίαν ἢ εὐθεῖαν τὴν γεγραμμένην οὐκ εὐθεῖαν οὖσαν. ὁ δὲ γεωμέτρης οὐδὲν συμπεραίνεται τῷ τήνδε εἶναι γραμμὴν ἣν αὐτὸς ἔφθεγκται, ἀλλὰ τὰ διὰ τούτων δηλούμενα.	A geometer does not suppose falsehoods, as some people asserted, saying that one should not use falsehoods, still a geometer speaks falsely when he calls one foot long a ‹line› that is not one foot long or calls straight a drawn ‹line› which is not straight. But a geometer does not conclude anything from the fact that the line which he has himself described is thus and so; rather, what is made clear by means of these.

The presence of the verb form ἐκτίθεσθαι in the first passage must not delude us into thinking[81] that the geometer's activity involved in the simile corresponds to the setting-out (ἔκθεσις) of the geometric objects involved in a proof. Such an identification purportedly fits Alexander's interpretation of the Aristotelian ἐκτίθεσθαι-procedure referred to in the passage as the use of denotative letters in schematic expositions of syllogistic.[82] In this reading, positing a line one foot long becomes something like a debased version of asserting ἔστω εὐθεῖα γραμμὴ ἡ ΓΔ "let ΓΔ be a straight line". But this reading has in its background the misguided idea that the denotative letters have a deictic function, and that they refer to the representation of a mathematical object in a diagram, so that general theorems are proved by referring to a particular diagram.

That this identification cannot stand is made clear by the other features of the line posited by the geometer in the passages above, namely, being straight and breadthless.[83] What Aristotle says, then, is that the *abstract* feature of being *exactly* one foot long is what is upset when a mathematician materially works out a geometric proof,[84] since no material straight line can be so, exactly as it cannot be straight or breadthless. To state tersely a point that might seem obvious but apparently was not (it is not even obvious today): a drawing cannot exactly represent a geometric object—and this has nothing to do with mathematical ἔκθεσις, in which a fully-fledged mathematical abstraction is deployed. As a consequence, the Aristotelian passage at *APr.* I.41 read above sets a simile, as is confirmed by the context, for the exemplification of a syllogistic scheme by means of concrete terms. Thus, Aristotle there simply outlines an analogy between his own procedure of ἐκτίθεσθαι—as is shown by the strategic position of the two verb forms, outside the mathematical horn of the analogy—and a mathematical activity that *is not* the activity carried out in the setting-out, but in the execution of the diagram.

[80] The polemical target is Protagoras (see *Metaph.* B.2, 997b35–998a4), as in the parallel passage in *Metaph.* N.2, 1089a21–26. We shall return on the verb ὑποτίθημι at the end of this Section.
[81] *Pace* the consensus of scholarship, see for instance Mignucci 1969, 495; Ross 1924, 476; Ross 1949, 541; Mendell 1998, 181–182; and, from a slightly different point of view, Smith 1989, 173. Nor should the presence of the verb ὑποτίθημι in the second passage delude us. This is simply an allusion to the suppositional mode of the clause, namely, to the presence of verb forms in the imperative.
[82] At *in APr.*, 379.14–380.27; this amounts to attaching a third meaning to the term ἔκθεσις.
[83] It goes without saying that no geometer would explicitly posit that a line is breadthless in a theorem. Nevertheless, being breadthless is an essential attribute of a line, since a line is defined as a "breadthless length" (I.def.2; the definition was known to Aristotle, who discusses it in an anti-Platonic perspective at *Top.* VI.6, 143b11–32). As a consequence, Aristotle rightly claims that a geometer is implicitly positing such a feature in each proof that involves lines.
[84] The line one foot long had the function of the reference line in the proto-theory of irrational lines outlined by Theodorus and Theaetetus at Plato, *Tht.* 147D. In Book X such a line is replaced, as we have seen, by the "expressible line", but they are both abstract objects.

The above discussion shows that, despite the terminological identity, the Aristotelian ἔκθεσις and the mathematical "setting-out" are not kindred notions:[85] it must in fact be excluded that ἔκθεσις was a technical term with a well-defined signification in mathematics, and that Aristotle borrowed both the term and a purportedly associated procedure.[86] I shall argue below that the inverse is more plausible: the use of the term ἔκθεσις to designate a specific part of a proposition is a late invention, concocted in philosophical quarters[87] on the basis of a perceived similarity with one of the Aristotelian uses seen above. I would even take it for almost certain that the key piece of Aristotelian doctrine to this effect was the text at *APr.* I.41 read above. Still, whoever coined the term with this passage in mind was unable to realize that a mathematical setting-out does not instantiate on the basis of an individual.

The doubt remains as to whether the setting-out got an early, alternative designation in mathematical writings—or at least whether a stable lexical constellation was associated with the mathematical activity of setting out as this is outlined in the present Section. I think that these questions can be answered in the positive by the following considerations, which draw again, directly on Aristotelian doctrines, only indirectly—and, as we shall see, sometimes desperately so—on Stoic evidence.[88] A final investigation into the mathematical lexicon will allow us to reach a conclusion which I deem satisfying even if not perfectly corroborated.

As we have seen in Sect. 1.5, Aristotle held that there are three kinds of principles of a deductive science. Since he more or less explicitly recognizes both definitions and "axioms" as principles, an interpretive crux has been debated among modern interpreters, namely, whether or not the postulates match any of the residual kinds introduced by Aristotle. An affirmative answer, grounded on T.L. Heath's popularization of a very influential paper by H.G. Zeuthen, was given in modern times by H.D.P. Lee:[89] Aristotle's ὑποθέσεις exactly correspond to the first three postulates, since the postulates posit the existence of the objects involved. Lee's view became an integral part of the folklore about pre-Euclidean mathematics, since the correspondence between existence and construction presents itself as "obvious" to modern eyes. However, well-argued criticisms have been levelled against this thesis: these either show that we have no grounds for supposing that the constructive postulates of the *Elements* were read, by Aristotle or by any ancient mathematician, as assumptions of existence,[90] or argue that the Aristotelian ὑποθέσεις most fittingly correspond to the local stipulations set out in the ἔκθεσις of a mathematical proposition.

The latter point was made by A. Gomez-Lobo. His interpretation is corroborated by the evidence adduced throughout this Section and in Sect. 3; it must be corrected on a major point, though: since the verb "to be" that often introduces the ἔκθεσις has a presential, not a copulative value, it is unnecessary to assign, as Gomez-Lobo does, a much-contrived "predicative value with ellipsis of

---

[85] That they are kindred notions is the position of all interpreters, in modern times since at least Einarson 1936.

[86] Recall also that to Aristotle (*Mem.* 449b30–450a7) it is only possible to think by means of instantial representations, and that he recognizes that this is one of the peculiarities of mathematical thinking.

[87] See below in this Section for further evidence.

[88] One must take cautiously the mentions of suppositions in well-known Platonic passages like *Phd.* 100A–101E, *Men.* 86D–87B, and *Rsp.* 510B–511E; these may well be local suppositions with a heuristic import. There is no serious secondary literature on the Platonic ὑποθέσεις. The celebrated beginning of the Hippocratic *VM* does not clarify the matter (in fact, it obscures it, for examples of what is "supposed" by the polemical target of the author of the treatise are principles such as hot and dry), unless in showing that ὑποθέσεις were not specific to mathematics or to philosophy; see the judicious discussion in Schiefsky 2005, 111–115 and 120–126.

[89] See Lee 1935, 114–117. As a matter of fact, these four pages are not much more than a cento of quotations from two of Heath's books. The article in question is Zeuthen 1896.

[90] See the clear discussion in Mueller 1991, 77–78.

the predicate"[91] to the same verb in the same syntactic configuration in Aristotelian passages such as *APo.* I.2, 72a18–24.[92]

To Aristotle, then, only definitions and general axioms can be principles of a mathematical *proof.* This does not mean that he did not recognize that constructive suppositions have a founding function in a mathematical *proposition*. The point, as I have shown,[93] is that what he calls λήμματα "assumptions" in a crucial, and generally neglected, passage at *Top.* I.1, 101a13–17, are necessary for, but only preliminary to, setting up the strictly deductive portion of a proposition, of which, as a consequence, they are not an integral part: and these λήμματα, in their context, can only mean "‹constructive› assumptions", that is, "constructive acts":

ἐκ τῶν οἰκείων μὲν τῇ ἐπιστήμῃ λημμάτων οὐκ ἀληθῶν δὲ τὸν συλλογισμὸν ποιεῖται· τῷ γὰρ ἢ τὰ ἡμικύκλια περιγράφειν μὴ ὡς δεῖ ἢ γραμμάς τινας ἄγειν μὴ ὡς ἂν ἀχθείησαν τὸν παραλογισμὸν ποιεῖται.	They[94] effect their syllogism from assumptions which are, on the one hand, appropriate to the science in question but, on the other, are not true—for they effect their paralogism either by describing the semicircles not as one should or by drawing certain lines not as they should be drawn.

If a construction is not an integral part of the deductive core of a proposition, the λήμματα on which it is grounded, either implicitly or explicitly, cannot be counted among the principles of a deductive science. Yet, the *Topica* passage attests to Aristotle's recognition that there are such "constructive assumptions", and it is of course not really important whether these λήμματα were explicitly stated as such at the beginning of some pre-Euclidean exposition of "Elements" or they simply figured as fixed formulaic expressions that characterize certain operations on geometric objects, which only in the *Elements* were enucleated and given the dignity of ἀρχαί "principles" or the status of basic constructive steps.

For these reasons, the tripartite division of principles found in the *Elements* and the one(s) proposed by Aristotle do not match, and in fact could not match. It is no surprise, then, that he does not mention the "postulates" we know from the mathematical tradition, and that he uses this term to characterize some principles on the basis of a dialectical criterion such as the degree of assent to them manifested by some learner (cf. Sect. 1.5). Nor can we assume that Aristotle, had he found the mathematical postulates stated as independent assumptions in some technical text, would have felt compelled to "replace" them with some other kind of principle, if he considered them irrelevant to the ἀπόδειξις: his classification of principles is simply grounded on different parameters.

However, given the obvious resonances of the Aristotelian denominations with the established mathematical lexicon, ancient interpreters had already tried to map the tripartite division(s) of principles proposed by Aristotle onto the tripartition attested in the *Elements*. In the exegetic tradition, the ὑποθέσεις came to designate generically the first principles of any science,[95] thereby including the axioms and the postulates. Proclus, instead, had no qualms in proposing an exact correspondence.[96] He applies the Aristotelian designations ὑποθέσεις, αἰτήματα, ἀξιώματα to the ἀρχαί of

[91] Gomez-Lobo 1977, 433; but contra see already Mendell 1986, 271–273.
[92] As no ancient mathematical principle overtly states existence (with one exception: the first and the third axiom in Archimedes, *Sph. cyl.* I, at *AOO* I, 6.2–5 and 6.11–14), whereas the suppositions in the ἔκθεσις exactly perform this function (of course, we must neglect the distinction between existential and presential value), these remarks undermine the main point in Malink 2017.
[93] See Acerbi 2013a, and cf. Mueller 1991, 77, on Aristotle regarding constructive assumptions as irrelevant to a *proof.*
[94] The grammatical subject is anyone proposing a false proof.
[95] See, for instance, Alexander, *in APr.*, 13.7–11.
[96] He did this at *iE*, 76.4–77.6 and 178.1–8.

the *Elements*. Expressly mentioning Aristotle as his source, Proclus justifies his mapping by showing that the three kinds of principle are in one-to-one correspondence with the three possible combinations of the two polar oppositions "self-evident" / "unknown" and "conceded by the learner" / "not conceded by the learner".[97] At the beginning of his commentary on the postulates, however, Proclus asserts that he has already dealt with "suppositions and terms, so they are called, in what has been previously said"—namely, when commenting on definitions. Proclus' incertitudes show the difficulties he had in establishing an exact correspondence between the Aristotelian principles and those of the *Elements*: he ends with identifying suppositions and definitions, against Aristotle's explicit statement of the contrary in *APo.* I.2.

If Aristotelian doctrines do not help us, let us turn to the evidence coming from Stoic logic. The ὑπόθεσις "supposition" was recognized by the Stoics as an independent λεκτόν "sayable", which is a mental "something" that roughly corresponds to our "meaning" (D.L. VII.65–68). Among sayables, prominently figure those that are αὐτοτελῆ "complete in themselves", namely, those that have a propositional structure and whose linguistic counterpart is a complete sentence. Important complete sayables are the ἀξιώματα "statements", whose linguistic counterparts are declarative sentences (ἀποφαντικά) or combinations of them by means of sentential connectors; the "statements" have a well-defined truth-value. Among non-declarative sayables, the linguistic counterparts of the Stoic ὑποθέσεις are characterized by an imperative without jussive connotation, which instead identifies another λεκτόν, namely, the προστακτικόν "imperative". As all sayables that are not statements, the suppositions are not truth-apt, giving rise so to speak to a fictitious world.[98]

Chrysippus wrote a series of treatises on suppositions and suppositional arguments (D.L. VII.196), collected in the fourth σύνταξις "division" of the λογικὸς τόπος πρὸς τοὺς λόγους καὶ τοὺς τρόπους "argumental corpus on arguments and modes";[99] the fourth title in this list is λόγοι ὑποθετικοὶ θεωρημάτων "Suppositional arguments of theorems"—possibly carrying a mathematical reference—the last title is περὶ ἐκθέσεων πρὸς Λεοδαμάντα "On setting-outs against/to Leodamas". One is reminded of Leodamas of Thasus, mentioned by Proclus, together with Archytas and Theaetetus, among those from whom "the theorems were increased and framed in a more scientific formulation". Again according to Proclus, Leodamas was among the first who applied the method of analysis and synthesis.[100] It is not impossible that Chrysippus addressed his own treatise to long-dead Leodamas. Diogenes Laertius (VII.66) also mentions the ὑποθετικόν "suppositional" in a list of "sayables", but the example falls into a lacuna.

As for late sources, examples of suppositions are found in Epictetus: "let it be night" and "let that you are lucky", in Ammonius: "let it be supposed that the Earth is the centre of the sphere of the Sun", and in the scholiast to *Int.*: "let it be supposed that the Earth has the ratio of a point to the Sun".[101] Simplicius reports an example of a Stoic suppositional argument devised to prove that outside the cosmos there is void;[102] the argument begins with the supposition: "let there be someone standing at the extreme of the Cosmos and extending a hand up". Within a list of forms of λόγος, Ammonius mentions "setting-out" and "supposition", and exemplifies the former by means of the

[97] The case of a self-evident yet not conceded principle of course does not apply—and a tripartite classification is the only one acceptable to a Neoplatonist. The example Proclus adduces of a ὑπόθεσις is the definition of a circle.
[98] See the capital study Bobzien 1997. The truth-bearers in Stoic logic are the statements, as seen. On truth-bearers in Aristotle, see Crivelli 2004, Part I, and Charles, Peramatzis 2016.
[99] On Stoic "modes" see Sects. 5.1.6 and 5.2.2.
[100] See *iE*, 66.14–18 and 211.18–23, respectively.
[101] In *Diss.* I.25.11 and 13, *in Int.*, 2.32, and ΣArist. *Int.*, 93b28, respectively. In *Diss.* I.7.22, Epictetus claims that "it is sometimes necessary to postulate [αἰτῆσαι] some supposition as the founding stone of the subsequent argument".
[102] He did this at *in Cael.*, 284.28–285.2.

ostensive phrase ἔστω εὐθεῖα γραμμὴ ἥδε "let a straight line be this one".[103] Ammonius ascribes the list to the Stoics, but we must be careful in assessing it as early Stoic doctrine. It is quite obvious, for instance, that the example in Ammonius owes very much to standard exegetic material that also surfaces in the following statement by Proclus (*iE*, 208.16–17) in his commentary on *El.* I.1:

μετὰ δὲ τὴν πρότασιν ἑξῆς ἡ ἔκθεσις· ἔστω ἡ δοθεῖσα εὐθεῖα πεπερασμένη ἥδε.	Immediately after the enunciation, the setting-out: let the given bounded straight line be this.

Despite what is intimated by the commentators, thus, we do not have reasons to think that the Stoics' setting-outs had anything to do with mathematical ἔκθεσις. More promising, instead, is the connection with Stoic "suppositions", as we shall see presently.

Before doing this, here is a last testimony that comes from philosophical sources. When referring to the meaning of "supposition" as ἀρχὴ ἀποδείξεως "principle of a proof",[104] Sextus Empiricus (*M* III.7–17) claims that a supposition is characterized by the assumption of a particular object as given and—as a consequence of this—by the use of the imperative δεδόσθω "let it be given", a verb form that is simply never met in the *Elements*.[105] No sources allude to the use of denotative letters in suppositions.

Let us now turn to mathematical evidence. When, in the course of a proof, reference is made to some assumption posited in the setting-out, this is marked by the verb ὑπόκειται "it has been supposed".[106] Conversely, the verb ὑπόκειμαι "to have been supposed" refers without exceptions to a supposition made in the setting-out or (9 occurrences out 50) in the construction. The following table presents the part of a proposition towards which such references are directed, whether affirmative or negative, within direct or indirect proofs. If a proposition is divided into cases or has a multiple enunciation, all references to the partial setting-outs are collected under the heading "main setting-out". I call "local" the setting-out that precedes an indirect argument, and that introduces new entities with respect to the main setting-out. The double slash precedes the additional material.

main setting-out	I.26, 29, 48, V.5, 6, 18, 19, VI.3, 5, 6 (*bis*), 7 (*ter*), 22, VII.2, 7, 20, 33, IX.10 (*bis*), 14, 30, 34, X.9por, 9/10, 21, 3, 38, 41/42, 47, XI.5, 16, 23 (*bis*), 34 (*bis*), 35, XII.4/5, XIII.2/3, 7 (*bis*) // X.13*alt*, 39*alt*
local setting-out	IX.12, 13, 18, 20, 36, X.16 (*bis*)
construction	IV.10, IX.20, X.33, XI.23, 23 (*bis*), 26 // X.54, 55[107]
supposition in indirect argument (spurious)	X.42, XII.12 // X.28/29II (*bis*)

The imperative ὑποκείσθω is found 8 times in the Euclidean corpus, 2 of which in the main text of the *Elements*. Let us read the occurrence in VI.7 (*EOO* II, 94.22–96.2 and 98.5–6)—the verb introduces, in a partial setting-out, the second additional condition on the angles of two triangles that must be proved to be similar. The supposition restates the liminal setting-out (which we also read) by making only the variant condition explicit:

[103] At *in Int.*, 2.31, and cf. ΣArist. *Int.*, 93b30–31, which gives the following as a paradigmatic ἐκθετικόν "setting-out-like" clause: "the geometers say setting out", apparently a meaningless epitome of a statement like "setting-out-like clause is what the geometers call setting-out".

[104] Sextus wrongly assumes, in his criticism, that a supposition is truth-apt.

[105] There is 1 occurrence in Apollonius, *Con.* II.4 (but this proposition is certainly spurious: see Sect. 2.4.1) and 13 occurrences in the Archimedean corpus, always introducing a setting-out, in *Sph. cyl.* I.6, II.4, *Con. sph.* 7, 19–20, *Spir.* 5–6 (*bis*), 7–9, *Aequil.* I.1.

[106] Recall that ὑπόκειμαι is a *praesens pro perfecto*.

[107] Strictly speaking, in X.54 and 55 immediate consequences of the construction are referred to.

ἔστω δύο τρίγωνα τὰ ΑΒΓ ΔΕΖ μίαν γωνίαν μιᾷ γωνίᾳ ἴσην ἔχοντα τὴν ὑπὸ ΒΑΓ τῇ ὑπὸ ΕΔΖ περὶ δὲ ἄλλας γωνίας τὰς ὑπὸ ΑΒΓ ΔΕΖ τὰς πλευρὰς ἀνάλογον, ὡς τὴν ΑΒ πρὸς τὴν ΒΓ οὕτως τὴν ΔΕ πρὸς τὴν ΕΖ, τῶν δὲ λοιπῶν τῶν πρὸς τοῖς Γ Ζ πρότερον ἑκατέραν ἅμα ἐλάσσονα ὀρθῆς.	Let there be two triangles, ΑΒΓ, ΔΕΖ, having one angle equal to one angle, ΒΑΓ to ΕΔΖ, and around other angles, ΑΒΓ, ΔΕΖ, the sides in proportion, as ΑΒ is to ΒΓ, so ΔΕ is to ΕΖ, and let first each of the remaining ‹angles› at Γ, Ζ together be less than a right ‹angle›.
ἀλλὰ δὴ πάλιν <u>ὑποκείσθω</u> ἑκατέρα τῶν πρὸς τοῖς Γ Ζ μὴ ἐλάσσων ὀρθῆς.	But now again, <u>let</u> each of those at Γ, Ζ <u>be supposed</u> not less than a right ‹angle›.

In other cases, suppositions are posited serving as principles: the imperative ὑποκείσθω introduces the assumptions in the *Optica* (both redactions) and, in the *Phaenomena*, the fundamental supposition that the Cosmos is spherical.[108]

In Apollonius we find the imperative within phrases like ὑποκείσθω τὰ μὲν ἄλλα τὰ αὐτά "let the others be supposed the same" or, ὑποκείσθω γὰρ τὰ αὐτά "in fact, let the same be supposed".[109] We also find 21 times the abbreviated phrase τῶν αὐτῶν ὑποκειμένων "the same being supposed", introducing additional cases that have received a proposition number at some point of the manuscript tradition. In every instance, the reference is to what is supposed in the main setting-out.

The Archimedean corpus offers a picture analogous to the picture seen in Euclid: there are occurrences in which the verb replaces the entire setting-out[110] or the construction (*Fluit.* II.10); occurrences in which the verb introduces or follows the basic assumptions of an entire treatise,[111] or additional suppositions within it,[112] or local suppositions (*Fluit.* I.8); or occurrences within the expression "let the others be supposed the same".[113] The same verb also makes reference to suppositions posited in the setting-out or in the enunciation (*Con. sph.* 19–20).[114]

So, the terminological record suggests that the term ὑπόθεσις denotes any supposition that grounds a proof or a series of proofs; this supposition has the status of a principle or of a simple assumption within a specific proposition.[115] In his discussion of I.6 as a converse of I.5, Proclus (*iE*, 252.5–24) contends that such deductive pairs interchange τὰ συμπεράσματα καὶ τὰ ὑποθέσεις "the conclusions and the suppositions". He takes the statement that the angles at the base are equal as the supposition of I.6, the logical subject of the proposition, namely, an isosceles triangle (!), as the supposition of I.5.

All in all, I feel entitled to conclude that one of the meanings of ὑπόθεσις attested in a significant part of the mathematical record is that of "specific supposition", whose main specimen are the suppositions posited in the setting-out. This suggests that the Stoic denomination ὑπόθεσις for a sayable characterized by the verb form in the imperative draws, by singling out a unifying syntactic feature, from a well-established mathematical terminology, and that the denomination ἔκθεσις for the setting-out is a later coinage.

[108] At *EOO* VII, 2.2 and 154.4, and VIII, 4.26, respectively.
[109] In *Con.* II.6, 49 (*ter*), III.7, 47–49, IV.5, 18.
[110] As in ὑποκείσθω τὰ εἰρημένα "let what has been said be supposed" in *Con. sph.* 17; other abbreviated phrases can be found in *Sph. cyl.* I.1 and 3.
[111] This happens in *Fluit.* I and *Aequil.* I—of this kind are the references to the celebrated heliocentric model of Aristarchus and to the Archimedean suppositions in *Ar.*
[112] In *Fluit.* I, just before proposition 8.
[113] In *Fluit.* II.8, and others akin to it, as in *Sph. cyl.* I.2por.
[114] 25 of the 40 occurrences are found in the treatises *Ar.* and *Fluit.*
[115] In a logical context, the antecedent of a conditional is called "supposition" by Philoponus, *in APr.*, 243.15–24.

4.2.1. *Determination*

In his commentary on Book I of the *Elements*, Proclus calls διορισμός "determination" the statement that immediately follows the setting-out and that states what must be proved or constructed. Such a clause is introduced by λέγω ὅτι "I claim that" in the case of theorems, by δεῖ δή "thus it is required" in the case of problems;[116] several problems do not have a determination (see also Sect. 4.3).[117] We have read examples of determinations, together with the setting-outs to which they are associated, at the beginning of Sect. 4.2; let us read two of them again; they are in propositions III.21 and IV.13 (*EOO* I, 220.18–21 and 306.23–25):

ἔστω κύκλος ὁ ΑΒΓΔ, καὶ ἐν τῷ αὐτῷ τμήματι τῷ ΒΑΕΔ γωνίαι ἔστωσαν αἱ ὑπὸ ΒΑΔ ΒΕΔ. λέγω ὅτι αἱ ὑπὸ ΒΑΔ ΒΕΔ γωνίαι ἴσαι ἀλλήλαις εἰσίν.	Let there be a circle, ΑΒΓΔ, and let there be angles, ΒΑΔ, ΒΕΔ, in a same segment, ΒΑΕΔ. I claim that angles ΒΑΔ, ΒΕΔ are equal to one another.
ἔστω τὸ δοθὲν πεντάγωνον ἰσόπλευρόν τε καὶ ἰσογώνιον τὸ ΑΒΓΔΕ. δεῖ δὴ εἰς τὸ ΑΒΓΔΕ πεντάγωνον κύκλον ἐγγράψαι.	Let there be a given pentagon both equilateral and equiangular, ΑΒΓΔΕ. Thus it is required to inscribe a circle in pentagon ΑΒΓΔΕ.

As there are "partial setting-outs", so there are "partial determinations": this happens if a proof is divided into cases[118] or if the enunciation of a proposition is the conjunction of several independent statements; these statements are proved in succession, and each of them is preceded by a suitable, "partial" determination.[119] Take for instance I.46, II.4, and the problems IV.6–7, 11–12, 15: in them, a quadrilateral or a polygon is proved to be regular by showing, first, that the sides are equal, then, that the angles also are.[120] Let us read the partial determination in IV.15 (*EOO* I, 316.7–9); it partly resumes the liminal determination, which states that the hexagon is equilateral and equiangular (I include the last step of the proof of the first of these statements):

ἰσόπλευρον ἄρα ἐστὶ τὸ ΑΒΓΔΕΖ ἑξάγωνον. λέγω δὴ ὅτι καὶ ἰσογώνιον.	Therefore the hexagon ΑΒΓΔΕΖ is equilateral. I now claim that it is also equiangular.

As this example shows, the partial determinations that introduce additional proofs are canonically introduced by λέγω δὴ ὅτι καί "I now claim that … also",[121] possibly curtailed by all possible combinations of missing δή or καί.

The problems may carry a double determination (see also Sect. 4.3):[122] the first determination resumes a part of the enunciation and is canonically introduced by δεῖ δή, the second follows the

[116] In Archimedes, *Sph. cyl.* I.5, we find 1 determination introduced by δεῖ οὖν "thus it is required" (*AOO* I, 18.14), the others are introduced by δεῖ δή. Archimedes systematically employed δεικτέον ὅτι (75 occurrences), with a modal connotation of necessity analogous to δεῖ δή, to introduce the determination of theorems. The only exception is in *Sph. cyl.* I–II, notoriously regularized by Isidorus of Miletus, where the canonical λέγω ὅτι dominates (there are 44 occurrences in the Archimedean corpus, of which 30 in this treatise, which exhibits only 11 δεικτέον "it must be proved").

[117] There is no determination in IV.10, VI.10, IX.18–19, X.27–35, 48–53, 85–90, XIII.18.

[118] This is quite rare, and sometimes of doubtful authenticity: see IX.19, XI.34, XIII.7.

[119] This happens for instance in I.34 or III.8–9.

[120] Always in this order, dictated by the canonical formulation of the conjoined predicate ἰσόπλευρόν τε καὶ ἰσογώνιον "both equilateral and equiangular" (cf. Sect. 5.3.5).

[121] Or by λέγω δὴ ὅτι οὐδέ "I now claim that … not … either" in the case of negative sentences, as in VIII.6.

[122] This happens, either as a liminal or as a partial determination, in I.9–12 22, 46, II.11, III.1, 17, IV.6–7, 11–12, 15 (*bis*), VII.2–3, 33, 34 (*bis*), 36 (*bis*), 39, VIII.2, 4 (*bis*), IX.18, 19 (*ter*), X.3, 4 (*bis*), 27–28, 32, 48–53, 85–90, XI.23, 26, XII.17, XIII.13–15, 16 (*bis*), 17 (*bis*).

construction, asserts that the object that results from it fits (a part of) the conditions of the problem, and is thereby introduced by λέγω ὅτι. Let us read I.11 (*EOO* I, 32.5–14)—the (instantiated) statement that reproduces the enunciation is the second determination:

ἔστω ἡ μὲν δοθεῖσα εὐθεῖα ἡ AB τὸ δὲ δοθὲν σημεῖον ἐπ' αὐτῆς τὸ Γ. δεῖ δὴ ἀπὸ τοῦ Γ σημείου τῇ AB εὐθείᾳ πρὸς ὀρθὰς γωνίας εὐθεῖαν γραμμὴν ἀγαγεῖν.	Let there be a given straight line, AB, and a given point on it, Γ. Thus it is required to draw from point Γ a straight line at right angles with straight line AB.
εἰλήφθω ἐπὶ τῆς ΑΓ τυχὸν σημεῖον τὸ Δ, καὶ κείσθω τῇ ΓΔ ἴση ἡ ΓΕ, καὶ συνεστάτω ἐπὶ τῆς ΔΕ τρίγωνον ἰσόπλευρον τὸ ΖΔΕ, καὶ ἐπεζεύχθω ἡ ΖΓ. λέγω ὅτι τῇ δοθείσῃ εὐθείᾳ τῇ AB ἀπὸ τοῦ πρὸς αὐτῇ δοθέντος σημείου τοῦ Γ πρὸς ὀρθὰς γωνίας εὐθεῖα γραμμὴ ἦκται ἡ ΖΓ.	Let a random point, Δ, be taken on ΑΓ, and let a ‹straight line›, ΓΕ, be set equal to ΓΔ, and let an equilateral triangle, ΖΔΕ, be constructed on ΔΕ, and let a ‹straight line›, ΖΓ, be joined. I claim that a straight line, ΖΓ, turns out to be drawn at right angles with a given straight line, AB, from a given point on it, Γ.

There are statements introduced by λέγω ὅτι and that are neither liminal nor partial determinations: these are the "local" determinations that precede a reduction to the impossible (cf. Sect. 5.2.1). It may happen that the local determination is so near to the liminal determination as to give the impression of being a part of it, as in IX.12 (*EOO* II, 362.17–22):

ἔστωσαν ἀπὸ μονάδος ὁποσοιδηποτοῦν ἀριθμοὶ ἀνάλογον οἱ Α Β Γ Δ. λέγω ὅτι ὑφ' ὅσων ἂν ὁ Δ πρώτων ἀριθμῶν μετρῆται, ὑπὸ τῶν αὐτῶν καὶ ὁ Α μετρηθήσεται.	Let there be as many numbers as we please in proportion from a unit, A, B, Γ, Δ. I claim that, by how many prime numbers Δ be measured, by the same A will also be measured.
μετρείσθω γὰρ ὁ Δ ὑπό τινος πρώτου ἀριθμοῦ τοῦ Ε. λέγω ὅτι ὁ Ε τὸν Α μετρεῖ.	In fact, let Δ be measured by some prime number, E. I claim that E measures A.

The denomination διορισμός for a specific part of a proposition is found for the first time in Proclus. As a matter of fact, the conditions of resolvability of a problem share this denomination and are also canonically introduced by the nexus δεῖ δή;[123] this was certainly the original signification. Proclus' discussion is in this respect surprising.[124] He dwells at length upon the first meaning of "determination", he only passes a couple of mentions about the other.[125] He does not offer any clarification of this obvious homonymy when commenting on I.22 (read in Sect. 4.2), the first proposition of the *Elements* that presents a "real" determination. Moreover, and contrary to what modern interpreters appear to assume, Proclus never mentions the statement that follows the setting-out *of a theorem*— namely, the statement formulated by a sentence that is introduced by λέγω ὅτι— as a διορισμός. His exposition might even suggest that "setting-out" and "determination" only exist as specific parts of *problems*.

The issue of the double signification of διορισμός was briefly dealt with by Eutocius in his commentary on Apollonius' *Conica*.[126] His goal was to explain the reference to the "determina-

[123] Of course, this nexus must be translated in the same way in both types of "determination".
[124] The text is at *iE*, 204.20–205.12.
[125] The mentions are at *iE*, 66.22–67.1 and 202.2–5. The characterizations of the determination in these passages are identical and must be ascribed to the same source.
[126] The two passages are at *AGE* II, 178.4–15, and I, 4.5–8, respectively.

tions" in the description Apollonius himself sketches of Book II of the *Conica*. Eutocius cites the complete *enunciation* of I.22 as an example of determination.[127]

The "real" determination is placed in fact within the enunciation of a problem;[128] it is expressly presented as a necessary condition since it is introduced by δεῖ δή; it is always (and must be) formulated as a constraint on the *givens* of the problem. The enunciations with a determination in the *Elements* are those of I.22, VI.28, and XI.23. Let us read again I.22 (*EOO* I, 52.15–18):

ἐκ τριῶν εὐθειῶν αἵ εἰσιν ἴσαι τρισὶ ταῖς δοθείσαις τρίγωνον συστήσασθαι. δεῖ δὴ τὰς δύο τῆς λοιπῆς μείζονας εἶναι πάντῃ μεταλαμβανομένας.

Construct a triangle from three straight lines that are equal to three given. Thus it is required that two, however permuted, are greater than the remaining one.

In the setting-out, the constraint is normally subsumed under the suppositions, as a participial construct or as a relative clause; a double occurrence of δεῖ δή is avoided (*EOO* I, 52.21–25):

ἔστωσαν αἱ δοθεῖσαι τρεῖς εὐθεῖαι αἱ Α Β Γ, ὧν αἱ δύο τῆς λοιπῆς μείζονες ἔστωσαν πάντῃ μεταλαμβανόμεναι, αἱ μὲν Α Β τῆς Γ αἱ δὲ Α Γ τῆς Β καὶ ἔτι αἱ Β Γ τῆς Α. δεῖ δὴ ἐκ τῶν ἴσων ταῖς Α Β Γ τρίγωνον συστήσασθαι.

Let there be three given straight lines, A, B, Γ, of which let two, however permuted, be greater than the remaining one, A, B of Γ and A, Γ of B and further B, Γ of A. Thus it is required to construct a triangle from ‹straight lines› equal to A, B, Γ.

In every problem that requires a determination in the *Elements*, the validity of the constraint is proved in a previous theorem: for I.22, VI.28, and XI.23, these are I.20, VI.27, and XI.20–21, respectively. Actually, XI.23 (*EOO* IV, 60.22–62.4) has a double determination:

ἐκ τριῶν γωνιῶν ἐπιπέδων, ὧν αἱ δύο τῆς λοιπῆς μείζονές εἰσι πάντῃ μεταλαμβανόμεναι, στερεὰν γωνίαν συστήσασθαι. δεῖ δὴ τὰς τρεῖς τεσσάρων ὀρθῶν ἐλάσσονας εἶναι.

Construct a solid angle from three plane angles two of which, however permuted, are greater than the remaining one. Thus it is required that the three ‹angles› be less than four right ‹angles›.

The above formulation makes only the constraint proved in XI.21 explicit as a determination. This constraint is "Every solid angle is contained by plane angles less than four right ‹angles›". The constraint dictated by XI.20, namely "If a solid angle be contained by three plane angles, any two, however permuted, are greater than the remaining one", is instead embodied in the enunciation. This choice can be explained if we read the enunciation of XI.22 (*EOO* IV, 58.5–9):

ἐὰν ὦσι τρεῖς γωνίαι ἐπίπεδοι, ὧν αἱ δύο τῆς λοιπῆς μείζονές εἰσι πάντῃ μεταλαμβανόμεναι, περιέχωσι δὲ αὐτὰς ἴσαι εὐθεῖαι, δυνατόν ἐστιν ἐκ τῶν ἐπιζευγνυουσῶν τὰς ἴσας εὐθείας τρίγωνον συστήσασθαι.

If there be three plane angles two of which, however permuted, are greater than the remaining one and equal straight lines contain them, it is possible to construct a triangle from the ‹straight lines› that join the equal straight lines.[129]

---

[127] In Apollonius's *Conica*, which contains just a handful of problems, determinations of this kind are only found in II.50 and 53: the determination in II.50 is introduced by δεήσει ἄρα "therefore it will be required", the determination in II.53 is introduced by δεῖ δή (*AGE* I, 290.21 and 310.22).

[128] On the denomination προσδιορισμός for such a kind of determination see Acerbi 2009a, 10–11.

[129] If the text is not corrupt, this is a typical example of cryptic mathematical style; one must understand "that join the extremities of the equal straight lines".

Propositions XI.20 and 22 are almost inverse to one another, even if XI.22 has the peculiar enunciation just read. Thus, XI.22 seems to have the function of detaching an accessory proof from the text of XI.23 and anticipating it. This entails eliminating the determination induced by XI.20 from the enunciation of XI.23. To this end, the determination must be transformed into an additional assumption, which must be presented as such in the part of the proof of XI.23 detached and anticipated in XI.22. The explicit modal connotation of theorem XI.22, whose emphasis on constructibility is in principle unnecessary, may be ascribed to the author's concern with remaining faithful to the typifying feature of a "real" determination, for this states the conditions of resolvability (underlined is the modal connotation!) of a problem.[130]

In the Greek mathematical corpus, problems that contain a "real" determination are seldom met;[131] moreover, most of the attested instances are quite naturally located at the end of the analysis in problems framed in the analysis-synthesis format. The strategy adopted in the *Elements* and, for instance, in *Con.* II.53,[132] of quite unnaturally anticipating the proof of the constraint, is thus forced by the purely synthetic format of these treatises.

Only two problems of the *Elements* are formulated as a search for their own conditions of resolvability: they are IX.18–19 (*EOO* II, 380.25–382.5 and 384.2–7). Let us read in succession their enunciations and setting-outs:[133]

δύο ἀριθμῶν δοθέντων ἐπισκέψασθαι, εἰ δυνατόν ἐστιν αὐτοῖς τρίτον ἀνάλογον προσευρεῖν.	Two numbers being given, investigate whether it is possible to find a third proportional to them.
ἔστωσαν οἱ δοθέντες δύο ἀριθμοὶ οἱ Α Β, καὶ δέον ἔστω ἐπισκέψασθαι εἰ δυνατόν ἐστιν αὐτοῖς τρίτον ἀνάλογον προσευρεῖν.	Let there be two given numbers, A, B, and let it be required to investigate whether it is possible to find a third proportional to them.
τριῶν ἀριθμῶν δοθέντων ἐπισκέψασθαι, πότε δυνατόν ἐστιν αὐτοῖς τέταρτον ἀνάλογον προσευρεῖν.	Three numbers being given, investigate when it is possible to find a fourth proportional to them.
ἔστωσαν οἱ δοθέντες τρεῖς ἀριθμοὶ οἱ Α Β Γ, καὶ δέον ἔστω ἐπισκέψασθαι πότε δυνατόν ἐστιν αὐτοῖς τέταρτον ἀνάλογον προσευρεῖν.	Let there be three given numbers, A, B, Γ, and let it be required to investigate when it is possible to find a fourth proportional to them.

These are number-theoretical *problems*, as is shown by the imperative in the enunciation and by the presence of δέον ἔστω "let it be required" in the setting-out/determination. A constraint for IX.18 is proved in IX.16; a constraint for IX.19 is proved in IX.17. Proposition IX.16 proves that "If two numbers be prime to one another, as the first is to the second, so the second will not be to some other". As a consequence, the two numbers in IX.18 must be non-prime to one another—this is the first constraint. However, the proof shows that there is a second constraint, namely, ὁ Α must measure the square on ὁ Β. This is also a sufficient condition, as is made clear by the proof of IX.18. Since a number that measures the square of a number cannot be prime to it, the second constraint implies the first, which is thereby superfluous. The present redaction of IX.18 may be rewritten in the format of a problem *cum* determination along the following lines:

[130] For a discussion of the relationships between XI.20 and 22, see Acerbi 2007, 390–392.
[131] "Real" determinations abounded in some lost works of the so-called analytic corpus, as Pappus attests to throughout Book VII of the *Collectio* and as we may gather from the Arabic translation of Apollonius' *De sectione rationis*. For the number-theoretical determinations in Diophantus' *Arithmetica* see just below.
[132] The theorem that proves the validity of the constraint is *Con.* II.52.
[133] On the use of πότε in IX.19 see Sect. 5.3.1. The text of the proof of IX.19 is corrupt; for a discussion, see Vitrac 1990–2001 II, 443, Acerbi 2007, 374–375, Vitrac, forthcoming.

*δύο ἀριθμῶν δοθέντων αὐτοῖς τρίτον ἀνάλογον προσευρεῖν. δεῖ δὴ τοὺς δοθέντας ἀριθμοὺς μὴ πρώτους πρὸς ἀλλήλους εἶναι καὶ τὸν πρῶτον τὸν ἀπὸ τοῦ δευτέρου μετρεῖν.
ἔστωσαν οἱ δοθέντες δύο ἀριθμοὶ μὴ πρῶτοι πρὸς πρὸς ἀλλήλους ὄντες οἱ Α Β, μετρείτω δὲ ὁ Α τὸν ἀπὸ τοῦ Β. δεῖ δὴ αὐτοῖς τρίτον ἀνάλογον προσευρεῖν.

Two numbers being given, find a third proportional to them. Thus it is required that the given numbers not be prime to one another and that the first measure the ‹square› on the second.
Let there be two given numbers, A, B, that are not prime to one another, and let A measure the ‹square› on B. Thus it is required to find a third proportional to them.

It is not said that the eccentric format of IX.18–19 is the sign of an "archaic" redaction. Our documentary record suggests that number-theoretical problems attained their standard format several centuries later, in Diophantus' *Arithmetica*. Problems IX.18–19 may thus well present "archaic" features without thereby belonging to the prehistory of the *Elements*.

As for the determinations in Diophantus, let us read *Ar*. I.27 (*DOO* I, 60.23–62.2):[134]

εὑρεῖν δύο ἀριθμοὺς ὅπως ἡ σύνθεσις αὐτῶν καὶ ὁ πολλαπλασιασμὸς ποιῇ δοθέντας ἀριθμούς.
δεῖ δὴ <u>τῶν εὑρισκομένων</u> τὸν ἀπὸ τοῦ ἡμίσεος τοῦ συναμφοτέρου τετράγωνον τοῦ ὑπ' αὐτῶν ὑπερέχειν τετραγώνῳ. ἔστι δὲ τοῦτο πλασματικόν.

Find two numbers in such a way that their composition and multiplication make given numbers.
Thus it is required that, <u>of the ‹numbers› found</u>, the square on the half of both together exceeds the ‹rectangle contained› by them by a square. But this is fictitious.

Problems in Diophantine number theory are submitted to two general requirements restricting the range of possible solutions. Diophantus admits in fact only positive quantities, expressible in numbers, as solutions. As a consequence, the determinations in the *Arithmetica* can take either the form of an inequality, when positivity of the solution is at issue, or the form of an identification of species (for instance, whether a number is a square or a cube), when rationality of the solution is secured, or in the same case, the form of a (negative) requirement of "congruence". As said, any such determination must set limitations on the givens of a problem in order for it to be solved. However, the above determination has been modified at some point of the tradition, in order to align it with the enunciation of *El*. II.5 (*EOO* I, 128.18–22):

ἐὰν εὐθεῖα γραμμὴ τμηθῇ εἰς ἴσα καὶ ἄνισα, τὸ ὑπὸ τῶν ἀνίσων τῆς ὅλης τμημάτων περιεχόμενον ὀρθογώνιον μετὰ τοῦ ἀπὸ τῆς μεταξὺ τῶν τομῶν τετραγώνου ἴσον ἐστὶ τῷ ἀπὸ τῆς ἡμισείας τετραγώνῳ.

If a straight line be cut into equals and unequals, the rectangle contained by the unequal segments of the whole ‹straight line› with the square on the ‹straight line› between the sections is equal to the square on the half.

Thus modified, the determination of *Ar*. I.27 is expressed in terms of the solutions sought, and for this reason it is an identity (the final clause, obviously a gloss of an intelligent reader, says that it is "fictitious"), which of course cannot impose any limitations on the solubility of the problem. Had the determination been formulated in terms of the given numbers, it would not have been empty, but then it would not have any longer preserved any connections with *El*. II.5.

[134] I have clarified the meaning of the adjective πλασματικός, which qualifies the determination, in Acerbi 2009a.

4.3. The Role of Constructions

The role of constructions in Greek mathematics is a notorious exegetic problem, which I shall treat only tangentially here. A clear discussion of the issue must be grounded on a typology that dispels possible ambiguities. My typology is fourfold. One must distinguish between: (a) constructions of mathematical objects; (b) problems of construction; (c) construction as a specific part of a mathematical proposition; (d) constructive acts. The four types of construction and their interactions can be summarily described as follows.

(a) *Constructions of mathematical objects*
A mathematical object is "constructed" when it is the result of a series of geometric or arithmetic operations that build on a set of assigned objects; this set may be empty (example: the construction of the very specific triangle in IV.10). Constructions of mathematical objects are normally[135] carried out in the pieces of mathematical writing called "problems of construction"—this is our type (b)—which in their turn usually resume, employ, and organize other constructions. As for their use, constructions in the present sense may be strictly functional to construct other mathematical objects (example: again the triangle in IV.10, functional to inscribe a pentagon in a circle—IV.11—which is in its turn functional to the construction of a regular icosahedron in XIII.16), or may have an obvious relevance as endpoints of complex, and possibly traditional, pieces of mathematical discourse (examples: constructions of the regular polyhedra, duplication of a cube, trisection of an angle, and quadrature of the circle).

It has been recently reaffirmed with force that constructions in the present sense are the internal engine of Greek mathematics,[136] in particular the three classical problems; more generally and more deeply, it is argued that Greek mathematics is constructive in essence. This historiographic position is questionable. (1) Many accounts about the constructive efforts of Greek geometers are biased by passage through two filters typical of Late Antiquity. The first is the ideology of the heuristic value of analysis (cf. Sect. 2.4.1); Pappus, *Coll.* VII.1, contends that the entire analytic corpus was composed to train one in solving problems of construction. The second filter is the compilatory attitude of the late commentators, our sole sources on the three classical problems. The very act of compiling solutions to the same problem creates a false impression of a mass effort. (2) The three classical problems did not receive equal attention: we know many solutions of the duplication problem, but far fewer are transmitted of the trisection problem, and only two for the quadrature of the circle: Archimedes' and the solution via quadratrix. The latter is fallacious, the former does not solve the problem on Greek standards (the circumference rectified using the spiral is not "given"). (3) In Greek mathematics prominence is accorded to systematic—for instance in the *Elements*, the *Conics* and the *Arithmetica*—and methodological aims, for instance in the entire Heronian corpus or in the "full immersions in analysis," as it were, constituted by the Apollonian analytic works. All of this is irreducible to a constructive endeavor. (4) Non-constructive assumptions are frequently made in Greek geometry.[137] An interesting example comes from the tract *On Isoperimetric Figures*.[138] The main theorem, namely, that a regular polygon is greater than any non-regular polygon isoperimetric

[135] But not exclusively: a counterexample to exclusivity are the constructions of the three conic sections in Apollonius, *Con.* I.11–13; we have read the enunciation of the first of these propositions in Sect. 4.1.
[136] This is the thesis of Knorr 1986.
[137] See Becker 1932 and 1936a, Mueller 1981, 27–29, 127–128, 139, 231–234, 263–264.
[138] Edition and commentary, with comparison of the extant versions, in Acerbi, Vinel, Vitrac 2010, 120–132.

to it and having the same number of sides, is proved by a sort of "local symmetrization": the maximal polygon is assumed to be neither equilateral nor equiangular; this assumption is driven to contradiction by showing how to construct, by making two adjacent sides or two angles equal, a greater isoperimetric polygon. The drawback of this clever argument lies in the fact that the existence of a maximal polygon is (implicitly) posited without a proof. Moreover, the process of "local symmetrization" is not effective: it cannot produce, starting from a given polygon, the maximal (regular) polygon in a predictable number of steps. The approach, then, is eminently non-constructive.

(b) *Problems of construction*

A problem of construction is a piece of mathematical writing that carries out a construction of a mathematical object—type (a) above. The structure of a problem of construction has been outlined in Sect. 1.1; more precisely, it comprises:[139]

i. An enunciation, whose logical subject is "someone" required to perform the construction of an object or to find it; the prescription is formulated in the aorist stem, thereby abstracting from any temporal connotation: the construction is a punctual operation, and for this reason is formulated as a constructive act—type (d) below.

ii. A setting-out that presents the assigned configuration; the only feature that demarcates the setting-out of a problem from the setting-out of a theorem is the presence, in the former, of the predicate "given" marking the objects on which the construction must be specifically carried out and that are so marked already in the enunciation; however, there are problems that do not feature "givens".[140]

iii. A determination introduced by δεῖ δή "thus it is required"; several problems do not have this "problematic" determination.[141]

iv. A construction of the required object, as well as some "auxiliary" constructions, according to the typology to be detailed in a moment: their set-theoretical union makes type (c) below. All such constructions are a sequence of conjoined constructive acts: type (d) below. No constructive act taken in isolation can a priori be said to belong to a problem or to a theorem.

v. A proof, which shows that the constructed object has the required properties ("construct <u>such-and-such</u> a triangle", as for instance in IV.10); it must be stressed that, before the end of the proof is reached, the object generated in the construction is just a "dumb" network of lines: the figure generated in the construction of I.46 is a square only at the very end of the proof. In problems more than in theorems, the boundary between the construction and the proof may not be clear-cut. The proof (or parts of it) may be introduced by a "theorematic" determination,[142] namely, the one opened by λέγω ὅτι "I claim that".[143]

vi. A conclusion, which may assume several forms, to be detailed in the next page.

---

[139] Cf. Sects. 4.1 (which includes a complete list of problems in the *Elements*), 4.2, and 4.2.1 for the first three items, Appendix A for a complete list of problems in the Greek mathematical corpus.

[140] The "givens" are absent in the enunciation of IV.10, X.10 (but here the verb is replaced by προτεθείσῃ), X.27–35, 48–53, 85–90, XIII.13–18.

[141] There is no determination in IV.10, VI.10, IX.18–19, X.27–35, 48–53, 85–90, XIII.18, a subset of which are the problems without "givens".

[142] A theorematic determination is sometimes required by specific features of the problem: in Book XIII, the enunciation includes both a problem and a theorem; in Book IV, some figures are first shown to be equilateral, then equiangular, and this second step requires a "partial" determination of the theorematic type (see Sect. 4.2.1).

[143] Theorematic determinations are absent in I.1–3, 23, 31, 42, 44–45, II.14, III.1, 25, 30, 33–34, IV.1–5, 8–10, 13–14, 16, VI.9–13, 18, 25, 28–30, X.10, 29–31, 33–35, XI.11–12, 26–27, XII.16, XIII.18. The form δεικτέον δή is found in X.49, 53. See also Sect. 4.2.1.

Let us thus see the forms that the conclusion of a problem of construction may assume. These are a general, non-instantiated conclusion, as in III.25 (*EOO* I, 230.8–9):[144]

κύκλου ἄρα τμήματος δοθέντος προσαναγέγραπται ὁ κύκλος.	Therefore, a segment of a circle being given, the circle turns out to be described out.

or an instantiated conclusion, as in I.2 (*EOO* I, 14.13–14):[145]

πρὸς ἄρα τῷ δοθέντι σημείῳ τῷ Α τῇ δοθείσῃ εὐθείᾳ τῇ ΒΓ ἴση εὐθεῖα κεῖται ἡ ΑΛ.	Therefore at a given point, A, a straight line, ΑΛ, turns out to be set equal to a given straight line, ΒΓ.

or an instantiated conclusion that includes a deductive step, as in I.1 (*EOO* I, 12.14–15):[146]

ἰσόπλευρον ἄρα ἐστὶ τὸ ΑΒΓ τρίγωνον, καὶ συνέσταται ἐπὶ τῆς δοθείσης εὐθείας πεπερασμένης τῆς ΑΒ.	Therefore triangle ΑΒΓ is equilateral, and turns out to be constructed on a given bounded straight line, AB.

or a conclusion that identifies an object as the required one, as in III.1 (*EOO* I, 168.9–10):[147]

τὸ Ζ ἄρα σημεῖον κέντρον ἐστὶ τοῦ ΑΒΓ.	Therefore point Z is the centre of ΑΒΓ.

or a procedural statement, as in IV.16 (*EOO* I, 320.9–12):

ἐὰν ἄρα ἐπιζεύξαντες τὰς ΒΕ ΕΓ ἴσας αὐταῖς κατὰ τὸ συνεχὲς εὐθείας ἐναρμόσωμεν εἰς τὸν ΑΒΓΔΕ κύκλον, ἔσται εἰς αὐτὸν ἐγγεγραμμένον πεντεκαιδεκάγωνον ἰσόπλευρόν τε καὶ ἰσογώνιον.	Therefore if joining BE, ΕΓ we adapt continuously straight lines equal to them in circle ΑΒΓΔΕ, an equilateral and equiangular pentadecagon will turn out to be inscribed in it.

There is no problematic conclusion in XII.16 and XIII.17–18.

The aspectual value of the perfect stem that characterizes most conclusions of problems and all constructive acts (see below) abstracts from temporal or durative connotations; the perfect stem refers to the *present* time of the subject (the patient, in this case, since the verb is always in the passive voice), while stressing that the present-time condition is the accomplished result of some past history: the constructed object is there, constructed, in its eternal present. Thus, the conclusion of a problem, while being a statement formulated by a sentence with the verb in the indicative, is not truth-apt, and has for this reason a metadiscursive connotation, which explicitly points to a form of knowledge.[148] For this reason, a constructive act refers to the conclusion of a problem as to its

[144] This kind of conclusion is also found in IV.2 (the Theonine manuscripts have it preceded by an instantiated conclusion, see *EOO* I, 274.18–19), 3, 5, 7–9, 11, 13–15, X.4, 27–28, XII.17, XIII.13, 14 (*bis*), 15, 16 (*bis*), 17. Recall that the problems in Book XIII require to perform two constructions.

[145] This kind of conclusion is also found in I.3, 9–12, 22–23, 31, 42, 44–45, II.11, 14, III.17, 30, 33–34, IV.1, 4, 6, 10, VI.9–13, 18, 25, 28–30, IX.18–19, X.3, 10, 29–35, X.85–90, XI.11–12, 23, 26–27, XIII.13, 15, 17.

[146] This kind of conclusion is also found in I.46, IV.3, 5, 7–9, 12, 15.

[147] It is required to find a point, and the proof is a reduction to the impossible. This kind of conclusion is also found in VII.2–3, 33–34, 36, 39, VIII.2, 4, X.48–53.

[148] I owe this insight to Beere, Morison, unpublished typescript. For this reason, the conclusion of a problem of construction has a strong presential connotation: "here it is, a square has finally been described on a given straight line".

own template, and does not "cite" such a conclusion word for word, as instead references to the enunciations of theorems can do: the transition from the indicative mood to the imperative mood in citations [see item (d) below and Sect. 2.3] highlights the fact that what is subsequently used in any application is the *actual* construction, and not a statement, a prescription, or the piece of operational knowledge that in some sense is validated by the complex "construction + proof" of a problem of construction.

(c) *Construction as a specific part of a proposition*

A construction is a specific part of a proposition (Sect. 1.1), possibly but not exclusively of a problem of construction—type (b) above. In the construction, the geometric or numeric configuration generated in the setting-out is completed with objects subsequently used in the proof. These objects may be required by the configuration assumed in the enunciation, or may be auxiliary objects, to be "discharged" in the proof, or may be objects to be constructed (in the case of problems). In the first two cases just mentioned, no justification is provided as to why such-and-such objects are introduced and not others. The construction is usually opened by a particle γάρ; as seen above (Sects. 1.1 and 4.2), this γάρ does not carry any explicative value: it is simply a scope particle.[149] The construction has a purely coordinative structure: it is made of a number of conjoined atomic steps, which apply one of the first three postulates, or constructions licensed by a previous problem; these are the "constructive acts"—type (d) below. The ordering in the sequence of constructive acts points to the fact that any act within the sequence produces an object that is apt to be operated upon by the subsequent constructive act: the temporal arrow is replaced by an operational arrow. Sometimes, deductive steps are inserted in the middle of a construction, so as to pinpoint features of the arrived at configuration that make the subsequent constructive act possible.

According to what has just been said, we may divide constructions into three categories:

1) Real auxiliary constructions: the involved geometric objects are used in the proof or in subsequent constructions but they are not mentioned in the setting-out. Let us take the construction of I.16 (*EOO* I, 42.13–16) as an example; only a triangle τὸ ΑΒΓ is assigned in the setting-out, one of whose sides ἡ ΒΓ is produced as far as τὸ Δ, a "something" that will not be used in the construction and that will simply remain a part of the denomination of two angles:

τετμήσθω ἡ ΑΓ δίχα κατὰ τὸ Ε, καὶ ἐπιζευχθεῖσα ἡ ΒΕ ἐκβεβλήσθω ἐπ' εὐθείας ἐπὶ τὸ Ζ, καὶ κείσθω τῇ ΒΕ ἴση ἡ ΕΖ, καὶ ἐπεζεύχθω ἡ ΖΓ, καὶ διήχθω ἡ ΑΓ ἐπὶ τὸ Η.	Let ΑΓ be bisected at E, and let BE be produced, once joined, in a straight line as far as Z, and let a ‹straight line›, EZ, be set equal to BE,[150] and let a ‹straight line›, ΖΓ, be joined, and let ΑΓ be drawn through as far as H.

The auxiliary constructive acts are discharged during the proof (see Sect. 2.3). Since these acts introduce objects whose nature is unpredictable on the basis of what precedes them in a proposition, they are the basic steps of a mathematical proposition that cannot be reduced to a mechanical procedure or to logical facts (but see Sect. 2.4.1).

[149] A formal proof of this is the absence of γάρ whenever the construction is opened by the presential verb form νενοήσθω, see the discussion in Sect. 4.2.

[150] The construction is bewildering, unless we keep in mind that the first occurrence of τὸ Ζ is just part of the name of the produced line. Integrating the first "a ‹straight line›" is crucial.

2) Constructions that make the geometric configuration implicitly assumed in the enunciation explicit: examples to a variable degree hybridized with the preceding category are II.1–8.[151] In particular, II.1 (*EOO* I, 120.1–4) entirely belongs to the present category—in it, a straight line ἡ ΒΓ cut at points τὰ Δ, Ε is given, and the relation to be proved involves the rectangles constructed on the segments that result from the two cuts:

ἤχθω γὰρ ἀπὸ τοῦ Β τῇ ΒΓ πρὸς ὀρθὰς ἡ ΒΖ, καὶ κείσθω τῇ Α ἴση ἡ ΒΗ, καὶ διὰ μὲν τοῦ Η τῇ ΒΓ παράλληλος ἤχθω ἡ ΗΘ διὰ δὲ τῶν Δ Ε Γ τῇ ΒΗ παράλληλοι ἤχθωσαν αἱ ΔΚ ΕΛ ΓΘ.	In fact, from B let a ‹straight line›, BZ, be drawn at right ‹angles› with ΒΓ, and let a ‹straight line›, BH, be set equal to A, and through H let a ‹straight line›, ΗΘ, be drawn parallel to ΒΓ and through Δ, Ε, Γ let ‹straight lines›, ΔΚ, ΕΛ, ΓΘ, be drawn parallel to BH.

3) Constructions required by the enunciation (therefore only in problems) and in which no auxiliary object is introduced; a paradigmatic example is I.46 (*EOO* I, 108.15–19), where it is required to ἀναγράψαι "describe" a square on a given straight line ἡ ΑΒ:

ἤχθω τῇ ΑΒ εὐθείᾳ ἀπὸ τοῦ πρὸς αὐτῇ σημείου τοῦ Α πρὸς ὀρθὰς ἡ ΑΓ, καὶ κείσθω τῇ ΑΒ ἴση ἡ ΑΔ, καὶ διὰ μὲν τοῦ Δ σημείου τῇ ΑΒ παράλληλος ἤχθω ἡ ΔΕ διὰ δὲ τοῦ Β σημείου τῇ ΑΔ παράλληλος ἤχθω ἡ ΒΕ.	Let a ‹straight line›, ΑΓ, be drawn at right ‹angles› with straight line AB from a point A on it, and let a ‹straight line›, ΑΔ, be set equal to AB, and through point Δ let a ‹straight line›, ΔΕ, be drawn parallel to AB and through point B let a ‹straight line›, BE, be drawn parallel to ΑΔ.

Among constructions must be included some suppositions typical of Book V, which normally consist in taking equimultiples, as in V.22 (*EOO* II, 62.1–4):

εἰλήφθω γὰρ τῶν μὲν Α Δ ἰσάκις πολλαπλάσια τὰ Η Θ, τῶν δὲ Β Ε ἄλλα, ἃ ἔτυχεν, ἰσάκις πολλαπλάσια τὰ Κ Λ, καὶ ἔτι τῶν Γ Ζ ἄλλα, ἃ ἔτυχεν, ἰσάκις πολλαπλάσια τὰ Μ Ν.	In fact, let equimultiples, H, Θ, be taken of A, Δ, and other, random, equimultiples, K, Λ, of B, E, and further other, random, equimultiples, M, N, of Γ, Z.

or those, fairly rare, which are found in the arithmetic Books, as in VIII.9 (*EOO* II, 294.23–27)—the iterative construction is made possible by VIII.2:

εἰλήφθωσαν γὰρ δύο μὲν ἀριθμοὶ ἐλάχιστοι ἐν τῷ τῶν Α Γ Δ Β λόγῳ ὄντες οἱ Ζ Η τρεῖς δὲ οἱ Θ Κ Λ, καὶ ἀεὶ ἑξῆς ἑνὶ πλείους ἕως ἂν ἴσον γένηται τὸ πλῆθος αὐτῶν τῷ πλήθει τῶν Α Γ Δ Β. εἰλήφθωσαν, καὶ ἔστωσαν οἱ Μ Ν Ξ Ο.	In fact, let two least numbers, Z, H, be taken that are in the ratio of A, Γ, Δ, B, three, Θ, K, Λ, and successively continually more by one, until their multiplicity become equal to the multiplicity of A, Γ, Δ, B. Let them be taken, and let them be M, N, Ξ, O.

or those, very simple, which we read in VII.37, VIII.4 and 15 (*EOO* II, 266.3–4, 278.26–27, and 312.24–314.2)—the second constructive act is licensed by VII.34:

[151] Propositions II.9–10 belong to the preceding category. Proposition II.14, "construct a square equal to a given rectilinear ‹figure›", is a hybrid between categories (1) and (3); as a matter of fact, in the non-trivial case the square required by the enunciation of this proposition is not even constructed!

ὁσάκις γὰρ ὁ Β τὸν Α μετρεῖ, τοσαῦται μονάδες ἔστωσαν ἐν τῷ Γ.	In fact, how many times B measures A, let there be so many units in Γ.
εἰλήφθω γὰρ ὁ ὑπὸ τῶν Β Γ ἐλάχιστος μετρούμενος ἀριθμὸς ὁ Η.	In fact, let the least number H measured by B, Γ be taken.
ὁ Γ γὰρ ἑαυτὸν πολλαπλασιάσας τὸν Ε ποιείτω ὁ δὲ Δ ἑαυτὸν πολλαπλασιάσας τὸν Η ποιείτω, καὶ ἔτι ὁ Γ τὸν Δ πολλαπλασιάσας τὸν Ζ ποιείτω ἑκάτερος δὲ τῶν Γ Δ τὸν Ζ πολλαπλασιάσας ἑκάτερον τῶν Θ Κ ποιείτω.	In fact, let Γ multiplying itself make E and let Δ multiplying itself make H, and further, let Γ multiplying Δ make Z and let Γ, Δ multiplying Z make Θ, K, respectively.

As noted at the beginning of Sect. 1.1, constructions do not necessarily follow immediately the setting-out or the determination (in both cases, either liminal or not), or more generally the fresh start of a new argument.[152] We also find constructions postponed in the middle of the proof; a pervasive dialectic between constructive acts and deductions is at work in some propositions of Book X (X.94 is an extreme example) or of Book XIII (as in XIII.16 and 17, which include a rich sequence of partial setting-outs).

Nor can the constructions be rigidly demarcated from the deductive steps, as we see in the case of IV.11 (*EOO* I, 298.18–300.5)—this is different from what happens in I.44 (see just below) because there is no "paraconditional" (see again below) between the two constructive clusters:

ἐκκείσθω τρίγωνον ἰσοσκελὲς τὸ ΖΗΘ διπλασίονα ἔχον ἑκατέραν τῶν πρὸς τοῖς Η Θ γωνιῶν τῆς πρὸς τῷ Ζ, καὶ ἐγγεγράφθω εἰς τὸν ΑΒΓΔΕ κύκλον τῷ ΖΗΘ τριγώνῳ ἰσογώνιον τρίγωνον τὸ ΑΓΔ ὥστε τῇ μὲν πρὸς τῷ Ζ γωνίᾳ ἴσην εἶναι τὴν ὑπὸ ΓΑΔ ἑκατέραν δὲ τῶν πρὸς τοῖς Η Θ ἴσην ἑκατέρᾳ τῶν ὑπὸ ΑΓΔ ΓΔΑ. καὶ ἑκατέρα ἄρα τῶν ὑπὸ ΑΓΔ ΓΔΑ τῆς ὑπὸ ΓΑΔ ἐστι διπλῆ. τετμήσθω δὴ ἑκατέρα τῶν ὑπὸ ΑΓΔ ΓΔΑ δίχα ὑπὸ ἑκατέρας τῶν ΓΕ, ΔΒ εὐθειῶν, καὶ ἐπεζεύχθωσαν αἱ ΑΒ ΒΓ ΓΔ ΔΕ ΕΑ.	Let an isosceles triangle, ZHΘ, be set out having each of the angles at H, Θ double of that at Z, and let a triangle, AΓΔ, equiangular to triangle ZHΘ be inscribed in circle ABΓΔE so has to be ΓAΔ equal to the angle at Z and those at H, Θ equal to AΓΔ, ΓΔA, respectively; therefore each of AΓΔ, ΓΔA is also double of ΓAΔ. Thus let AΓΔ, ΓΔA be bisected by straight lines ΓE, ΔB, respectively, and let ‹straight lines›, AB, BΓ, ΓΔ, ΔE, EA, be joined.

What happens in I.44 (*EOO* I, 102.15–104.5) is worth a short discussion:

συνεστάτω τῷ Γ τριγώνῳ ἴσον παραλληλόγραμμον τὸ ΒΕΖΗ ἐν γωνίᾳ τῇ ὑπὸ ΕΒΗ, ἥ ἐστιν ἴση τῇ Δ, καὶ κείσθω ὥστε ἐπ' εὐθείας εἶναι τὴν ΒΕ τῇ ΑΒ, καὶ διήχθω ἡ ΖΗ ἐπὶ τὸ Θ, καὶ διὰ τοῦ Α ὁποτέρᾳ τῶν ΒΗ ΕΖ παράλληλος ἤχθω ἡ ΑΘ, καὶ ἐπεζεύχθω ἡ ΘΒ. καὶ ἐπεὶ εἰς παραλλήλους τὰς ΑΘ ΕΖ εὐθεῖα ἐνέπεσεν ἡ ΘΖ, αἱ ἄρα ὑπὸ ΑΘΖ ΘΖΕ γωνίαι δυσὶν ὀρθαῖς εἰσιν ἴσαι· *αἱ ἄρα*	In an angle EBH that is equal to Δ let a parallelogram, BEZH, be constructed equal to triangle Γ, and let it be set so as to be BE in a straight line with AB, and let ZH be drawn through as far as Θ, and through A let a ‹straight line›, AΘ, be drawn parallel to either of BH, EZ, and let a ‹straight line›, ΘB, be joined. And since a straight line, ΘZ, fell[153] on parallels AΘ, EZ, therefore the angles AΘZ, ΘZE are equal to two right

[152] The presence of independent subproofs is necessary if a division into cases is required by the geometric configuration or by a proof by reduction to the impossible.

[153] The aorist ἐνέπεσεν "fell" is required because this configurational state of affairs is not directly governed by a construction licensed by a postulate or by a problem; see below under item (d); it remains that what happens is a constructive act. The same reason justifies the present stem συμπιπτέτωσαν "let them meet" in the subsequent construction.

ὑπὸ ΒΘΗ ΗΖΕ δύο ὀρθῶν ἐλάσσονές εἰσιν· αἱ δὲ ἀπὸ ἐλασσσόνων ἢ δύο ὀρθῶν εἰς ἄπειρον ἐκβαλλόμεναι συμπίπτουσιν· αἱ ΘΒ ΖΕ ἄρα ἐκβαλλόμεναι συμπεσοῦνται. ἐκβεβλήσθωσαν καὶ συμπιπτέτωσαν κατὰ τὸ Κ, καὶ διὰ τοῦ Κ σημείου ὁποτέρα τῶν ΕΑ ΖΘ παράλληλος ἤχθω ἡ ΚΛ, καὶ ἐκβεβλήσθωσαν αἱ ΘΑ ΗΒ ἐπὶ τὰ Λ, Μ σημεῖα.

‹angles›; *therefore ΒΘΗ, ΗΖΕ are less than two right ‹angles›*; and ‹straight lines› from less than two right ‹angles› meet once unboundedly produced; therefore ΘΒ, ΖΕ once produced will meet. Let them be produced, and let them meet at Κ, and through point Κ let a ‹straight line›, ΚΛ, be drawn parallel to either of ΕΑ, ΖΘ, and let ΘΑ, ΗΒ be produced as far as points Λ, Μ.

Apparently, there is a series of deductive steps opened by a paraconditional[154] in the middle of the construction or, conversely, a series of constructive acts is placed in the middle of the proof, after the "anaphora" that canonically opens the proof (there are only deductive steps after the quote). But this reading is unwarranted: there is just one deductive step (in italics above) inside a very long construction; as said, this step has the function of identifying a feature of the arrived at configuration that makes the subsequent constructive act possible. The point is that what precedes and follows the italicized sentence are two applications of I.post.5 (its contrapositive first, then the postulate itself, with concise quote): qua postulate it counts as a constructive act (the aorist ἐνέπεσεν proves this, see below), but, as a consequence of its declarative structure, it can only be discharged if it is included in a paraconditional or in a deductive sequence (concise citation serving as coassumption, and subsequent conclusion).

(d) *Constructive acts*

Constructive acts are atomic geometric or arithmetic operations undergone by one or more mathematical objects, each of which serves as a syntactic subject but as a semantic patient. These acts are formulated by single sentences with the verb in the imperative. They are the independent units of meaning of which a construction—type (c) above—is made; they summarize the construction of a mathematical object—type (a) above—; their formulation depends on the primary occurrence of the construction of type (a) within a problem of construction—type (b) above.

All basic constructive acts are formulated in the passive perfect imperative, thus: ἐπεζεύχθω ἡ ΑΒ "let a ‹straight line› ΑΒ be joined".[155] Any object introduced in the construction is designated by an indefinite noun phrase; as in the clause just read, such a designation is very often left understood. Denotative letters are adjoined as appositions to the designation: they are the names of the indefinite designations (see Sect. 3.3). The use of the perfect stem refers[156] to the stylistic practice canonically adopted in the conclusion of problems; it emphasizes the value of ἐνεστὼς συντελικός "accomplished present" of such a stem, abstracting from temporal connotations (Sect. 1.1). The imperative sets constructive acts in the "suppositional mode" of Stoic logic:[157] constructive acts do not have a truth-value, and in fact they are, on a par with the suppositions that figure in the setting-out, discharged in the course of the proof (Sect. 2.3). It is of capital importance to realize that no constructive act operates on given objects, nor does it produces given objects as output: the predicate "given" appears nowhere in the first three postulates; in every problem of construction the given objects are named by means of a lettered designation in the setting out, and under this name

[154] This is the statement introduced by καὶ ἐπεὶ "and since"; see Sects. 1.1, 4.4, and 5.3.2.
[155] Special verbs or special contexts may not require the perfect: see below the examples taken from VII.37 and VIII.15.
[156] And it does so in a strong sense: see just below and Sect. 2.3.
[157] See Bobzien 1997 and Sect. 4.2.

they are operated upon in the construction (check any constructive act in this Section, or in the paradigmatic matrices discussed in Sect. 2.3).

Every constructive act is expressed by an invariant formula, whose template is the instantiated conclusion of the relevant problem (see Sect. 2.3). In the case of the postulates, however, this is impossible, since they are not instantiated. In this instance, the "principle of the first occurrence" provides the necessary template. Let us read first I.post.1–3 (*EOO* I, 8.7–12)—they are introduced by a verb whose meaning has a strong directive connotation:

ᾐτήσθω ἀπὸ παντὸς σημείου ἐπὶ πᾶν σημεῖον εὐθεῖαν γραμμὴν ἀγαγεῖν,	Let it be required to draw a straight line from any point to any point,
καὶ πεπερασμένην εὐθεῖαν κατὰ τὸ συνεχὲς ἐπ' εὐθείας ἐκβαλεῖν,	and to produce a bounded straight line continuously in a straight line,
καὶ παντὶ κέντρῳ καὶ διαστήματι κύκλον γράφεσθαι.	and that a circle can be described with any centre and radius.

Note the second-aorist infinitive in the first two postulates: this stem abstracts from any aspectual or temporal connotation, very much as the aorist infinitive that figures in the enunciations of problems: any basic constructive step is a punctual operation. Note also that γράφω "to describe" in the third postulate does not have a second aorist: and in fact a present infinitive is employed.[158] These constructive acts, along with I.1, are all together at work for the first time in I.2 (*EOO* I, 12.24–14.3)[159]—let me stress again: all verb forms are in the perfect stem (underlined), all new mathematical objects are explicitly designated by indefinite noun phrases (italics):

<u>ἐπεζεύχθω</u> γὰρ ἀπὸ τοῦ Α σημείου ἐπὶ τὸ Β σημεῖον *εὐθεῖα* ἡ ΑΒ, καὶ <u>συνεστάτω</u> ἐπ' αὐτῆς *τρίγωνον ἰσόπλευρον* τὸ ΔΑΒ, καὶ <u>ἐκβεβλήσθωσαν</u> ἐπ' εὐθείας ταῖς ΔΑ ΔΒ *εὐθεῖαι* αἱ ΑΕ ΒΖ, καὶ κέντρῳ μὲν τῷ Β διαστήματι δὲ τῷ ΒΓ *κύκλος* <u>γεγράφθω</u> ὁ ΓΗΘ.	In fact, from point A to point B <u>let</u> *a straight line*, AB, <u>be drawn</u>, and <u>let</u> *an equilateral triangle*, ΔAB, <u>be constructed</u> on it, and <u>let</u> *straight lines*, AE, BZ, <u>be produced</u> in a straight line with ΔA, ΔB, and with centre B and radius BΓ <u>let</u> *a circle*, ΓHΘ, <u>be described</u>.

The main difference between the formulations just read and those in the postulates comes from the use of the verb ἐπιζεύγνυμι "to join" instead of ἄγω "to draw" in the applications of the first postulate;[160] otherwise the formulations conform to a canonical matrix expression (see Sect. 2.3). Such a difference of use is rigidly adhered to throughout the ancient mathematical corpus:[161] ἐπιζεύγνυμι generates straight lines by joining two points (namely, by using the first postulate);

[158] I take the meaning to be passive and the present stem to carry a modal connotation of possibility (cf. also the presence of παντί "any"): see Sect. 5.2.

[159] Postulate 2 is not used in I.1. I take I.2 for simplicity's sake, but of course the "principle of the first occurrence" must also be applied to I.1 in the case of I.post.1 and 3. We have already read the construction of I.2 in Sect. 2.5.

[160] The term διάστημα "radius" is employed in the third postulate and in its citations for grammatical reasons: a prepositional phrase like ἐκ τοῦ κέντρου "radius" (lit. "a ‹straight line› from the centre") without the preposed article, as it is required in the applications of the third postulate, cannot be declined, but a dative is required: see Federspiel 2005. The expression ἐκ τοῦ κέντρου without preposed article can only be a grammatical subject or a nominal complement of the copula, both of which are necessarily in the nominative.

[161] See Federspiel 2002, 137–147, for a detailed study. Exceptions are the first postulate, 3 dubious occurrences in Archimedes (in the enunciations of *Sph. cyl.* I.9–10 and in the construction of *Aequil.* II.8), and a fair number in Apollonius' *Conica*, in particular in participial expressions contained in the enunciation of 15 propositions (I.1, II.1, 29, 34, III.44–45, 47, IV.1, 4–6, 9, 13, 15, 18).

ἄγω generates straight lines whose function is specified, normally by means of prescriptions on their direction or on their relation to other geometric objects whose function is specified (parallelism, perpendicularity, tangency, section, etc.): it is an all-purpose verb, and is met in many constructive acts. A "drawn" straight line often passes διά "through" one or more points. The preposition is occasionally prefixed to the verb: διάγω "to draw through", which, however, is not a mere stylistic variant of ἄγω διά.

The use of the verb διάγω is in fact interesting and deserves a digression (see also Sect. 5.1.3). In the Euclidean corpus, this verb is mainly employed in the perfect imperative; it is often synonymous with ἐκβάλλω "to produce", within applications of the second postulate in which the endpoint of the produced straight line is *mentioned*, as in I.16 we have read above.[162] Since this endpoint is very often left undetermined, the constructive act is also underdetermined.[163] Such residual degrees of freedom are made explicit when the straight line "drawn through" passes through one point, but no other point on the line is specified,[164] as we may check by reading again our paradigmatic proposition III.2 (*EOO* I, 168.24–170.2):

μὴ γάρ, ἀλλ' εἰ δυνατόν, πιπτέτω ἐκτὸς ὡς ἡ ΑΕΒ, καὶ εἰλήφθω τὸ κέντρον τοῦ ΑΒΓ κύκλου καὶ ἔστω τὸ Δ, καὶ ἐπεζεύχθωσαν αἱ ΔΑ ΔΒ, καὶ διήχθω ἡ ΔΖΕ.

In fact not, but, if possible, let it fall outside as AEB, and let the centre of circle ABΓ be taken and let it be Δ, and let ‹straight lines›, ΔA, ΔB, be joined, and let a ‹straight line›, ΔZE, be drawn through.

The letter Z is here introduced for the first time in the proof. The letter E is only part of the name of straight line ἡ AEB, but here it is not the name of a point: it becomes the name of a point only when it is identified as the intersection of straight lines ἡ AEB and ἡ ΔZE; exactly for this reason the text uses διάγω. Otherwise it would have read:

*καὶ ἐπεζεύχθωσαν αἱ ΔΑ ΔΒ ΔΕ, καὶ ἡ ΔΕ τεμνέτω τὸν ΑΒΓ κύκλον κατὰ τὸ Ζ.

And let ‹straight lines›, ΔA, ΔB, ΔE, be joined, and let ΔE cut the circle ABΓ at Z.

The only (partial) exception to such a connotation of indeterminacy of διάγω occurs when the verb is used for planes "drawn through" one or more straight lines, as in the following extract from proposition XI.13 (*EOO* IV, 36.12–15):[165]

εἰ γὰρ δυνατόν, ἀπὸ τοῦ αὐτοῦ σημείου τοῦ Α τῷ ὑποκειμένῳ ἐπιπέδῳ δύο εὐθεῖαι αἱ ΑΒ ΑΓ πρὸς ὀρθὰς ἀνεστάτωσαν ἐπὶ τὰ αὐτὰ μέρη, καὶ διήχθω τὸ διὰ τῶν ΒΑ ΑΓ ἐπίπεδον.

In fact, if possible, from a same point, A, let two straight lines, AB, AΓ, be erected on the same side at right ‹angles› with the underlying plane, and let the plane through BA, AΓ be drawn through.

---

[162] There are 30 occurrences of this kind in the main text of the *Elements*; they are in XII.17 (present participle), VI.3 (passive aorist participle), I.16, 20–21, 44, II.3, 11, III.1, 9–10, 15, 20, 25, 31, IV.15, VI.9, 26, XII.16–17, XIII.1, 9–12, 14 (perfect imperative).

[163] But the second postulate does not mention any endpoint, either.

[164] There are 18 occurrences of διάγω with this meaning in the main text of the *Elements*, in III.8, 32 (passive aorist subjunctive), XI.20 (passive aorist participle), III.1, 2, 5, 6, 8, 23, 32, IV.3, XI.2, 4, 11, 31, XIII.2 (perfect imperative). The indefinite character of the construction is made prominent by the systematic presence of determiners of indefiniteness or of arbitrariness (see Sect. 5.1.2–3); the only exceptions are in III.2, 23, XI.2, 20, 31, XIII.2.

[165] The other two occurrences are in XI.5 and 7. The exception is partial because in XI.7 the plane is drawn through one straight line only, which does not completely determine the plane, whereas two straight lines do.

I now present a typology of the verbs that in the *Elements* formulate constructive acts, and more generally configurational or numeric states of affairs; the lists do not contain verbs with metadiscursive connotation or that are not operative (like εἰμί "to be" or ἔχω "to have"); each item is followed, normally in its first occurrence (some verbs figure in more than one type), by the total number of its occurrences in the whole of the *Elements*; this completes the lexicographic survey of Sect. 1.5 (the underlined verbs are not terminative).

(A) States of affairs and constructions licensed by definitions: ἀναστρέφω "to convert", said of ratios, V.def.16 (23); ἀντιπέπονθα "to happen to be in inverse relation", said of regions, VI.def.2 (50); ἀπέχω "to be distant", said of straight lines in a circle, III.def.4–5 (13); βέβηκα "to happen to stand", said of angles upon an arc of a circumference, III.def.9 (21); διαιρέω "to divide", said of ratios, V.def.15 (146); ἐγγράφω "to inscribe", said of rectilinear figures in a circle and vice versa, IV.def.1, 3, 5 (88); ἐναρμόζω "to adapt", said of straight lines to a circle, IV.def.7 (10); ἐφάπτομαι "to be tangent", said of straight lines to a circle and of circles to one another, III.def.2–3 (92); κέκλιμαι "to happen to be inclined", said of straight lines to a plane, XI.def.7 (1); περιγράφω "to circumscribe", said of rectilinear figures in a circle and vice versa, IV.def.2, 4, 6 (57); περιέχομαι "to be contained", said of parallelograms by their sides, II.def.1; πολλαπλασιάζω "to multiply", said of numbers, VII.def.16 (271); σύγκειμαι "to be compounded", said of ratios, VI.def.5 (162);[166] συντίθημι "to compound", said of ratios, V.def.14; ταράσσομαι "to be perturbed", said of ratios, V.def.18 (10); τέτμημαι (ἄκρον καὶ μέσον λόγον) "to be cut" (scil. in extreme and mean ratio), said of straight lines, VI.def.3.

(B) Constructive acts licensed by postulates: γράφω "to describe", said of circles, I.post.3 (58); ἐκβάλλω "to produce", said of straight lines, I.post.2 (69); ἐπιζεύγνυμι "to join", said of straight lines, I.post.1 (ἄγω in the postulate) (291).

(C) Constructive acts licensed by problems: ἄγω "to draw", said of straight lines that are parallel to one another, I.31, perpendicular to one another, I.11–12, or to a plane, XI.11, or tangent to a circle, III.17 (231); ἀναγράφω "to describe", said of squares, I.46, of rectilinear regions similar to another rectilinear region, VI.18, of parallelepipeds, XI.27 (84); ἀνίστημι "to erect", said of straight lines perpendicular to a plane, XI.12 (55); ἀφαιρέω "to remove", said of straight lines from another straight line, I.3, VI.9, of segments from a circle, III.34 (138); γράφω "to describe", said of segments of a circle on a straight line, III.33; ἐγγράφω "to inscribe", said of rectilinear figures in a circle and vice versa, IV.2, 4, 6, 8, 11, 13, 15, 16, of polygons in a circle, XII.16, of polyhedra in concentric spheres, XII.17; ἐναρμόζω "to adapt", said of straight lines to a circle, IV.1; (προσ)ευρίσκω "to find", said of points, III.1 (λαμβάνω "to take" in the applications), of straight lines in proportion, VI.11–13, of numbers, VII.2–3, 33–34, 36, 39, VIII.2, 4, IX.18–19, of irrational lines, X.27–35, 48–53, 85–90 (105, of which 32 with prefix); κεῖμαι "to be set", said of straight lines equal to another straight line, I.2 (τίθημι "to set" in the enunciation) (136); παραβάλλω "to apply", said of regions to a straight line, I.44, VI.28–30 (170); περιγράφω "to circumscribe", said of rectilinear figures in a circle and vice versa, IV.3, 5, 7, 9, 12, 14; περιλαμβάνω "to comprehend", said of regular polyhedra with a sphere, XIII.13–18 (26); προσαναγράφω "to describe out", said of circles from a segment of a circle, III.25 (5); συνίστημι "to construct", said of triangles, I.1, 22–23, of parallelograms equal to a triangle or to a rectilinear region, I.42, 45, of squares equal to a rectilinear region, II.14, of triangles, IV.10, of regions similar to a region and equal to another, VI.25,

[166] All compound forms of κεῖμαι serve as passive of the corresponding forms of τίθημι (βάλλω in one instance).

of solid angles, XI.23, 26, of regular polyhedra, XIII.13–18 (126); τέμνω "to cut", said of angles, I.9, of straight lines, I.10, VI.10, of straight lines that are cut in extreme and mean ratio, II.11, VI.30, of arcs of a circumference, III.30 (395).

(D) Items that are not licensed by any of the previous, but figure in them or in theorems (relevant occurrences are provided): ἄγω "to draw", said of lines through points, passim; ἀναπληρόω "to complete", said of parallelograms, XII.2 (1); ἀνίστημι "to erect", said of solids, XII; ἀποκαθίστημι "to return", said of the lines that generate the boundary of solids of revolution, XI.def.14, 18, 21 (7); ἀπολαμβάνω "to cut off", said of lines from another line, I.def.18, III.def.9–10 and passim (21); ἅπτομαι "to touch", said of lines, passim (50); γράφω "to describe", said of semicircles, VI.13, X, XIII; δέχομαι "to admit of", said of segments of circles with respect to an angle, III.def.11 (11); διάγω "to draw through", said of lines and of planes, passim (see also above) (57); διαιρέω "to divide", said of numbers, VII.def.6–7 and passim (146); διαλείπω "to leave out", said of numbers, IX.8–10 (17); διαφέρω "to differ", said of numbers, VII.def.7, of regions, X (17); δύναμαι "to be worth", said of straight lines, X (342); ἔκκειμαι "to be set out", said of any sort of objects (see Sect. 4.2) (159); ἐκτίθημι "to set out", said of numbers IX.36, of the edges of the regular polyhedra, XIII.18 (3); ἐμπεριέχω "to contain", said of solids with respect to another solid, XII (6); ἐμπίπτω "to fall on", said of straight lines on a straight line, I.post.5 and passim (86); ἔρχομαι "to pass", said of straight lines through a point, III.12, 33 (6); ἐφίστημι "to stand", said of straight lines on a straight line, XI.def.5, XI.4 (41); ἵστημι "to be set up", said of straight lines on a straight line, I.def.10 (9); καταγράφω "to describe completely", said of various configurations left to be completed, II, VI, X, XIII (18); κατασκευάζω "to construct", said of various configurations (abridged constructions), passim (32); κεῖμαι "to lie", said of lines with respect to a straight line, I.def.4, of points, I.def.15; μένω "to stand still", said of the axes of solids of revolution, XI.def.14, 18, 21 (12); μεταλαμβάνομαι "to be permuted", said of straight lines, I.17, of angles, XI.20 (19); νοέω "to conceive", said of points and solids (see Sect. 4.2) (10); παράκειμαι "to be applied", said of regions to a straight line, X (37); παραλλάσσω "to fall beside", said of straight lines, I.8, of segments of a circle, III.24 (2); παρεμπίπτω, said of straight lines between lines, III.16 (6); περαίνομαι "to be bounded", said of lines, I.1 (13); περατοῦμαι "to be bounded", said of lines by lines or points, I.def.17, XI.def.17 (2); περιάγομαι "to rotate", said of the lines that generate the circular bases of a cylinder, XI.def.23 (1); περιέχω / –ομαι "to contain / be contained", said of straight lines and an angle and vice versa, I.def.9, of regions by their boundary, I.def.14–15, 18–19, I.cn.9, III.def.6–10, XI.9–13, 23, 25–28 and passim, of numbers, VII.def.19–20 and passim (362); περιλαμβάνω "to comprehend", said of solids of revolution by their own boundary, XI.def.14, 18, 21 (26); περιφέρομαι "to rotate", said of the lines that generate the boundary of solids of revolution, XI.def.14, 18, 21 (10); πίπτω "to fall on", said of points within a region, III.25, of straight lines on a point, III.11, of straight lines on a straight line, passim, of straight lines inside or outside a region, III–IV and passim (43); προσαναπληρόω "to complete", said of circles, III.25 (1); προσεκβάλλω "to produce", said of straight lines, I.5, 16, 32, III.2 (10); προσαρμόζω, "to fit", said of straight lines compounding the irrationals by removal, X (82); προσπίπτω "to fall on", said of straight lines on a line, I.post.5 and passim (42); στρέφω "to turn", said of solids of revolution around their axes (3); συμβάλλω "to meet", said of straight lines with straight lines, planes, or surfaces, passim (18); συμπίπτω "to meet", said of straight lines and of planes, I.def.23 and passim (35); (συμ)πληρόω "to complete", said of parallelograms, of parallepipeds, of cylinders, XII.15 (33); συντίθημι "to compound", said of magnitudes, V–VI, X, of straight lines, X, XIII, of numbers, VII–IX; τέμνω "to cut", said of regions or of lines, I.def.17 and passim (61); τίθημι "to set", said of points on a point, I.4, 8, III.24

(5); ὑπερβάλλω "to exceed", said of parallelograms, VI.29 (5); ὑπόκειμαι "to underlie", said of planes, XI.def.3 and XI–XIII passim ("to be supposed" see Sect. 4.2) (140); ὑποτείνω "to extend under", said of straight lines under angles or arcs, I.18–19, III.29 (68); φέρομαι "to move", said of the lines that generate the boundary of solids of revolution, XI.def.14, 18, 21 (7); ψαύω, "to touch", said of polygons and of polyhedra with respect to a circle and a sphere, respectively, XII.16–18 (9).

(E) Operations, relations and predicates of various kinds, mainly transversal to genera of objects: ἀνθυφαιρέω "to subtract / remove in turn", VII.1 and X.2–3 (6); ἀφαιρέω "to subtract", said of numbers, "to remove", said of any other object, I.cn.3 and passim; γίγνομαι "to result", "to yield", "to come to be", said of any kind of object, most notably of third or fourth proportionals, passim (127); δίδομαι "to be given", said of any given object, passim (317); διπλασιάζομαι "to be doubled", said of numbers, IX.32, 34 (6); ἐλλείπω "to fall short", said of magnitudes, V.def.5, of parallelograms, VI.28 (58); ἐμπίπτω "to fall", said of numbers as mean proportionals between other numbers, VIII–IX; ἐφαρμόζω "to coincide", said of geometric objects, I.cn.7 (32); καταμετρέω "to measure out", said of magnitudes and of numbers, V.def.1–2, VII.def.3–5 (13); λαμβάνω "to take" (κατα– 3 occurrences), said of any kind of object, in particular of random points on a line, of the centre of a circle, passim, of (equi)multiples, V, of third and fourth proportionals, VI, of GCD and LCM of numbers, of minimal representatives of ratios, VII–VIII, of mean proportionals, X (180); (κατα– / περι– / ὑπο)λείπομαι "to remain", said of any remainder after a removal or a subtraction, passim [(13 + 2 + 7) + 31]; μετρέω "to measure", said of magnitudes and numbers, VII.def.8–15 and passim (809); ποιέω "to make", said whenever a configuration (in particular in X) or, most frequently, a number is the output of an operation, VII–IX (629); πολλαπλασιάζω "to take multiples", said of magnitudes, V.def.4; VII–IX; πρόσκειμαι "to be added", said of any object, passim (60); προσλαμβάνω "to take in addition", said of straight lines to a straight line, XIII (5); προστίθημι "to add", said of any object, I.cn.2, 4 (10); σύγκειμαι "to be compounded", said of numbers, VII.def.2; συγκρίνω "to compare", said of the edges of the regular polyhedra, XIII.18 (1); τυγχάνω "to happen", said of objects taken "at random" (see Sect. 5.1.2) (98); ὑπερέχω "to exceed", said of magnitudes, V.def.5, 7, of any kind of object, passim (110).

This list highlights the huge number of operational notions that are not defined in the *Elements*—not even implicitly, as for instance by means of a problem of construction. The list also highlights that the operatory lexicon of the *Elements* is very rich, and in fact locally redundant: check the number of synonyms in the specific sublexica of intersection, of completion of figures, and the bewildering abundance of verbs that identify the remainder of a removal or of a subtraction. All these redundancies involve verbs with prefixed prepositions. In general, note the fair number of verbs with double prefix: one of them specifies the verb and hence the type of the operation, the other marks the degree to which the operation is carried out, usually κατα– (completely) or προσ– (further); see for instance προσεκβάλλω versus the canonical ἐκβάλλω, all subsequent occurrences of the former being citations of the primary occurrence in I.16. As is well known, some prefixes may carry a double connotation: παρα– suggests adaptation to a substrate but also a missed action; ἀνα– suggests something being "erected" but also something being "completed". On the other hand, homologous yet different operations may be designated by the same verb, see the lexicon of "composition" of ratios and of numbers. Finally, verb forms that suppose non-constructive yet geometric states of affairs usually are in the active or middle voice, present imperatives with stative value; a case in point is the rich operational sublexicon of intersection.

It must be once more stressed that the enunciations of problems are formulated with the verb in the imperative:[167] the emphasis is placed on the realization of the construction, not on the possibility of its realization.[168] Concerns of constructibility and of existence, which could easily be implemented by phrases such as δυνατόν ἐστι or simply ἔστι "it is possible",[169] are absent.[170] The only problem in the *Elements* whose enunciation appears to assert constructibility is XI.22; we have read it in Sect. 4.2.1. This is only a surface reading, however: as seen, the formulation of the enunciation of XI.22 is theorematic; therefore, its modal connotation is strictly functional to the application of XI.22 in the subsequent problem.[171]

An explicit emphasis on constructibility is placed in the παράδοξα "surprising" problems of some Erykinos, described by Pappus.[172] Let us read two enunciations (*Coll.* III.60 and 72):[173]

ἐν παντὶ τριγώνῳ, πλὴν τοῦ ἰσοπλεύρου καὶ ἰσοσκελοῦς τοῦ τὴν βάσιν ἐλάσσονα τῆς πλευρᾶς ἔχοντος, <u>δυνατόν ἐστι</u> συσταθῆναί τινας ἐπὶ τῆς βάσεως ἐντὸς δύο εὐθείας ἴσας ταῖς ἐκτὸς ὁμοῦ λαμβανομέναις.	In every triangle, with the exceptions of the equilateral and of the isosceles having the base less than the side, <u>it is possible</u> that on the base some[174] two straight lines be constructed inside equal to the external ones taken together.
δοθέντος παραλληλογράμμου χωρίου <u>δυνατόν ἐστιν</u> εὑρεῖν ἕτερον παραλληλόγραμμον ὥστε αὐτὸ μὲν τὸ ἐπιταχθὲν μέρος εἶναι τοῦ δοθέντος ἑκάστην δὲ πλευρὰν ἑκάστης πολλαπλασίαν κατὰ τὸν δοθέντα ἀριθμόν.	Given a parallelogrammic region, <u>it is possible</u> to find another parallelogram so as to be a prescribed part of the given one, and each side a multiple of each according to a given number.

The modal connotation of the παράδοξα problems is, however, connected with the fact that they admit infinitely many solutions. Pappus repeatedly insists on this point; he even seems to present Erykinos' elaborations as an example of a systematic investigation into a group of problems that exhibit this very characteristic.

As is to be expected, modal formulations are found in the Archimedean corpus much more often than in other authors.[175] Note, however, that the enunciation of *Sph. cyl.* I.5[176] is formulated with an infinitive, and that in *Con. sph.* 7–9 no construction is performed; in all remaining occurrences, the problem formulated with modal connotation has again infinitely many solutions.[177]

[167] Note also the directive δεῖ δή "thus it is required" that introduces the determination of problems.
[168] A complete, commented list of the problems in the Greek mathematical corpus—with emphasis on existential issues—is presented in Appendix A.
[169] See Posidonius' fragment quoted in Sect. 1.5.
[170] As seen in Sect. 4.2, Zeuthen 1896 proposed to regard some constructions as existence proofs. See the criticisms in Knorr 1983, who, however, adopts a tendentious reading of Zeuthen's thesis.
[171] Modal connotations, but with other goals, can also be found in the number-theoretical "problems" IX.18–19; see again the discussion in Sect. 4.2.1.
[172] Erykinos is cited by name at *Coll.* III.59; the extract from his treatise is *Coll.* III.60–73, less likely III.58–74, as usually assumed. The theorem in *Coll.* III.58 is repeated almost verbatim by Proclus (*iE*, 327.8–24) and, in a slightly modified form, by Eutocius in his commentary on some of the assumptions in Archimedes' *Sph. cyl.* I (*AOO* III, 12.18–14.4). Eutocius seems to draw other material from the same corpus: see the whole passage that deals with the assumptions (ibid., 6.4–14.30). Erykinos' problems are a variation on the theme of *El.* I.21 (see also *Coll.* V.10–11): "If two straight lines be constructed inside a triangle on one of its sides starting from its extremities, the constructed ‹straight lines› will be less than the remaining two sides of the triangle, and will contain a greater angle" (*EOO* I, 50.4–8).
[173] At Hultsch 1876–78, 106.10–13 and 126.19–23, respectively. The first enunciation keeps a strong diorismatic connotation; Pappus explains (*Coll.* III.62) why the construction is impossible for the excluded triangles.
[174] Note the determiner of indefiniteness "some" (see Sect. 5.1.3), a further sign that the solution is not unique.
[175] In *Sph. cyl.* I.2–4, 6 (*bis*), *Con. sph.* 7–9, 19–20, *Spir.* 3–9, 21–23, *Aequil.* I.6–7, *Quadr.* 20por.
[176] The modal connotation excepted, this enunciation is strictly analogous to the three enunciations of *Sph. cyl.* I.6.
[177] Note also the explicitly existential supposition in assumptions 1 and 3 of *Sph. cyl.* I, *AOO* I, 6.2–5 and 6.11–14, respectively (cf. Sect. 4.2).

Coming back to the *Elements*, the existential statement that opens X.def.3 is in fact a distortion of the enunciation of theorem X.10: it is certainly a late, and clumsy, addition to the text, which has originated a messy manuscript tradition.

The issue of the uniqueness of the solution to a problem of construction is never thematized as such in Greek mathematics. A small constellation of stylistic solutions was devised to cope with this problem; they may sometimes appear contrived to our eyes. Proposition I.7 (*EOO* I, 24.12–16) denies the subsistence (and not the possibility of the subsistence, even if the future has an inherent modal connotation) of two different configurations constructed from the same givens—in our language, this is the uniqueness of a triangle of given sides (I.7 is of course a theorem; its proof proceeds by reduction to the impossible; it states and proves the core insight of the SSS criterion of congruence of triangles—which is proposition I.8—as an independent theorem):

ἐπὶ τῆς αὐτῆς εὐθείας δύο ταῖς αὐταῖς εὐθείαις ἄλλαι δύο εὐθεῖαι ἴσαι ἑκατέρα ἑκατέρᾳ οὐ συσταθήσονται πρὸς ἄλλῳ καὶ ἄλλῳ σημείῳ ἐπὶ τὰ αὐτὰ μέρη τὰ αὐτὰ πέρατα ἔχουσαι ταῖς ἐξ ἀρχῆς εὐθείαις.

On a same straight line, two other straight lines respectively equal to the same two straight lines will not be constructed at different points on the same side having the same extremes as the original straight lines.

The same formulation can be found in XI.13 (*EOO* IV, 36.9–11)—this is the uniqueness of the perpendicular to a plane from a given point; again, the verb of the principal clause is in the future,

ἀπὸ τοῦ αὐτοῦ σημείου τῷ αὐτῷ ἐπιπέδῳ δύο εὐθεῖαι πρὸς ὀρθὰς οὐκ ἀναστήσονται ἐπὶ τὰ αὐτὰ μέρη.

From a same point, two straight lines will not be erected on the same side at right ‹angles› with a same plane.

or in III.16 (*EOO* I, 208.8–11)—this is the uniqueness of the tangent to a circumference from a given point on it:

ἡ τῇ διαμέτρῳ τοῦ κύκλου πρὸς ὀρθὰς ἀπ' ἄκρας ἀγομένη ἐκτὸς πεσεῖται τοῦ κύκλου, καὶ εἰς τὸν μεταξὺ τόπον τῆς τε εὐθείας καὶ τῆς περιφερείας ἑτέρα εὐθεῖα οὐ παρεμπεσεῖται.

A ‹straight line› drawn at right ‹angles› with the diameter of a circle from an extremity ‹of its› will fall outside the circle, and another straight line will not interpolate in the place between both the straight line and the circumference.

Uniqueness finds an explicit formulation in Book X, when it is proved that some irrational lines are univocally determined by their mode of generation. This happens in X.42–47 and X.79–84; we read X.42 (*EOO* III, 120.21–22):

ἡ ἐκ δύο ὀνομάτων κατὰ ἓν μόνον σημεῖον διαιρεῖται εἰς τὰ ὀνόματα.

A binomial can be divided into the names at one point only.

The "determinations" of these theorems, however, are formulated as a negative statement. This fact radicalizes one of the instances in which the complex setting-out + determination is fairly different from the enunciation. Let us read what happens in X.42 (*EOO* III, 120.23–27):[178]

[178] Note the aberrant presence of the deductive step that closes the setting-out.

ἔστω ἐκ δύο ὀνομάτων ἡ ΑΒ διῃρημένη εἰς τὰ ὀνόματα κατὰ τὸ Γ· αἱ ΑΓ ΓΒ ἄρα ῥηταί εἰσι δυνάμει μόνον σύμμετροι. λέγω ὅτι ἡ ΑΒ κατ' ἄλλο σημεῖον οὐ διαιρεῖται εἰς δύο ῥητὰς δυνάμει μόνον συμμέτρους.	Let there be a binomial, AB, that turns out to be divided into the names at Γ; therefore ΑΓ, ΓΒ are expressibles commensurable in power only. I claim that AB cannot be divided at another point in two expressibles commensurable in power only.

Not all problems in the *Elements* require to *construct* an object. Some of them require to εὑρεῖν or προσευρεῖν "find" the object, thereby carrying a strong existential connotation. Such problems are prominent in number theory and in the theory of irrational lines,[179] whose objects share features that make them "less generic" than standard geometric entities. The same explanation applies to the εὑρεῖν-constructions that can be read in the geometric Books (III.1 and VI.11–13). The propositions of Book X are further characterized by the absence of "given" objects on which the construction has to be performed: for instance, one simply reads εὑρεῖν τὴν δευτέραν ἀποτομήν "Find a second apotome" (X.86, at *EOO* III, 258.14).

A peculiar characteristic of Book I is the initial string of problems of construction I.1–3.[180] The presence of these three propositions seems dictated by foundational concerns directed by requirements of deductive economy.[181] In this perspective, it is significant that I.1 asks to construct an equilateral triangle, even if an isosceles triangle would have perfectly served the scope.[182] However, if we are not willing to introduce in I.1 geometric objects other than a given straight line, only special isosceles triangles are available,[183] and this requires less immediate constructions than the construction performed in I.1. Thus, the choice of the equilateral triangle is also induced by requirements of deductive and constructive economy.

Some of the constructions at the beginning of Book I, in particular I.2–3, admit of several geometric configurations, depending on the relative positions of the objects involved. Here as elsewhere in the *Elements*, only one of the configurations is treated, since the proof associated with almost any alternative configuration is identical in every respect to the proposed proof, including the assignments of denotative letters.[184] Thus, what introduces a spuriously particularizing character is the diagram *actually drawn*; this induces in its turn the feeling that operating a distinction of cases is necessary. Collecting the alleged "missing cases" was the hobby-horse of ancient commentators; it became an enumerative obsession in Eutocius' commentary on Apollonius' *Conica*.[185] Some of the many configurations of I.2–3 which Proclus (*iE*, 224.5–232.9) compiles γυμνασίας ἕνεκα "for exercise" are totally fictitious; Proclus himself notes that the Euclidean proof fits all these cases, but nevertheless goes on to present them. The omitted case of I.7 (see just above), namely, the case in which the two triangles are not the one inside the other, is not of this kind. Proclus (*iE*, 262.7–263.4) offers a simple proof that applies the second part of the enunciation of I.5, which shows that also the angles under the base of an isosceles triangle are equal.

---

[179] They are VII.2–3, 33–34, 36, 39, VIII.2, 4, IX.18–19, X.3–4, 10, 27–35, 48–53, 85–90.

[180] The beginning of Book I comprises: I.1, construction of an equilateral triangle of given side; I.2, application of a given segment to a given point; I.3, removal of a given segment from a given segment; I.4, SAS congruence criterion; I.5–6, isosceles triangles have the angles at the base equal, and vice versa; I.7–8, SSS congruence criterion (I.7 has the role outlined above); I.9–10, bisection of a given segment and of a given angle; I.11–12, perpendicular to a given segment from a given point, lying on it or not.

[181] On foundational themes in Greek mathematics see Acerbi 2010b.

[182] So Hero *apud* an-Nayrīzī, in Tummers 1994, 39.4–41.19; Proclus (*iE*, 218.12–219.17) wrongly compiles this part.

[183] For instance, the isosceles triangle whose side is double of the base.

[184] These alternative configurations must be carefully kept distinct from "limiting cases" such as the one occurring in proposition III.2 (see Sect. 1.1).

[185] See Acerbi 2012, 175–176 and 200, on this.

4.4. ANAPHORA

A proof normally starts with a joint reference to one or more mathematical states of affairs supposed in the setting-out or in the construction and to some consequence of them; the consequence is drawn thanks to the implicit "intermediation" of a previously proved theorem. This part of a proposition—yet it is not a *specific* part of a proposition independent of the proof—has recently been baptized "anaphora" by M. Federspiel,[186] since it "refers back" to what has been supposed in the setting-out or in the construction. This is the part of the proof in which the suppositions are "discharged" in the most explicit way.

From the syntactic point of view, the "anaphora" is a system comprising a causal subordinate clause, introduced by ἐπεί "since" and with a verb form in the present or perfect indicative (the perfect stem is used whenever a direct reference is made to a constructive act), followed by a principal clause with a verb form in the indicative or in the imperative, possibly introduced by ἄρα "therefore" (see the discussion in Sect. 5.3.2). The causal subordinate, which I shall call the "antecedent", contains the reference, the principal clause (the "consequent") contains its consequence. The complex subordinate + principal clause itself may take the form of an instantiated reference to the theorem that carries out the intermediation.

If the "anaphora" is located at the very beginning of the proof, the subordinant ἐπεί is usually completed by scope particles like οὖν "then" or γάρ "in fact"; an *incipit* like καὶ ἐπεί "and since" is also frequent (cf. Sect. 5.3.2). Let us read the liminal "anaphorae" of VII.26 (*EOO* II, 242.4–6)—reference to the setting-out, intermediation by VII.24,

ἐπεὶ γὰρ ἑκάτερος τῶν Α Β πρὸς τὸν Γ πρῶτός ἐστιν, καὶ ὁ ἐκ τῶν Α Β ἄρα γενόμενος πρὸς τὸν Γ πρῶτος ἔσται.	In fact, since each of A, B is prime to Γ, therefore the ‹number› resulting from A, B will also be prime to Γ.

of IV.4 (*EOO* I, 278.16–21)—double reference to immediate consequences of the construction, coordinated by δέ, consequent introduced for this reason by δή "thus" (cf. Sects. 5.3.2 and 5.3.6), and therefore no intermediation,[187]

καὶ ἐπεὶ ἴση ἐστὶν ἡ ὑπὸ ΑΒΔ γωνία τῇ ὑπὸ ΓΒΔ ἐστὶ δὲ καὶ ὀρθὴ ἡ ὑπὸ ΒΕΔ ὀρθῇ τῇ ὑπὸ ΒΖΔ ἴση, δύο δὴ τρίγωνά ἐστι τὰ ΕΒΔ ΖΒΔ τὰς δύο γωνίας ταῖς δυσὶ γωνίαις ἴσας ἔχοντα καὶ μίαν πλευρὰν μιᾷ πλευρᾷ ἴσην τὴν ὑποτείνουσαν ὑπὸ μίαν τῶν ἴσων γωνιῶν κοινὴν αὐτῶν τὴν ΒΔ.	And since an angle, ΑΒΔ, is equal to an ‹angle› ΓΒΔ and a right ‹angle›, ΒΕΔ, is also equal to a right ‹angle›, ΒΖΔ, thus there are two triangles, ΕΒΔ, ΖΒΔ, having two angles equal to two angles and one side (the one extending under one of the equal angles) common to them, ΒΔ, equal to one side.

and of V.7 (*EOO* II, 22.16–18)—references to the construction and to the setting-out, again coordinated by δέ; no intermediation:

[186] See Federspiel 1995 and 1999, but he only refers to subordinates in the perfect stem and contends—*pace* Proclus—that the "anaphora" is a specific part of a proposition. This is just one more instance of the occupational neurosis of professional linguists called "compulsive normativity".

[187] But the formulation of the consequent prepares for the intermediation to come in the next step: the consequent comprises in fact an instantiated citation of the antecedent of the conditional in the enunciation of proposition I.26 (this is the ASA-AAS congruence criterion of triangles); in the citation, note the absence, with respect to the template, of the syntagm ἑκατέραν ἑκατέρᾳ "respectively"; cf. Sect. 4.5.4.

ἐπεὶ οὖν ἰσάκις ἐστὶ πολλαπλάσιον τὸ Δ τοῦ Α καὶ τὸ Ε τοῦ Β ἴσον δὲ τὸ Α τῷ Β, ἴσον ἄρα καὶ τὸ Δ τῷ Ε.

Then since Δ is equimultiple of A and E of B and A is equal to B, therefore Δ is also equal to E.

If the verb in the antecedent of the "anaphora" is in the perfect stem, the formulaic character of the instantiated reference to the intermediating theorem is particularly marked, as we see on the examples of VI.11 (*EOO* II, 108.7–9)—intermediation by VI.2,

ἐπεὶ οὖν τριγώνου τοῦ ΑΔΕ παρὰ μίαν τῶν πλευρῶν τὴν ΔΕ ἦκται ἡ ΒΓ, ἀνάλογόν ἐστιν ὡς ἡ ΑΒ πρὸς τὴν ΒΔ, οὕτως ἡ ΑΓ πρὸς τὴν ΓΕ.

Then since a ‹straight line›, ΒΓ, turns out to be drawn parallel to one of the sides ΔΕ of a triangle, ΑΔΕ, in proportion, as ΑΒ is to ΒΔ, so ΑΓ is to ΓΕ.

and of II.11 (*EOO* I, 152.21–24)—coordinant δέ, intermediation by II.4:

ἐπεὶ γὰρ εὐθεῖα ἡ ΑΓ τέτμηται δίχα κατὰ τὸ Ε πρόσκειται δὲ αὐτῇ ἡ ΖΑ, τὸ ἄρα ὑπὸ τῶν ΓΖ ΖΑ περιεχόμενον ὀρθογώνιον μετὰ τοῦ ἀπὸ τῆς ΑΕ τετραγώνου ἴσον ἐστὶ τῷ ἀπὸ τῆς ΕΖ τετραγώνῳ.

In fact, since a straight line, ΑΓ, turns out to be bisected at E and ZA turns out to be added to it, therefore the rectangle contained by ΓΖ, ΖΑ with the square on AE is equal to square on EZ.

In the last example, note the *praesens pro perfecto* πρόσκειται, as usual with all forms of κεῖμαι.

In particular instances, an imperative is employed in the consequent of the "anaphora": we read it at the beginning of the proof of III.8 (*EOO* I, 186.6–7), where a constructive act is performed—reference to an immediate consequence of the construction:

καὶ ἐπεὶ ἴση ἐστὶν ἡ ΑΜ τῇ ΕΜ, κοινὴ προσκείσθω ἡ ΜΔ.

And since AM is equal to EM, let ΜΔ be added as common.

A reference to a constructive act is the most frequent kind of reference in clauses featuring an imperative in the consequent. However, examples of "anaphorae" that open the construction and refer to the setting-out can be found, as in I.18 (*EOO* I, 46.7–14). This is followed by the two "anaphorae" that initialize the proof (the second anaphora is devised to justify a coassumption introduced by δέ), in which the construction itself and a consequence of it are employed, respectively—intermediating theorems I.16 and I.5; note also the quite infrequent (and possibly spurious) postposed antecedent, introduced by ἐπεὶ καί "since … also":

ἐπεὶ γὰρ μείζων ἐστὶν ἡ ΑΓ τῆς ΑΒ, κείσθω τῇ ΑΒ ἴση ἡ ΑΔ, καὶ ἐπεζεύχθω ἡ ΒΔ.

In fact, since ΑΓ is greater than AB, let a ‹straight line›, ΑΔ, be set equal to AB, and let a ‹straight line›, ΒΔ, be joined.

καὶ ἐπεὶ τριγώνου τοῦ ΒΓΔ ἐκτός ἐστι γωνία ἡ ὑπὸ ΑΔΒ, μείζων ἐστὶ τῆς ἐντὸς καὶ ἀπεναντίον τῆς ὑπὸ ΔΓΒ· ἴση δὲ ἡ ὑπὸ ΑΔΒ τῇ ὑπὸ ΑΒΔ – ἐπεὶ καὶ πλευρὰ ἡ ΑΒ τῇ ΑΔ ἐστιν ἴση –· μείζων ἄρα καὶ ἡ ὑπὸ ΑΒΔ τῆς ὑπὸ ΑΓΒ.

And since an angle, ΑΔΒ, is external to a triangle, ΒΓΔ, it is greater than an internal and opposite one ΔΓΒ; and ΑΔΒ is equal to ΑΒΔ—since a side AB is also equal to ΑΔ—; therefore ΑΒΔ is also greater than ΑΓΒ.

An "anaphora" is not necessarily located at the beginning of the proof; the reference in such non-liminal "anaphorae" can point to results proved within the ongoing proof or in previous theorems. Typical of non-liminal "anaphorae" is the introductory syntagm πάλιν ἐπεί "again, since": it is usually found in replicas of similar "anaphorae" formulated previously. Let us see examples of non-liminal "anaphorae" within the proofs of II.9 and II.4. In II.9, the two "anaphorae" introduced by πάλιν ἐπεί initialize deductive sequences strictly parallel to those that precede them, and which are introduced by καὶ ἐπεί. Let us read first a long extract from II.9 (*EOO* I, 144.3–25)—no intermediation; note the homologous postposed explanations (between dashes below; cf. Sects. 1.4 and 4.5.3) of the second δέ-conjunct in the antecedent of the first two "anaphorae":

καὶ ἐπεὶ ἡ ὑπὸ ΗΕΖ ἡμίσειά ἐστιν ὀρθῆς ὀρθὴ δὲ ἡ ὑπὸ ΕΗΖ – ἴση γάρ ἐστι τῇ ἐντὸς καὶ ἀπεναντίον τῇ ὑπὸ ΕΓΒ –, λοιπὴ ἄρα ἡ ὑπὸ ΕΖΗ ἡμίσειά ἐστιν ὀρθῆς· ἴση ἄρα ἐστὶν ἡ ὑπὸ ΗΕΖ γωνία τῇ ὑπὸ ΕΖΗ· ὥστε καὶ πλευρὰ ἡ ΕΗ τῇ ΗΖ ἐστιν ἴση. πάλιν ἐπεὶ ἡ πρὸς τῷ Β γωνία ἡμίσειά ἐστιν ὀρθῆς ὀρθὴ δὲ ἡ ὑπὸ ΖΔΒ – ἴση γὰρ πάλιν ἐστὶ τῇ ἐντὸς καὶ ἀπεναντίον τῇ ὑπὸ ΕΓΒ –, λοιπὴ ἄρα ἡ ὑπὸ ΒΖΔ ἡμίσειά ἐστιν ὀρθῆς· ἴση ἄρα ἡ πρὸς τῷ Β γωνία τῇ ὑπὸ ΔΖΒ· ὥστε καὶ πλευρὰ ἡ ΖΔ πλευρᾷ τῇ ΔΒ ἐστιν ἴση. καὶ ἐπεὶ ἴση ἐστὶν ἡ ΑΓ τῇ ΓΕ, ἴσον ἐστὶ καὶ τὸ ἀπὸ ΑΓ τῷ ἀπὸ ΓΕ· τὰ ἄρα ἀπὸ τῶν ΑΓ ΓΕ τετράγωνα διπλάσιά ἐστι τοῦ ἀπὸ ΑΓ· τοῖς δὲ ἀπὸ τῶν ΑΓ ΓΕ ἴσον ἐστὶ τὸ ἀπὸ τῆς ΕΑ τετράγωνον – ὀρθὴ γὰρ ἡ ὑπὸ ΑΓΕ γωνία –· τὸ ἄρα ἀπὸ τῆς ΕΑ διπλάσιόν ἐστι τοῦ ἀπὸ τῆς ΑΓ. πάλιν ἐπεὶ ἴση ἐστὶν ἡ ΕΗ τῇ ΗΖ, ἴσον καὶ τὸ ἀπὸ τῆς ΕΗ τῷ ἀπὸ τῆς ΗΖ· τὰ ἄρα ἀπὸ τῶν ΕΗ ΗΖ τετράγωνα διπλάσιά ἐστι τοῦ ἀπὸ τῆς ΗΖ τετραγώνου· τοῖς δὲ ἀπὸ τῶν ΕΗ ΗΖ τετραγώνοις ἴσον ἐστὶ τὸ ἀπὸ τῆς ΕΖ τετράγωνον· τὸ ἄρα ἀπὸ τῆς ΕΖ τετράγωνον διπλάσιόν ἐστι τοῦ ἀπὸ τῆς ΗΖ.

And since HEZ is half of a right ‹angle› and EHZ is a right ‹angle›—for it is equal to the internal and opposite one EΓB—, therefore EZH as a remainder is half of a right ‹angle›; therefore angle HEZ is equal to EZH; so that a side EH is also equal to HZ. Again, since the angle at B is half of a right ‹angle› and ZΔB is a right ‹angle›—for again it is equal to the internal and opposite one EΓB—, therefore BZΔ as a remainder is half of a right ‹angle›; therefore the angle at B is equal to ΔZB; so that a side ZΔ is also equal to a side ΔB. And since AΓ is equal to ΓE, the ‹square› on AΓ is also equal to ΓE; therefore the squares on AΓ, ΓE are double of that on AΓ; and to those on AΓ, ΓE is equal the square on EA—for angle AΓE is a right ‹angle›—therefore that on EA is double of that on AΓ. Again, since EH is equal to HZ, that on EH is equal to that on HZ; therefore the squares on EH, HZ are double of the square on HZ; and to the squares on EH, HZ is equal the square on EZ; therefore the square on EZ is double of that on HZ.

There are four "anaphorae" in the above extract: the first two of them apply a result just proved in the same proof and one proved in theorem I.29,[188] respectively; the last two "anaphorae" employ an equality posited in the construction and the equality arrived at at the end of the deductive chain initialized by the first of the four "anaphorae", respectively. In II.4 (*EOO* I, 126.23–25), instead,

καὶ ἐπεὶ ἴσον ἐστὶ τὸ ΑΗ τῷ ΗΕ καί ἐστι τὸ ΑΗ τὸ ὑπὸ τῶν ΑΓ ΓΒ – ἴση γὰρ ἡ ΗΓ τῇ ΓΒ –, καὶ τὸ ΗΕ ἄρα ἴσον ἐστὶ τῷ ὑπὸ ΑΓ ΓΒ.

And since AH is equal to HE and AH is the ‹rectangle contained› by AΓ, ΓB—for HΓ is equal to ΓB—therefore HE is also equal to that by AΓ, ΓB.

the first condition in the antecedent of the paraconditional refers to I.43 without the intermediation of a reference to geometric states of affairs posited in the construction; the second condition, coordinated by καί, is an immediate consequence (which is recognized as such in the postposed explanation) of an equality proved in the first part of the proof.

[188] The reference to I.29 is justified by a postposed explanation in the form of instantiated partial citation.

4.5. Proof

A proof (ἀπόδειξις) is a connected sequence of inferences whose conclusion is a statement to be proved and in which new deductive material is fed in by means of references to states of affairs previously proved or supposed,[189] or assumed as principles. This deductive process is initialized by identifying some properties of the geometric configuration generated at the end of the construction as relevant for the statement to be proved. Within the proof, a prominent role is played by the logic of relations and of predicates: this is the "form" of the deductive machine.

4.5.1. *The logic of relations*

The discussion of the logic of relations in this long Section is divided into four subsections:

1) Relations and relational syllogisms in ancient logical doctrines, in particular as witnessed to in Aristotle, in Galen, and in some Aristotelian commentators (Sects. 4.5.1.1 and 4.5.1.4).
2) General remarks on predicates and relations in the Greek mathematical corpus (4.5.1.2).
3) The identifying criterion of a relation: the position of the relational operator (4.5.1.3).
4) Interactions between relations and deductive structure: these interactions are transitivity, symmetry, composition, and stability (4.5.1.4).

4.5.1.1. Aristotle and Galen on relations

Relations as logical entities and their properties fell within the range of ancient dialectical doctrines only tangentially and in extremis. Aristotle's discussions of the relatives, in *Cat.* 7,[190] *Top.* IV.4 and VI.8, and, as an item of a summary philosophical lexicon, in *Metaph.* Δ.15, simply point out the relational peculiarities of a number of predicates ("polyadic predicates"), even if some evolution can be detected from the exposition in *Cat.* 7 to the item in the philosophical lexicon of *Metaph.* Δ.15. The exposition in *Cat.* 7 simply lists single *relata* under the category of (mathematical) relatives; these are entities seen from the perspective of their satisfying a specific predicate, like τὸ διπλάσιον "what is double".[191] The real focus of the chapter is in fact on the issues of convertibility and of co-subsistence of relatives.[192] The exposition in *Metaph.* Δ.15 mentions pairs of *relata* in their reciprocal relationships, by means of expressions like

1) διπλάσιον πρὸς ἥμισυ "double to half", or τριπλάσιον πρὸς τριτημόριον "triple to third part", or in general πολλαπλάσιον πρὸς πολλοστημόριον "multiple to multiple parts" (these are called κατ' ἀριθμόν "numeric" relatives);
2) τὸ τμητικὸν πρὸς τὸ τμητόν "the cutter to what can be cut" (relatives κατὰ δύναμιν ποιητικὴν καὶ παθητικήν "according to an active or a passive capacity");

[189] These are the suppositions included in the setting-out or in the construction that precede the proof.
[190] On the discussion in *Cat.* 7 see most recently Duncombe 2015. For relations in Plato, see Scaltsas 2016.
[191] The context (see for instance 6b11–14) makes it obvious that the articles are not mere "citation" articles [that is, corresponding to our quotation marks (cf. Sect. 3.2.1), which would entail a mention of a true relative, like "the double"], even if Aristotle's wording is not always clear-cut.
[192] List of relatives at 6a36–b14; contrary relatives at 6b15–27; conversion (verb ἀντιστρέφω) at 6b28–7b14; co-subsistence (ἅμα εἶναι) at 7b15–8a12; discussion of an aporia at 8a13–b24.

3) τὸ μετρητὸν πρὸς τὸ μέτρον "measurable to measure" (the former is said to be a relative τῷ ἄλλο πρὸς αὐτὸ λέγεσθαι "because another item is said as related to it").

In *Metaph.* Δ.15, less specific *relata* are also mentioned, such as τὸ ἴσον καὶ ὅμοιον καὶ ταὐτό "what is equal, similar, and identical",[193] which is assigned to the genus of numeric relatives.[194] Mentions are also found of entities that derive from a single predicate, like τὸ ἀδύνατον "what is impossible", very much in the style of the list in the *Categories*. Thus, in Aristotle there is no thematization or study of relations as logical entities, as instead he does with his syllogistic in the case of predicates. Likewise, no thematization of relations as logical entities can be found in the Stoic doctrine of the relatives, whose main testimony is a passage in Simplicius.[195]

Chapters XVI–XIX of Galen's *Institutio logica* contain the first outline of a logical theory of relations. Galen refers to deductions that specifically pertain to relations as τρίτον εἶδος συλλογισμῶν "a third species of syllogism", and calls them κατὰ τὸ πρός τι συλλογισμοί "relational syllogisms". He repeatedly asserts that their formulations assume suitable καθολικά "general" or γενικά "generic" axioms as templates, and that this fact validates relational syllogisms as demonstrative inferences.[196] Galen also claims that relational syllogisms are particularly useful in arithmetic and in logistic;[197] he corroborates his claim by setting out a number of mathematical examples,[198] which can be distributed in three broad categories (it is noteworthy that, in most of his examples, Galen uses ordinals as schematic letters, very much as in the Stoic "modes"—see Sect. 5.1.6):

(*a*) Deductions by transitivity of equality (*Inst. Log.* I.2–3 and XVI.6). They are validated by the general axiom enunciated as *El.* I.cn.1, which Galen quotes, along with a mention of its being applied by "Euclid in the first theorem" (*Inst. Log.* XVI.6).

(*b*) Deductions by composition of ratios (*Inst. Log.* XVI.1–3 and 9), like "$A = 2B$; $B = 2C$: therefore $A = 4C$". They are validated by an axiom that ranges over every multiple or submultiple, but of which Galen only provides formulations that are restricted to well-defined (sub)multiples, as for instance in *Inst. Log.* XVI.3:

εἰ γὰρ ὅδε τις ὁ ἀριθμὸς τοῦδέ τινος εἴη τριπλάσιος, τοῦ δὲ τριπλασίου πάλιν ἕτερος εἴη τριπλάσιος, ἐνναεπλάσιος ἂν εἴη ὁ μείζων ἀριθμὸς τοῦ ἐλάττονος, καὶ ἀναστρέψαντί σοι πάλιν ὁ ἐλάττων τοῦ μείζονος ἔνατον ἔσται μέρος.	If some specific number were triple of some specific number, and again another were triple ‹of the triple›, the greater number would be nine times the lesser, and again, if you convert,[199] the lesser will be a ninth part of the greater.

[193] See *Metaph.* Δ.15, 1021a8–14 and 1021b6–8, where the relational character of the abstract nouns ἰσότης "equality" and ὁμοιότης "similitude" is said to derive from the relational character of the associated entities.

[194] Aristotle also asserts that a species of a genus known to be a relative is a relative.

[195] At *in Cat.*, 165.32–166.29. See Mignucci 1988 and Menn 1999 for discussions.

[196] See *Inst. Log.* XVI.5, 10, 12, XVII.1, XVIII.1, 6, 8. A possible contribution by Posidonius is mentioned at *Inst. Log.* XVIII.8; see Sect. 2.1 for a discussion of *Inst. Log.* XVI.5–10, a text that is crucial to understand the entire Greek demonstrative practice.

[197] At *Inst. Log.* XVI.1, 6 and XVIII.5; the received text does not mention geometry.

[198] Neither in Galen nor in Alexander (see below) mathematical inferences make all of the relational syllogisms. For instance, Galen presents examples of inferences involving τὸ μᾶλλον "what is more" (*Inst. Log.* XVI.12), like "the virtue of the better is worthier of choice; and soul is better than body; therefore, the virtue of the soul is worthier of choice than the virtue of the body", or relations that are converse of one another (XVI.10), like "being the father / son of". Still, he does not offer any serious discussion of converse relations.

[199] What is meant here is the "conversion" of a relation, not of a ratio. I have not translated "by conversion" (ἀναστρέψαντι is the standard mathematical marker of this operation on ratios) because of the presence of the dative of interest σοι.

To this category can be likened the inferences by addition and removal of equals from equals (XVI.4 and 7–8). These are validated by the general axioms enunciated in *El.* I.cn.2–3.[200]

(*c*) Deductions by particularization (*Inst. Log.* XVIII.5–7), called ἀνὰ λόγον "proportional" and of the kind "*A*:*B*::*C*:*D*; *A*:*B*::2:1: therefore *C*:*D*::2:1".[201] Galen is right in maintaining that these are not a species of category (*a*), for two equivalent ratios are not equal but the same: so transitivity of equality does not apply. Such inferences are validated by the much-contrived general axiom ὧν ὁ αὐτὸς ὁ λόγος καθόλου, τούτων καὶ οἱ κατὰ μέρος λόγοι πάντες οἱ αὐτοί "of what the ratio is in general the same, all particular ratios of these will also be the same".

As is clear from the examples, the generality of the "general axiom" is variable: Galen does not bother about providing a version of the axiom in (*b*) valid for a specific (sub)multiple; the axiom in (*c*) is not simply I.cn.1 straightforwardly reformulated for identity—as we would have expected—but an axiom fabricated as a backward calque of the particular inference at issue.

We may well wonder whether Galen might have intended to use relational syllogisms as a first step towards a systematic rethinking of the conditions of validity of an inference. He seems to corroborate this hypothesis when he asserts that (almost)[202] all syllogisms διὰ τὴν τῶν ἐπιτεταγμένων[203] αὐτοῖς καθολικῶν ἀξιωμάτων πίστιν ἔχουσι τὴν σύστασιν "derive their formulation from the validating power of the general axioms that are superordinate to them" (see Sect. 2.1 for my translation).[204] This shows that Galen's proposal goes beyond an analysis of the logic of relations: thus, the qualifier "relational" for the "third species" points to an accidental feature of such syllogisms—their essential feature is simply that they cannot fit the inferential schemes regarded as canonical by the prevailing logical schools. And in fact, the discussion of non-relational inferences such as "you say 'it is day'; but you say the truth: therefore it is day"[205] extends over the entire chapter XVII of the *Institutio logica*. I conclude that Galen's short exposition,[206] even if it contains germs of a potentially subversive approach, has been overestimated by modern scholarship.[207]

Galen was not alone in his attempt at taming relational inferences. Alexander, followed as usual by a whole school of commentators,[208] excluded from the class of well-formed (Aristotelian) syllogisms such arguments with particular premises as require an additional, generic, premise in order validly to conclude. He shows how to transform these arguments so as to recover a correct syllogistic form: assume the additional premise as a major, σύστειλον "merge" the particular premises of the original argument into a single coassumption and assume it as a minor.[209] The ill-formed arguments belong to the class of ἀμεθόδως περαίνοντες "unmethodically concluding" arguments;[210]

[200] In this instance, Euclid is not mentioned.
[201] It is current practice to write *A*:*B*::*C*:*D* and not *A*:*B* = *C*:*D* because two ratios are not said to be "equal", but "the same".
[202] But Galen's σχεδόν "almost" at *Inst. Log.* XVII.1 is certainly nothing more than a precautionary modulation.
[203] See Sect. 5.1.6 for the "inverse" technical term ὑποτάττω "to subordinate".
[204] Galen worries about pointing out (*Inst. Log.* XVII.7, and I.5 before it) that the meaning of ἀξίωμα he assumes is ἐξ αὑτοῦ πιστὸν λόγον "self-validating argument"—and not "statement", as in Stoic logic.
[205] This is an "unmethodically concluding argument" according to Alexander. See below for the relevant passages.
[206] See for instance *Inst. Log.* XVII.1.
[207] See most recently the discussion, which includes Alexander's contribution but seems to me to miss the main point of Galen's proposal, in Barnes 2007, 419–447 (see also Barnes 1993). For a more sympathetic assessment, see Morison 2008, 105–113. On Galen's logic, see also Bobzien 2004.
[208] See Alexander, *in APr.*, 21.28–22.23, 68.21–69.4, and 344.9–346.6, *in Top.*, 14.18–15.14; [Ammonius], *in APr.*, 70.1–71.6; Philoponus, *in APr.*, 36.10–11 and 321.7–322.18; [Themistius], *in APr.*, 121.20–123.8.
[209] The operation of "merging" does not simply amount to taking the conjunction of the particular premises, but entails a reformulation of them (an "alignment") so as to fit the formulation of the additional premise. See Sect. 2.1 for detail.
[210] Galen mentions these arguments only at the end of his treatise (*Inst. Log.* XIX.6); he establishes no link with relational syllogisms, and even declares that the unmethodically concluding arguments are "superfluous". See Bobzien 1999, 152–155, for a first orientation on such arguments.

they are variously ascribed to the Stoics or to νεώτεροι "moderns".[211] Among such arguments, the standard inference by transitivity of equality is adduced as the most representative example. I shall further discuss these passages in Sect. 4.5.1.4 below.

4.5.1.2. Relations and predicates

In Greek mathematics, relations are used more extensively than predicates.[212] There are two main reasons for this: first, what must be proved in a theorem is normally that some relation (equality, similitude, parallelism, etc.) holds between suitable elements of a specific geometric configuration; second, relations may enjoy properties—like transitivity—that make them particularly suited to interact with inferences. There is just one treatise in Greek mathematics that deals exclusively with the logic of a specific class of predicates: the *Data*.[213]

Since ancient Greek logic does not offer sustained discussions of what a relation is, we are confronted with a non-trivial interpretive problem: to identify what counts for a linguistic unit employed in Greek mathematics to be possibly classified as a relation. Of course, this problem cannot be solved simply by projecting back a list of logical objects that, to our eyes, *obviously are* relations. Nor can we restrict the investigation into relevant properties of relations to those properties that, to our eyes, *obviously are* relevant properties, as for instance transitivity or symmetry. In other words, we need an independent criterion to identify the extension of the predicate "relation" in Greek mathematics. To do this, let us first give a few necessary conditions for a sub-inferential linguistic unit to formulate a relation. Such a linguistic unit must display:

1) "empty places" where to locate the "terms", in the form of names, of noun phrases, or of denotative or numeral letters;
2) a verbal unit (frequently copula + adjective serving as the nominal complement of the copula) with the function of "relational operator";[214]
3) a linguistic form that is invariable or that admits of variants within a well-defined and restricted lexical and syntactic range;
4) an active interaction with inferential structures (the "logic of relations"), possibly giving prominence to properties that we recognize as typical of a relation, such as transitivity.

Such conditions are not sufficient. In the next Section, we shall see that other features of a linguistic unit crucially count for identifying it as a relation: position and ordering of the terms and of the relational operator; variants of wording that prove decisive when properties such as transitivity or symmetry interact with a deduction; correlation between linguistic form and position within a proof; correlation between linguistic form and specific mathematical transformations undergone within specific inferences. By taking these features into account, I shall formulate a "fundamental criterion" for a linguistic unit to count as a relation. Moreover, since there is no "instruction

[211] It is not clear whether the "moderns" can always be identified with the Stoics, nor whether all forms of arguments mentioned by Galen or by Alexander can be ascribed to them; see Barnes 1990, 68–75.
[212] On the Stoic conception of predicates, see most recently Bobzien, Shogry 2020.
[213] See the detailed discussion in Sect. 2.4.1, and Acerbi 2011a.
[214] Writing "relational operator" before defining what a relation is is just a lexical shortcut, for which the benevolent reader will forgive me.

manual" for writing Greek mathematics, the search for identifying criteria and for regulating principles must refer to mathematical practice, and hence cannot but be grounded on the statistical distribution of independent linguistic patterns.[215]

Since the basic unit of meaning in the formulaic system of the Greek mathematical idiolect is the statement, and since such an idiom is eminently generative (see Sects. 2.2–3), relations might be seen even where there is none. Consider in fact *n*-entries formulaic expressions such as the first application of the construction of I.23 in I.24 (*EOO* I, 58.7–9):[216]

συνεστάτω πρὸς τῇ ΔΕ εὐθείᾳ καὶ τῷ πρὸς αὐτῇ σημείῳ τῷ Δ τῇ ὑπὸ ΒΑΓ γωνίᾳ ἴση ἡ ὑπὸ ΕΔΗ.	On straight line ΔΕ and at point Δ on it let an ‹angle›, ΕΔΗ, be constructed equal to angle ΒΑΓ.

If this linguistic unit fits requirements (1) to (3) above thanks to its matrix structure (Sect. 2.3), it does not fit requirement (4)—nor does it fit the "fundamental criterion" for a linguistic unit to count as a relation, as we shall see in a moment. And in fact, such a formula gives expression to a constructive act, and a constructive act is deductively sterile.

Let me summarize in advance the results of my discussion: the table below sets out the main predicates and relations to be found in the *Elements* and in the *Data*. With one obvious exception, I have understood the verb form whenever this is "is" or "are". Note—and this will prove crucial in Sects. 4.5.1.3–4—that some relations admit of a double formulation, as for instance "*A* is similar to *B*" or "*A* and *B* are similar to one another".

Predicates

geometric	numeric	theory of irrational lines[217]
A given (in magnitude etc.)	*A* prime, compound, perfect	*A* expressible (lines and regions)
A equiangular, equilateral, even-sided	*A* square, cube, plane, solid	*A* irrational (*idem*)
	A odd, even, even times even	

Binary relations

general	geometric	general or numeric
	A similar to *B*	
A is *B*	*A* in a straight line with *B*	
A equal to *B*	*A* parallel to *B*	*A* multiple of *B*
A greater (less) than *B*	*A* equiangular with *B*	ratio of *A* to *B*
A greater (less) than *B* by a given magnitude	*A* homologous to *B*	*A* and *B* have a ratio
	A about the same diameter as *B*	*A* part/s of *B*
A greater (less) than *B* than in ratio by a given magnitude	*A* equal and similar to *B*[218]	*A* measures *B*
	A coincides with *B*	*A* and *B* prime to one another
A commensurable with *B* (in length, in power, etc.)	*A* is worth more than *B* by the square on a straight line (in)commensurable with itself	*A* and *B* similar plane or solid

[215] This may be difficult to swallow, but think that grammatical rules are (and cannot but be) established exactly in this way, and this entails a fair, and ineliminable, degree of circularity in the approach (see my *Liminalia*).
[216] See Sect. 2.3 on the matrix structure of constructions.
[217] The typical predicates of the theory of irrational lines are relations of (in)commensurability in which one of the terms is saturated by a fixed straight line or by a region, called "expressible": $irr(*)_e \equiv inc(e,*)$. Thus, a straight line can only be irrational with respect to a reference line; cf. Sect. 2.4.1 for this procedure of partial "saturation" of a relation.
[218] The putative relation "*A* similar and placed similarly to *B*" only figures within constructive acts and does not satisfy the fundamental criterion formulated in the next Section.

Quaternary relations
A to *B* as *C* to *D*
A to *B* has a duplicate / triplicate ratio than *C* to *D*
A and *B* equimultiple of *C* and *D*, respectively (*two forms*)
A of *B* same part/s that *C* of *D*
A measures *B* and *C* measures *D* an equal ‹number› of times

Quaternary relations can always be reduced to nested binary relations; the fundamental criterion will apply only to binary relations. For instance, proportionality among magnitudes reduces to identity of ratios. Numeric proportionality is instead grounded on the quaternary relation "being the same part/s that", which in its turn results from the composition of identity (the "nesting" relation) and of the binary relation "being part/s of" (the "nested" relation). Linguistically, structures of this kind are marked by the presence of adverbs like ἰσάκις "equi–", if the nesting relation is equality, or by determinations of identity.[219] Conversely, no object is identified by a binary relation—to repeat what should be a commonplace: a Greek ratio is not a mathematical object—and this is one of the reasons why differences of formulation are highly significant.

No ternary relations are found in Greek mathematics. Candidates could be "*A* is a common measure of *B* and *C*", "*A* is a multiple of *B* according to *C*", or "*A* measures *B* according to *C*". Actually, any of the terms in them is univocally determined by the other two. Therefore, "*A* measures *B* according to *C*" is not a ternary relation—and in fact we shall see that it does not fit the fundamental criterion—but a binary relation "*A* measures *B*" saturated by inserting the name of the result of the operation in the formulation: it is just a designation of the mathematical object *C*.

4.5.1.3. The fundamental criterion: the position of the relational operator

Let us now come to the fundamental criterion of identification of a two-place linguistic item as a binary relation. Since the sole ground for such an investigation are textual data, I assume a peculiar *syntactic* feature of the relations used in Greek mathematics as their exclusive characterization; this feature, if added to the necessary conditions (1) – (3) above, makes them sufficient to identify a two-place linguistic item as a binary relation:[220]

If, in all or in most occurrences, the verbal unit of a two-place linguistic item that satisfies conditions (1) – (3) above is placed in an external position with respect to the pair of terms, this linguistic item formulates a binary relation.[221]

If the linguistic complex "copula + nominal complement" is divided, what counts as the relational operator is the complement. The criterion entails that the terms in a linguistic item that formulates a relation must be arranged as a strictly contiguous pair, unless prevailing grammatical constraints intervene, as for instance the insertion of particles like ἄρα or δέ, whose position within a

[219] About identity belonging to the categories of relatives, see *Metaph.* Δ 15, 1021a8–14.

[220] As in many other cases, the syntactic feature central to the criterion is a recurrent pattern in Greek prose style; its significance in the mathematical idiolect comes from its very high statistical frequency.

[221] When this does not happen, I shall say that the relational operator "is in an internal position etc." or that it "divides the pair of terms". In all premises of an Aristotelian syllogism, the operator (the verb forms ὑπάρχει "belongs" or κατηγορεῖται "is predicated of", possibly negated) is external to the pair of terms, with the additional feature that one of the two term-places is endowed with a suitable quantity operator, in the dative or in the genitive, according to the verb employed: in the dative, these quantity operators are παντί "to all", οὐδενί "to no", τινι "to some".

sentence is rigidly regulated. As said, the formulation of the fundamental criterion obviously entails that conditions (1) to (3) above are automatically satisfied;[222] it is instead *a fact* that condition (4) is always satisfied whenever the fundamental criterion is.

Thus, what formulates a relation looks like the following linguistic item, which is typical of the "language of the givens":[223]

τὸ Α τοῦ Β δοθέντι μεῖζόν ἐστιν ἢ ἐν λόγῳ A than B is greater than in ratio by a given ‹magnitude›

The contiguous pair of terms is τὸ Α τοῦ Β; the relational operator is the entire syntagm δοθέντι μεῖζόν ἐστιν ἢ ἐν λόγῳ "is greater than in ratio by a given ‹magnitude›". It turns out, as is already clear from the table set out in the previous Section, that no linguistic complex that formulates a relation according to the fundamental criterion fails to express a "relation" according to modern standards. Conversely, the fundamental criterion allows us to exclude the expression of specific geometric configurations from the class of relations, even if they are fully-fledged relations on modern standards: for in such expressions the verbal unit happens *regularly* to divide the pair of terms.[224] These expressions are for instance "*A* is perpendicular to *B*", "*A* is at right angles with *B*", "*A* is orthogonal to *B*",[225] or "*A* is on *B*", "*A* is the centre of *B*", "*A* is tangent to *B*". And in fact, these geometric states of affairs are (this is a mere *statistical fact*) deductively sterile, unlike for instance the strictly analogous relation "*A* is parallel to *B*".

So, equality is a relation because its linguistic expression satisfies conditions (1) – (3) and because we *most frequently* read ἴση ἐστὶν ἡ ΑΒ τῇ ΓΔ "is equal AB to ΓΔ" instead of ἡ ΑΒ ἴση ἐστὶ τῇ ΓΔ "AB is equal to ΓΔ". There are three reasons why this does not happen in *all* occurrences.[226]

(*a*) The first reason is obvious: the Greek mathematical idiom is a (highly contrived) subset of Greek natural language; it is not a symbolic language, in which some expressions are well-formed and others are not. But if this is true, a distribution of occurrences in which some linguistic patterns are greatly prevailing can only be explained by adherence to a practice regarded as canonical. And, as I try to show throughout this book, lexical and stylistic conventions, when rigidly adhered to, have an actual mathematical import.

(*b*) The second reason is that the position of the relational operator strictly correlates with the position of the relation within a specific part of a mathematical proposition:[227] it is *a fact* that the relational operator is in an external position when the deductive machine is on, namely, within a proof; on the contrary, its position can be (and usually is) internal to the pair of terms within the enunciation, and as a consequence within the setting-out—that is, when the deductive machine is off. And it is *a fact*—a most amazing feature of Greek mathematical style—that, whenever the end of the proof is being approached, that is, whenever the linguistic complex that will constitute the

---

[222] As for condition 3, it must apply to every linguistic form of a relation.

[223] This relation means that *A* minus the given magnitude has a given ratio to *B*. This is a cryptic relation, which prompted the Byzantine scholar John Chortasmenus to jot down an abusive scholium against Diophantus: Acerbi 2013b. Some translations in this Section will be ungrammatical, since they will try to preserve syntactic features of Greek language that are not idiomatic in English.

[224] They are counted as "relations" in a modern sense in the table of the definitions of the *Elements* in Sect. 1.5.

[225] The first formulation is specific to orthogonality in a plane, the third to orthogonality in the space.

[226] If the formulation is not unique, we have to require that the fundamental criterion be satisfied by just one formulation, even if it is normally satisfied by every formulation.

[227] For simplicity's sake, I shall relax the distinction between linguistic level and logical level from now on.

conclusion / enunciation starts taking its shape, the relational operator shifts to the internal position: the attraction of the formulaic pole of the (instantiated) conclusion, and hence of the enunciation, is stronger than the attraction of the formulaic pole of the proof. This entails that condition (4) above is an essential requisite for a linguistic item to be a relation: in a sense, a relation is linguistically deactivated when it is not included in a proof.

But why is the relational operator placed in an internal position in the enunciation? Because the mathematical objects are not designated there by letters, but by (sometimes quite long) noun phrases: the relational operator—which in an enunciation can only be deductively neutralized, as we have just seen—has the sole function of forestalling ambiguities in the identification of the terms. And why is the relation operator placed in an external position in the proof? Because of saliency, for it is the sole linguistic unit of the relation that interacts with the deductive chain: it is so to speak the handle that allows linking a relation to another in a deductive chain; the handle is "retractable" and disappears outside strictly deductive contexts.

The evolution of the position of the relational operator is well represented in proposition II.8 (*EOO* I, 138.2–142.6), here preceded by its diagram—the numbers of the steps with internal operator are in boldface, those with external operator are in italics; the "English" translation in the third column is purely symbolic (see also Sect. 2.4.1):[228]

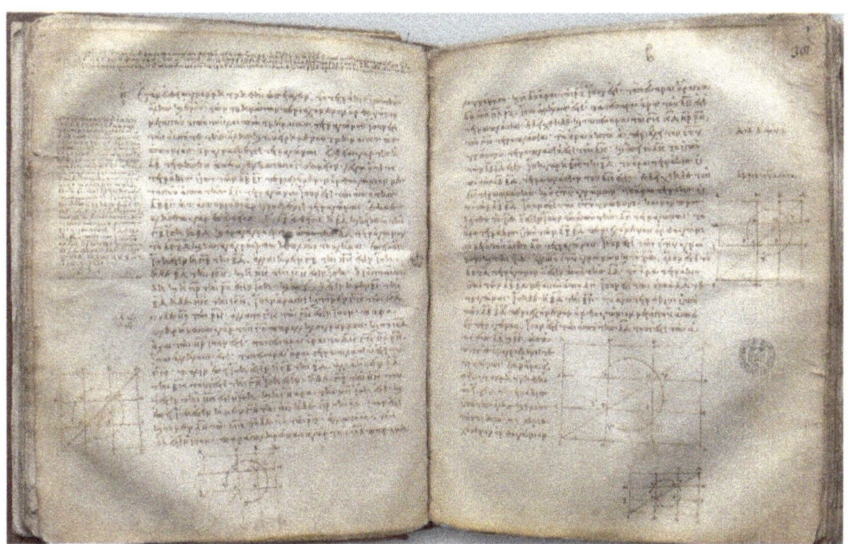

The diagrams attached to *El.* II.8 in Bodl. Dorv. 301, ff. 37v–38r
four diagrammatic scholia (three of which in the hand of the main copyist) try to de-oversymmetrize the original diagram; the rightmost diagram carries the indication "this is the most accurate"

[228] The last constructive act of step 3 does not refer to the diagram, but to the geometric configuration. Despite the fact that σχῆμα is the "‹geometric› figure", whereas the "diagram" is the καταγραφή, the verb καταγράφω formulates a standard constructive act (16 occurrences in the main text of the *Elements*, all in the form καταγεγράφθω, in II.7, 8, VI.27–29, X.91–96, XIII.1–5; 2 occurrences in *Data* 58–59). The figure is said to be described completely "twice" because it is a double replica of the figure in II.7. The fact that the correlation between the denotative letters and the mathematical objects is made clear only in the proof is a notational problem (which hinges on the incidental fact that part of the construction is omitted) and does not prove anything about the deductive role of the diagram: to repeat, the diagram is traced on the basis of the written text, and cannot have an inferential import.

1	ἐὰν εὐθεῖα γραμμὴ τμηθῇ ὡς ἔτυχεν, τὸ τετράκις ὑπὸ τῆς ὅλης καὶ ἑνὸς τῶν τμημάτων περιεχόμενον ὀρθογώνιον μετὰ τοῦ ἀπὸ τοῦ λοιποῦ τμήματος τετραγώνου ἴσον ἐστὶ τῷ ἀπό τε τῆς ὅλης καὶ τοῦ εἰρημένου τμήματος ὡς ἀπὸ μιᾶς ἀναγραφέντι τετραγώνῳ.	*aRb*
2	εὐθεῖα γάρ τις ἡ ΑΒ τετμήσθω ὡς ἔτυχεν κατὰ τὸ Γ σημεῖον. λέγω ὅτι τὸ τετράκις ὑπὸ τῶν ΑΒ ΒΓ περιεχόμενον ὀρθογώνιον μετὰ τοῦ ἀπὸ τῆς ΑΓ τετραγώνου ἴσον ἐστὶ τῷ ἀπὸ τῆς ΑΒ ΒΓ ὡς ἀπὸ μιᾶς ἀναγραφέντι τετραγώνῳ.	
3	ἐκβεβλήσθω γὰρ ἐπ' εὐθείας τῇ ΑΒ εὐθεῖα ἡ ΒΔ, καὶ κείσθω τῇ ΓΒ ἴση ἡ ΒΔ, καὶ ἀναγεγράφθω ἀπὸ τῆς ΑΔ τετράγωνον τὸ ΑΕΖΔ, καὶ καταγεγράφθω διπλοῦν τὸ σχῆμα.	*aRb*
4	ἐπεὶ οὖν ἴση ἐστὶν ἡ ΓΒ τῇ ΒΔ	*Rab*
5	ἀλλὰ ἡ μὲν ΓΒ τῇ ΗΚ ἐστιν ἴση	*abR*
6	ἡ δὲ ΒΔ τῇ ΚΝ,	*abR*
7	καὶ ἡ ΗΚ ἄρα τῇ ΚΝ ἐστιν ἴση.	*abR*
8	διὰ τὰ αὐτὰ δὴ καὶ ἡ ΠΡ τῇ ΡΟ ἐστιν ἴση.	*abR*
9	καὶ ἐπεὶ ἴση ἐστὶν ἡ ΒΓ τῇ ΒΔ	*Rab*
10	ἡ δὲ ΗΚ τῇ ΚΝ,	*Rab*
11	ἴσον ἄρα ἐστὶ καὶ τὸ μὲν ΓΚ τῷ ΚΔ	*Rab*
12	τὸ δὲ ΗΡ τῷ ΡΝ·	*Rab*
13	ἀλλὰ τὸ ΓΚ τῷ ΡΝ ἐστιν ἴσον – παραπληρώματα γὰρ τοῦ ΓΟ παραλληλογράμμου –·	*abR*
14	καὶ τὸ ΚΔ ἄρα τῷ ΗΡ ἴσον ἐστίν·	*abR*
15	τὰ τέσσαρα ἄρα τὰ ΔΚ ΓΚ ΗΡ ΡΝ ἴσα ἀλλήλοις ἐστίν·	–
16	τὰ τέσσαρα ἄρα τετραπλάσιά ἐστι τοῦ ΓΚ.	*aRb*
17	πάλιν ἐπεὶ ἴση ἐστὶν ἡ ΓΒ τῇ ΒΔ	*Rab*
18	ἀλλὰ ἡ μὲν ΒΔ τῇ ΒΚ, τουτέστι τῇ ΓΗ, ἴση	*abR*
19	ἡ δὲ ΓΒ τῇ ΗΚ, τουτέστι τῇ ΗΠ, ἐστιν ἴση,	*abR*
20	καὶ ἡ ΓΗ ἄρα τῇ ΗΠ ἴση ἐστίν.	*abR*
21	καὶ ἐπεὶ ἴση ἐστὶν ἡ μὲν ΓΗ τῇ ΗΠ	*Rab*
22	ἡ δὲ ΠΡ τῇ ΡΟ,	*Rab*
23	ἴσον ἐστὶ καὶ τὸ μὲν ΑΗ τῷ ΜΠ	*Rab*
24	τὸ δὲ ΠΛ τῷ ΡΖ·	*Rab*
25	ἀλλὰ τὸ ΜΠ τῷ ΠΛ ἐστιν ἴσον – παραπληρώματα γὰρ τοῦ ΜΛ παραλληλογράμμου –·	*abR*
26	καὶ τὸ ΑΗ ἄρα τῷ ΡΖ ἴσον ἐστίν·	*abR*
27	τὰ τέσσαρα ἄρα τὰ ΑΗ, ΜΠ, ΠΛ, ΡΖ ἴσα ἀλλήλοις ἐστίν·	–
28	τὰ τέσσαρα ἄρα τοῦ ΑΗ ἐστι τετραπλάσια·	*abR*
29	ἐδείχθη δὲ καὶ τὰ τέσσαρα τὰ ΓΚ ΚΔ ΗΡ ΡΝ τοῦ ΓΚ τετραπλάσια·	*abR*
30	τὰ ἄρα ὀκτώ, ἃ περιέχει τὸν ΣΤΥ γνώμονα, τετραπλάσιά ἐστι τοῦ ΑΚ.	*aRb*
31	καὶ ἐπεὶ τὸ ΑΚ τὸ ὑπὸ τῶν ΑΒ, ΒΔ ἐστιν	*abR*
32	– ἴση γὰρ ἡ ΒΚ τῇ ΒΔ –,	*Rab*
33	τὸ ἄρα τετράκις ὑπὸ τῶν ΑΒ ΒΔ τετραπλάσιόν ἐστι τοῦ ΑΚ·	*aRb*
34	ἐδείχθη δὲ τοῦ ΑΚ τετραπλάσιος καὶ ὁ ΣΤΥ γνώμων·	*aRb*
35	τὸ ἄρα τετράκις ὑπὸ τῶν ΑΒ, ΒΔ ἴσον ἐστὶ τῷ ΣΤΥ γνώμονι.	*aRb*
36	κοινὸν προσκείσθω τὸ ΞΘ, ὅ ἐστιν ἴσον τῷ ἀπὸ τῆς ΑΓ τετραγώνῳ·	*aRb*
37	τὸ ἄρα τετράκις ὑπὸ τῶν ΑΒ, ΒΔ περιεχόμενον ὀρθογώνιον μετὰ τοῦ ἀπὸ ΑΓ τετραγώνου ἴσον ἐστὶ τῷ ΣΤΥ γνώμονι καὶ τῷ ΞΘ·	*aRb*
38	ἀλλὰ ὁ ΣΤΥ γνώμων καὶ τὸ ΞΘ ὅλον ἐστὶ τὸ ΑΕΖΔ τετράγωνον, ὅ ἐστιν ἀπὸ τῆς ΑΔ·	*aRb*
39	τὸ ἄρα τετράκις ὑπὸ τῶν ΑΒ ΒΔ μετὰ τοῦ ἀπὸ ΑΓ ἴσον ἐστὶ τῷ ἀπὸ ΑΔ τετραγώνῳ·	*aRb*

40	ἴση δὲ ἡ ΒΔ τῇ ΒΓ·	*Rab*
41	τὸ ἄρα τετράκις ὑπὸ τῶν ΑΒ, ΒΓ περιεχόμενον ὀρθογώνιον μετὰ τοῦ ἀπὸ ΑΓ τετραγώνου ἴσον ἐστὶ τῷ ἀπὸ τῆς ΑΔ, τουτέστι τῷ ἀπὸ τῆς ΑΒ καὶ ΒΓ ὡς ἀπὸ μιᾶς ἀναγραφέντι τετραγώνῳ.	*aRb*
42	ἐὰν ἄρα εὐθεῖα γραμμὴ τμηθῇ ὡς ἔτυχεν, τὸ τετράκις ὑπὸ τῆς ὅλης καὶ ἑνὸς τῶν τμημάτων περιεχόμενον ὀρθογώνιον μετὰ τοῦ ἀπὸ τοῦ λοιποῦ τμήματος τετραγώνου ἴσον ἐστὶ τῷ ἀπό τε τῆς ὅλης καὶ τοῦ εἰρημένου τμήματος ὡς ἀπὸ μιᾶς ἀναγραφέντι τετραγώνῳ, ὅπερ ἔδει δεῖξαι.	*aRb*

This text calls for a number of remarks. First, step 6 does not contain a relational operator, which is simply constructed ἀπὸ κοινοῦ with step 5; the same for steps 10, 12, 22, 24. Second, step 16 is an exception to the external-operator rule.[229] Still, after the lengthy subinference in steps 17–28, the same statement as in step 16 is retrieved with external operator in step 29, where it is a coassumption (see Sect. 4.5.5); step 16, instead, is the conclusion of a partial argument, which already contains *in nuce* the linguistic structure of the statement to be proved in II.8. In fact, step 30, which for this reason is the formulaic consequence of step 16, starts the sequence of statements with an internal operator that gradually evolve to shape the instantiated conclusion (I shall say that it starts being "aligned" with it). The only exceptions in the sequence are steps 31–32 and 40, in which the external position of the relational operator is justified by the fact that they are coassumptions that refer to suppositions posited in the construction (including the part of it that is omitted).

(*c*) A discussion of steps 15 and 27 brings to the fore the third reason for there not being a unique formulation of every specific relation. In steps 15 and 27 the form is Α Β Γ Δ ἴσα ἀλλήλοις ἐστίν "A, B, Γ, Δ are equal to one another". Two covariant reasons justify this form. First: there are four terms equal to one another, and the formulation "is equal * to #" has only two entries. Second: since the standard formulation is not available, the relation must conform to the formulaic pole of the primary occurrence of the equality relation in the inference rule I.cn.1 (*EOO* I, 10.2), which serves as a template for transitivity of equality and includes a reciprocal pronoun:

τὰ τῷ αὐτῷ ἴσα καὶ ἀλλήλοις ἐστὶν ἴσα.	Items equal to a same item are also equal to one another.

Note also (see Sect. 4.5.1.4 for a fuller discussion) that the formulation of equality with the reciprocal pronoun ἀλλήλοις makes the symmetry of the relation of equality explicit, but hides, as far as linguistic self-evidence is concerned, the fact of being suited to an inference by transitivity: and in fact, steps 15 and 27 are the conclusions of trivial one-premise inferences.

Add to these the remarks to be made about step 3. The adjective ἴση is in an internal position because it is included in an instantiated citation of I.2: this is a constructive act, not a relation. The ordering of the terms is the same as that of the (instantiated) conclusion of I.2, even if, in the citation, the verb form is preposed, as is usual in constructive acts. As a reminder of this phenomenon (already discussed in Sect. 2.3), let us read in parallel the conclusion of I.2 (*EOO* I, 14.13–14) and the normal form of its application:[230]

[229] But recall again that the "exceptions" do not constitute a problem in non-formalized languages.

[230] The reference to the point to which the given straight line must be attached (underlined syntagm) is omitted in the application since it is obvious from the assignments of letters.

πρὸς ἄρα τῷ δοθέντι σημείῳ τῷ Α τῇ δοθείσῃ εὐθείᾳ τῇ ΒΓ ἴση εὐθεῖα κεῖται ἡ ΑΛ. κείσθω τῇ ΑΒ ἴση ἡ ΒΓ

To the attraction of the constructive formula κείσθω τῇ ΑΒ ἴση ἡ ΒΓ we must also ascribe the internal position of the relational operator in the frequent instances of "inversion" between nominative and dative in non-constructive sentences like τῇ ΑΒ ἴση ἐστὶν ἡ ΓΔ, some examples of which we read in a long extract from II.10 (*EOO* II, 150.3–19):

τὰ ἄρα ἀπὸ τῶν ΕΓ ΓΑ τετράγωνα διπλάσιά ἐστι τοῦ ἀπὸ τῆς ΓΑ τετραγώνου· τοῖς δὲ ἀπὸ τῶν ΕΓ ΓΑ ἴσον ἐστὶ τὸ ἀπὸ τῆς ΕΑ· τὸ ἄρα ἀπὸ τῆς ΕΑ τετράγωνον διπλάσιόν ἐστι τοῦ ἀπὸ τῆς ΑΓ τετραγώνου. πάλιν ἐπεὶ ἴση ἐστὶν ἡ ΖΗ τῇ ΕΖ, ἴσον ἐστὶ καὶ τὸ ἀπὸ τῆς ΖΗ τῷ ἀπὸ τῆς ΖΕ· τὰ ἄρα ἀπὸ τῶν ΗΖ ΖΕ διπλάσιά ἐστι τοῦ ἀπὸ τῆς ΕΖ· τοῖς δὲ ἀπὸ τῶν ΗΖ ΖΕ ἴσον ἐστὶ τὸ ἀπὸ τῆς ΕΗ· τὸ ἄρα ἀπὸ τῆς ΕΗ διπλάσιόν ἐστι τοῦ ἀπὸ τῆς ΕΖ· ἴση δὲ ἡ ΕΖ τῇ ΓΔ· τὸ ἄρα ἀπὸ τῆς ΕΗ τετράγωνον διπλάσιόν ἐστι τοῦ ἀπὸ τῆς ΓΔ· ἐδείχθη δὲ καὶ τὸ ἀπὸ τῆς ΕΑ διπλάσιον τοῦ ἀπὸ τῆς ΑΓ· τὰ ἄρα ἀπὸ τῶν ΑΕ ΕΗ τετράγωνα διπλάσιά ἐστι τῶν ἀπὸ τῶν ΑΓ ΓΔ τετραγώνων· τοῖς δὲ ἀπὸ τῶν ΑΕ ΕΗ τετραγώνοις ἴσον ἐστὶ τὸ ἀπὸ τῆς ΑΗ τετράγωνον· τὸ ἄρα ἀπὸ τῆς ΑΗ διπλάσιόν ἐστι τῶν ἀπὸ τῶν ΑΓ ΓΔ.

Therefore the squares on ΕΓ, ΓΑ are double of the square on ΓΑ; and to those on ΕΓ, ΓΑ is equal that on ΕΑ; therefore the square on ΕΑ is double of the square on ΓΑ. Again, since ΖΗ is equal to ΕΖ, that on ΖΗ is also equal to that on ΖΕ; therefore those on ΗΖ, ΖΕ are double of that on ΕΖ; and to those on ΗΖ, ΖΕ is equal that on ΕΗ; therefore that on ΕΗ is double of that on ΕΖ; and ΕΖ is equal to ΓΔ; therefore the square on ΕΗ is double of that on ΓΔ; and that on ΕΑ was also proved double of that on ΑΓ; therefore the squares on ΑΕ, ΕΗ are double of the squares on ΑΓ, ΓΔ; and to the squares on ΑΕ, ΕΗ is equal the square on ΑΗ; therefore that on ΑΗ is double of those on ΑΓ, ΓΔ.

The phenomenon of "inversion" itself is explained by adherence to the disposition of the terms that is required by the normal form of an inference by transitivity, which amounts to an application of the principle of saliency (see the next Section). Note that, as in II.8 the operator τετραπλάσιος "quadruple" divides the terms (step 30), so, in II.10 just read, the operator διπλάσιος "double" is in an internal position because it belongs to a germinal statement of the conclusion.[231]

By conversion, according to the principle of the syntactic template explained in Sect. 2.3, we must expect that a relation be invariantly formulated when it derives from the canonical, instantiated form of a definition or of the enunciation of a proposition. Such a canonical form emerges according to two mechanisms.

(1) The canonical form of the relation derives from a problem of construction. Let us recall the path (clarified in detail at the end of Sect. 2.2) leading from the constructive act "being produced continuously in a straight line with" to the standard form of the relation "being in a straight line with"; the path goes from the first, non-instantiated formulation in I.post.2, through the first application of I.post.2 as a constructive act in I.2, to the first formulation of the relation as an instantiated statement in the "determination" of I.14 (*EOO* I, 8.9–10, 12.26–14.1, and 38.12–13):

[231] Most relational operators in this deductive chain are internal to the pair of terms because of reason (*b*) above: attraction of the formulaic pole of the conclusion (the proof is approaching its end).

καὶ πεπερασμένην εὐθεῖαν κατὰ τὸ συνεχὲς ἐπ' εὐθείας ἐκβαλεῖν.	And to produce a bounded straight line continuously in a straight line.
ἐκβεβλήσθωσαν ἐπ' εὐθείας ταῖς ΔΑ ΔΒ εὐθεῖαι αἱ ΑΕ ΒΖ.	Let straight lines, AE, BZ, be produced in a straight line with ΔA, ΔB.
λέγω ὅτι ἐπ' εὐθείας ἐστὶ τῇ ΓΒ ἡ ΒΔ.	I claim that BΔ is in a straight line with ΓB.

The two occurrences of the relation "being in a straight line with" with internal operator which we read, in proposition I.14 (*EOO* I, 38.14–15) itself, inside the local setting-out associated with the reduction to the impossible, can be explained by the contingent requirement of emphasizing the straight line to be produced—and which is the object of the reduction—by preposing its mention to the verb form:

εἰ γὰρ μή ἐστι τῇ ΒΓ ἐπ' εὐθείας ἡ ΒΔ, ἔστω τῇ ΓΒ ἐπ' εὐθείας ἡ ΒΕ.	In fact, if BΔ is not in a straight line with BΓ, let BE be in a straight line with ΓB.

We thus see that the pragmatic requirement of saliency (a stylistic resource that pertains to expressivity) may be stronger than the logico-formulaic constraint of keeping the relational operator in an external position, if the two requirements conflict.

A second, interesting example is provided by the path leading to the standard form of the relation "being parallel to".[232] There is a bifurcation between the formulation of the constructive act in I.32, rigidly formulaic since it adheres to the conclusion of I.31, and the formulation of the relation in the determination of I.27, in fact a canonical transformation of the primary reciprocal[233] form contained in the enunciation of I.27 itself.[234] It goes without saying that the verbal unit divides the two terms in the formulation of the constructive act, while it is in an external position in the relation. Let us read in this order the four passages in I.31, 32, and 27 enunciation and setting-out (*EOO* I, 76.10–12, 76.24–25, 66.18–20, and 66.23–24):

διὰ τοῦ δοθέντος ἄρα σημείου τοῦ Α τῇ δοθείσῃ εὐθείᾳ τῇ ΒΓ παράλληλος εὐθεῖα γραμμὴ ἦκται ἡ ΕΑΖ.	Therefore through a given point, A, a straight line, EAZ, turns out to be drawn parallel to a given straight line, BΓ.
ἤχθω γὰρ διὰ τοῦ Γ σημείου τῇ ΑΒ εὐθείᾳ παράλληλος ἡ ΓΕ.	In fact, through point Γ let a ‹straight line›, ΓE, be drawn parallel to straight line AB.
ἐὰν εἰς δύο εὐθείας εὐθεῖα ἐμπίπτουσα τὰς ἐναλλὰξ γωνίας ἴσας ἀλλήλαις ποιῇ, παράλληλοι ἔσονται ἀλλήλαις αἱ εὐθεῖαι.	If a straight line falling on two straight lines make the alternate angles equal to one another, the straight lines will be parallel to one another.
λέγω ὅτι παράλληλός ἐστιν ἡ ΑΒ τῇ ΓΔ.	I claim that AB is parallel to ΓΔ.

(2) The canonical form of the relation directly derives from a principle, as for instance the definitions of a part and of a multiple in VII.def.3 and 5 (*EOO* II, 184.6–7 and 184.9–10):[235]

---

[232] Cf. the discussions of I.31 in Sects. 2.3 and 3.3.
[233] Namely, with the pronoun ἀλλήλων. This transformation is the same as the transformation undergone by "equality", which we shall discuss in the following Section.
[234] I.31 also requires that the point through which the parallel is drawn is mentioned.
[235] On these notions, see also the discussion in Sect. 5.3.4.

μέρος ἐστὶν ἀριθμὸς ἀριθμοῦ ὁ ἐλάσσων τοῦ μείζονος, ὅταν καταμετρῇ τὸν μείζονα.	A number is a part of a number, the less of the greater, whenever it measure the greater out.
πολλαπλάσιος δὲ ὁ μείζων τοῦ ἐλάσσονος, ὅταν καταμετρῆται ὑπὸ τοῦ ἐλάσσονος.	And the greater is a multiple of the less, whenever it be measured out by the less.

In a definition, that which is to be defined is in a liminal position and without the article, as is customary: it results that the relational operators "part" and "multiple" are automatically external to the pair of terms in the relation.[236] The form so established directly becomes, by the usual mechanism of rigid adherence to the primary occurrence, the standard form.

However, when a relation is employed in inhomogeneous mathematical contexts, different still equivalent formulations can be used in order to emphasize complementary features of the link between the terms. This phenomenon is well exemplified by the equivalent formulations ὁ Α τὸν Β μετρεῖ "A measures B" and ὁ Α τοῦ Β μέρος ἐστίν "A is a part of B"[237] of the binary relation "being a part of", and by the equivalent formulations of the associated quaternary relation ἰσάκις ὁ Α τὸν Β μετρεῖ καὶ ὁ Γ τὸν Δ "A measures B and Γ ‹measures› Δ an equal ‹number› of times" and ὁ Α τοῦ Β τὸ αὐτὸ μέρος ἐστὶν ὅπερ ὁ Γ τοῦ Δ "A of B is really the same part that Γ of Δ".

Some of the relations listed in the table of the previous Section are the one the inverse of the other. In particular, this is the case for the relations that formulate the notions of "part" and of "multiple". As we have seen above, in VII.def.3 and 5 this fact is directly, and simply, borne out by the inversion of the qualifiers "greater" / "less" in the denominations of the terms and by the transition from the active to the passive voice of the verb form. However, the two forms of the quaternary relation of equimultiplicity[238] are not backed up by homologous forms of the relation of equipartition. The two formulations of equipartition are in fact linguistically inhomogeneous:

equimultiplicity	equipartition
ἰσάκις ἐστὶ πολλαπλάσιον τὸ Α τοῦ Β καὶ τὸ Γ τοῦ Δ	ἰσάκις ὁ Α τὸν Β μετρεῖ καὶ ὁ Γ τὸν Δ
ἔστι τὰ Α Γ τῶν Β Δ ἰσάκις πολλαπλάσια	ὁ Α τοῦ Β τὸ αὐτὸ μέρος ἐστὶν ὅπερ ὁ Γ τὸν Δ

The reason for this polarization is that a technical term like "parts" (pl.) is not available on the right column, because it has an independent meaning (see Sect. 5.3.4).

The relational operator in the second just-read formulation of equipartition divides the four terms.[239] More generally, in some quaternary relations, the operator can be found in an internal position even in rigidly canonical formulations: for this reason the fundamental criterion is restricted to binary relations. As a matter of fact, the sample of such quaternary relations is small; a short discussion allows clarifying the phenomenon. The internal position of the operator in the relation "*A* to *B* has a duplicate / triplicate ratio than *C* to *D*" is forced by the primary occurrence in the template, namely, V.def.9–10, of which we read V.def.9 (*EOO* II, 4.11–13):

ὅταν δὲ τρία μεγέθη ἀνάλογον ᾖ, τὸ πρῶτον πρὸς τὸ τρίτον διπλασίονα λόγον ἔχειν λέγεται ἤπερ πρὸς τὸ δεύτερον.	Whenever three magnitudes be in proportion, the first to the third is said to have a duplicate ratio than to the second.

236 Note the two parallel, and rigidly separated, pairs ἀριθμὸς ἀριθμοῦ | ὁ ἐλάσσων τοῦ μείζονος.
237 The equivalence is in fact posited in VII.def.3 read above.
238 See the next Section for a detailed discussion.
239 As to the position in the first form, recall that the relational operator is not the verb form, but the adverb ἰσάκις.

The point is that, for duplicate ratios, there does not exist a compact formulation that solely comprises terms, like the formulation of "simple" proportionality "as *A* is to *B*, so *C* is to *D*". Moreover, if we compare the definition just read with the second formulation of equipartition, we realize that both of them "mention" the compounding binary relation: the "ratio" in V.def.9 (underlined above), "being a part of" in the second formulation of equipartition—and the same happens in expressions that contain the sequences "having the same ratio as" or "having a greater ratio than", in which the relational operator also divides one pair of terms from the other. What happens in all these cases, then, is simply that the formulation "dismantles" the quaternary structure of the relation in its binary components: there is no longer any quaternary relation around. This phenomenon can be better perceived in a specific correlative formulation of equipartition that features a relative subordinate clause. A similar correlative structure exists for the equimultiples; we find both of them at work side by side in VII.5 (*EOO* II, 202.11–15):

<table>
<tr><td>ὁσαπλασίων ἄρα ἐστὶν ὁ ΒΓ τοῦ Α, τοσαυταπλασίων ἐστὶ καὶ συναμφότερος ὁ ΒΓ ΕΖ συναμφοτέρου τοῦ Α Δ· ὃ ἄρα μέρος ἐστὶν ὁ Α τοῦ ΒΓ, τὸ αὐτὸ μέρος ἐστὶ καὶ συναμφότερος ὁ Α Δ συναμφοτέρου τοῦ ΒΓ ΕΖ.</td><td>Therefore whichever multiple is ΒΓ of Α, such a multiple is also ΒΓ, ΕΖ, both together, of Α, Δ, both together; therefore whichever part is Α of ΒΓ, the same part is also Α, Δ both together, of ΒΓ, ΕΖ, both together.</td></tr>
</table>

Again, there is no quaternary relation here: the formulation simply amounts to recognizing that the compounding binary relations are identical. In the same way, the εἰλήφθω-constructions of equimultiples that are so frequent in Book V do not have a relational character, and in fact the relational operator is in an internal position. We see this on the example of V.3 (*EOO* II, 6.9–14), which includes two relations in good order:

<table>
<tr><td>καὶ εἰλήφθω τῶν Α Γ ἰσάκις πολλαπλάσια τὰ ΕΖ ΗΘ. λέγω ὅτι ἰσάκις ἐστὶ πολλαπλάσιον τὸ ΕΖ τοῦ Β καὶ τὸ ΗΘ τοῦ Δ.</td><td>And let there be taken ΕΖ, ΗΘ equimultiple of Α, Γ. I claim that ΕΖ is equimultiple of Β and ΗΘ of Δ.</td></tr>
<tr><td>ἐπεὶ γὰρ ἰσάκις ἐστὶ πολλαπλάσιον τὸ ΕΖ τοῦ Α καὶ τὸ ΗΘ τοῦ Γ [...]</td><td>In fact, since ΕΖ is equimultiple of Α and ΗΘ of Γ [...]</td></tr>
</table>

4.5.1.4. Interactions between relations and the deductive machine: transitivity, symmetry, stability

The crucial role of lexicographic features is not limited to what I have outlined in the previous Section. These features also bear on the transformations undergone by the relations. There are four main types of such a transformation:

1) Inferences by transitivity, by far the commonest type.[240]
2) Inferences by symmetry, to be found in Book V only.
3) Composition of relations.

[240] A particular case of inferences by transitivity are the replacements of terms, sometimes introduced by τουτέστι "that is". None of these clauses is a priori suspect of authenticity.

4) Operations that modify the terms of a relation while keeping the relation between the resulting terms: a typical example are the transformations of ratios in proportion theory (V.def.12–18): alternation, inversion, composition, division, conversion, “through an equal”, and “through an equal in perturbed proportion”.

TYPES (1) AND (2): TRANSITIVITY AND SYMMETRY

Some of the relations listed above are assumed or proved to be transitive (**type 1**) in the *Elements*: equality in I.cn.1, parallelism in I.30 (straight lines in a same plane) and XI.9 (straight lines not in a same plane), sameness of ratio (that is, proportionality) in V.11, similitude of rectilinear figures in VI.21 (but the authenticity of this proposition is doubtful), commensurability in X.12.[241] All the enunciations that formulate results of transitivity have the same form; let us read I.30 (*EOO* I, 74.4–5) as an example (I.cn.1 and V.11 will be read just below):

αἱ τῇ αὐτῇ εὐθείᾳ παράλληλοι καὶ ἀλλήλαις εἰσὶ παράλληλοι.	‹Straight lines› parallel to a same straight line are also parallel to one another.

Thus, the general form of all transitivity statements is as follows:

τὰ τῷ αὐτῷ *R* καὶ ἀλλήλοις εἰσὶ *R*.	items *R* to a same item are also *R* to one another

This syntactic form serves as a meta-template for all transitivity statements; it reflects exactly their common logical form, that is, their common matrix structure.

The setting-outs + determinations that correspond to the above statements are typically formulated as follows (I.30, at *EOO* I, 74.6–7):

ἔστω ἑκατέρα τῶν ΑΒ ΓΔ τῇ ΕΖ παράλληλος. λέγω ὅτι καὶ ἡ ΑΒ τῇ ΓΔ ἐστι παράλληλος.	Let each of AB, ΓΔ be parallel to EZ. I claim that AB is also parallel to ΓΔ.

How does the principle of the template work in this case? The only propositions in which all steps of an inference by transitivity are made explicit are I.1–2; in both cases, equality is the relation at issue. The instantiated inference is formulated in such a way as to keep a form which is as near as possible to the non-instantiated formulation of the basic transitivity-rule for equality. Let us read I.1 (*EOO* I, 12.8–12), accompanied by a symbolic transcription:

ἴση ἐστὶν ἡ ΒΓ τῇ ΒΑ·	ΒΓ is equal to BA;	Rab
ἐδείχθη δὲ καὶ ἡ ΓΑ τῇ ΑΒ ἴση·	and ΓA was also proved equal to AB;	cbR
ἑκατέρα ἄρα τῶν ΓΑ ΓΒ τῇ ΑΒ ἐστὶν ἴση·	therefore each of ΓA, ΓB is equal to AB;	$(c\wedge a)bR$
τὰ δὲ τῷ αὐτῷ ἴσα καὶ ἀλλήλοις ἐστὶν ἴσα·	and items equal to a same item are also equal to one another;[242]	I.cn.1
καὶ ἡ ΓΑ ἄρα τῇ ΓΒ ἐστὶν ἴση.	therefore ΓA is also equal to ΓB.	caR

[241] Equimultiplicity is a quaternary relation and therefore admits of several forms of “transitivity”; see the discussion later in this Section.

[242] I.cn.1 is the statement of an inference rule; for this reason it is applied as a coassumption in arguments like this.

The third step of this inference, which amounts to "merging" the first two steps into a single statement that contains a form of the pronominal indefinite adjective ἑκάτερος "each of", is often neglected in recent discussions of the argument in I.1, but it is necessary if the fact has to be made manifest that the form of (a part of) the basic transitivity-rule—namely, as it is formulated in I.cn.1: τὰ τῷ αὐτῷ ἴσα "items equal to a same item"—is preserved; this is what I call "alignment" of a formula to (a part of) a general principle that serves as a template.[243]

Actually, the linguistic form of I.cn.1 and of its kin listed above is tricky and deserves a discussion. For I.cn.1 is the statement of an inference rule and therefore it contains three relations, two of which—merged in the noun phrase τὰ τῷ αὐτῷ ἴσα, wherefrom the plural—serve as premises and one—the noun phrase featuring a reciprocal pronoun τὰ ἀλλήλοις ἴσα—as the conclusion. Therefore, the plural ἴσα "equal" has different functions. In τὰ ἀλλήλοις ἴσα, it marks a formulation of equality I shall call "symmetric" on account of both morphology (the two terms are named as a plurality, in the nominative) and logic, for this formulation makes the symmetry of the relation of equality manifest; an instantiated symmetric formulation as a statement reads thus: τὰ Α Β ἴσα ἀλλήλοις ἐστίν. On the other hand, the plural ἴσα in τὰ τῷ αὐτῷ ἴσα simply points to the fact that two relations are merged, each of which has the "asymmetric" (as far as morphology is concerned: the two terms are named separately, nominative vs. dative) form "* to # is equal", an instantiated asymmetric formulation of which is τὸ Α τῷ Β ἴσον ἐστίν. Since both the symmetric formulation and the asymmetric formulation figure in I.cn.1, either of them must (and any of them may) serve as a template for all instantiated formulations of equality. The adopted template is the most "natural" one: the symmetric formulation requires the additional presence of the pronoun ἀλλήλοις and does not adequately separate the two terms, since they are named in the plural. On account of this, it is quite obvious that the asymmetric formulation has to be adopted as a template.[244]

Now, merging the two equality relations that serve as premises in the transitivity-rule I.cn.1 by means of the noun phrase τὰ τῷ αὐτῷ ἴσα also dictates the form in which the terms must be arranged in any instantiation of the two premises as independent clauses: the form of the instantiation must be "*abR*; *cbR*", whose essential features are (*i*) the terms are not divided by the relational operator; (*ii*) the common term has the same position in the two premises. The first feature is our definition of a "relation". The second feature deserves a long digression.

The first two statements in the proof of I.1 read above have the form *Rab*; *cbR*; the conclusion has the form *caR*. The resulting inferential scheme "*Rab*; *cbR*: therefore *caR*" has the common term in the same position in the two premises; I shall call it a "homological" scheme. A scheme in which the two occurrences of the common term are adjacent, as in modern practice and as in "*Rab*; *bcR*: therefore *acR*", I shall call "chiastic". There are in all four possible dispositions of the terms in the premises of a transitivity inference, as shown by the table in the following page:[245]

---

[243] The importance of the step was rightly pointed out by Alexander in his attempt at dressing an argument by transitivity in syllogistic clothes: A is equal to B will be syllogistically inferred from Γ is equal to B and A is equal to Γ "when, after adding the general premise saying 'items equal to a same item are also equal to one another', we merge what has been assumed as two premises into one single premise, which has the same meaning as the two: this is 'A and Γ are equal to the same (in fact to B)'" (*in APr.*, 344.16–19). The reference to the fully-fledged inference in I.1 is clear, even if not explicit. Cf. Sects. 2.1 and 4.5.1.1 on this.

[244] The asymmetric formulation is even adopted in inferences as the one above, where the last step instantiates the symmetric conclusion of the inference rule by means of an asymmetric form.

[245] Inversions of the terms in the conclusion will only be relevant in specific instances.

abR; *bcR*: therefore *acR* *baR*; *cbR*: therefore *acR*	chiastic disposition of the terms
abR; *cbR*: therefore *acR* *baR*; *bcR*: therefore *acR*	homological disposition of the terms

It is crucial to realize that the arrangement of the terms in I.1 is not induced by contextual constraints. The straight lines that are proved to be equal in the preliminary steps are in fact radii of the same circle: therefore, the order in which they appear in the equality statements is not dictated by any particular mathematical requirement. I conclude that the choice of setting the terms according to a homological scheme is deliberate. The statistical analysis I shall expound in points (*a*) – (*d*) below corroborates this conclusion.[246] Moreover, the common term is more frequently placed in first position in both premises, "*Rab*; *Rac*: therefore *Rcb*", than in second position.[247] The relational operator *R* is always external to the pair of terms; its linguistic expression is the same both in the premises and in the conclusion; it is more often preposed to the terms than postposed, a fact that is readily explained by saliency.

I now carry out a statistical analysis of the disposition of the terms in transitivity inferences. The samples for similitude and commensurability are too restricted and do not provide significant data. Therefore, I shall discuss only equality, parallelism, and notions of proportion theory.

(*a*) *Transitivity of equality*[248]

The results for transitivity of equality are set out in the table below.[249] The first row lists the Books of the *Elements* in order, the second row gives the number of occurrences of homological / chiastic arrangements. For instance, the first column means that, in Book I, transitivity of equality is handled 8 times using the homological arrangement, 5 times following the chiastic arrangement. The next-to-last column gives the same numbers for all alternative proofs. The sum total for the *Elements* is 76/35. It is to be noticed that among the chiastic arrangements the occurrences are reckoned of the infrequent scheme "$b = a$; $c = b$: therefore $a = c$".

I	II	III	IV	V	VI	VII	VIII	IX	X	XI	XII	XIII	alt.	tot.
8/5	5/2	12/1	6/4	0/1	3/7	6/0	0/0	8/3	4/2	1/2	2/2	13/3	8/3	76/35

(*b*) *Transitivity of parallelism*

Transitivity of parallelism is applied a handful of times. Homological arrangements are found in XI.10, 13, and XII.17, as well as in *Data* 28. The first occurrence presents an almost fully-fledged inference, very much in the style of I.1 seen above. The only chiastic arrangement is in IV.7.

[246] In Acerbi 2009b, which I closely follow here, I have applied these results to show that transitivity cannot "explain" the fact that Aristotle calls the first figure syllogisms "perfect".
[247] The conclusion is less frequently provided in the form *Rbc*, namely, with non-inverted terms.
[248] In certain instances (a strict minority, as far as I have been able to check), a particular ordering of the terms may fit the mathematical context better than another. This would introduce a bias in our statistical sample. If the sample is sufficiently large, however, the effects of such contextual considerations will become marginal or, if not, will tend to cancel out, and we can be reasonably confident that what we find as the preponderant formulation in mathematical texts was perceived as the most suitable way to express inferences by transitivity.
[249] The analysis of chained applications of transitivity is less straightforward, since it frequently happens that some steps are understood, and we cannot determine with certainty the ordering of the terms in such steps. See also below on this.

(*c*) *Transitivity of proportionality*

On account of their abstract character, notions tied to proportion theory (elaborated in Book V and applied in some of the subsequent Books) enjoy a higher degree of formulaic standardization than those typical of other Books of the *Elements*. The most representative notion of general proportion theory for which transitivity holds is, of course, the relation of proportionality itself. The homological arrangement used in handling transitivity is adhered to with remarkable consistency in this case too. To this claim V.11 (*EOO* II, 34.4–9), where transitivity of proportionality is proved, seems to constitute a notable exception. The enunciation, which has the typical format seen above, is in fact instantiated in the setting-out by means of a chiastic arrangement:

οἱ τῷ αὐτῷ λόγῳ οἱ αὐτοὶ καὶ ἀλλήλοις εἰσὶν οἱ αὐτοί.	‹Ratios› that are the same with a same ratio are also the same with one another.
ἔστωσαν γὰρ ὡς μὲν τὸ Α πρὸς τὸ Β οὕτως <u>τὸ Γ πρὸς τὸ Δ</u>, ὡς δὲ <u>τὸ Γ πρὸς τὸ Δ</u> οὕτως τὸ Ε πρὸς τὸ Ζ. λέγω ὅτι ἐστὶν ὡς τὸ Α πρὸς τὸ Β οὕτως τὸ Ε πρὸς τὸ Ζ.	In fact, as A is to B, so let <u>Γ be to Δ</u>, and, as <u>Γ is to Δ</u>, so let E be to Z. I claim that, as A is to B, so E is to Z.

Still, a part of the manuscripts of the Arabic translation of the *Elements* carries the homological pattern;[250] and, whenever V.11 is applied, this is by far the preferred format: the sum total of the occurrences is in fact 36/10.[251] No theorem analogous to V.11 is proved for numbers, yet transitivity of proportionality is repeatedly applied in the arithmetic Books. The homological arrangement is followed almost everywhere: the sum total is 28/2.[252] The following table summarizes these data:

I	II	III	IV	V	VI	VII	VIII	IX	X	XI	XII	XIII	tot.
0/0	0/0	0/0	0/0	5/0	14/2	5/0	23/2	0/0	4/3	1/1	12/4	0/0	64/12

The very *condition* of proportionality formulated in the *definiens* of V.def.5 was conceived as a relation for which transitivity holds.[253] This fact is used where we would expect to find it, namely, in the proof of V.11. The symmetry of the condition of proportionality is used in V.13, where the inequality required in the condition is checked in inverted order[254]—and it is at any rate secured by the very formulation of V.def.5.[255]

[250] See Engroff 1980, 213.

[251] This happens in V.16 (*bis*), 18, 23 (*bis*), VI.3, 5–6, 18–20 (*bis*), 22 (*bis*), 23 (*bis*), 24–26, X.27–28, 68, 113, XI.17, XII.2, 4, 5, 9 (*bis*), 11, 12 (*ter*), 14, 18 (*bis*). Chiastic arrangements can be found only in VI.7, 19; X.31–32, 112, XI.33, XII.2, 4–5, 11. Occurrences of chained inferences in the case of transitivity of proportionality are in V.23, VI.1, 2 (*bis*), 14 (*bis*), 15 (*bis*), 20, X.25, 93, 98–100, XI.31, 33 (*bis*), 34 (*bis*), 37 (*bis*), XII.1, 4, 11, 15 (*bis*).

[252] In VII.14, 17, 19 (*bis*), 34, VIII.2 (*quater*), 4, 5 (*bis*), 8 (*bis*), 9, 10 (*quinquies*), 11, 12 (*bis*), 18 (*bis*), 19 (*ter*). The only exceptions are in VIII.20–21.

[253] I have discussed this problem in Acerbi 2003a, 193–197; I shall not retrieve the discussion here. The text of V.def.5 can be read in Sects. 5.1.6 and 5.3.4. I am referring to V.def.5 in its "bare" formulation, that is, abstracting from the fact that it identifies a relation that can be designated by the single term "proportion" and that can be written "as *A* is to *B*, so *C* is to *D*", a form that is compact but opaque as to its correspondence with the operative part of the definition (the same incongruity we find for instance in the definition of "equi<u>lateral</u> tri<u>angle</u>").

[254] See the deductive steps at *EOO* II, 40.13–18. This peculiarity has been regarded as a logical fallacy (Mueller 1981, 149 n. 13). The phenomenon does not arise in Gerard's Arabo-Latin translation, since this exhibits the same formulation both in the definition and in the applications (on the fact that the *definiens* of V.def.5 is slightly different from the formulation of its applications see Acerbi 2003a, 196–197).

[255] Reflexivity is a metamathematical requirement; we may hardly expect to find it in Greek mathematics.

(d) *Transitivity of equimultiplicity*

The second notion of general proportion theory for which transitivity holds is equimultiplicity.[256] It is a quaternary relation that, within deductions, appears only in Book V. As for all quaternary relations (see Sect. 4.5.1.2), equimultiplicity is made of two nested relations: the relation "A is a multiple of B" and equality (of the intended multiples). Thus, A, C are equimultiple (ἰσάκις πολλαπλάσια) of B, D iff $A = mB$ and $C = mD$ for some integer m[257]—which I shall write $(A,C) = m(B,D)$. The notion of "equimultiplicity" is not defined in the *Elements*. This can be explained by the fact that the term πολλαπλάσιος "multiple" is defined in V.def.2, whereas the adverb ἰσάκις in ἰσάκις πολλαπλάσιος "equimultiple" retains its current meaning.

The notion of "equimultiplicity" is studied for its own sake in Book V: the series of results in V.1–6 amount to a little, independent theory: if V.1, 2, 4 are crucially used later in proportion theory, V.3 is only applied in the proof of V.4, and V.5, 6 are not employed in Book V—nor are they applied, a fortiori, in the rest of the *Elements*. Some propositions in proportion theory are strictly correlate, beyond identity of formulation (with an asterisk), with those in the equimultiplicity-theory string V.1–6:

equimultiple theory	1	2	3	4*	5*	6
proportion theory	12	18	22	15*	19*	17

Contrary to what happens with the propositions on numeric partition and proportion in Book VII, none of the propositions in the first row is used to prove its proportion-theoretical correlate.

My symbolic rewriting of equimultiplicity $(A,C) = m(B,D)$ is partly misleading, for it gives prominence to only one of the two possible couplings of the four terms, namely, the coupling in which the two multiples on the one side and the two base magnitudes on the other form different pairs. Actually, Greek mathematical language is richer than this, as it sets out and employs both couplings, identifying them by means of different formulations. I shall say that "is equimultiple A of B and C of D" is in Form I, whereas "A, C of B, D are equimultiple" is in Form II. In Form I, the abstract linguistic pattern is $R[mB,B][mD,D]$; in Form II, it is $[mB,mD][B,D]S$, where R and S symbolize the linguistic expressions of the relations at issue:[258]

Form I	ἰσάκις ἐστὶ πολλαπλάσιον τὸ Α τοῦ Β καὶ τὸ Γ τοῦ Δ	is equimultiple[259] A of B and C of D	$R[mB,B][mD,D]$
Form II	ἔστι τὰ Α Γ τῶν Β Δ ἰσάκις πολλαπλάσια	A, C of B, D are equimultiple	$[mB,mD][B,D]S$

My symbolic rewriting of equimultiplicity $(A,C) = m(B,D)$ is faithful to Form II only. As is often the case with quaternary relations, equimultiplicity is symmetric only if particular permutations of the terms are performed. Such a requirement is nicely met by the two forms, both by means of the actual disposition of the terms, by means of the differentiation induced by the different cases (nominative or genitive), and, in Form II, by the use of the plural.

[256] The peculiar logic of the complementary relation "A of B is the same part/s that C of D" is discussed in Sect. 5.3.4.
[257] The sign nB must be read as the repeated addition of B according to the units in n, with $n > 1$.
[258] In the case of Form I, the symbol [,] denotes in its turn a relation, namely, "being a multiple of"; in the case of Form II, the same symbol simply denotes an ordered pair. I have also replaced A with mB and C with mD.
[259] The construction with two subjects and the verb form in the singular is typical of Greek language: see the remarks in Humbert 1960, 73–74.

Equimultiplicity for fixed *m* is transitive: if $(A,C) = m(B,D)$ and $(E,C) = m(F,D)$, then $(A,E) = m(B,F)$.[260] This is a straightforward consequence of transitivity of equality: the identity of the second members of the pairs of relata in the first two equimultiplicity statements—the premises—guarantees that the same multiple is at work in both,[261] and therefore this feature transfers to the third equimultiplicity statement—the conclusion. However, Form II could not enter into such an inference, since the four magnitudes involved are set in an order that is unsuitable to allow replacement of linguistic complexes. Only Form I is suited to fit into the homological transitivity scheme—and in fact, where transitivity of equimultiplicity is employed, namely, in V.5 (*bis*), 6, 8, 17 (*bis*), only Form I is applied, and in all occurrences the arrangement of the (pairs of) terms is homological. In the arithmetic Books, a similar pattern is at work in VII.7 (*bis*), where the relation is "being the same part that".

Form II is invariably used in other contexts: (*i*) in the construction before a proof; (*ii*) inside a proof, whenever the proportionality condition of V.def.5 is made explicit;[262] (*iii*) at the very end of a proof, when an already proved or assumed equimultiplicity statement is retrieved. In none of these contexts, which are almost invariably tied to V.def.5 and give rise to the majority of the overall occurrences, do we find chains of inferences by transitivity of equimultiplicity.[263] A general reckoning in Book V gives the following figures: V.11–13, 22–23 have Form II only; V.2, 5 and 15 have Form I only; V.3–4, 6–8, 16–17 have both. The other theorems do not use the equimultiples. In V.16, the only occurrence of Form I is within a citation of V.15, in whose setting-out we do find Form I. The other references to V.15 (they are in V.16 and 23) are in Form II, following the "local" use of the host propositions.

This discussion shows that there are two formulations of equimultiplicity because any of them is better suited than the other to fit into a well-defined mathematical context: compartmented use has an exact counterpart in a differential formulation.

The statistical analysis carried out so far allows answering the following questions. How do we recognize that an argument is (or can be labelled as) an inference by transitivity? and how can we construe the Ancients recognizing this very state of affairs? Which is the less biased way of projecting back such a seemingly transparent notion as transitivity into ancient texts? Were the Ancients able to formulate an inference by transitivity independently of the actual arrangement of the terms? Surprising as it may seem, evidence that comes from mathematical texts quite definitely suggests that no, they were not able to do that (or, better said, they simply did not do that), and the evidence is particularly telling *exactly* because the relations involved in the above survey are symmetric. Their very property of symmetry, in fact, entails that the order of the terms in the premises of an inference by transitivity is immaterial. In other words, and since the sample is reasonably large, we should expect a statistical zero as far as the prominence is concerned of some among the four possible configurations of terms *aRb–bRc*, *bRa–cRb*, *aRb–cRb*, *bRa–bRc*.

But this definitely does not happen. Since we are only interested in the arrangements of the terms and since all uninstantiated enunciations of the transitivity rules for binary relations have the

[260] This rule is never proved. In proposition V.3, it is shown how to compound relations of equimultiplicity with fixed terms and different multiples: if $(A,C) = n(B,D)$ and $(E,F) = m(A,C)$, then $(E,F) = k(B,D)$, for some *k*—note that the theorem does not specify that $k = nm$. This form of transitivity is different from the form considered here (we shall deal with it presently), but the mere fact that *some* form of transitivity is at issue forces the enunciation of the equimultiplicity statement to be in Form I.

[261] We must not forget that, unlike my symbolic transcription, any equimultiplicity statement is so worded as precisely to avoid mentioning the common multiple.

[262] The existence of Form II can be explained by the fact that it fits into the formulation of V.def.5 better than Form I.

[263] See V.4–5, 8 and 17; V.6 has Form I in the proof, Form II elsewhere.

same formulation, we are allowed to add all the data from the *Elements* set out under points (*a*) – (*d*) above and to collect them in the same statistical sample. The result 151/49 for the partition of the occurrences calls for an explanation in terms of a practice directed by an intentionally applied criterion, precisely because it is utterly unlikely as a statistical fluctuation: the probability that it is the outcome of an unbiased random distribution of the common terms is less than 1 over 10^7. All of this should not surprise us. If we exclude the denotative letters and some signs in Diophantus, Greek mathematicians used only natural language. We are, instead, deeply accustomed to using symbolic notations. Our perception of what pertains to *logical form* cannot be assumed to coincide with their perception. If only natural language is used, regular linguistic patterns are what is at one's disposal in order to recognize, or even to create, kinds of argument whose validity does not depend on the actual terms set out in them (singular terms, schematic letters, etc.), that is, arguments that are valid *in virtue of a certain form*: the solution that was given to the problem of creating a stable linguistic format was also, even if not exclusively, aimed at producing a recognizable logical form. It is not surprising, then, that seemingly trivial features such as the ordering of the terms can reasonably be taken to replace, in the perspective of the ancient Greek mathematicians, more articulated and far-reaching notions as the one we designate with the noun "transitivity"—and the criterion to be followed to single out a mostly suitable disposition of the terms is, again, adherence to the template of the archetypal non-instantiated formulations (read again the text of I.1).

All mathematical relations considered so far are symmetric, namely, *aRb* if and only if *bRa*. When the relation is not symmetric, the chiastic arrangement is necessary in a schematic presentation of transitivity, if the same relation has to be kept in both premises. For instance, we currently write "$a > b$; $b > c$: therefore $a > c$" (chiastic arrangement of the terms and the same relation in both premises), and not "$a > b$; $c < b$: therefore $a > c$" (homological arrangement of the terms but different relations in the premises). The second formulation is totally correct but would appear quite eccentric. That this was not at all eccentric to ancient Greek eyes we shall presently see. In the *Elements* there are a few inferences by transitivity of "greater than". The seemingly natural chiastic scheme "$a > b$; $b > c$: therefore $a > c$" is attested in X.44 and XI.21 only.[264] The eccentric homological scheme "$a > b$; $a < c$: therefore $c > b$" is instead found in I.21 (*bis*), XI.23 (conclusion stated as $b < c$) and in the alternative proofs to X.1 and to XII.17.[265] What is important is that, in both occurrences in I.21 (*EOO* I, 50.22–24, and cf. the subsequent inference at 52.5–8), the premise I have written as "$a < c$" does not contain the sign "less than" in the text, but "greater than" with the two terms interchanged:[266]

αἱ ΓΕ ΕΒ ἄρα τῶν ΓΔ ΔΒ μείζονές εἰσιν· *ἀλλὰ τῶν* *ΒΕ ΕΓ* μείζονες ἐδείχθησαν αἱ ΒΑ ΑΓ· πολλῷ ἄρα αἱ ΒΑ ΑΓ τῶν ΒΔ ΔΓ μείζονές εἰσιν.	Therefore ΓΕ, ΕΒ are greater than ΓΔ, ΔΒ; but greater than *ΒΕ, ΕΓ* were proved ΒΑ, ΑΓ; therefore ΒΑ, ΑΓ are much greater than ΒΔ, ΔΓ.

Note, as seen above, that the "inversion" of the terms within the coassumption (the terms ΒΕ, ΕΓ, in italics in the quote, are thereby strongly marked for saliency) entails the displacement of the relational operator in an internal position. In XI.23, instead, the inference aims at proving a "less

[264] As well as in the alternative proof of XI.21; this is not an independent occurrence since the relevant part of the proof is repeated without changes.
[265] Add to these occurrences a scheme of the kind "$a > b$; $a = c$: therefore $c > b$" in XIII.8.
[266] Such "inversions" are quite common in the case of symmetric relations, most notably equality. They are marked by the fact that the term in the dative precedes the term in the nominative. As we have seen in Sect. 4.5.1.3, most of the inversions appear to serve the purpose of setting the terms in a homological arrangement.

than" inequality, as is clear from the context, and the premise with interchanged terms is in fact the first one, a disposition that fits very well its resulting from a sum of "greater than" inequalities. Therefore, not even in the case of a non-symmetric relation adherence to a chiastic ordering was felt as a necessary feature of the inference. On the contrary, adherence to the "standard" lexicographic ordering of the terms must be considered—at least as far as mathematical practice is concerned and if the converse of the relation at issue exists—a stronger regulative criterion than keeping the relation fixed throughout the inference.

Late Aristotelian commentators were not eager to see connections between transitivity and syllogismhood, even if the obviousness of some typical inferences by transitivity and their current and widespread application in mathematical arguments clearly puzzled them (see also Sects. 2.1 and 4.5.1.1). They appear to have been pretty casual in their expositions as far as the ordering of the terms was concerned. Alexander (*in APr.*, 344.9–346.6) excludes from the class of well-formed syllogisms all those deductions that need an additional premise. Among the examples, the transitivity inference for equality is almost invariably adduced. Alexander writes it as "$c = b$; $a = c$: therefore $a = b$",[267] which indeed does fit neither a homological transitivity scheme nor a "standard" chiastic one, namely, with the common term in the middle. The same example can also be found in Galen, *Inst. Log.* XVI.6:[268] after a quotation of the general statement of transitivity of equality in the form we find in I.cn.1, reference is made to the basic inference in I.1, which is formulated, by introducing ordinals in the Stoic fashion (see Sects. 5.1.6 and 5.2.2), as

ἐπεὶ γὰρ τὰ τῷ αὐτῷ ἴσα καὶ ἀλλήλοις ἴσα ἐστίν, δέδεικται δὲ τὸ πρῶτόν τε καὶ τὸ δεύτερον τῷ τρίτῳ ἴσον, ἑκατέρῳ αὐτῶν ἴσον ἂν εἴη οὕτω τὸ πρῶτον.	since items equal to a same item are also equal to one another, and both the first and the second turn out to be proved equal to the third, the third would in this way be equal to each of them.

A less frequently adduced argument has "having the same parents" as the transitive relation. In the same sample of authors as the sample for equality, we find the argument formulated three times in the scheme "aRb; bRc: therefore aRc" and once in the scheme "aRb; cRb: therefore aRc".[269]

An interesting point in the commentators' analysis of transitivity arguments is their identifying them as second figure syllogisms with particular premises.[270] This is to some extent surprising: since both particular premises and relations of equality do convert *simpliciter*, a transitivity argument for equality would fit into every figure. Therefore, a deliberate choice was made by Alexander

[267] This example is at *in APr.*, 344.14–15 (cf. Sect. 2.1). For a list of the occurrences of the transitivity inference for equality in the ancient exegetic literature, see Barnes 1990, 68–71. The schemes followed by some Aristotelian commentators are as follows: Alexander, *in Top.*, 14.21–22: "$a = b$; $b = c$: therefore $a = c$" (the second clause is in the form "$c = b$" in Par. gr. 1874 and in the Aldine); [Ammonius], *in APr.*, 70.11–12: "$a = b$; $b = c$: therefore $a = c$"; Philoponus, *in APr.*, 36.10–11 and 321.10–11, and [Themistius], *in APr.*, 121.23–24: "$a = b$; $c = b$: therefore $a = c$". It is fairly obvious, however, that the accounts are not independent of one another both in the choice of the example and in the way the problems such inferences raise are discussed. Alexander may well be their common source: Philoponus mentions him as his own source at *in APr.*, 321.8. However, we may well concede that the ordering of the terms could be unintentionally changed by each single commentator or by copyists (see the mentioned variant reading).

[268] At *Inst. Log.* I, Galen twice offers an argument by transitivity of equality applied to three men called Theon, Dion, and Philon. The whole passage is corrupt, but the scheme seems to be "$a = b$; $c = b$: therefore $a = c$" in the first occurrence and "$b = a$; $b = c$: therefore $a = c$" in the second. Here, too, the general statement of transitivity of equality is quoted.

[269] The first three at Alexander, *in APr.*, 344.32–34 and 345.6–7, and at [Themistius], *in APr.*, 122.3–4, the fourth at Philoponus, *in APr.*, 321.22–24 (the second and fourth occurrence are schemes of the invalid syllogism where "being siblings" replaces "having the same parents").

[270] And this leads immediately to the conclusion that the transitivity arguments are not syllogistically valid. The premises are indeterminate (in fact they are singular), and it is a standard Aristotelian doctrine (*APr.* I.4, 26a29–30) that indeterminate premises are equivalent to particular premises.

and his colleagues, and I surmise that the second figure was chosen looking at the standard ordering of the transitivity inferences, namely, the homological order. But why the third figure has not been selected? Maybe because the standard examples of syllogisms invalid by form were second figure syllogisms, as for instance the notorious fallacy put forward by Caeneus and reported by Aristotle in *APo.* I.12.

A very interesting symmetry argument, which shows that universal negative premises convert *simpliciter* on the basis of the manifest symmetry of the relation formulated by ἀποζεύγνυμι "to be disjoined from", is ascribed to Eudemus and Theophrastus by Alexander, *in APr.*, 31.4–10 and 34.13–15, and, drawing from Alexander and in a less neat way as usual, by Philoponus, *in APr.*, 48.11–18. Let us read Alexander—note the use of the perfect stem:

τὴν δὲ δεῖξιν οὕτως ποιοῦνται· κείσθω τὸ Α κατὰ μηδενὸς τοῦ Β· εἰ κατὰ μηδενός, ἀπέζευκται τοῦ Β τὸ Α καὶ κεχώρισται· τὸ δὲ ἀπεζευγμένον ἀπεζευγμένου ἀπέζευκται· καὶ τὸ Β ἄρα παντὸς ἀπέζευκται τοῦ Α· εἰ δὲ τοῦτο, κατὰ μηδενὸς αὐτοῦ.

They shape their proof as follows: let A be set ‹to be said› of no B; if of none, A turns out to be disjoined and separated from B; but what turns out to be disjoined turns out to be disjoined from what turns out to be disjoined: therefore B also turns out to be disjoined from every A; and if this, ‹it is said› of none of them.

Finally, two allusions to the symmetry of a relation in technical contexts are worth a mention. The first is in Proclus, *iE*, 373.5–23 (commentary on I.30—transitivity of parallelism), where he remarks that "the geometer was used to prove, in the arguments about relations [ἐν τοῖς περὶ τῶν σχέσεων λόγοις], the identity pervading all items that have the same relation to a same item" and claims that not all relations are transitive, even if a sufficient condition for this is their being symmetric (!). Proclus uses ἀντιστρέφουσι συνωνύμως "synonymically convert" for "being symmetric" and adduces equality, similitude, identity, and παράλληλος θέσις "parallel position" as examples—the same examples as in Aristotle, *Metaph.* Δ.15, and in the same order—arguing that "parallelism is a similitude of position, so to speak". The second allusion is in Cleomedes, *Cael.* I.1.24 (I.1.228–232): on the example of the Antipodeans, he distinguishes σχέσεις "relations" that ἀντιστρέφουσι "convert" and those that do not.

Types (3) and (4): Composition and Stability under Operations

Interesting examples of composition of relations (**type 3**) come from Book V. Let us start with V.3 (*EOO* II, 12.2–7), where it is shown how to compound equimultiplicity relations with different multiples—it is a species of transitivity, as suggested by the fact that the enunciation is in Form I; note also the δι' ἴσου "through an equal" phrase that marks the operation on equimultiples:[271]

ἐὰν πρῶτον δευτέρου ἰσάκις ᾖ πολλαπλάσιον καὶ τρίτον τετάρτου ληφθῇ δὲ ἰσάκις πολλαπλάσια τοῦ τε πρώτου καὶ τρίτου, καὶ δι' ἴσου τῶν ληφθέντων ἑκάτερον ἑκατέρου ἰσάκις ἔσται πολλαπλάσιον τὸ μὲν τοῦ δευτέρου τὸ δὲ τοῦ τετάρτου.

If first of second is equimultiple and third of fourth and there be taken equimultiples of the third and fourth, through an equal, of the ‹multiples› respectively taken, will also be equimultiple the one of the second, the other of the fourth.

[271] For the use of ordinals see Sect. 5.1.6. In the first sentence, the relational operator is in an internal position, and of course there remains in the setting-out; the same happens in V.2. In the proof of V.3, however, the relational operator is moved to external position, with the sole exception of the instantiated citation of V.2 (validation by template is a stronger regulating principle than adherence to the formulaic environment).

Thus, if $(A,C) = m(B,D)$ and $(E,Z) = n(A,C)$, then $(E,Z) = k(B,D)$ for some k—the theorem does not assert that $k = nm$. This is a form of transitivity that acts on the multiples, but the consistent use of Form I for the first premise and of Form II for the second premise makes it impossible to determine whether the arrangement is chiastic or homological.

The proposition of proportion theory that correlates with V.3 is V.22, which proves the stability of the δι' ἴσου transformation under transition to identical ratios. Let us read the setting-out of proposition V.22 (*EOO* II, 60.22–27):

ἔστω ὁποσαοῦν μεγέθη τὰ Α Β Γ καὶ ἄλλα αὐτοῖς ἴσα τὸ πλῆθος τὰ Δ Ε Ζ σύνδυο λαμβανόμενα ἐν τῷ αὐτῷ λόγῳ, ὡς μὲν τὸ Α πρὸς τὸ Β οὕτως τὸ Δ πρὸς τὸ Ε, ὡς δὲ τὸ Β πρὸς τὸ Γ οὕτως τὸ Ε πρὸς τὸ Ζ. λέγω ὅτι καὶ δι' ἴσου ἐν τῷ αὐτῷ λόγῳ ἔσται.

Let there be as many magnitudes as we please, A, B, Γ, and others equal to them in multiplicity, Δ, E, Z, taken two and two together in a same ratio, as A is to B, so Δ is to E, and as B is to Γ, so E is to Z. I claim that they will also be in a same ratio through an equal.

Thus, if A:B::D:E and B:C::E:Z, then A:C::D:Z; to this theorem refers Galen's example in *Inst. Log.* XVI.1–3 and 9. The δι' ἴσου phrase marks the transition from ratios A:B and B:C to ratio A:C, and from ratios D:E and E:Z to ratio D:Z, an operation that we might simply read as the multiplication of the two pairs of ratios.[272] The inference fits very neatly into a scheme of transitivity, but in this case the canonical disposition of the terms is chiastic.

The fact that the inference above is identified by an additional qualification such as δι' ἴσου turns out not to be accidental, as is shown by other occurrences of the syntagm in the *Elements*.[273] The first occurrence of use is in proposition V.20 (*EOO* II, 56.6–10), a theorem which is a preliminary step to the proof of V.22:

ἐὰν ᾖ τρία μεγέθη καὶ ἄλλα αὐτοῖς ἴσα τὸ πλῆθος σύνδυο λαμβανόμενα καὶ ἐν τῷ αὐτῷ λόγῳ δι' ἴσου δὲ τὸ πρῶτον τοῦ τρίτου μεῖζον ᾖ, καὶ τὸ τέταρτον τοῦ ἕκτου μεῖζον ἔσται, κἂν ἴσον, ἴσον, κἂν ἔλαττον, ἔλαττον.

If there be three magnitudes and others equal to them in multiplicity taken two and two together and in a same ratio and the first be greater than the third through an equal, the fourth will also be greater than the sixth, and if equal, equal, and if less, less.

The enunciation states that, if A:B::D:E and B:C::E:Z and δι' ἴσου $A > C$, then $D > Z$, and the same for $A = C$ and $A < C$. The syntagm is placed in the antecedent of the conditional, not in the consequent as in V.22. In such a position, it cannot, of course, mark an operation on ratios: it simply stresses the fact that, of the four terms in the sequence A, B, B, C, only the extremes are taken into account, whereas the "equal" middle terms are not.[274]

The definition of the "through an equal" operation in V.def.17 (*EOO* II, 6.7–13) confirms that this was the general meaning of δι' ἴσου:

[272] In ancient terms, this is the operation of composition of ratios: see Acerbi 2018 on the long history of this notion.

[273] Most of the occurrences in the ancient corpus are related to applications of V.22–23, or of their extensions to inequalities between ratios; see Aujac 1986 for a survey. The operation in V.23 is a modification of the operation in V.22, in which the disposition of the terms in the second pair of ratios is E:D – Z:E. Accordingly, the qualification δι' ἴσου is further qualified by the expression ἐν τεταραγμένῃ ἀναλογίᾳ "in perturbed proportion".

[274] The other occurrences of seemingly incongruous use that Aujac adduces, namely, Archimedes, *Aequil.* I.12, and Pappus, *Coll.* III.96, do not stand a close scrutiny. The involved operations on ratios are not univocally determined, and the δι' ἴσου phrase can be read as a genuine reference to *El.* V.22.

<u>δι᾽ ἴσου</u> λόγος ἐστὶ πλειόνων ὄντων μεγεθῶν καὶ ἄλλων αὐτοῖς ἴσων τὸ πλῆθος σύνδυο λαμβανομένων καὶ ἐν τῷ αὐτῷ λόγῳ, ὅταν ᾖ ὡς ἐν τοῖς πρώτοις μεγέθεσι τὸ πρῶτον πρὸς τὸ ἔσχατον, οὕτως ἐν τοῖς δευτέροις μεγέθεσι τὸ πρῶτον πρὸς τὸ ἔσχατον· ἢ ἄλλως· λῆψις τῶν ἄκρων καθ᾽ ὑπεξαίρεσιν τῶν μέσων.	A ratio <u>through an equal</u> is, there being several magnitudes and others equal to them in multiplicity taken two and two together and in a same ratio, whenever, as in the first magnitudes the first is to the last, so in the second magnitudes the first be to the last; or otherwise: taking the extremes by removing the middles.

The text is corrupt and conflates a true definition and a reformulation of theorem V.22, but a satisfactory reading can be reconstructed quite easily.[275] It is enough to combine Gerard's Arabo-Latin translation[276]

proportio que equalitas nominatur est proportio quarundam extremitatum ad alias cum fuerint quantitates plures duabus, et fuerint cum eis alie quantitates secundum earum numerationem in proportione una, et remote fuerint equaliter que sunt in medio.

with Gerard's report, immediately subsequent to this text, of Thābit ibn Qurra claiming that he had found a definition substantially identical to the first definition in V.def.17 *in alia scriptura*.[277] This allows to reconstruct a definition of the δι᾽ ἴσου operation on ratios certainly nearer to the original than the definition attested in the Greek manuscripts:

*δι᾽ ἴσου λόγος ἐστὶ πλειόνων ὄντων μεγεθῶν καὶ ἄλλων αὐτοῖς ἴσων τὸ πλῆθος σύνδυο λαμβανομένων καὶ ἐν τῷ αὐτῷ λόγῳ, λῆψις τῶν ἄκρων καθ᾽ ὑπεξαίρεσιν τῶν μέσων.	A ratio through an equal is, there being several magnitudes and others equal to them in multiplicity taken two and two together and in a same ratio, taking the extremes by removing the middles.

Thus, the *definiens* simply reads "taking the extremes by removing the middles". The interest in the common middle term is further borne out by the very expression δι᾽ ἴσου, where the understood noun is very likely ὅρου "term".[278] This and the formulation of V.def.17 show that the syntagm δι᾽ ἴσου was intended to capture a unique feature of the schematic representation of such inferences by transitivity as are characterized by a disposition with contiguous middle terms: in principle, then, the marker δι᾽ ἴσου is not specific to dispositions of terms in a proportion, as the applications in V.3 and in V.20 confirm.[279] The creation of an ad hoc syntagm to single out only this subspecies of inference by transitivity suggests, once again, that the chiastic arrangement was not perceived as the canonical disposition of terms among those that fit a transitivity scheme.

---

[275] In the same way, V.def.18 contains the enunciation of V.23. On these definitions, see again Aujac 1986, and Federspiel 2006a and 2006b.

[276] The text is at Busard 1984, c. 118.38–43.

[277] The presence of alternative proofs and definitions in the *Elements* is very often the result of collation of isolated variant readings: see the discussion in Vitrac 2004.

[278] And not δι᾽ ἴσου διαστήματος "through an equal interval", an expression that introduces an unwelcome interference with the meaning of διάστημα as "ratio", a meaning that is amply attested in the ancient corpus: see for instance the entire *Sectio canonis* and the discussion at Porphyry, *in Harm.*, 91.4–95.23. On the meaning of διάστημα and λόγος in Aristotle and in the mathematical tradition see Ugaglia 2016.

[279] Add, in the case of the enunciations of V.20–21, the fact that the δι᾽ ἴσου phrase figures in the *antecedent* of the conditional.

Transformations of **type (3)** allow a smooth transition to the main form of interaction between a relation and the deductive structure: this is the stability of a relation under suitable manipulations of the terms; these give rise to "non-logical" inferential schemes. The δι' ἴσου composition in V.22 can be seen as a specific instance of an operation that secures a form of stability to identity of ratios. To this category belong all operations on ratios mentioned at the very beginning of this Section; they are in fact examples of **type (4)** transformations. To the same category also belongs the composition of relations of equimultiplicity, whose theory, as we have seen, is developed in V.1–6. Here is the "mathematical content" of these propositions, written in symbolic form:[280]

1) if $(A,C, \ldots) = n(B,D, \ldots)$, then $(A,A + C + \ldots) = n(B,B + D + \ldots)$;
2) if $(A,C) = n(B,D)$ and $(E,F) = m(B,D)$, then $(A + E,C + F) = k(B,D)$;
3) if $(A,C) = n(B,D)$ and $(E,F) = m(A,C)$, then $(E,F) = k(B,D)$;
4) if $A{:}B{::}C{:}D$, then, for every n and m, $nA{:}mB{::}nC{:}mD$;[281]
5) if $(A,C) = n(B,D)$ with $C, D < A, B$, respectively, then $(A - C,A) = n(B - D,B)$;
6) if $(A,C) = n(B,D)$ and $(E,F) = m(B,D)$ with $E, F < A, C$, respectively, then $A - E = B$ and $C - F = D$ or $(A - E,C - F) = k(B,D)$.

Now, here are the corresponding propositions of proportion theory (see also above), in which identity of ratios is proved to be stable under:

i) sum of antecedents and of consequents:
 if $A{:}B{::}C{:}D{::}E{:}F\ldots$, then $(A + C + E \ldots){:}(B + D + F\ldots){::}A{:}B$ (V.12);
ii) composition: if $A{:}B{::}C{:}D$, then $(A + B){:}B{::}(C + D){:}D$ (V.18);
iii) "through an equal": if $A{:}B{::}C{:}D$ and $B{:}E{::}D{:}F$, then $A{:}E{::}C{:}F$ (V.22);
iv) transition to equimultiples: $A{:}B{::}nA{:}nB$, for any n (V.15);
v) removal of antecedents from antecedents and of consequents from consequents: if $A{:}B{::}C{:}D$ with $C, D < A, B$, respectively, then $(A - C){:}(B - D){::}A{:}B$ (V.19);
vi) division: if $A{:}B{::}C{:}D$, then $(A - B){:}B{::}(C - D){:}D$ (V.17).[282]

To these we must add the stability of identity of ratios under

vii) transitivity: if $A{:}B{::}C{:}D$ and $C{:}D{::}E{:}F$, then $A{:}B{::}E{:}F$ (V.11);
viii) "mixed" transitivity: if $A{:}B{::}C{:}D$ and $C{:}D > E{:}F$, then $A{:}B > E{:}F$ (V.13);
ix) alternation: if $A{:}B{::}C{:}D$, then $A{:}C{::}B{:}D$ (V.16);
x) conversion, if $A{:}B{::}C{:}D$, then $A{:}(A - B){::}C{:}(C - D)$, and inversion, if $A{:}B{::}C{:}D$, then $B{:}A{::}D{:}C$ (stated as obvious in the spurious porisms to V.19 and to V.7, respectively);
xi) "through an equal" in perturbed proportion: if $A{:}B{::}D{:}F$ and $B{:}E{::}C{:}D$, then $A{:}E{::}C{:}F$ (V.23);[283]
xii) sum of antecedents in different proportions with equal consequents: if $A{:}B{::}C{:}D$ and $E{:}B{::}F{:}D$, then $(A + E){:}B{::}(C + F){:}D$ (V.24).

---

[280] If we are interested in the stability of a relation, it is superfluous to specify that, for instance, $k = n + m$ in V.2, or $k = nm$ in V.3, or $k = n - m$ in V.6 (since a unit is not a number, the case $n - m = 1$ must be treated separately).
[281] This is the first application of V.def.5.
[282] A ratio is standardly conceived as greater-to-less.
[283] In V.22–23, the terms in the proportions that act as premises are placed in such a way that the same final proportion results in both theorems.

Finally, let us see the "interactions" between (in)equalities of terms and identity of ratio:

xiii) $A{:}B{::}C{:}B$ and $B{:}A{::}B{:}C$ iff $A = C$ (V.7, 9);
xiv) $A{:}B > C{:}B$ and $B{:}A < B{:}C$ iff $A > C$ (V.8, 10);
xv) if $A{:}B{::}C{:}D$ and $A > C$, then $B > D$, and the same for "equal to" and "less than" (V.14);
xvi) "through an equal": if $A{:}B{::}C{:}D$ and $B{:}E{::}D{:}F$ and $A > E$, then $C > F$, and the same for "equal to" and "less than" (V.20);
xvii) "through an equal" in perturbed proportion: if $A{:}B{::}D{:}F$ and $B{:}E{::}C{:}D$, and $A > E$, then $C > F$, and the same for "equal to" and "less than" (V.21).

Thus, the theory of Book V almost entirely consists (I have listed 24 propositions out of 25) in proving that the identity of ratios is stable under suitable operations on the ratios themselves. The manner of proof is always the same, and entails applying V.def.5 in the two directions:[284] a relation of proportionality is assumed; V.def.5 is applied in one direction as a necessary condition to proportionality; if the transformations of the equimultiples induced by the transformation that is assumed to operate on the ratios (such as alternation, etc.) do not introduce further constraints on the initial choice of the multiples, the genericity of the initial choice is preserved, so that V.def.5 can finally be applied as a sufficient condition, proportionality being thus restored on the ratios resulting from the performed operation. That no further constraints are introduced is shown in the crucial propositions V.1–2 and 4—the very kernel of the whole theory—supplemented by a suitable form of transitivity of equimultiplicity, left unproved.

More generally, transformations of **types (1) – (4)** always figure in theorems that prove results of stability. That a deep concern with stability was the prime mover of the investigations into the properties of relations is corroborated by the following, quick survey of principles and theorems that pertain to relations and that involve stability:[285]

(1) Results that concern equality, included among the common notions that preface Book I: stability under addition and removal of equal magnitudes (I.cn.2–3), and similarly for stability under doubling and halving (5–6), preceded by the stability of inequality under addition of equal magnitudes (4).[286] Recall Galen's examples in *Inst. Log.* XVI.4 and 7–8, which we have read in Sect. 2.1 (but see also Sect. 4.5.1.1).

(2) The proportion theory expounded in Book VII is independent of that of Book V; it does not even aim at proving the numeric counterpart of all results proved for magnitudes (an easy task, at that). The following table sets out the correlative propositions:

V	1	5	19	12	16	22
VII	5–6	7–8	11	12	13	14

[284] When this does not happen: the specific assumptions of V.7 trivialize the requirement of genericity and make the check of the definition immediate; V.15 does not use equimultiples to check V.def.5. In V.8 and V.13, V.def.7 is applied.
[285] We may add the stability of the operation of cutting a straight line in extreme and mean ratio under the addition of the greater segment (XIII.5).
[286] The fact that I.cn.4–6 are spurious only confirms that issues of stability were regarded as crucial.

Very much as in V.1–6, our little "theory of equimultiplicity", the initial string VII.4–10 studies the notion of "part/s". In VII.4, it is proved that either of the relations "being part/s" holds for any ordered pair of numbers. Three paired propositions follow, namely, VII.5–6, 7–8, 9–10: they prove, for numbers that are "part/s" of another, the relation from which will derive, according to the definition, the homonymous relation of numeric proportion theory: in the order, sum of antecedents and of consequents, removal of antecedents from antecedents and of consequents from consequents, alternation.[287] The associated propositions of numeric proportion theory are, in this order, VII.12, 11, 13. Proposition VII.14 proves the property "through an equal". In VII.15, a particular form of alternation is proved applied to the quaternary relation "measuring the same ‹number› of times".[288]

(3) Results of primality theory in VII.23–30. Relative primality is stable under transition to: a part of one of the assumed coprime numbers (23); the square of one of the coprime numbers (25); the pairwise product of pairs of coprime numbers (26); the squares or the cubes of coprime numbers (27); their sum, to any of the assumed coprime (28). Finally, the product of numbers that are severally prime to a third is also prime to this number (24), a prime is prime relative to any number it does not measure (29), and, if a prime measures the product of two numbers, then it also measures at least one of the two assumed numbers (30).

(4) Stability of the relation of commensurability under identity of ratio (X.11) and under addition (15); stability of the relation of incommensurability under transition to magnitudes commensurable with the assumed magnitudes (13) and under addition (16); stability of the relation "being worth more by the square on a straight line commensurable with itself" under transition to magnitudes in the same ratio as the assumed magnitudes (14); stability of the class of medial lines and medial regions under commensurability (23 and porism); stability of the six irrationals generated by addition (66–70) and of the six irrationals generated by removal (103–107) under commensurability. In general, the theory of irrational lines in Book X can be seen as a search for minimal classes of straight lines stable under two operations: the formation of rectangular regions contained by straight lines and the parabolic application of a region on a straight line.

(5) In the *Data*,[289] propositions 10–11 and 13–21: detailed analysis of the relations "being greater than in ratio by a given" (= **A**) and "being in a given ratio" (= **B**). Stability of **A** under the operations of "composition" and "conversion" of ratios (10–11); transitivity under composition of **A** and **B** (13). If given magnitudes are compounded with (14) or are removed from (15) magnitudes in relation **B**, the resulting magnitudes are either in relation **A** or in relation **B** to one another; if a given magnitude is compounded with either of magnitudes in relation **B**, while another given magnitude is removed from the other, only relation **A** obtains (16). If two magnitudes are in relation **A** with respect to a same magnitude, they are either in relation **A** or in relation **B** to one another (17). If a same magnitude is in relation **A** with respect to two magnitudes, these are either in relation **A** or in relation **B** to one another (18). Transitivity of relation **A** (19). If magnitudes in relation **B** are removed from (20) or compounded with (21) two given magnitudes, the resulting magnitudes are either in relation **A** or in relation **B** to one another.

---

[287] Of course, no alternation can be proved in equimultiple theory.
[288] The form is particular because one of the terms is a unit.
[289] On a metalevel, all *Data*-theorems prove stability of the predicate "being given" under suitable transformations.

4.5.2. *Metamathematical markers: potential and analogical proofs, references to the obvious, optative mood, personal verb forms*

A small constellation of linguistic tools has the function of curtailing deductive sequences or proofs, whenever these are obvious or reproduce almost word-for-word sequences that have already been fully unfolded.[290] Three categories can be distinguished:

1) References to the obvious, normally introduced by δῆλον "clear" and its variants δηλαδή "clearly" and δηλονότι "it is clear that", or by φανερόν "manifest". In the *Elements*, the references to the obvious never call into play properties of the diagram.
2) Analogical proofs, introduced by διὰ τὰ αὐτὰ δή (δέ) "for the (very) same ‹reasons›".[291]
3) Potential proofs, introduced by references to the similarity of the deduction: the most frequent form is ὁμοίως δὴ δείξομεν "very similarly we shall prove" followed by a that-clause.

A detailed analysis of these three categories is as follows.

(1) The references to the obvious are infrequent and must be regarded a priori as interpolations. In the whole of the *Elements*, δῆλον and its kin exhibit the following occurrences:

δῆλον	III.25, X.9/10, 44, 111, *alt*VI.6, *alt*X.23, 38
δηλαδή	III.25 (*bis*), X.4, 10, XI.3, XII.4/5, 17, XIII.15, *alt*X.18, 23
δηλονότι	IV.8, VI.7, X.47

An author like Archimedes presents 222 references to the obvious featuring the stem δηλ–; further 87 items are associated with φανερόν.[292]

A πόρισμα "porism" is an unproved statement, which is declared φανερόν "manifest" on the grounds of a proposition just proved. A porism is normally introduced by ἐκ δὴ τούτου φανερόν ὅτι "thus it is manifest from this that", but there are exceptions:[293] stylistic variants are found in the porisms to IV.5, IX.11,[294] X.9,[295] X.114;[296] a potential proof is found in the first porism to VI.20;[297] nothing at all in the porism to VI.20[298] and in proposition XII.17. The following table collects the distribution, in the main text of the *Elements*, of the occurrences of φανερόν outside porisms; this predicate obviously serves as a stylistic marker of the arithmetic Books and of Book X:

[290] A style strongly connotated by metamathematical markers with non-trivial deductive import characterizes Diophantus' *Arithmetica*; this style is thoroughly studied in Acerbi 2011e, sect. 7.1. The prefatory epistles of Archimedes' and Apollonius' treatises contain major pieces of information about metamathematical and foundational issues, as we have partly seen in Sect. 3.4: see the discussion in Vitrac 2008.
[291] The denominations "analogical proof" and "potential proof" are first introduced in Vitrac 1990–2001 IV, 46.
[292] Most references to the obvious in Archimedes open a proved statement or even the conclusion of a proof: therefore, they are a stylistic trait. There are only 7 occurrences in Apollonius.
[293] Many porisms are certainly spurious.
[294] Phrase καὶ φανερὸν ὅτι "and it is manifest that".
[295] Phrase καὶ φανερὸν ἐκ τῶν δεδειγμένων ἔσται ὅτι "and it will be manifest from what has been proved that".
[296] Phrase καὶ γέγονεν ἡμῖν καὶ διὰ τούτου φανερὸν ὅτι "and it has resulted manifest to us from this that"—this is particularly aberrant, because of the presence of the dative of interest ἡμῖν.
[297] Phrase ὡσαύτως δὲ καὶ ἐπὶ τῶν ὁμοίων τετραπλεύρων δειχθήσεται ὅτι "and likewise it will be also proved for the similar quadrilaterals that".
[298] However, this ends with ὥστε καὶ καθόλου φανερὸν ὅτι "so that it is also in general manifest that".

III	31, 35
VII	2
VIII	6, 9, 15, 21
IX	3, 13, 15, 32–34
X	3, 4, 13 (*bis*), 16/17, 18, 28/29, 42, 44, 54, 57–59, 115
XI	20, 22, 34
XII	3, 17

Note the unusual postpositive formulation in VIII.6 (*EOO* II, 288.12–13):

ὅτι μὲν οὖν οἱ Α Β Γ Δ Ε ἑξῆς ἀλλήλους οὐ μετροῦσιν, φανερόν – οὐδὲ γὰρ ὁ Α τὸν Β μετρεῖ.

Then now, that A, B, Γ, Δ, E do not successively measure one another, it is manifest—for A does not measure B either.

We also find this formulation in IX.32–34 and, the reference to the obvious being replaced by δέδεικται "it has been proved", in IX.9 (*bis*).

(2) and (3) The analogical proofs are introduced by διὰ τὰ αὐτὰ δή (δέ) "for the (very) same ‹reasons›".[299] This is a short-range and lightly-loaded anaphoric tool that introduces an inferential unit virtually identical to the just preceding inferential unit. Longer range and heavier deductive load is assigned to a potential proof, usually introduced by ὁμοίως δὴ δείξομεν "very similarly we shall prove" followed by a that-clause.[300] Some variant readings (wrongly) replace the particle δή with δέ,[301] set the verb in the passive future,[302] or feature only the adverb ὁμοίως.[303] The different range of the two kinds of abbreviated proof can be best appreciated by comparing the analogical proof in III.21 (*EOO* I, 220.24–222.4)

καὶ ἐπεὶ ἡ μὲν ὑπὸ ΒΖΔ γωνία πρὸς τῷ κέντρῳ ἐστὶν ἡ δὲ ὑπὸ ΒΑΔ πρὸς τῇ περιφερείᾳ καὶ ἔχουσι τὴν αὐτὴν περιφέρειαν βάσιν τὴν ΒΓΔ, ἡ ἄρα ὑπὸ ΒΖΔ γωνία διπλασίων ἐστὶ τῆς ὑπὸ ΒΑΔ. διὰ τὰ αὐτὰ δὴ ἡ ὑπὸ ΒΖΔ καὶ τῆς ὑπὸ ΒΕΔ ἐστι διπλασίων.

And since angle BZΔ is at the centre and BAΔ at the circumference and they have a same arc BΓΔ as base, therefore angle BZΔ is double of BAΔ. For the very same ‹reasons› BZΔ is also double of BEΔ.

with potential proofs like the one occurring in V.6 (*EOO* II, 20.11–22.2)—the long, unfolded proof unusually ends with a conditional introduced by ὥστε "so that":

ἔστω γὰρ πρότερον τὸ ΗΒ τῷ Ε ἴσον. λέγω ὅτι καὶ τὸ ΘΔ τῷ Ζ ἴσον ἐστίν.
κείσθω γὰρ τῷ Ζ ἴσον τὸ ΓΚ. ἐπεὶ ἰσάκις ἐστὶ πολλαπλάσιον τὸ ΑΗ τοῦ Ε καὶ τὸ ΓΘ τοῦ Ζ ἴσον δὲ τὸ μὲν ΗΒ τῷ Ε τὸ δὲ ΚΓ τῷ Ζ, ἰσάκις ἄρα ἐστὶ πολλαπλά-

In fact, let first HB be equal to E. I claim that ΘΔ is also equal to Z.
In fact, let a ‹magnitude›, ΓK, be set equal to Z. Since AH of E and ΓΘ of Z is equimultiple and HB is equal to E and KΓ to Z, therefore AB of E and

[299] There are 150 occurrences in the whole of the *Elements*. There is 1 occurrence of διὰ ταὐτὰ τοίνυν "now then, for the same ‹reasons›", in I.21 (*EOO* I, 52.4). The particle τοίνυν occurs twice in Apollonius (*Con.* I.41 and 52), but 9 times in Autolycus and 17 in Archimedes (1 in *Sph. cyl.* I, 7 in *Con. sph.*, 6 in *Spir.*, 1 in *Aequil.* I, 1 in *Ar.*, 1 in *Fluit.* I).
[300] There are 86 occurrences in the whole of the *Elements*.
[301] This mistake is statistically irrelevant.
[302] There are 19 occurrences, in propositions I.15, 27, III.37, IV.3, 12 (*bis*), 13 (*bis*), 14, 15 (*bis*), IX.19, X.1, 31, XI.18, XII.5–6, XIII.16–17.
[303] There are 8 occurrences, in III.25, IV.15–16, X.13/14, 21/22, 68, XII.4, XII.4*alt*.

σιον τὸ ΑΒ τοῦ Ε καὶ τὸ ΚΘ τοῦ Ζ· ἰσάκις δὲ ὑπόκειται πολλαπλάσιον τὸ ΑΒ τοῦ Ε καὶ τὸ ΓΔ τοῦ Ζ· ἰσάκις ἄρα ἐστὶ πολλαπλάσιον τὸ ΚΘ τοῦ Ζ καὶ τὸ ΓΔ τοῦ Ζ. ἐπεὶ οὖν ἑκάτερον τῶν ΚΘ ΓΔ τοῦ Ζ ἰσάκις ἐστὶ πολλαπλάσιον, ἴσον ἄρα ἐστὶ τὸ ΚΘ τῷ ΓΔ· κοινὸν ἀφῃρήσθω τὸ ΓΘ· λοιπὸν ἄρα τὸ ΚΓ λοιπῷ τῷ ΘΔ ἴσον ἐστίν· ἀλλὰ τὸ Ζ τῷ ΚΓ ἐστιν ἴσον· καὶ τὸ ΘΔ ἄρα τῷ Ζ ἴσον ἐστίν· ὥστε εἰ τὸ ΗΒ τῷ Ε ἴσον ἐστίν, καὶ τὸ ΘΔ ἴσον ἔσται τῷ Ζ.	ΚΘ of Ζ is equimultiple; and ΑΒ of Ε and ΓΔ of Ζ has been supposed equimultiple; therefore ΚΘ of Ζ and ΓΔ of Ζ is equimultiple. Then since each of ΚΘ, ΓΔ of Ζ is equimultiple, therefore ΚΘ is equal to ΓΔ; let a ‹magnitude›, ΓΘ, be removed as common; therefore ΚΓ as a remainder is equal to ΘΔ as a remainder; but Ζ is equal to ΚΓ; therefore ΘΔ is also equal to Ζ; so that if ΗΒ is equal to Ε, ΘΔ will also be equal to Ζ.
ὁμοίως δὴ δείξομεν ὅτι, κἂν πολλαπλάσιον ᾖ τὸ ΗΒ τοῦ Ε, τοσαυταπλάσιον ἔσται καὶ τὸ ΘΔ τοῦ Ζ.	Very similarly we shall prove that, even if ΗΒ be multiple of Ε, such a multiple will also be ΘΔ of Ζ.

Sometimes, a potential proof is induced by peculiarities of the enunciation, as in I.17 (*EOO* I, 44.9–13 and 44.19–22), where a tricky syntagm (see Sect. 4.2 for the translation) has expressly required to exhaust all combinations:

παντὸς τριγώνου αἱ δύο γωνίαι δύο ὀρθῶν ἐλάσσονές εἰσι πάντῃ μεταλαμβανόμεναι.	Two angles of every triangle, however permuted, are less than two right ‹angles›.
ἔστω τρίγωνον τὸ ΑΒΓ. λέγω ὅτι τοῦ ΑΒΓ τριγώνου αἱ δύο γωνίαι δύο ὀρθῶν ἐλάττονές εἰσι πάντῃ μεταλαμβανόμεναι	Let there be a triangle, ΑΒΓ. I claim that two angles of triangle ΑΒΓ, however permuted, are less than two right ‹angles›.
αἱ ἄρα ὑπὸ ΑΒΓ ΒΓΑ δύο ὀρθῶν ἐλάσσονές εἰσιν. ὁμοίως δὴ δείξομεν ὅτι καὶ αἱ ὑπὸ ΒΑΓ ΑΓΒ δύο ὀρθῶν ἐλάσσονές εἰσι καὶ ἔτι αἱ ὑπὸ ΓΑΒ ΑΒΓ.	Therefore ΑΒΓ, ΒΓΑ are less than two right ‹angles›. Very similarly we shall prove that ΒΑΓ, ΑΓΒ are also less than two right ‹angles›, and further ΓΑΒ, ΑΒΓ.

Other metamathematical markers are the following.

(*a*) The optative mood is employed as a marginal residue,[304] in particular within a formulaic conditional clause that marks a division into cases of a problem; the first occurrence in the *Elements* is found in II.14 (*EOO* I, 160.9–17):[305]

[304] But the optative mood is highly frequent in *Optica* redaction **B** (12 occurrences, against 2 in *Optica* **A**). Occurrences of optative in Apollonius are in *Con.* I.7, II.46 (the same conditional clause as in *El.* II.14), and III.4, at *AGE* I, 26.25, 266.18, 324.23. The optative is more frequent in Archimedes, for instance in *Sph. cyl.* I.32, *AOO* I, 118.2; I.34, 126.10; *Con. sph.* 32, 444.16; *Aequil.* II.4–5, *AOO* II, 176.11 and 180.2; *Ar.*, *AOO* II, 232.18, 234.20, 244.5–6, 256.31; *Quadr.* 5, *AOO* II, 270.9; 6, 272.24; 14, 286.21 and 288.18; 15, 292.19; 16, 298.9, *Fluit.* II.2, *AOO* II, 350.3. As a matter of fact, in some technical texts we find inferences systematically drawn with the optative in the conclusion: a beautiful example are the arguments in [Aristotle], *LI*. Compare the extracts from Galen's *Inst. Log.* read in Sects. 2.1 and 4.5.1.1.

[305] The other occurrences of such a conditional clause are in IV.1, VI.28, VII.31 (*bis*), 32, XI.11. A similar clause is found once in Archimedes, *Sph. cyl.* I.5; in the same author, we find references to the assigned problem as an ἐπίταγμα "prescription" in *Sph. cyl.* I.2 (*bis*), 3, 4, and in *Con. sph.*, proem. This noun also characterizes the several "prescriptions" of a complex problem in Pappus, most notably in Book VII of the *Collectio* (34 occurrences). The verb ἐπιτάττω "to prescribe" is ubiquitous in Diophantus' *Arithmetica*, since it presents the specific (and paradigmatic) values of numbers that are assigned in the enunciation. A meaning similar to that attested in Diophantus is in *El.* VI.9 (*bis*), where a "prescribed" part of a straight line must be removed from it, in VIII.2 (*ter*), where it is asked to find least numbers "as many as one prescribes" in continuous proportion, and in Hero, *Metr.* III.6 and 21–23; see also Pappus, *Coll.* III.27 (*ter*), 66 (*bis*), 67, 72 (*bis*), 74, IV.74 (an expression similar to that in *El.* VIII.2, and the same in *Coll.* III.67), VIII.25, 26. In Pappus, a ratio is frequently said to be "prescribed".

τῷ δοθέντι εὐθυγράμμῳ ἴσον τετράγωνον συστήσασθαι.	Construct a square equal to a given rectilinear ‹figure›.
ἔστω τὸ δοθὲν εὐθύγραμμον τὸ Α. δεῖ δὴ τῷ Α εὐθυγράμμῳ ἴσον τετράγωνον συστήσασθαι.	Let there be a given rectilinear ‹figure›, A. Thus it is required to construct a square equal to the rectilinear ‹figure› A.
συνεστάτω γὰρ τῷ Α εὐθυγράμμῳ ἴσον παραλληλόγραμμον ὀρθογώνιον τὸ ΒΔ. εἰ μὲν οὖν ἴση ἐστὶν ἡ ΒΕ τῇ ΕΔ, γεγονὸς ἂν εἴη τὸ ἐπιταχθέν – συνέσταται γὰρ τῷ Α εὐθυγράμμῳ ἴσον τετράγωνον τὸ ΒΔ.	In fact, let a rectangular parallelogram, BΔ, be constructed equal to the rectilinear ‹figure› A. Then if BE is equal to EΔ, that which was prescribed would have come to be—for a square BΔ turns out to be constructed equal to the rectilinear ‹figure› A.

The two propositions of Book VII that contain the formulaic expression γεγονὸς ἂν εἴη τὸ ἐπιταχθέν are theorems, not problems as II.14 is. These propositions also present marked stylistic peculiarities, among which a formulation of a recursive procedure featuring the noun ἐπίσκεψις "investigation"; this noun figures again in some proofs in redaction **b** of Book XII (see below).[306] Let us read the text of VII.31 (*EOO* II, 250.17–252.2); it is a reduction to the impossible:[307]

καὶ εἰ μὲν πρῶτός ἐστιν ὁ Γ, γεγονὸς ἂν εἴη τὸ ἐπιταχθέν. εἰ δὲ σύνθετος, μετρήσει τις αὐτὸν ἀριθμός. τοιαύτης δὴ γινομένης ἐπισκέψεως ληφθήσεταί τις πρῶτος ἀριθμὸς ὃς μετρήσει τὸν πρὸ ἑαυτοῦ, ὃς καὶ τὸν Α μετρήσει – εἰ γὰρ οὐ ληφθήσεται, μετρήσουσι τὸν Α ἀριθμὸν ἄπειροι ἀριθμοί, ὧν ἕτερος ἑτέρου ἐλάσσων ἐστίν, ὅπερ ἐστὶν ἀδύνατον ἐν ἀριθμοῖς.	And if Γ is prime, that which was prescribed would have come to be. And if it is compounded, some number will measure it. Then such an investigation coming to be some prime number will be taken that will measure the one before itself, which will also measure A—for if it will not be taken, unboundedly many numbers, each of which is less than the other, will measure number A, which is really impossible in numbers.

The Theonine manuscripts, apparently concerned with the partition between theorems and problems, modify in VII.31–32 the above clause to δῆλον ἂν εἴη τὸ ζητούμενον "what is sought would be clear", in order to align it with the standard formulation. Add to this formula the isolated clauses εἰ τύχοι "random" in X.72, certainly spurious (see Sect. 5.1.2), and the occurrences of εἴη in X.def.4 and XI.34, the latter within a postposed explanation that is certainly an interpolation.[308]

(*b*) Verb forms in the first person—which I shall call "personal"—are extremely rare, even if they are not necessarily suspected of inauthenticity.[309] The only obvious exceptions to such a claim of rareness are of course the formulae λέγω ὅτι, which introduces the determination of theorems, and ὁμοίως δὴ δείξομεν, which introduces a potential proof (for the former, see Sect. 4.2.1).

The occurrences of verb forms in the first person plural in the main text of the *Elements* and of the *Data* are set out, distributed by termination, in the following table:[310]

---

[306] In XII.9 (*bis*), 10–11, *EOO* IV, 402.29–403.1, 403.30–404.1, 405.12–14, 409.22–25. Add to these the occurrences of ἐπισκέπτομαι "to investigate" in the enunciation and in the setting-out of IX.18–19 (cf. Sect. 4.2.1).

[307] I insert the integration of the Theonine manuscripts in my translation. The soundness of the variant is confirmed by the indirect tradition: see Busard 1983, 217.509–510, and Busard 1984, c. 182.17–20. Cf. also Heiberg's perplexities about the text of **P** in *EOO* II, 250.24 *app.*

[308] At *EOO* III, 218.21, 4.1; IV, 108.2, respectively.

[309] For their high frequency in Diophantus see Acerbi 2011e, introduction and sect. 7.1.3.

[310] The occurrences of δείξομεν "we shall prove" outside ὁμοίως δὴ/(δὲ) δείξομεν are marked with an asterisk.

–ωμεθα	III.25
–αμεθα	*Data* def.1–2[311]
–ουμεθα	X.62
–αμεν	X.28/29II, 32/33, 44
–υμεν	XII.10 (*bis*)
–ωμεν	IV.15por, 16 (*bis*), VI.20porII, IX.34 (*bis*), X.28/29II, XI.1, 23/24, 26 (*bis*), XII.2 (*ter*), 3, 4, 10 (*sexties*), 11 (*bis*), 16, 17 (*ter*), XIII.11, 13, 15, 16 (*quater*), 17
–ομεν	IV.15por (*bis*), 16por (*bis*), V.8, VI.22/23*, IX.34 (*bis*), X.10, 41/42*, XI.23/24*, XII.2, 10 (*bis*), 11, 12*, 12, 16, XIII.18*

Prima facie, the verb forms in the first person are a stylistic marker of the stereometric Books, because much of the others figure in patently spurious material. Actually, this phenomenon also affects Books XI–XIII, for inauthentic segments of text also contain the occurrences in III.25, V.8,[312] X.44,[313] XI.1, 26 (*bis*), XII.2 (*ter*), 3, 10 (*sexties*), 11 (*bis*),[314] XII.4.[315] Add to these all metamathematical remarks interpolated in a proof, such as ἐμάθομεν γάρ "for we learned it" in X.10, ὁμοίως δὴ τοῖς προτέροις ἐπιλογιούμεθα "we shall infer very similarly to those before" in X.62, and ὡς ἔμπροσθεν ἐδείκνυμεν "as we previously proved" in XII.10.[316] To one and the same reviser I ascribe the 7 occurrences of ἐπιζεύξωμεν "we join" in Book XIII. Support to my contention: all occurrences figure in the antecedent of a conditional; for two of them (the only occurrence in XIII.11 and the second occurrence in XIII.16), the consequent of the same conditional carries the metadiscursive verb form συνάγεται "one deduces";[317] two occurrences are included in postposed explanations introduced by ἐπειδήπερ (XIII.13 and the first occurrence in XIII.16); the constructive step that includes one of the occurrences is justified by a postposed explanation like διὰ τὴν ὁμοιότητα "because of the similitude" (the third occurrence in XIII.16); the remaining two occurrences (XIII.15 and the fourth occurrence in XIII.16) must be questioned as to their authenticity because they have the same form as the others and pertain to the same statements.[318]

Probably authentic instances of personal verb forms are instead the double occurrence in IV.16,[319] a construction whose redaction is particularly unconstrained, the occurrences in XII.16–17[320] and XIII.17–18,[321] and the four occurrences in IX.34 (*EOO* II, 404.2–19), probably the "less instantiated" proposition of the *Elements*. Let us read this theorem in its entirety (please do not ask me to add a diagram):[322]

---

311 This is the only occurrence of a first-person verb form in the *Data*; we have seen its import in Sect. 2.4.1.
312 These are additional cases, almost certainly spurious.
313 This is included in a useless remark, which also contains a reference to the obvious and the quasi-hapax καθ' ὑπόθεσιν "by supposition" (the only other occurrence is in X.47); I surmise that it must be ascribed to the same reviser as of lemma X.41/42 and of the beginning of X.42.
314 Most of these occurrences in Book XII are postposed explanations introduced by ἐπειδήπερ; cf. Sect. 4.5.3.
315 This is an appendage that offers a generalization not required by the enunciation. The listed occurrences are at *EOO* I, 230.4; II, 30.6 (περαίνομεν τὴν ἀπόδειξιν "we conclude the proof"); III, 126.3; IV, 8.21, 80.15 (*bis*), 142.13 and 142.24 (*bis*), 154.23–24, 188.2, 188.26–190.1 (*ter*), 192.14–17 (*bis*), 198.4–5 (*bis*), 162.9, respectively.
316 See *EOO* III, 32.15–16, 190.17; IV, 188.20 and 192.13, respectively.
317 This form is found again in V.25, within a part of the proof that has certainly been rewritten, and in X.28/29II.
318 They are, in the order given in the text, at *EOO* IV, 280.15, 310.2–3; 292.10, 308.18; 310.21; 302.14–15, 310.27–28.
319 We read it at *EOO* I, 320.26.
320 The first occurrence in XII.17, within a postposed explanation introduced by ἐπειδήπερ and containing the hapax ἐπινοήσωμεν, is certainly spurious (*EOO* IV, 228.20). The second occurrence in the same proposition (form νοήσωμεν) belongs to a segment of text that has undergone a rewriting that has reduced the number of personal forms from 3 to 2; in redaction **b** (= XII.16) we find twice the personal form ἕξομεν (*EOO* IV, 234.17 and 421.1).
321 These are at *EOO* IV, 322.11 and 334.18.
322 For the iterative meaning of ἀεί "continually" in mathematical texts, see Federspiel 2004. On iterative ἀεί in philosophical texts see Ugaglia 2009. (But this is the only meaning compatible with a non-linear conception of Time; just read Thucydides to find plenty of iterative ἀεί.) The enunciation of IX.34 should definitely be stated for even numbers.

ἐὰν ἀριθμὸς μήτε τῶν ἀπὸ δυάδος διπλασιαζομένων ᾖ μήτε τὸν ἥμισυν ἔχῃ περισσόν, ἀρτιάκις τε ἄρτιός ἐστι καὶ ἀρτιάκις περισσός.	If a number neither be of those doubled from a dyad nor have its half odd, it is both even times even and even times odd.
ἀριθμὸς γὰρ ὁ Α μήτε τῶν ἀπὸ δυάδος διπλασιαζομένων ἔστω μήτε τὸν ἥμισυν ἐχέτω περισσόν. λέγω ὅτι ὁ Α ἀρτιάκις τέ ἐστιν ἄρτιος καὶ ἀρτιάκις περισσός.	In fact, let a number, A, neither be of those doubled from a dyad nor have its half odd. I say that A is both even times even and even times odd.
ὅτι μὲν οὖν ὁ Α ἀρτιάκις ἐστὶν ἄρτιος, φανερόν – τὸν γὰρ ἥμισυν οὐκ ἔχει περισσόν. λέγω δὴ ὅτι καὶ ἀρτιάκις περισσός ἐστιν. ἐὰν γὰρ τὸν Α τέμνωμεν δίχα καὶ τὸν ἥμισυν αὐτοῦ δίχα καὶ τοῦτο ἀεὶ ποιῶμεν, καταντήσομεν εἴς τινα ἀριθμὸν περισσὸν ὃς μετρήσει τὸν Α κατὰ ἄρτιον ἀριθμόν. εἰ γὰρ οὔ, καταντήσομεν εἰς δυάδα, καὶ ἔσται ὁ Α τῶν ἀπὸ δυάδος διπλασιαζομένων, ὅπερ οὐχ ὑπόκειται· ὥστε ὁ Α ἀρτιάκις περισσός ἐστιν· ἐδείχθη δὲ καὶ ἀρτιάκις ἄρτιος· ὁ Α ἄρα ἀρτιάκις τε ἄρτιός ἐστι καὶ ἀρτιάκις περισσός, ὅπερ ἔδει δεῖξαι.	Then now, that A is even times even, it is manifest—for it does not have its half odd. I now claim that it is also even times odd. In fact, if we bisect A and its half and do this continually, we shall come upon some odd number that will measure A according to an even number. In fact, if not, we shall come upon a dyad, and A will be among those doubled from a dyad, which has really not been supposed; so that A is even times odd; and it was also proved even times even; therefore A is both even times even and even times odd, which it was really required to prove.

This text preludes the most interesting occurrences of personal verb forms. They figure in the canonical, non-instantiated formulation of the iterative procedure (in the technical meaning defined in Sect. 1.2) of the method of exhaustion (cf. Sect. 5.1.3); they contain the personal verb form καταλείψομεν "we shall leave out"; they are further characterized by a series of circumstantial participles that act as modifiers of the (implicit) operating subject. Let us read the first occurrence in proposition XII.10 (*EOO* IV, 190.9–14):

τέμνοντες δὴ τὰς ὑπολειπομένας περιφερείας δίχα καὶ ἐπιζευγνύντες εὐθείας καὶ ἀνιστάντες ἐφ' ἑκάστου τῶν τριγώνων πρίσματα ἰσοϋψῆ τῷ κυλίνδρῳ καὶ τοῦτο ἀεὶ ποιοῦντες καταλείψομέν τινα ἀποτμήματα τοῦ κυλίνδρου ἃ ἔσται ἐλάττονα τῆς ὑπεροχῆς ᾗ ὑπερέχει ὁ κύλινδρος τοῦ τριπλασίου τοῦ κώνου.	Thus, bisecting the arcs that remain and joining straight lines and erecting on each of the triangles prisms with the same height as the cylinder and doing this continually we shall leave some segments of the cylinder that will be less than the excess by which the cylinder exceeds the triple of the cone.

This procedure is found again, with suitable adaptations, in XII.2, 10–12, 16.[323] Only the last occurrence is backed up by redaction **b** of the Greek text.[324] Redaction **b**, however, also presents the clause in proposition XII.5, in a passage in which the main redaction has only an abbreviated formula in the passive voice.[325] In redaction **b**, the above formula, which must be regarded as a later normalization resulting from an interference with the procedural code, is replaced by a simple genitive absolute, which, however, displays a decidedly metamathematical connotation: τοιαύτης δὴ

[323] In *EOO* IV, 144.6–10, 190.13–19, 198.18–220.2, 206.11–16, 226.18–21, respectively.

[324] This is XII.15 **b**, *EOO* IV, 416.27–30: verb form καταλήψομεν "we shall take".

[325] Compare *EOO* IV, 397.23–27: verb form λήψομεν "we shall take", and *EOO* IV, 166.6–10: verb form λειφθῶσι "they have remained".

γινομένης ἀεὶ ἐπισκέψεως "thus such an investigation coming about continually"; we have already read this clause in VII.31.[326]

The circumstantial participles are characteristic of such clauses as we have just read a specimen: 21 out of the 26 occurrences in the main text of the *Elements* are found in the 5 propositions of Book XII listed above.[327] Redaction **b** of this Book only keeps the two occurrences in XII.16, but adds one in XII.5. As for the other instances of circumstantial participles, those in IV.16 and V.8 are found in the above-mentioned arguments in which also figure verb forms in the first person. Add the spurious occurrences in IX.13[328] and in XII.12; in XII.12, two participles separate δείξομεν from the canonically adjoined ὁμοίως δή.[329] As is to be expected, circumstantial participles that act as modifiers of the operating subject are absent in the *Data*.

4.5.3. *Postposed arguments*

A "postposed explanation" is an explicative clause that follows the statement it is intended to explain. The postposed explanations, whose size ranges from short syntagms to complete sentences, exhibit four peculiarities:

1) They interrupt the "natural" deductive ordering, namely, the ordering that proceeds "forward" (for "backward" metadeductive ordering in analysis, see Sect. 2.4.1).
2) They are often marked by the presence of intensive causal subordinants,[330] such as ἐπείπερ "since ... really", ἐπειδή "since ... quite", ἐπειδήπερ "since ... really quite" (cf. Sect. 5.3.2);[331] by formulaic clauses like διὰ τό "because" + infinitive; or, in the commonest form (a bit more than 200 times in the *Elements*), by explicative γάρ "for".[332]
3) They often include non-instantiated citations of theorems. The lack of instantiation introduces a surplus of generality, which is out of place within a proof.
4) Their mathematical import is frequently very poor or simply irrelevant.

[326] In **b** XII.2, 9 (*bis*), 10, 11, in *EOO* IV, 392.10–12, 402.29–403.1 and 403.30–404.1, 405.12–14, 409.22–25. In the first occurrence the noun is διαίρεσις "division" and the adverb is missing.
[327] Four participles are repeated, identical and in the same order, in XII.10 (*bis*), 11, 12.
[328] At *EOO* II, 368.22; δεικνύντες πάλιν "by proving again" within a potential proof.
[329] At *EOO* IV, 210.22–26, to be compared with **b** XII.10, ibid., 407.10.
[330] The use of ἐπειδή may also result from Atticizing habits: Federspiel 2008b, 539. This would only confirm that the authenticity of the clauses introduced by this subordinant is doubtful.
[331] The only occurrence of εἴπερ "since ... really" in the whole of the *Elements* is in the spurious material that completes X.9por, *EOO* III, 28.17; other occurrences are in *Optica* **A** 32 (likewise spurious) and **B** preface (*bis*), at *EOO* VII, 58.10, 144.7, and 150.22, respectively. On εἴπερ, see Wakker 1994, 315–329; on the particle περ in Homer, see the thorough, and much theoretically-oriented, study Bakker 1988 (on the fact that περ actually follows Wackernagel's law, see Ruijgh 1990, 216). Apart from the nearly 700 occurrences of ὅπερ in the QED clause, in the "reference to the impossible" in a reduction (see Sect. 5.2.1), and in some specific formulae (see Sect. 5.3.4), ὅσπερ "which ... really" figures in III.25 (*bis*), VII.6 (*bis*), 8 (*sexties*), 11, 12, 20 (all in the formula ὁ EB τοῦ ZΔ τὸ αὐτὸ μέρος ἐστὶν ἢ μέρη, ἅπερ ὁ AB τοῦ ΓΔ "EB of ZΔ is really the same part or parts that AB of ΓΔ"; for the "parts", see Sects. 4.5.1.3 and 5.3.4), X.9, 44, XIII.4; on ὅσπερ, see Monteil 1963, 160–172, and Wakker 1994, 320 n. 40. The other intensive morph of the relative pronoun, ὅστις "which" (for which see Monteil 1963, 124–159), can be found in I.def. 4 (straight line), 7 (plane), 17 (diameter), 23 (parallel straight lines), I.42; III.def.2 (straight line tangent to a circle), 3 (tangent circles), III.16, XI.38 *vulgo*, XII.15 and 15**b**, 17, XIII.2*alt*. It is obvious that the intensive form is well suited to definitions. The relative pronoun ὅς itself exhibits nearly 700 occurrences.
[332] On regressive γάρ (in this function, a "push" particle that temporarily self-embeds the discourse at a lower level), in fact a typical resource of Greek prose style, see Sicking, van Ophuijsen 1993, 20–25, De Jong 1997, Bakker 2009b, and Netz 2001 in an Aristotelian context. Apollonius Dyscolus has a long section of his *On connectors* on regressive γάρ: *GG* II.1.1, 239.9–241.29; his account shows the difficulty the ancient grammarians had in categorizing scope particles.

All these trivializing features are actualized in the following postposed explanation, drawn from proposition IV.15 (*EOO* I, 314.12–17)—we are clumsily explained for what reason the angles of an equilateral triangle are equal:

ἰσόπλευρον ἄρα ἐστὶ τὸ ΕΗΔ τρίγωνον· καὶ αἱ τρεῖς ἄρα αὐτοῦ γωνίαι αἱ ὑπὸ ΕΗΔ ΗΔΕ ΔΕΗ ἴσαι ἀλλήλαις εἰσίν – ἐπειδήπερ τῶν ἰσοσκελῶν τριγώνων αἱ πρὸς τῇ βάσει γωνίαι ἴσαι ἀλλήλαις εἰσίν, καί εἰσιν αἱ τρεῖς τοῦ τριγώνου γωνίαι δυσὶν ὀρθαῖς ἴσαι.

Therefore triangle ΕΗΔ is equilateral: therefore its three angles ΕΗΔ, ΗΔΕ, ΔΕΗ are also equal to one another—since the angles at the base of isosceles triangles are really quite equal to one another, and the three angles of the triangle are equal to two right ‹angles›.

Instances of nested postposed explanations are sometimes found. They may belong to different species, as in XII.8 (*EOO* IV, 178.24–28):

ὡς δὲ τὸ ΒΗΜΛ στερεὸν πρὸς τὸ ΕΘΠΟ στερεόν, οὕτως ἡ ΑΒΓΗ πυραμὶς πρὸς τὴν ΔΕΖΘ πυραμίδα – ἐπειδήπερ ἡ πυραμὶς ἕκτον μέρος ἐστὶ τοῦ στερεοῦ διὰ τὸ καὶ τὸ πρίσμα, ἥμισυ ὂν τοῦ στερεοῦ παραλληλεπιπέδου, τριπλάσιον εἶναι τῆς πυραμίδος.

And, as solid ΒΗΜΛ is to solid ΕΘΠΟ, so pyramid ΑΒΓΗ is to pyramid ΔΕΖΘ—since the pyramid is really quite a sixth part of the solid because the prism, which is half of the parallelepipedal solid, is also triple of the pyramid.

A postposed explanation may even contain an entire deduction with nested conditional, as in XII.2 (*EOO* IV, 142.9–17)—this yields a juxtaposition of connectors, an aberrant syntax, and the final repetition of the statement to be explained:[333]

τὸ δὴ ἐγγεγραμμένον τετράγωνον μεῖζόν ἐστιν ἢ τὸ ἥμισυ τοῦ ΕΖΗΘ κύκλου – ἐπειδήπερ, ἐὰν διὰ τῶν Ε Ζ Η Θ σημείων ἐφαπτομένας εὐθείας τοῦ κύκλου ἀγάγωμεν, τοῦ περιγραφομένου περὶ τὸν κύκλον τετραγώνου ἥμισύ ἐστι τὸ ΕΖΗΘ τετράγωνον· τοῦ δὲ περιγραφέντος τετραγώνου ἐλάττων ἐστὶν ὁ κύκλος· ὥστε τὸ ΕΖΗΘ ἐγγεγραμμένον τετράγωνον μεῖζόν ἐστι τοῦ ἡμίσεως τοῦ ΕΖΗΘ κύκλου.

Thus the square that turns out to be inscribed is greater than the half of circle ΕΖΗΘ—since, if through points Ε, Ζ, Η, Θ we draw straight lines tangent to the circle, square ΕΖΗΘ is really quite half of the square circumscribed to the circle; and the circle is less than the circumscribed square; so that the square that turns out to be inscribed is greater than the half of circle ΕΖΗΘ.

A comparison with the Arabo-Latin tradition suggests that most of the postposed explanations introduced by intensive forms of ἐπεί "since" are interpolations: for instance, none of the 38 occurrences of ἐπειδήπερ "since ... really quite" in the main text of the *Elements* is attested in the indirect tradition.[334] We thus have every reason to think that a very small, original sample of postposed arguments has gradually been enriched. Several scholarly actions may have contributed to the

[333] Note also the personal verb form ἀγάγωμεν "we draw" and the identical denotative letters that name two different objects, a circle and a square.

[334] We find them in III.16por, IV.3, 15, V.8 (*bis*), 12, VI.28, X.9por, 23por, 28/29I, 73, XI.1, 8, 26, 33, XII.2 (*bis*), 3, 7por, 8, 10 (*ter*), 11–12, 17 (*quater*), 18, XIII.8, 13 (*ter*), 16, 17 (*bis*), 18. Two further occurrences can be found in the alternative proofs, and others were relegated by Heiberg in the critical apparatus. Add the occurrence in *Data* 49, and the two instances of ἐπείπερ "since ... really" in VI.19por and XI.33por.

phenomenon: later revisions, collation of exemplars operated by the copyists, or, simply and most likely, marginal annotations that have found their way into the text.[335]

4.5.4. *Instantiated and non-instantiated citations of theorems*

We have seen in Sect. 2.2 that instantiated and non-instantiated citations of previous results constitute the long-range anaphoric structure of the *Elements*. In particular, the formulation of a constructive act conforms to a template that coincides either with one of the postulates or with the (instantiated) conclusion of the problem that licenses the constructive act. This Section presents some examples of citations of theorems; the template is in this case the enunciation.[336] The examples are graded according to the degree of conformity to the original. As is to be expected, such citations are mostly found in paraconditionals; as a consequence, the syntactic structure of the original statement is kept almost unchanged, if this was in conditional form.

Let us start with a word-for-word citation that is certainly spurious, as is amply confirmed by the striking, "bookish" initial reference to the theorem referred to;[337] this citation is found in proposition XII.2 (*EOO* IV, 144.10–16):

ἐδείχθη γὰρ ἐν τῷ πρώτῳ θεωρήματι τοῦ δεκάτου βιβλίου ὅτι, δύο μεγεθῶν ἀνίσων ἐκκειμένων, ἐὰν ἀπὸ τοῦ μείζονος ἀφαιρεθῇ μεῖζον ἢ τὸ ἥμισυ καὶ τοῦ καταλειπομένου μεῖζον ἢ τὸ ἥμισυ καὶ τοῦτο ἀεὶ γίγνηται, λειφθήσεταί τι μέγεθος ὃ ἔσται ἔλασσον τοῦ ἐκκειμένου ἐλάσσονος μεγέθους.

For it was proved in the first theorem of the tenth book that, two unequal magnitudes being set out, if from the greater a ‹magnitude› greater than the half be removed and from the remainder one greater than the half and this come about continually, some magnitude will have remained that will be less than the lesser magnitude set out.

The SAS criterion of congruence of triangles has a complex enunciation and a widespread application. The several forms of its citations deserve a detailed discussion. Let us read the template in I.4 (*EOO* I, 16.9–16):[338]

ἐὰν δύο τρίγωνα τὰς δύο πλευρὰς ταῖς δυσὶ πλευραῖς ἴσας ἔχῃ ἑκατέραν ἑκατέρᾳ καὶ τὴν γωνίαν τῇ γωνίᾳ ἴσην ἔχῃ τὴν ὑπὸ τῶν ἴσων εὐθειῶν περιεχομένην, καὶ τὴν βάσιν τῇ βάσει ἴσην ἕξει, καὶ τὸ τρίγωνον τῷ τριγώνῳ ἴσον ἔσται, καὶ αἱ λοιπαὶ γωνίαι ταῖς λοιπαῖς γωνίαις ἴσαι ἔσονται ἑκατέρα ἑκατέρᾳ, ὑφ᾽ ἃς αἱ ἴσαι πλευραὶ ὑποτείνουσιν.

If two triangles have two sides equal to two sides, respectively, and the angle contained by the equal straight lines equal to the angle, they will also have the base equal to the base, and the triangle will be equal to the triangle, and the remaining angles, those under which the equal sides extend, will be equal to the remaining angles, respectively.

---

[335] For a detailed analysis of the phenomenon of interpolations in Greek mathematical texts, see Vitrac 1990–2001 IV, 32–71. On ancient and Byzantine recensions of Greek mathematical texts, see Acerbi 2016.

[336] Recall that the general conclusion of a theorem, if it is present, is identical to the enunciation.

[337] There are a handful of such references in the *Elements*, see Sect. 4.5.5. As for the inauthenticity of this reference, see Vitrac 1990–2001 IV, 266 n. 15.

[338] Two remarks on the translation: the Greek article that precedes the numerals "two" is a standard feature of the mathematical code, and says nothing as to the definite character of the expression; the final relative clause only determines αἱ λοιπαὶ γωνίαι. On the latter point see Federspiel 2003, 332–333.

The first citation of the enunciation of theorem I.4 in I.5 keeps almost all linguistic units of the original. The exceptions are: the two occurrences of "sides" are omitted and replaced by denotative letters; the reference to the equal angles contained by the equal sides is simplified since such angles coincide. On the other hand, the syntactic structure is radically modified. The original, very long, conditional statement features two conjuncts in the antecedent and three conjuncts in the consequent. In the citation, this sentential structure is transformed into a deduction that comprises a paraconditional, a coassumption, and a conclusion. The conclusion embodies the three conjuncts that figure in the consequent of the original conditional;[339] the coassumption contains the second of the conjuncts that figure in the antecedent of the same conditional.

The presence of the paraconditional in the citation of I.4 in I.5 is motivated by the phenomenon of "alignment" I have discussed in Sects. 4.5.1.3–4. The point is that, in I.5, the two equalities between the relevant sides of the triangle are stated separately: AZ = AH in the construction; AB = AΓ in the setting-out, as this is one of the assumptions of the theorem.[340] But the enunciation of I.4 subsumes the two equalities under one and the same statement, namely ἐὰν δύο τρίγωνα τὰς δύο πλευρὰς ταῖς δυσὶ πλευραῖς ἴσας ἔχῃ ἑκατέραν ἑκατέρᾳ "if two triangles have two sides equal to two sides, respectively". Therefore, a deductive step is needed that "aligns" the two-statement form of the equalities in I.5 with the one-statement form of the template. The result is a hybrid paraconditional, whose consequent is introduced by δή "thus", the particle that marks sentences formulating statements that obviously derive from the assigned conditions (cf. Sects. 5.3.2 and 5.3.6). In the citation, which we finally read below, note the long segment of text without denotative letters and the two occurrences of ἑκατέρα ἑκατέρᾳ "respectively";[341] this syntagm is devised to make the pairs of terms ordered pairs and cannot be eliminated (I.5, *EOO* I, 20.15–23):

ἐπεὶ οὖν ἴση ἐστὶν ἡ μὲν AZ τῇ AH ἡ δὲ AB τῇ AΓ, δύο <u>δὴ</u> αἱ ZA AΓ δυσὶ ταῖς HA AB ἴσαι εἰσὶν <u>ἑκατέρα ἑκατέρᾳ</u>· καὶ γωνίαν κοινὴν περιέχουσι τὴν ὑπὸ ZAH· βάσις ἄρα ἡ ZΓ βάσει τῇ HB ἴση ἐστίν, καὶ τὸ AZΓ τρίγωνον τῷ AHB τριγώνῳ ἴσον ἔσται, <u>καὶ αἱ λοιπαὶ γωνίαι ταῖς λοιπαῖς γωνίαις ἴσαι ἔσονται ἑκατέρα ἑκατέρᾳ, ὑφ' ἃς αἱ ἴσαι πλευραὶ ὑποτείνουσιν</u>, ἡ μὲν ὑπὸ AΓZ τῇ ὑπὸ ABH ἡ δὲ ὑπὸ AZΓ τῇ ὑπὸ AHB.

Then since AZ is equal to AH and AB to AΓ, <u>thus</u> two ‹sides›, ZA, AΓ, are equal to two ‹sides›, HA, AB, <u>respectively</u>; and they contain a common angle, ZAH; therefore base ZΓ is equal to base HB, and triangle AZΓ will be equal to triangle AHB, <u>and the remaining angles, those under which the equal sides extend, will be equal to the remaining angles, respectively</u>, AΓZ to ABH and AZΓ to AHB.

Similarly shaped is the citation of I.8 in I.9.[342] Because of the absence of any long segment of text without denotative letters (underlined above), this might seem to be less exactly a word-for-word reference than the reference to I.4 in I.5 just read. As a matter of fact, the two original enunciations have undergone the same linguistic transformations. Let us read the enunciation of proposition I.8 (*EOO* I, 26.13–17) and its first citation in I.9 (*EOO* I, 30.1–5)—note again the persistence of the distributive syntagm ἑκατέρα ἑκατέρᾳ:

---

[339] In this case the only linguistic transformation from the template to the citation amounts to adding the denotative letters.
[340] This happens at *EOO* I, 20.12–13 and 20.6–7, respectively.
[341] See also the second example discussed in Sect. 4.4.
[342] Proposition I.8 is the SSS criterion of congruence of triangles.

ἐὰν δύο τρίγωνα τὰς δύο πλευρὰς ταῖς δύο πλευραῖς ἴσας ἔχῃ ἑκατέραν ἑκατέρᾳ, ἔχῃ δὲ καὶ τὴν βάσιν τῇ βάσει ἴσην, καὶ τὴν γωνίαν τῇ γωνίᾳ ἴσην ἕξει τὴν ὑπὸ τῶν ἴσων εὐθειῶν περιεχομένην.

If two triangles have two sides equal to two sides, respectively, and also have the base equal to the base, they will also have the angle contained by the equal straight lines equal to the angle.

ἐπεὶ γὰρ ἴση ἐστὶν ἡ ΑΔ τῇ ΑΕ κοινὴ δὲ ἡ ΑΖ, δύο δὴ αἱ ΔΑ ΑΖ δυσὶ ταῖς ΕΑ ΑΖ ἴσαι εἰσὶν ἑκατέρα ἑκατέρᾳ· καὶ βάσις ἡ ΔΖ βάσει τῇ ΕΖ ἴση ἐστίν· γωνία ἄρα ἡ ὑπὸ ΔΑΖ γωνίᾳ τῇ ὑπὸ ΕΑΖ ἴση ἐστίν.

In fact, since ΑΔ is equal to ΑΕ and ΑΖ is common, thus two ‹sides›, ΔΑ, ΑΖ, are equal to two ‹sides›, ΕΑ, ΑΖ, respectively; and base ΔΖ is equal to base ΕΖ; therefore angle ΔΑΖ is equal to angle ΕΑΖ.

One of the most concise citations of the enunciation of I.4 is in XI.4 (*EOO* IV, 14.5–9):

καὶ ἐπεὶ δύο αἱ ΑΕ ΕΔ δυσὶ ταῖς ΓΕ ΕΒ ἴσαι εἰσὶ καὶ γωνίας ἴσας περιέχουσιν, βάσις ἄρα ἡ ΑΔ βάσει τῇ ΓΒ ἴση ἐστίν, καὶ τὸ ΑΕΔ τρίγωνον τῷ ΓΕΒ τριγώνῳ ἴσον ἔσται· ὥστε καὶ γωνία ἡ ὑπὸ ΔΑΕ γωνίᾳ τῇ ὑπὸ ΕΒΓ ἴση ἐστίν.

And since two ‹sides›, ΑΕ, ΕΔ, are equal to two ‹sides›, ΓΕ, ΕΒ, and they contain equal angles, therefore base ΑΔ is equal to base ΓΒ, and triangle ΑΕΔ will be equal to triangle ΓΕΒ; so that angle ΔΑΕ is also equal to angle ΕΒΓ.

A beautiful example of a word-for-word citation, still completely instantiated, is the reference to the enunciation of V.2 in V.3 (*EOO* II, 12.23–14.3)—to retrieve the template, it is enough to omit the denotative letters:

ἐπεὶ οὖν πρῶτον τὸ ΕΚ δευτέρου τοῦ Β ἰσάκις ἐστὶ πολλαπλάσιον καὶ τρίτον τὸ ΗΛ τετάρτου τοῦ Δ ἔστι δὲ καὶ πέμπτον τὸ ΚΖ δευτέρου τοῦ Β ἰσάκις πολλαπλάσιον καὶ ἕκτον τὸ ΛΘ τετάρτου τοῦ Δ, καὶ συντεθὲν ἄρα πρῶτον καὶ πέμπτον τὸ ΕΖ δευτέρου τοῦ Β ἰσάκις ἐστὶ πολλαπλάσιον καὶ τρίτον καὶ ἕκτον τὸ ΗΘ τετάρτου τοῦ Δ.

Then since first, ΕΚ, of second, Β, and third, ΗΛ, of fourth, Δ, is equimultiple, and fifth, ΚΖ, of second, Β, and sixth, ΛΘ, of fourth, Δ, is also equimultiple, therefore first and fifth compounded, ΕΖ, of second, Β, and third and sixth, ΗΘ, of fourth, Δ, will also be equimultiple.

Since the original enunciations of the relevant theorems of Book V are very short, all operations on ratios must be regarded as citations of them. However, conformity to the original formulation may vary: the range is from a literal citation, as in the first retrieval of the enunciation of V.17 in proposition V.18 (*EOO* II, 52.12–15),

καὶ ἐπεί ἐστιν ὡς τὸ ΑΒ πρὸς τὸ ΒΕ οὕτως τὸ ΓΔ πρὸς τὸ ΔΗ, συγκείμενα μεγέθη ἀνάλογόν ἐστιν· ὥστε καὶ διαιρεθέντα ἀνάλογον ἔσται.

And since, as ΑΒ is to ΒΕ, so ΓΔ is to ΔΗ, compounded magnitudes are in proportion; so that they will also be in proportion divided.

to the canonical format for this kind of reference: this is the standard shortcut that identifies the operation, as for instance the relational dative διελόντι "by division" that marks V.17, a proposition that is cited in such concise a way in X.14 (*EOO* III, 42.10–14):[343]

---

[343] The standard shortcuts identifying these operations are: ἀνάπαλιν "by inversion" (V.7por), ἐναλλάξ "by alternation" (V.16), διελόντι "by division" (V.17), συνθέντι "by composition" (V.18), ἀναστρέψαντι "by conversion" (V.19por), δι' ἴσου "through an equal" (V.22). So, there are three relational datives and three adverbs, one of which is a prepositional expression. The three adverbs are found in the enunciations in which the operations are first introduced. The same is the

ἔστιν ἄρα ὡς τὰ ἀπὸ τῶν Ε Β πρὸς τὸ ἀπὸ τῆς Β οὕτως τὰ ἀπὸ τῶν Δ Ζ πρὸς τὸ ἀπὸ τῆς Δ· διελόντι ἄρα ἐστὶν ὡς τὸ ἀπὸ τῆς Ε πρὸς τὸ ἀπὸ τῆς Β οὕτως τὸ ἀπὸ τῆς Ζ πρὸς τὸ ἀπὸ τῆς Δ.

therefore, as the ‹squares› on E, B are to that on B, so those on Δ, Z are to that on Δ; therefore, by division, as that on E is to that on B, so that on Z is to that on Δ.

It may happen that the stem of a verb form is changed when the enunciation is cited. Let us read the citation of II.5 in III.35 (*EOO* I, 258.20–23), a case in which the original passive aorist in the antecedent of the conditional is changed to a perfect stem in the antecedent of the paraconditional:

ἐπεὶ οὖν εὐθεῖα ἡ ΑΓ τέτμηται εἰς μὲν ἴσα κατὰ τὸ Η εἰς δὲ ἄνισα κατὰ τὸ Ε, τὸ ἄρα ὑπὸ τῶν ΑΕ ΕΓ περιεχόμενον ὀρθογώνιον μετὰ τοῦ ἀπὸ τῆς ΕΗ τετραγώνου ἴσον ἐστὶ τῷ ἀπὸ τῆς ΗΓ.

Then since a straight line, AΓ, turns out to be cut in equal ‹segments› at H and in unequal ones at E, therefore the rectangle contained by AE, EΓ with the square on EH is equal to that on HΓ.

A few lines later in III.35 (*EOO* I, 258.26–27), we may read a reference to I.47, in its typical, very compressed, form:

ἀλλὰ τοῖς μὲν ἀπὸ τῶν ΕΗ ΗΖ ἴσον ἐστὶ τὸ ἀπὸ τῆς ΖΕ.

But equal to those on EH, HZ is that on ZE.

4.5.5. *Assumptions and coassumptions*

A complete demonstrative argument normally begins with a paraconditional and ends with a conclusion, marked by the presence of ἄρα "therefore". The core of the argument is a connected sequence of atomic inferences assumption–coassumption–conclusion;[344] within this core, the conclusion of a self-contained deductive step serves as the assumption of the subsequent step. As a consequence, if we exclude the liminal paraconditional, fresh deductive material can only be fed into a proof by means of the coassumptions. There are in fact no one-premise arguments in mathematics: the conclusion is always drawn in virtue of a primary assumption and of a previous result that supplements it (see Sect. 4.5.1.1). It may happen—and it does frequently happen—that the reference to the previous result is absent: but this only means that it is understood. The conclusion of such an enthymematic inference is preferably introduced by ὥστε "so that", even if instances of two consecutive ἄρα are not infrequent (see Sect. 5.3.6).

Most coassumptions are introduced by a coordinative δέ "and".[345] This is also the main use of the conjunction ἀλλά "but". Peculiar forms of coassumption are the analogical and potential proofs, as well as minimal constructive acts like those introduced by ἀφῃρήσθω "let it be removed" and by προκείσθω "let it be added". It may happen that the reference to a result previously supposed or proved within the ongoing proof is marked by the presence of ὑπόκειται "it has been supposed", ἐδείχθη "it was proved", or the like. A part of these species we read at work in III.35 (*EOO* I, 258.24–260.13)—the coassumptions are underlined:

case for the dative ἀναστρέψαντι; a different verb form is instead present in the other two cases: these are διαιρεθέντα and συντεθέντα, respectively, which qualify the magnitudes involved in the operation.

[344] Cf. D.L. VII.76, where Crinis' definition of a λόγος "argument" is reported: τὸ συνεστηκὸς ἐκ λήμματος καὶ προσλήψεως καὶ ἐπιφορᾶς "what is made of an assumption, a coassumption, and a conclusion"; see also below and Sect. 5.1.6, where we shall read the entire passage.

[345] For a general assessment of δέ as "boundary-marker", see Bakker 1993.

τὸ ἄρα ὑπὸ τῶν ΑΕ ΕΓ μετὰ τῶν ἀπὸ τῶν ΗΕ ΗΖ ἴσον ἐστὶ τοῖς ἀπὸ τῶν ΓΗ ΗΖ· ἀλλὰ τοῖς μὲν ἀπὸ τῶν ΕΗ ΗΖ ἴσον ἐστὶ τὸ ἀπὸ τῆς ΖΕ, τοῖς δὲ ἀπὸ τῶν ΓΗ ΗΖ ἴσον ἐστὶ τὸ ἀπὸ τῆς ΖΓ· τὸ ἄρα ὑπὸ τῶν ΑΕ ΕΓ μετὰ τοῦ ἀπὸ τῆς ΖΕ ἴσον ἐστὶ τῷ ἀπὸ τῆς ΖΓ· ἴση δὲ ἡ ΖΓ τῇ ΖΒ· τὸ ἄρα ὑπὸ τῶν ΑΕ ΕΓ μετὰ τοῦ ἀπὸ τῆς ΕΖ ἴσον ἐστὶ τῷ ἀπὸ τῆς ΖΒ. διὰ τὰ αὐτὰ δὴ καὶ τὸ ὑπὸ τῶν ΔΕ ΕΒ μετὰ τοῦ ἀπὸ τῆς ΖΕ ἴσον ἐστὶ τῷ ἀπὸ τῆς ΖΒ· ἐδείχθη δὲ καὶ τὸ ὑπὸ τῶν ΑΕ ΕΓ μετὰ τοῦ ἀπὸ τῆς ΖΕ ἴσον τῷ ἀπὸ τῆς ΖΒ· τὸ ἄρα ὑπὸ τῶν ΑΕ ΕΓ μετὰ τοῦ ἀπὸ τῆς ΖΕ ἴσον ἐστὶ τῷ ὑπὸ τῶν ΔΕ ΕΒ μετὰ τοῦ ἀπὸ τῆς ΖΕ· κοινὸν ἀφῃρήσθω τὸ ἀπὸ τῆς ΖΕ· λοιπὸν ἄρα τὸ ὑπὸ τῶν ΑΕ ΕΓ περιεχόμενον ὀρθογώνιον ἴσον ἐστὶ τῷ ὑπὸ τῶν ΔΕ ΕΒ περιεχομένῳ ὀρθογωνίῳ.

therefore the ‹rectangle contained› by AE, EΓ with the ‹squares› on HE, HZ is equal to those on ΓH, HZ; but that on ZE is equal to those on EH, HZ, that on ZΓ is equal to those on ΓH, HZ; therefore that by AE, EΓ with that on ZE is equal to that on ZΓ; and ZΓ us equal to ZB; therefore that by AE, EΓ with that on EZ is equal to that on ZB. For the very same ‹reasons› that by ΔE, EB with that on ZE is also equal to that on ZB; and that by AE, EΓ with that on ZE was also proved equal to that on ZB; therefore that by AE, EΓ with that on ZE is equal to that by ΔE, EB with that on ZE; let that on ZE be removed as common; therefore the rectangle contained by AE, EΓ as a remainder is equal to the rectangle contained by ΔE, EB.

The internal reference introduced by ἐδείχθη points to a result proved a couple of lines before.

Let us see the use of ἀλλά in detail. In the main text of the *Elements* there are 400 occurrences, with the following distribution:

	I	II	III	IV	V	VI	VII	VIII	IX	X	XI	XII	XIII	tot.
# prop.	48	14	37	16	25	33	39	27	36	115	39	18	18	465
% # signs	7.6	3.3	6.9	3.4	4.9	7.6	5.7	5	5.1	26.1	9	8.3	6.9	100
ἀλλά	28	20	29	10	11	47	8	19	32	102	37	39	18	400
% ἀλλά	7	5	7.25	2.5	2.75	11.75	2	4.75	8	25.5	9.25	9.75	4.5	100

Even if the particle does not occur in one context only, we may regard it as a (negative) stylistic marker for Books V and, especially, VII; otherwise, its frequency is approximately proportional to the frequency of signs in each Book, that is, to the deductive density.

As said, the primary context of use of ἀλλά is introducing coassumptions. In a minority of cases, it is followed by the emphatic particle μήν "of course";[346] the nexus thereby assumes the form that seems recommended in Stoic logic.[347] Negative coassumptions can be introduced by οὐδὲ μήν "nor of course".[348] If the context requires it, ἀλλά is accompanied by an adverbial καί.[349] In most cases, however, ἀλλά is directly followed by an article, or, when it precedes a relation of proportionality, is part of the syntagm ἀλλ' ὡς "but, as".[350] The main exception to these two basic nexuses is the expression ἀλλ' εἰ δυνατόν "but if possible" (see Sect. 5.2.1 for detail); the other exceptions are

[346] There are 38 occurrences in the whole of the *Elements*, in propositions I.4, I.19, 25, III.16, V.10 (*bis*), VI.8, 20, VII.19 (*bis*), 24, 30, VIII.5, 19 (*bis*), IX.3, 12 (*quater*), 13 (*ter*), 15, 18 (*bis*), 19, 36, X.6, 41/42, XI.4, XII.6, **b**2 (*bis*), 10–11, XIII.4, 10, 17, 18, XII.4*alt*, and moreover also in **b**XII.2 (*bis*), 10–11, There are 2 occurrences in the *Data*, in propositions 51 and 68*alt*.

[347] Cf. again D.L. VII.76, in Crinis' definition of a λόγος "argument". I qualified with "seems" since two examples of an argument (in fact, a first indemonstrable: see Sect. 5.2.2) are adduced: the first has δέ in the coassumption (as most of the examples of a first indemonstrable in ancient sources), the second has ἀλλὰ μήν. It is true that the second example is the τρόπος "mode" of a first indemonstrable (see Sects. 5.1.6 and 5.2.2 for this notion).

[348] In I.19, 25, III.16, V.10 (*bis*), XIII.18. The only occurrence of μήν that does not accompany another particle is in X.51.

[349] There are 49 occurrences in the whole of the *Elements*, 18 in the *Data*.

[350] There are 71 occurrences in the main text of the *Elements*.

coassumptions in which ἀλλά is followed by a verb[351] or by disparate syntagms like συναμφότερ– in V.8, X.17–18, μείζονα καθ' ὑπόθεσιν "greater by supposition" in X.44, 47. The only noun that directly follows ἀλλά is πυραμίς "pyramid" in XII.3. To conclude this overview, note the two strictly parallel coassumptions in V.11 and 17: they are formulated ἀλλὰ εἰ ὑπερεῖχε "but if it exceeded", so that they contain two of the scarce occurrences of verb forms in the imperfect.[352]

The secondary context of use of ἀλλά can broadly be regarded as coassumptive: this is the nexus ἀλλὰ δή "but now", which introduces the further cases of a multi-layered proof. This may happen either because several cases are required by a multiple enunciation, as in I.26, or because a partition into cases is necessary since the proof sets out several geometric configurations. In the main text of the *Elements* there are 42 occurrences of ἀλλὰ δή;[353] they are distributed in the following way between the two types just described (I neglect the occurrence in X.18/19; with asterisk are the forms ἀλλὰ δὴ πάλιν):

multiple enunciation	I.26*, III.3, 14, VI.2–3, 7*, 14–17, 22, VIII.15, 17, IX.9–10, X.9 (*bis*), 11, 15–17, XI.37, XII.9, 15, XIII.7
several configurations	III.33 (*bis*), 36, IV.5 (*bis*), V.8, IX.18 (*bis*), 19* (*ter*), 20, X.71 (*ter*), 72, XII.15

Let us discuss now the coassumptions that have the form of references to results proved earlier in a proof. As for those that contain forms of ὑπόκειμαι "to have been supposed", let us get again a look at the table set out in Sect. 4.2, where the specific parts of a proposition are listed to which such coassumptions referer, whether affirmative or negative, within direct or indirect proofs. To repeat the table caption, if a proposition is divided into cases or has a multiple enunciation, all references to the partial setting-outs are collected under the heading "main setting-out". The "local setting-out" is the setting-out that precedes an indirect argument, and which introduces new entities with respect to the main setting-out. An asterisk marks the occurrences within direct proofs; the double slash precedes the additional material.

main setting-out	I.26, 29, 48*, V.5*, 6*, 18, 19*, VI.3*, 5*, 6 (*bis*)*, 7 (*ter*), 22*, VII.2, 7*, 20, 33, IX.10 (*bis*), 14, 30, 34, X.9por, 9/10, 21*, 37*, 38*, 41/42, 47, XI.5, 16, 23 (*bis*), 34 (*bis*), 35*, XII.4/5*, XIII.2/3, 7*(*bis*) // X.13, 39
local setting-out	IX.12, 13, 18, 20, 36, X.16 (*bis*)
construction	IV.10*, IX.20, X.33*, XI.23, 23* (*bis*), 26* // X.54*, 55*
suppositions in indirect argument (spurious)	X.42, XII.12 // X.28/29II (*bis*)

In the canonical practice, then, the verb ὑπόκειμαι refers without exceptions to a supposition introduced in the setting-out or in the construction.

The statements that are characterized by the presence of forms of δείκνυμι "to prove" refer to results proved within the ongoing proof or, less frequently, in previous propositions. The following tables set out the distribution of their occurrences, according to verb form, particle or term that accompany the coassumption, "range" of the anaphora (internal range if nothing follows, otherwise the target-proposition is indicated in brackets). Let us see first the less frequent forms of δείκνυμι:

[351] The verbs are "to be" in XI.34 (*bis*), XII.15, XIII.7, παραλλάσσω "to fall beside" in I.8, τέμνω "to cut" in IV.13, δύναμαι "to be worth" in the spurious X.9por, where we also find ἀλλ' ἁπλῶς "but in general".

[352] For instance, forms of the imperfect of the verb "to be" only occur in I.19 (*bis*), 25 (*bis*), III.37, V.17, VIII.2, XII.2 (*bis*). They all are found within coassumptions, the last two of which are introduced by ἀλλά.

[353] There are 3 occurrences in *Data* 10–11, 44.

	δή (καί)	ἀλλά	πάλιν	other clauses
δεικνύντες			IX.13	
ἐδείκνυμεν				XII.10 (XII.2)
ἐδείχθαμεν				X.44 (X.41/42)
ἐδείχθησαν	I.13, 15, X.55, 67	I.21		

	ὅτι	reduction	other clauses
δέδεικται	VI.1, 33, IX.9 (8, *bis*, postposed verb), 18 (16), 19 (17), X.9/10 (VIII.26), 18/19 (9por), XI.25, XII.13, **b** XII.3 (XI.39), 12	**b**XII.2, 5, 10–11	X.111, XIII.17 (XI.38), *Data* 5

Some of the above clauses are obvious interpolations, like ὡς ἔμπροσθεν ἐδείκνυμεν "as we previously proved" in XII.10 and ὡς ἐπάνω ἐδείξαμεν "as we proved above" in X.44, or the participial nexus δεικνύντες πάλιν "proving again" in IX.13. In the second table above, note the construct δέδεικται + participle in X.111, ὅπερ ἀδύνατον δέδεικται "which has really been proved impossible" in **b** XII.2, 5, 10–11, τοῦτο γὰρ δέδεικται ἐν τῷ παρατελεύτῳ θεωρήματι τοῦ ἑνδεκάτου βιβλίου "for this has been proved in the penultimate theorem of the eleventh book" in XIII.17, ὡς δέδεικται "as has been proved" in *Data* 5.

Let us now pass to the most frequent form, namely, ἐδείχθη. In the table below, an asterisk marks occurrences of ἐδείχθη δὲ ὅτι "and it was proved that"; the row headed "parac." records occurrences within paraconditionals:

δέ (καί)	I.1, 2, 5 (*bis*), 7, 19*, 25*, 30, 32–34, 46, II.4, 8 (*bis*), 10, III.4–6, 10, 12, 13* (*bis*), 35, IV.6–7, 11–12, 15, V.10* (*bis*), 19, 23, VI.3, 10, 18, 20por (*bis*), 32, VII.8, VIII.5, 9–10, 18, IX.8, 34, X.5–6, 28/29II*, 33, 53, XI.15, 17, 20, 23*-24, XII.2*, 5*, 10*-11*, 12* (*bis*), 17–18*, XIII.1, 2 (*bis*), 6, 7 (*bis*), 9–10, 14 (*bis*), 15–17, **b**XII.2*, 5*, 9*, 10 (*bis*), 10*, 11*, 17*, *Data* 42
ὅτι	III.16 (postposed verb), X.28/29I
ἀλλά	I.21, III.15, 25, IV.15, VI.8, 24, 28, VII.10, VIII.19, XI.31, XII.3, XIII.3–4, 8–9, 13
other	III.32, VI.7,19, VIII.19, IX.36, X.44, 91–92, 107, XI.4 (*bis*), 18, 20, 35, XII.2, 5, 7, 18, XIII.11
parac.	I.5, IV.12, V.19, VI.1, 4, 23–24, X.9por, 93 (*ter*), 94–95, 96 (*ter*), XI.4 (*bis*), XII.12, XIII.16, 18
reduction	III.8, IV.4, 8, 13, X.84, XI.2, XII.2, 5, 11–12, 18

Of course, the last two rows do not list coassumptions.[354] The heading "other" includes the following clauses: relative clause in III.32, X.44, XI.20; clause introduced by καί, and therefore postposed verb, in VI.7 (ἐδείχθη + participle), IX.36, XI.4 (*bis*), 18, 35; clause introduced by δέ, and postposed verb, in X.92, XII.7 (*bis*), XIII.11. Certainly spurious are the decidedly metamathematical turns of phrase in VI.19 (ἐπείπερ ἐδείχθη ὡς "since it was really proved that"), VIII.19 (ὡς ἐν τῷ πρὸ τούτου θεωρήματι ἐδείχθη "as it was proved in the theorem before the present one"), X.91 (ὡς ἐν τοῖς ἔμπροσθεν ἐδείχθη "as it was proved in the previous ones"), 107 (ὡς ἐδείχθη "as it was proved"), XII.2 (ἐδείχθη γὰρ ἐν τῷ πρώτῳ θεωρήματι τοῦ δεκάτου βιβλίου "for it was proved in the first theorem of the tenth book"), XII.5, 18 (ὡς ἔμπροσθεν ἐδείχθη "as it was previously proved"). Note finally that, given the standard practice of understanding the verb "to be", a form ἐδείχθη can be inserted in a clause without modifying its syntax, a fact that makes this verb form suitable for punctual interpolations.

[354] On the presence of coordinative ἀλλά within the antecedent of a paraconditional, see Sect. 5.3.2 and the remarks in Federspiel 2008b, 543–545.

Forms of the participle προδεδειγμένος "what has previously been proved" are an obvious stylistic marker of Book X.[355] Unevenly distributed are also the occurrences of forms of the participle προειρημένος "previously said".[356] The participle προκείμενος "proposed" is again restricted to Book X;[357] it has the sole metamathematical function of curtailing some setting-outs, as we may see in the following extract from X.39 (*EOO* III, 114.4–11):

ἐὰν δύο εὐθεῖαι δυνάμει ἀσύμμετροι συντεθῶσι ποιοῦσαι τὸ μὲν συγκείμενον ἐκ τῶν ἀπ' αὐτῶν τετραγώνων ῥητὸν τὸ δ' ὑπ' αὐτῶν μέσον, ἡ ὅλη εὐθεῖα ἄλογός ἐστιν· καλείσθω δὲ μείζων.	If two straight lines incommensurable in power be compounded making the ‹region› compounded of the squares on them expressible and the one ‹contained› by them medial, the whole straight line is irrational; let it be called major.
συγκείσθωσαν γὰρ δύο εὐθεῖαι δυνάμει ἀσύμμετροι αἱ ΑΒ ΒΓ ποιοῦσαι τὰ προκείμενα. λέγω ὅτι ἄλογός ἐστιν ἡ ΑΓ.	In fact, let two straight lines incommensurable in power, ΑΒ, ΒΓ, be compounded making what has been proposed. I claim that ΑΓ is irrational.

Other metamathematical phrases that contain verbs with prefix προ– are in X.4por (τὸ πόρισμα προχωρεῖ "the porism proceeds") and X.41/42 (δείξομεν ἤδη προεκθέμενοι λημμάτιον τοιοῦτον "we shall prove by previously setting out such a little lemma"); note also the aberrant QED-like formulae in X.53/54 (ἃ προέκειτο δεῖξαι "which it had been proposed to prove") and XI.23por (ὅπερ προέκειτο ποιῆσαι "which it had really been proposed to do"). Internal to the object language are instead the sparse occurrences of forms of the participle προτεθείς "previously assigned". This is connected with the peculiar formulation of the theorem on the unboundedness of the class of prime numbers IX.20 (*ter*) and, of course, with the contexts of the theory of irrational lines in which an expressible is "previously assigned", as in X.def.3 (*bis*), def.4, X.10 (*ter*).

"Predictions" on proofs to be carried out are made by the other forms of δείκνυμι; after all, this is the role played by the verb in potential proofs (whose occurrences are here excluded):

δείκνυται	X.def.3
δείξομεν	VI.22/23, 41/42, XI.23/24, XII.12, XIII.18
δειχθήσεται	I.16, 47, IV.12, 14, VI.20porI, X.32, 64, XIII.13
δειχθήσονται	X.44

The phrases that include δειχθήσεται "it will be proved" are for the most part canonical clauses of potential proofs with a syntagm inserted immediately after the liminal ὁμοίως (δή): a genitive absolute in I.16 and 47, τῷ πρὸ τούτου "to that before the present one" in IV.14, πάλιν "again" in X.32, γάρ "for" in X.64. There remain the aberrant occurrences in IV.12 (διὰ τὰ αὐτὰ δειχθήσεται "for the same ‹reasons› it will be proved": it is an analogical proof, not a potential proof!), VI.20porI,[358] XIII.13 (ὡς ἑξῆς δειχθήσεται "as it will be subsequently proved"). In X.44 we find the only occurrence of an analogical proof introduced by the sequence κατὰ τὰ αὐτὰ δή "according to the very same ‹reasons›".

[355] In X.55, 56, 59, 62, 63, 65; the form δεδειγμένος occurs in X.9, 58, προδέδεικται in XIII.17.
[356] They can be found in II.3 (*bis*), X.19, 20, 77, 82–83, 84 (*bis*), XI.36 (*ter*), XIII.16, 17 (*bis*), 18. In II.3 and XI.36, the participle is found in the enunciation; this explains also the other occurrences in the same propositions.
[357] In X.39, 40, 41 (*bis*), 47, 76–78, 83 (*bis*), 110, plus 3 occurrences in the alternative proofs.
[358] Clause ὡσαύτως δὲ καί; the adverb ὡσαύτως "likewise" is also found in dubious potential proofs as those in V.8, 20porI, X.23por, XII.15. In V.15 (*bis*), 16, 23, VI.1, its presence is instead induced by the enunciation of V.15 (which is the template of these citations), where it has a well-defined mathematical meaning.

5. THE LOGICAL SYNTAX

This part of the book mainly studies the use of such linguistic units as have a logical import: these are all sorts of connectors and coordinants, qualifiers, adverbia, and noun phrases. The longest Section of this part (Sect. 5.1) treats the manyfold ways generality is explicitly implemented, in a mathematical proposition, by means of suitable determiners; in the same perspective, I shall also discuss the use of the article, and a possibly parallel linguistic structure in Stoic logic. A Section on modals (5.2) will allow us to see how indirect proofs are framed in Greek mathematics. The formation of typical non-simple statements by means of subordinants and coordinants is thoroughly analysed in Sects. 5.3.1 (conditional clauses), 5.3.2 ("paraconditional" clauses, namely, systems made of a causal subordinate and a principal clause), 5.3.4 (disjunctions), and 5.3.5 (conjunctions). In these Sections, frequent parallels are drawn between the mathematical practice and logical or grammatical doctrines. Similar parallels are also drawn in the residual Sections of this part, which deal with negation (5.3.3) and with syllogistic connectors (5.3.6).

5.1. Quantification; implicit and explicit generality

Galen, at *Inst. Log.* XII.5–8, is the only ancient Greek author who describes—with some philosophical overload, as becomes clear in his final remark—the several ways a general enunciation is formulated in a mathematical text:[1]

5 κατὰ συμβεβηκὸς τότε φαίνονταί τινες ἀποφάνσεις τε καὶ δείξεις εἶναι κατὰ μέρος· ὡς γὰρ πρὸς τὴν περὶ παντὸς τριγώνου δεῖξίν τε καὶ ἀπόφανσιν ὅτι δυοῖν ὀρθαῖς ἴσας ἔχει τὰς τρεῖς γωνίας, ἐπὶ μέρους δόξειεν ἂν εἶναι πρότασις λέγουσα μὴ "πᾶν τρίγωνον", ἀλλ' "ἔνια τὰς πρὸς τῇ βάσει γωνίας ἴσας ἀλλήλαις ἔχειν". **6** οὕτως μὲν οὖν ῥηθὲν οὔπω διωρισμένην οὐδ' ἐπιστημονικὴν ἔχει τὴν ἀπόφανσίν τε καὶ γνῶσιν, ἐκείνως δὲ ἐπιστημονικήν τε καὶ καθόλου "πᾶν ἰσοσκελὲς τρίγωνον τὰς πρὸς τῇ βάσει γωνίας ἴσας ἀλλήλαις ἔχει". **7** συνήθης δὲ τοῖς Ἕλλησι λέξις ἐστὶ καὶ ἡ διὰ τῆς τῶν ἀριθμῶν προτάξεως ἐνδεικνυμένη τὸ καθόλου καὶ ἡ ἄνευ τούτων· ὡσαύτως γὰρ ἔστιν εἰπεῖν "πᾶν ἰσοσκελὲς τρίγωνον τὰς πρὸς τῇ βάσει γωνίας ἴσας ἀλλήλαις ἔχει" ὡς "τὰ ἰσοσκελῆ τρίγωνα τὰς πρὸς τῇ βάσει γωνίας ἴσας ἀλλήλαις ἔχει". **8** καὶ μέντοι καὶ κατὰ τὸν ἑνικὸν ἀριθμὸν ἔθος ἐστὶ τοῖς Ἕλλησιν ἑρμηνεύειν τὰ οὕτω λεγόμενα, καὶ διαφέρει γε οὐδέν, εἰ "τὰ ἰσοσκελῆ τρίγωνα"

5 Some demonstrative enunciations sometimes appear, because of incidental features, to be particular: for, compared to the demonstrative enunciation about every triangle, that it has the three angles equal to two right ‹angles›, the enunciation would seem particular saying not that "every triangle" but that "some have the angles at the base equal to one another". **6** Worded in this way, then, neither the enunciation nor the associated knowledge are well defined or scientific, whereas, worded in that way: "every isosceles triangle has the angles at the base equal to one another", they are both scientific and general. **7** Still, it is customary for the Greeks both the wording intimating generality by preposing multiplicity and the wording without it: for to say "every isosceles triangle has the angles at the base equal to one another" is tantamount to say "the isosceles triangles have the angles at the base equal to one another". **8** But there is more: it is also customary for the Greeks to express what has been said in the singular, and there is no difference at

[1] But see also, in another context, Aristotle, *Int.* 7, 17a35–b16. I skip the first part of *Inst. Log.* XII.5, which is corrupt. I take it that the rest of the text, integrations included, is sound and conveys Galen's thought. Recall also Posidonius' text we have read in Sect. 1.5.

F. Acerbi, *The Logical Syntax of Greek Mathematics*, Sources and Studies in the History of Mathematics and Physical Sciences,
https://doi.org/10.1007/978-3-030-76959-8_5

λέγουσιν "ἅπαντα τὰς πρὸς τῇ βάσει γωνίας ἴσας ἀλλήλαις ἔχειν" ἢ "τὸ ἰσοσκελὲς τρίγωνον"· εἰς γὰρ τὸ εἶδος ἀποβλέποντες τὸ ἅπασι τοῖς κατὰ μέρος ὑπάρχον ὡς περὶ ἑνὸς εἰκότως ποιοῦνται τὴν ἀπόφανσιν· καὶ γάρ ἐστιν ὡς εἶδος ἕν.	all whether they say that "all isosceles triangles have the angles at the base equal to one another" or "the isosceles triangle": for looking at a form that belongs to all particulars, they rightly shape the enunciation as if ‹it were› about one—and in fact, as a form it is one.

Galen's remarks about what is "customary for the Greeks" point to a remarkable phenomenon: Greek mathematics seldom uses quantification. For about thirty theorems of the *Elements* have a quantified enunciation; all the others express generality by referring to one of the following items:

a) classes of objects taken as pluralities;
b) generic representatives of a specific class;
c) pairs or *n*-ples of objects.

Let us read one example for each category in this order; a quantified statement is also included, and it is placed first in the list of quotes: the examples are taken from propositions I.16, I.47, I.6, and I.15 (*EOO* I, 42.6–8, 110.10–13, 22.19–21, and 40.6–7):[2]

παντὸς τριγώνου μιᾶς τῶν πλευρῶν προσεκβληθείσης ἡ ἐκτὸς γωνία ἑκατέρας τῶν ἐντὸς καὶ ἀπεναντίον γωνιῶν μείζων ἐστίν.	One side of every triangle being produced, the external angle is greater than each of the internal and opposite angles.
ἐν τοῖς ὀρθογωνίοις τριγώνοις τὸ ἀπὸ τῆς τὴν ὀρθὴν γωνίαν ὑποτεινούσης πλευρᾶς τετράγωνον ἴσον ἐστὶ τοῖς ἀπὸ τῶν τὴν ὀρθὴν γωνίαν περιεχουσῶν πλευρῶν τετραγώνοις.	In right-angled triangles, the square on the side extending under the right angle is equal to the squares on the sides containing the right angle.
ἐὰν τριγώνου αἱ δύο γωνίαι ἴσαι ἀλλήλαις ὦσιν, καὶ αἱ ὑπὸ τὰς ἴσας γωνίαις ὑποτείνουσαι πλευραὶ ἴσαι ἀλλήλαις ἔσονται.	If two angles of a triangle be equal to one another, the sides extending under the equal sides will also be equal to one another.
ἐὰν δύο εὐθεῖαι τέμνωσιν ἀλλήλας, τὰς κατὰ κορυφὴν γωνίας ἴσας ἀλλήλαις ποιοῦσιν.	If two straight lines cut one another, they make the vertical angles equal to one another.

The last three enunciations show that generality was not expressed by means of additional linguistic units, as quantifiers are, but simply, and most frequently, by the absence of constraints that could limit the validity of the result. Conversely, whenever such additional linguistic units are present, they convey a logical added value that must be carefully analyzed in order to understand its origin and its aims. This analysis also involves a diachronic dimension: a scrutiny of the textual tradition of the *Elements* shows in fact that the determiners of generality are more likely to be added, by later revisers of the Euclidean text, than items in other lexical categories are, thereby generating a phenomenon of overdetermination of generality.[3] For this reason, in this Section I shall extensively use the Arabo-Latin tradition of the *Elements* as a witness to textual layers different from, and possibly earlier than, the textual layers the Greek tradition gives us access to—with the

[2] English language makes a crucial article disappear from the enunciation of I.47.
[3] A major exception to this trend, connected with a peculiar feature of Arabic language, will be described in Sect. 5.1.3.

caveat that the Arab revisers were apparently affected by the same overgeneralizing neurosis, even if their attention was very often focused on segments of text complementary to those affected by revisions in the Greek line of tradition. As a consequence, I shall assume the following as a working hypothesis: *whenever the direct and the indirect tradition diverge as to the presence of determiners of generality, the less adulterated text is the one in which such determiners are absent.*

The following typology lists the linguistic items used in Greek mathematics to make generality explicit.[4] The typology also sets up the terminology I shall use in this Section, listing at the same time its subsections. I shall generically designate such linguistic items "determiners of generality":

- Quantifiers: forms of the pronominal adjectives πᾶς or ἅπας "all", "every", "any" (Sect. 5.1.1).
- Determiners of arbitrariness: forms of the verb τυγχάνω "to happen" (5.1.2).[5]
- Determiners of indefiniteness: forms of the adjective τις "some" (5.1.3).
- Generalizing qualifiers: forms of the adjectives ὁποιοσοῦν "whichever", ὁποσοιδηποτοῦν, ὁσοιδηποτοῦν, and ὁποσοιοῦν "as many as we please" (all only in the plural). Of the same kind are explicit mentions of σχήματα "figures" or of εἴδη "forms" (5.1.4).[6]
- Articles, whose presence seemingly aims at limiting generality (5.1.5).
- Ordinal numbers acting as dummy letters or as variables (5.1.6).

The final Sect. 5.1.7 will briefly discuss the use of indefinite conditionals in Stoic logic.

5.1.1. *Quantifiers*

The table below sets out the propositions of the *Elements* whose enunciation is quantified;[7] an asterisk marks the propositions enunciated in conditional form; the bracketed numbers refer to propositions that contain forms of πᾶς or ἅπας that do not convey generality to the enunciation:[8]

I	16–20, (22), 32, 43
III	16
V	(1*), (12*)
VI	24, 27
VII	4, 29, 31–32
IX	8*, 9*, 10*, 20, (35*)
XI	2, 18*, 21
XII	3, (4*), 7, 10, **b**3, 6, 9

[4] I do not include syntactic devices like the conditional form of an enunciation; this will be treated in Sect. 5.3.
[5] The literal translation "to happen" will always be replaced by adverbial or adjectival "(at) random".
[6] These mentions amount to generalizing on the number of sides.
[7] The 4 occurrences of σύμπας "sum total" in IX.36 have nothing to do with generality. I include the propositions of redaction **b** of Book XII; recall that XII.6 is absent in redaction **b**, so that XII.7 = XII.6 **b**, etc. Archimedes quantifies his enunciations much more often than the *Elements* does: in the corpus of his writings there are 98 theorems enunciated in non-conditional form, 36 of which are quantified: these are *Sph. cyl.* I.13–15, 18, 33–34, 42, 44, II.2; *Circ.* 1 and 3; *Con. sph.* 4–5, 18, 21, 25, 27, 29, 31; *Aequil.* I.9–10, 13–15, II.4, 8, 10; *Quadr.* 17 and 24, *Fluit.* I.2; *Meth.* 2, 4, 6–8, 10. In *Sph. cyl.* I and II, all theorems enunciated in the prefatory epistles are quantified.
[8] Take for example V.12 (*EOO* II, 36.10–13): ἐὰν ᾖ ὁποσαοῦν μεγέθη ἀνάλογον, ἔσται ὡς ἓν τῶν ἡγουμένων πρὸς ἓν τῶν ἑπομένων οὕτως <u>ἅπαντα</u> τὰ ἡγούμενα πρὸς <u>ἅπαντα</u> τὰ ἑπόμενα "If there be as many magnitudes as we please in proportion, as one of the antecedents is to one of the consequents, so <u>all</u> antecedents will be to <u>all</u> consequents". The determiner "all" simply stands for the sum of the items mentioned.

Some definitions and postulates must be added to the list: these are I.def.15, I.post.1, post.3, post.4, II.def.1–2, VI.def.4, and XI.def.3.11 (*EOO* I, 4.9–13, 8.7–8, 8.11–14, 118.2–8, II, 72.11–12, and IV, 2.5–7, 4.10–15), which we read now and which I shall discuss just below:[9]

I.def.15	κύκλος ἐστὶ σχῆμα ἐπίπεδον ὑπὸ μιᾶς γραμμῆς περιεχόμενον πρὸς ἣν ἀφ' ἑνὸς σημείου τῶν ἐντὸς τοῦ σχήματος κειμένων πᾶσαι αἱ προσπίπτουσαι εὐθεῖαι ἴσαι ἀλλήλαις εἰσίν.	A circle is a plane figure contained by one line such that all the straight lines falling on it from one point among those lying inside the figure are equal to one another.
post. 1	ἠτήσθω ἀπὸ παντὸς σημείου ἐπὶ πᾶν σημεῖον εὐθεῖαν γραμμὴν ἀγαγεῖν.	Let it be required to draw a straight line from any point to any point.
post. 3	καὶ παντὶ κέντρῳ καὶ διαστήματι κύκλον γράφεσθαι.	And that a circle can be described with any centre and radius.
post. 4	καὶ πάσας τὰς ὀρθὰς γωνίας ἴσας ἀλλήλαις εἶναι.	And that all right angles be equal to one another.
II.def.1	πᾶν παραλληλόγραμμον ὀρθογώνιον περιέχεσθαι λέγεται ὑπὸ δύο τῶν τὴν ὀρθὴν γωνίαν περιεχουσῶν εὐθειῶν.	Every rectangular parallelogram is said to be contained by two of the straight lines containing a right angle.
II.def.2	παντὸς δὲ παραλληλογράμμου χωρίου τῶν περὶ τὴν διάμετρον αὐτοῦ παραλληλογράμμων ἓν ὁποιονοῦν σὺν τοῖς δυσὶ παραπληρώμασι γνώμων καλείσθω.	Of every parallelogrammic region, let one whichever of the parallelograms about its diagonal with the two complements be called gnomon.
VI.def.4	ὕψος ἐστὶ παντὸς σχήματος ἡ ἀπὸ τῆς κορυφῆς ἐπὶ τὴν βάσιν κάθετος ἀγομένη.	The[10] height of every figure is the ‹straight line› drawn from the vertex perpendicular to the base.
XI.def.3	εὐθεῖα πρὸς ἐπίπεδον ὀρθή ἐστιν, ὅταν πρὸς πάσας τὰς ἁπτομένας αὐτῆς εὐθείας καὶ οὔσας ἐν τῷ ὑποκειμένῳ ἐπιπέδῳ ὀρθὰς ποιῇ γωνίας.	A straight line is orthogonal to a plane whenever it make right angles with all the straight lines that touch it and that are in the underlying plane.
XI.def.11	στερεὰ γωνία ἐστὶν ἡ ὑπὸ πλειόνων ἢ δύο γραμμῶν ἁπτομένων ἀλλήλων καὶ μὴ ἐν τῇ αὐτῇ ἐπιφανείᾳ οὐσῶν πρὸς πάσαις ταῖς γραμμαῖς κλίσις. Ἄλλως. στερεὰ γωνία ἐστὶν ἡ ὑπὸ πλειόνων ἢ δύο γωνιῶν ἐπιπέδων περιεχομένη μὴ οὐσῶν ἐν τῷ αὐτῷ ἐπιπέδῳ πρὸς ἑνὶ σημείῳ συνισταμένων.	A solid angle is the inclination with respect to all lines ‹contained› by more than two lines that touch one another and that are not in a same surface. Otherwise. A solid angle is the one contained by more than two plane angles that are not in a same plane constructed at one point.

We are interested in the definitions in which the *definiens* refers to "all" objects of a given class; these are I.def.15 and XI.def.3.[11] The question is: how can we establish whether an object fits any

[9] All forms of πᾶς below are of course non-articular. For a fresh discussion see Bakker 2009a, 250–252.
[10] The definite article is required in English since there is only one straight line that can serve as a height according to the definition. The absence of a definite article determining the term to be defined (when this is a noun) is a characteristic phenomenon of Greek language; it bears the name of *Definitionsstil*. It simply means that the *definiendum* is taken to be the nominal complement of the copula.
[11] The first definition in XI.def.11 is spurious; the transmitted text is the result of contamination: just recall that the *Elements* always refers to the second definition when the notion of "solid angle" is employed: cf. XI.20–21, 23, 26–27,

of these definitions, if infinitely many items must be checked? The answer requires a detailed discussion. Let us start with I.def.15.

In the *Elements*, the definition of a circle is always used as a necessary condition: a circle is assigned; as a consequence, either its radii are equal or some point lies on the circle. The definition is also used as a necessary condition when a circle that passes through some points is claimed to pass through others.[12] Thus, the *Elements* never sets out to check, according to I.def.15, that some unspecified line is a circle.[13] This problem, however, was formulated and solved in a number of peculiar ways in the rest of the Greek mathematical corpus. A first formulation can be found in *Data*, def. 6 (*EOO* VI, 2.13–15)—this definition posits, in the "language of the givens", the existence and the uniqueness of a circle with given centre and given radius; its exact constructive counterpart is I.post.3:

τῇ θέσει δὲ καὶ τῷ μεγέθει κύκλος δεδόσθαι λέγεται οὗ δέδοται τὸ μὲν κέντρον τῇ θέσει ἡ δὲ ἐκ τοῦ κέντρου τῷ μεγέθει.	A circle is said to be given in position and in magnitude of which the centre is given in position and the radius in magnitude.

To say it otherwise: a point that is the extremity of a segment of given length whose other extremity is fixed lies on a given circumference—or, worded as a locus theorem in the way Charmandrus did according to Pappus, *Coll.* VII.24 (cf. Sect. 2.4.1):

ἐὰν εὐθείας τῷ μεγέθει δεδομένης τὸ ἓν πέρας ᾖ δεδομένον, τὸ ἕτερον ἅψεται θέσει δεδομένης περιφερείας κοίλης.	If one extremity of a straight line given in magnitude be given, the other will touch a concave arc given in position.
ἐὰν ἀπὸ δύο δεδομένων σημείων κλασθῶσιν εὐθεῖαι δεδομένην περιέχουσαι γωνίαν, τὸ κοινὸν αὐτῶν σημεῖον ἅψεται θέσει δεδομένης περιφερείας κοίλης.	If from two given points straight lines be inflected containing a given angle, their common point will touch a concave arc given in position.

The second enunciation reformulates as a locus theorem the property of the circle proved in *El.* III.21.[14] Since a locus theorem has a proof and establishes both the existence and the uniqueness of the locus-curve, Charmandrus *proved* something that in the *Elements* and in the *Data* is only assumed, namely, that both the property in I.def.15 and the property in III.21 are necessary and sufficient conditions for a line to be a circle—and Charmandrus apparently did this by eliminating any reference to infinitely many objects. We shall clarify this important point in a moment.

36; see Vitrac 1990–2001 IV, 82–83, for a discussion. The pseudo-Heronian *Definitiones* do not contain a definition of a solid angle.

[12] This deductive step, characterized by a specific formulaic expression, like ὁ ἄρα κέντρῳ τῷ Ε διαστήματι δὲ ἑνὶ τῶν ΑΕ ΕΒ ΕΓ κύκλος γραφόμενος ἥξει καὶ διὰ τῶν λοιπῶν σημείων "therefore the circle described with centre E and radius one of AE, EB, EΓ will also pass through the other points" (III.25, in *EOO* I, 228.17–19), is found in III.25, 33, IV.4–5, 8–9, 13–14, XII.17, XIII.13–16. In the propositions of Book XIII, the same expression refers to a sphere that passes through some points (of course according to the sphere's definition in the *Elements*).

[13] What we find is the proof that three equal radii of a circle suffice to determine its centre (III.9), or the construction that, given an arc of a circumference, shows how to draw the entire circumference that contains the given arc (III.25). A remark on III.9; its enunciation is: "If a point be taken inside a circle and from the point more than two equal straight lines fall on the circle, the taken point is the centre of the circle". It is misleading to claim that this theorem proves that a circle is uniquely determined by three points on it: the enunciation does not even refer to any intersection between the circle and the straight lines, and hence to any point on the circle.

[14] Enunciation of III.21: "In a circle, the angles in a same segment are equal to one another". A "property" of a specific geometric figure is a *necessary* condition for a generic figure to be exactly that specific figure.

Before doing this, let us return to I.def.15, which some authors do apply to checking whether a line is a circle. In Theodosius, *Sph.* I.1, it is proved that a section of a sphere by a plane is a circle.[15] If the plane passes through the centre of the sphere, the fact that the radii of the sphere are equal immediately entails that the section is a circle that has the same centre as the sphere.[16] If the plane does not pass through the centre of the sphere, Theodosius first determines a point that will eventually work as the centre of the circle, then proves that two straight lines (they are not expressly qualified as "generic") drawn from that point as far as the section of the sphere by the plane are equal. The result is generalized to all straight lines drawn from that point as far as the section by means of a standard logico-stylistic resource: the "potential proof" (see Sect. 4.5.2). Let us read how (*Sph.* I.1, 4.19–21)—for a harmless logical fallacy is present:[17]

ὁμοίως δὴ δείξομεν ὅτι καὶ <u>πᾶσαι</u> αἱ ἀπὸ τοῦ Ε πρὸς τὴν ΑΒΓ γραμμὴν προσπίπτουσαι ἴσαι ἀλλήλαις εἰσίν· ἡ ἄρα ΑΒΓ γραμμὴ κύκλου περιφέρειά ἐστιν, ἧς κέντρον τὸ Ε.	Very similarly we shall prove that <u>all</u> the ‹straight lines› that from E fall on line ΑΒΓ are also equal to one another; therefore line ΑΒΓ is a circumference of a circle, whose centre is E.

Now, that *all* straight lines drawn from point τὸ Ε to line ἡ ΑΒΓ are equal is exactly what is required to prove, according to I.def.15: the potential proof can only be a step towards proving this result, but cannot contain a statement of it. The correct formulation is: "Very similarly we shall prove that all the *other* ‹straight lines› that from E fall on line ΑΒΓ are also equal to one another".

Be that as it may, the Greek geometers might have adopted two strategies to frame a proof that involves a check on infinitely many objects:

1) To check a finite number of objects, qualifying them as "generic" from the very outset.
2) To check a finite number of objects, possibly without qualifying them as "generic", the extension to the whole multiplicity being handed over to a potential proof (possibly without the fallacy just seen).

The Greek geometers normally adopted the second strategy. Another example is in Apollonius, *Con.* I.4, where he proves that a section of a cone parallel to the base circle is again a circle. He does this again by means of the kind of quasi-fallacious potential proof just seen, with the additional, harmless, quirk of requiring from the very outset that a point taken on the section be arbitrary.[18]

A totally different strategy is adopted by Apollonius in *Con.* I.5 and II.48, in which he proves again that a particular conic section is a circle. To do this, he applies the result that the property proved in various ways in *El.* II.14, III.35, and VI.8 and 13 is also a sufficient condition,[19] once a straight line is identified as a diameter, for a line to be a circle. The property states that, if a straight

[15] Theodosius employs the Euclidean definition of the circle, and a definition of the sphere that begins σφαῖρά ἐστι σχῆμα στερεόν "a sphere is a solid figure", what follows being a calque of the Euclidean definition of the circle itself. The *Elements* offers a different, kinematic, definition of the sphere, in XI.def.14.

[16] Theodosius quotes Euclid's definition of the circle and his own definition of the sphere but omits the quantifier.

[17] The deductive fallacy is shared by *Con.* I.4–5 and *El.* XI.4 and 18, which we shall read presently.

[18] The quirk is at *AGE* I, 16.2; the syntagm is τι σημεῖον "some point": see Sect. 5.1.3. Adding the determiner of indefiniteness "some" amounts to a quirk because the subsequent generalization by means of a potential proof does not require making the genericity of the assumed objets explicit.

[19] That this property is also a sufficient condition is proved by Pappus among the lemmas he presents for completing the deductive structure of *Con.* I, by Serenus in *Sect. cyl.* 4, and two centuries later by Eutocius in his commentary on *Con.* I.5 (these are *Coll.* VII.237, *Opuscula*, 16.2–18, and *in Con.* I.5, *AGE* II, 208.17–210.7, respectively). I have discussed the associated doxography in Acerbi 2012c, 155–156.

line is drawn from a point on a circle perpendicular to any of its diameters, the square on the perpendicular is equal to the rectangle contained by the two segments of the diameter cut off by the perpendicular itself.[20] The proofs in *Con.* I.5 and II.48 are similar, with one crucial difference: only *Con.* I.5 ends with a potential proof, after a check of the said property performed on one single point τὸ Θ;[21] in II.48, instead, the conclusion is simply stated after the condition is proved to hold for two different points τὸ Γ and τὸ Λ. Let us read the two final arguments of *Con.* I.5 and II.48 (*AGE* I, 20.2–7 and 272.28–274.2) one after the other:[22]

καὶ τὸ ὑπὸ τῶν ΚΖ ΖΗ ἄρα ἴσον ἐστὶ τῷ ἀπὸ τῆς ΖΘ. ὁμοίως δὴ δειχθήσονται καὶ πᾶσαι αἱ ἀπὸ τῆς ΗΘΚ γραμμῆς ἐπὶ τὴν ΗΚ ἠγμέναι κάθετοι ἴσον δυνάμεναι τῷ ὑπὸ τῶν τμημάτων τῆς ΗΚ· κύκλος ἄρα ἐστὶν ἡ τομὴ οὗ διάμετρος ἡ ΗΚ.	Therefore the ‹rectangle contained› by KZ, ZH is also equal to the ‹square› on ZΘ. Very similarly all the ‹straight lines› drawn from line HΘK perpendicular to HK will also be proved to be worth equal to the ‹rectangle contained› by the segments of HK; therefore the section is a circle whose diameter is HK.
ἴσον ἄρα τὸ μὲν ἀπὸ ΓΡ τῷ ὑπὸ ΜΡΝ τὸ δὲ ἀπὸ ΛΣ τῷ ὑπὸ ΜΣΝ· κύκλος ἄρα ἐστὶν ἡ ΛΓΜ γραμμή.	Therefore that on ΓP is equal to that by MPN and that on ΛΣ to that by MΣN; therefore line ΛΓM is a circle.

As he does in *Con* I.4, Apollonius qualifies point τὸ Θ in I.5 as an arbitrary point on the section;[23] in II.48, instead, the two representative points τὸ Γ and τὸ Λ are simply introduced as the extremities of two perpendiculars dropped on the section. Thus, Apollonius' argument in II.48 shows that marking a point on the line to be checked as arbitrary or framing a potential proof is only a matter of style: just take two points (and maybe simply one) and check the property; since the chosen point is assumed to be generic simply because it is not asserted that it is a special point, the proof will apply to all points. The arbitrariness of the choice may be simply indicated by the absence of the article (σημεῖον "a point").

By using the example of the circle, I have alluded twice to the major technical challenge of stating—and proving—necessary and sufficient conditions for a line to be a well-defined and specific curve. In ancient technical jargon, identifying any such condition amounts to identifying a σύμπτωμα "characteristic property" of the curve. A σύμπτωμα is in fact a constant relation of equality or proportionality that holds among variable or invariant geometric elements attached to a generic line: speaking boldly and in modern terms, this corresponds to the "equation" of a curve. An ἀρχικόν "principal" characteristic property was usually provided together with the very definition of a curve, but the characteristic property was carefully differentiated from the definition, which was normally set out in generative or "mechanical" terms (cf. Sect. 3.4). The best known example of such an approach is Apollonius, *Con.* I.11–13: at the very beginning of each of these enunciations, the single conic sections are defined in generative terms ("cut a cone by a plane in such-and-such a way"); the associated principal characteristic property is stated at the end of each enunciation.[24] Properties derived from the ἀρχικὸν σύμπτωμα by means of chains of equalities or identities

[20] In *Con.* I.21, Apollonius generalizes this property to every central conic.
[21] Eutocius (*AGE* II, 208.7–10) reads the potential proof in his copies of the *Conics* and transcribes it in his commentary; this does not prove that the potential proof is original, but makes it likely.
[22] Note, in *Con.* I.5, the idiom of "powers" typical of Book X of the *Elements*.
[23] This happens at *AGE* I, 18.13; the syntagm is again τι σημεῖον "some point".
[24] The practice of carefully distinguishing the definition and the characteristic property is described as a matter of course by Pappus at *Coll.* III.20, and, accordingly, endorsed in his own survey of special curves at *Coll.* IV.30–59.

were perceived as (and in algebraic terms obviously are) characteristic as well; in the case of the parabola, for instance, this is the way *Con.* I.20 deduces the property that the squares on the ordinates are proportional to the associated abscissas.

A special kind of mathematical proposition, with peculiar stylistic features, was conceived to store characteristic properties of known lines: these are the locus theorems, of which Charmandrus' loci read above are a very simple example. The enunciation of a locus theorem sets out exactly a univocal relation between a set of geometric constraints imposed upon a point (summarized in the σύμπτωμα), and a curve the point is said ἅψεσθαι "to touch", that is, to lie on—so, no need to refer to infinitely many points, no need to qualify the point as arbitrary: and no need to suppose that a line is made of points.[25] In other terms, if a point satisfies the constraints, then it belongs to a well-determined curve, and vice versa. This amounts to proving both the existence and the uniqueness of the locus-curve. The peculiar demonstrative format of a locus theorem guarantees that the relation set out in its enunciation is in fact a necessary and sufficient condition for a line to be a well-defined and specific curve (see Sect. 2.4.1).[26]

But what about dealing with characteristic properties outside the format of locus theorems? Let us take again the example of conic sections: the ἀρχικὰ συμπτώματα associated with the three conic sections are proved to be necessary conditions in *Con.* I.11–13—that is: *if* a line is such-and-such a conic section, *then* the line has such-and-such a property. Can the converse be proved? Yes, it can, and very easily, but Apollonius waits until *Con.* I.52–60 to prove that, for given values of the parameters, the principal συμπτώματα of the three conic sections are also sufficient conditions—that is: *if* such-and-such a property holds for a certain line, *then* the line is such-and-such a conic section. What is more, Apollonius definitely does not state the enunciations in this way, and fills in a subtle lacuna in these proofs later on, with the proofs in *Con.* VI.1–3. Let us see what kind of subtle lacuna there is in *Con.* I.52–60; this will lead us to the technical core of the proof that the συμπτώματα of the three conic sections are also sufficient conditions. Let us take the example of the parabola. Its principal characteristic property is that, *once a straight line is identified as a diameter*, the square on any ordinate associated with that diameter is equal to the rectangle contained by the abscissa associated with that ordinate and an invariant straight line called ὀρθία "upright side".[27] Now, the lacuna in the sufficiency proof for the parabola[28] consists in assuming that a parabola with a given parameter will pass through any point for which the characteristic property, with the same parameter and referred to the same diameter, holds. As is clear, this lacuna amounts to taking for granted the uniqueness of the curve associated with an assigned characteristic property. It is not a harmless lacuna: it is simply the technical core of the sufficiency proof. The proofs in Book VI are not expressly devised to fill the gap, and do that rather incidentally.

Yet, besides being filled in Book VI, there is a simple reason for there being such a lacuna in Apollonius' proof: the uniqueness proof *in the said sense* is immediate. Let us read how Eutocius settles the issue in his commentary on *Con.* I.5 (*AGE* II, 208.10–15)—his proof refers to a circle and to the property proved in *El.* II.14, III.35, and VI.8 and 13:[29]

---

[25] It is hardly necessary to recall that a major Aristotelian insight (argued at length in *Ph.* VI, in order to refute the Zenonian paradoxes) is that a line is not composed of points.

[26] On Chrysippus likening mathematical loci to Platonic Forms (so Geminus *apud* Proclus, *iE*, 395.13–18, in the comment on I.35) see Caston 1999, 199.

[27] This line is traditionally called *latus rectum*. The upright side is the form parameter of the parabola as seen from a specific diameter: upright sides associated with different diameters of the same parabola are different, see *Con.* I.49.

[28] Read *Con.* I.53, *AGE* I, 164.7–12.

[29] See also the spurious postposed explanation (cf. Sect. 4.5.3) in XIII.13 (*EOO* IV, 292.9–12).

καὶ δυνατὸν μέν ἐστιν ἐπιλογίσασθαι τοῦτο διὰ τῆς εἰς ἀδύνατον ἀπαγωγῆς. εἰ γὰρ ὁ περὶ τὴν ΚΗ γραφόμενος κύκλος οὐχ ἥξει διὰ τοῦ Θ σημείου, ἔσται τὸ ὑπὸ τῶν ΚΖ ΖΗ ἴσον ἤτοι τῷ ἀπὸ μείζονος τῆς ΖΘ ἢ τῷ ἀπὸ ἐλάσσονος, ὅπερ οὐχ ὑπόκειται.	It is also possible to infer this by reduction to the impossible. For if the circle described about KH will not pass through point Θ, the ‹rectangle contained› by KZ, ZH will be equal either to the ‹square› on a ‹straight line› greater than ZΘ or to that on a lesser one, which has really not been supposed.

In short: the uniqueness of the curve is secured in a trivial way by the fact that the characteristic property is an equality, for the equality would not obtain for the same kind of curve if this did not pass through the intended point. It remains to explain my italicized qualification "in the said sense" in the previous paragraph. The point is the following. As recalled above, "Greek" lines are not defined as the extension of a property that operates a selection within the widest genus of "lines": they are defined individually by a variety of generative techniques (cf. Sect. 3.4).[30] For this reason, the form of a line is never determined *directly* by a characteristic property, but further qualifications are required. In particular, a characteristic property selects a subspecies out of a well-defined class of lines: for instance, within the class of conic sections, a characteristic property determines a specific section. What a characteristic property says, then, is that a unique line of assigned form and with assigned parameters (cf. the italicized clause "once a straight line is identified as a diameter" above) will pass through a suitable point in the plane. For this reason, the species of the line is a given of the problems in *Con.* I.52–60, so that different proofs must be framed for parabola, hyperbola, and ellipsis.—All these complications disappear in the format of locus theorems.

So far for checking *El.* I.def.15. Definition XI.def.3 allows highlighting other logical and textual subtleties. Let us discuss its application in the relevant direction; this is found in XI.4: the other proofs that a straight line is orthogonal to a plane depend on this proposition. Let us first read the enunciation and the setting-out of XI.4 (*EOO* IV, 12.18–25):

ἐὰν εὐθεῖα δύο εὐθείαις τεμνούσαις ἀλλήλας πρὸς ὀρθὰς ἐπὶ τῆς κοινῆς τομῆς ἐπισταθῇ, καὶ τῷ δι' αὐτῶν ἐπιπέδῳ πρὸς ὀρθὰς ἔσται.	If a straight line stand on the common section at right ‹angles› with two straight lines cutting one another, it will also be at right ‹angles› with the plane through them.
εὐθεῖα γάρ <u>τις</u> ἡ ΕΖ δύο εὐθείαις ταῖς ΑΒ ΓΔ τεμνούσαις ἀλλήλας κατὰ τὸ Ε σημεῖον ἀπὸ τοῦ Ε πρὸς ὀρθὰς ἐφεστάτω. λέγω ὅτι ἡ ΕΖ καὶ τῷ διὰ τῶν ΑΒ ΓΔ ἐπιπέδῳ πρὸς ὀρθάς ἐστιν.	In fact, let <u>some</u> straight line, EZ, stand from E at right ‹angles› with two straight lines, AB, ΓΔ, cutting one another at point E. I claim that EZ is also at right ‹angles› with the plane through AB, ΓΔ.

The proof is long but simple. An arbitrary straight line ἡ ΗΕΘ is taken in the plane identified by straight lines αἱ ΑΒ, ΓΔ, and passing through their intersection τὸ Ε; it is then proved that ἡ ΗΕΘ is also perpendicular to the straight line ἡ ΕΖ that is supposed to be perpendicular to αἱ ΑΒ, ΓΔ from τὸ Ε. The arbitrariness of ἡ ΗΕΘ is made explicit by τις "some", and is further strengthened by ὡς ἔτυχεν "at random" in the construction.[31] Now, the Greek and the Arabo-Latin tradition diverge as to how to formulate the transition, required by XI.def.3, from straight line ἡ ΗΕΘ being

[30] The only exception are homeomeric lines, for which see again Acerbi 2010a.

[31] The determiner of arbitrariness is absent in the indirect tradition, and the same for τυχόντος that qualifies point τὸ Ζ on the straight line outside the plane. See Sects. 5.1.2–3 on these markers.

perpendicular to ἡ ZE to *all* straight lines in the intended plane being perpendicular to ἡ ZE. Let us read first the final part of the Greek proof of XI.4 (*EOO* IV, 16.5–16):

ὀρθὴ ἄρα ἑκατέρα τῶν ὑπὸ HEZ ΘEZ γωνιῶν· ἡ ZE ἄρα πρὸς τὴν HΘ τυχόντως διὰ τοῦ E ἀχθεῖσαν ὀρθή ἐστιν. ὁμοίως δὴ δείξομεν ὅτι ἡ ZE καὶ πρὸς πάσας τὰς ἁπτομένας αὐτῆς εὐθείας καὶ οὔσας ἐν τῷ ὑποκειμένῳ ἐπιπέδῳ ὀρθὰς ποιήσει γωνίας· εὐθεῖα δὲ πρὸς ἐπίπεδον ὀρθή ἐστιν, ὅταν πρὸς πάσας τὰς ἁπτομένας αὐτῆς εὐθείας καὶ οὔσας ἐν τῷ αὐτῷ ἐπιπέδῳ ὀρθὰς ποιῇ γωνίας· ἡ ZE ἄρα τῷ ὑποκειμένῳ ἐπιπέδῳ πρὸς ὀρθάς ἐστιν· τὸ δὲ ὑποκείμενον ἐπίπεδόν ἐστι τὸ διὰ τῶν AB ΓΔ εὐθειῶν· ἡ ZE ἄρα πρὸς ὀρθάς ἐστι τῷ διὰ τῶν AB ΓΔ ἐπιπέδῳ.

Therefore each of the angles HEZ, ΘEZ is a right ‹angle›; therefore ZE is orthogonal to HΘ, which is randomly drawn through E. Very similarly we shall prove that ZE will also make right angles with all the straight lines that touch it and that are in the underlying plane; and a straight line is orthogonal to a plane whenever it make right angles with all the straight lines that touch it and that are in a same plane; therefore ZE is at right ‹angles› with the underlying plane; and the underlying plane is the one through straight lines AB, ΓΔ; therefore ZE is at right ‹angles› with the plane through AB, ΓΔ.

And now let us read the versions in Adelard I and Gerard; the correspondence between the denotative letters is obvious:[32]

Adelard I

duorum itaque angulorum *nbt* et *nbk* uterque rectus. quare *nb* super *kt* iuxta duos angulos rectos. sicque manifestum quia omnis linea producta ex *b* super superficiem duarum linearum *gd* et *zh* producet angulum rectum ex *nb*. erit igitur *nb* perpendicularis super superficiem *gdhz*.

Gerard

ergo quisquis duorum angulorum *hbt* et *hbk* est rectus, ergo *hb* est perpendicularis super *tk*. et similiter ostenditur, quod omnis linea a puncto *b* in superficie linearum *gd ez* propterea continet cum *hb* angulum rectum. ergo linea *hb* est perpendicularis erecta super superficiem *gd*; *ez*.

The Greek proof is clearly longer and far more contrived than the Latin proofs. In the Greek proof, we may note the unusual, almost non-instantiated potential proof (underlined), the complete citation of XI.def.3 that follows this potential proof, and the last three steps, which only the useless reference to the "underlying plane" in the quote of XI.def.3 makes necessary.[33] And further: in the part of the proof that is not transcribed above, there are 4 references to previous proofs marked by ἐδείχθη in 7 lines and a likewise useless analogical proof,[34] and all of them are absent in the indirect tradition. We must conclude that the Greek proof is the result of a punctilious revision; the outcome, flawless but cumbersome and inelegant, makes too many steps explicit in its attempt at "saturating" the deductive structure.

[32] Busard 1983, 302.106–110, and Busard 1984, c. 340.48–53, respectively. Earlier in both texts, we find a reference to the definition of right angle that does not figure in the Greek tradition. As already noted, different revisers perceive different deductive gaps.

[33] The Greek redaction also contains the determiner of arbitrariness in adverbial form τυχόντως "randomly" (underlined in the text), which in the entire Greek mathematical corpus exhibits just another occurrence (Theon, *in Alm. I.2*, *iA*, 333.3). Recall also the two forms of τυγχάνω that figure in the first part of the proposition and that are likewise absent in the indirect tradition.

[34] See *EOO* IV, 14.22–16.1 and 14.18–19, respectively. For ἐδείχθη as a marker, see Sect. 4.5.5.

This and countless other cases corroborate the working hypothesis that the presence of some or many determiners of generality is the result of layers of revisions, whose aim is to produce a more perspicuous text (cf. Sect. 5.1).

So much for the definitions. The propositions that carry a quantifier in the enunciation do not deserve a detailed discussion; I just note that there are no salient mathematical features that discriminate between the propositions that are quantified and those that are not.[35] If we exclude the enunciations in conditional form, a common trait of the quantified enunciations is to state a property of a most general genus of geometric objects: all triangles (I.16–20, 32, XI.2); all parallelograms (I.43, VI.24); all solid angles (XI.21); all pyramids (XII.3), prisms (7), and cones (10); all numbers (VII.4 and 32), prime numbers (29), and composite numbers (31). Add to these two maximum results,[36] in which quantification has specific aims. It remains that the sample is limited, that almost every such theorem might also be enunciated without the quantifier, and that a phenomenon of stylistic attraction may have suggested formulating in the same way the theorems in the small group I.17–20, or the pairs I.16/I.32 and I.43/VI.24. The proofs of the quantified enunciations cannot be distinguished from those of the non-quantified enunciations.

Three scattered remarks close this Section.

In III.16 and XI.2 a property is shown to apply to a class of objects by proving that no objects of the intended classes do not satisfy it.[37] As said, VI.27 (*EOO* II, 158.13–19) is a maximum theorem preliminary to the theory of application of areas:

πάντων τῶν παρὰ τὴν αὐτὴν εὐθεῖαν παραβαλλομένων παραλληλογράμμων καὶ ἐλλειπόντων εἴδεσι παραλληλογράμμοις ὁμοίοις τε καὶ ὁμοίως κειμένοις τῷ ἀπὸ τῆς ἡμισείας ἀναγραφομένῳ μέγιστόν ἐστι τὸ ἀπὸ τῆς ἡμισείας παραβαλλόμενον παραλληλόγραμμον ὅμοιον ὂν τῷ ἐλλείμματι.	Of all the parallelograms applied to a same straight line and falling short by parallelogrammic forms both similar and similarly placed to the one described on the half ‹of the straight line›, the greatest is the parallelogram applied to the half ‹and› that is similar to the defect.

The proof is a model of sobriety: no quantification is added, as well as no determiners of generality: a parallelogram is taken that fulfils the constraints but that is not constructed on the half of the assigned straight line, and it is shown that this parallelogram is less than the parallelogram constructed on the half of the assigned straight line.

Finally, let us read the enunciation and the setting-out of XI.18 (*EOO* IV, 48.2–7):

ἐὰν εὐθεῖα ἐπιπέδῳ τινὶ πρὸς ὀρθὰς ᾖ, καὶ πάντα τὰ δι' αὐτῆς ἐπίπεδα τῷ αὐτῷ ἐπιπέδῳ πρὸς ὀρθὰς ἔσται.	If a straight line be at right ‹angles› with some plane, all the planes through it will also be at right ‹angles› with the same plane.
εὐθεῖα γάρ τις ἡ AB τῷ ὑποκειμένῳ ἐπιπέδῳ πρὸς ὀρθὰς ἔστω. λέγω ὅτι καὶ πάντα τὰ διὰ τῆς AB ἐπίπεδα τῷ ὑποκειμένῳ ἐπιπέδῳ πρὸς ὀρθάς ἐστιν.	In fact, let some straight line, AB, be at right ‹angles› with the underlying plane. I claim that all the planes through AB will also be at right ‹angles› with the underlying plane.

[35] In Book VII, the only three theorems that present a general conclusion happen to have a quantifier in the enunciation. As for the Arabo-Latin tradition, Gerard's source quantifies all theorems of Book XII!

[36] These are III.16 and VI.27; for VI.27 see just below.

[37] Take XI.2 as an example: the proof that any two intersecting straight lines belong to one single plane amounts to proving that no two intersecting straight lines can have a part in a plane and another outside it.

The Greek proof has the same structure as XI.4 discussed above, and the same textual problems. XI.def.3 is invoked in the same almost-non-instantiated form we have seen in XI.4. A non-instantiated citation of XI.def.4 is also provided that makes the presence of the immediately subsequent ἐδείχθη-reference to a proof necessary (XI.18, *EOO* IV, 48.20–23 = XI.def.4, IV, 2.8–11):

καὶ ἐπίπεδον πρὸς ἐπίπεδον ὀρθόν ἐστιν, ὅταν αἱ τῇ κοινῇ τομῇ τῶν ἐπιπέδων πρὸς ὀρθὰς ἀγόμεναι εὐθεῖαι ἐν ἑνὶ τῶν ἐπιπέδων τῷ λοιπῷ ἐπιπέδῳ πρὸς ὀρθὰς ὦσιν.	And a plane is orthogonal to a plane whenever the straight lines drawn in one of the planes at right ‹angles› with the common section of the planes be at right ‹angles› with the remaining plane.

All of this is absent in the indirect tradition, which also lacks one of the two forms of the verb τυγχάνω “to happen” we find in the Greek text. The missing verb form (for which see the next Section) is contained in the potential proof that closes the proof of XI.18 (*EOO* IV, 50.2–4)—needless to say, this is worded in the quasi-fallacious form we have become accustomed to:

ὁμοίως δὴ δειχθήσεται καὶ πάντα τὰ διὰ τῆς AB ἐπίπεδα ὀρθὰ τυγχάνοντα πρὸς τὸ ὑποκείμενον ἐπίπεδον.	Very similarly all the planes through AB will also be proved to happen to be orthogonal to the underlying plane.

On the other hand, the Arabo-Latin proof seems to be inconclusive, since it ends by proving, again by means of a quasi-fallacious potential proof, that the plane through the intended line is orthogonal to the underlying plane—*gd* is the intersection of the plane through the intended line with the underlying plane:[38]

Adelard I	Gerard
eodemque modo ostendemus quia omnes linee producte a linea *gd* supra duos angulos rectos in superficie *abgd* erunt supra superficiem assignatam.	et similiter ostenditur quod omnis perpendicularis producta ex linea *gd* in superficie *abgd* est perpendiculariter erecta super superficiem datam.

However, it remains true that the plane through the intended line is declared from the very outset to be arbitrary[39]—this we do not find in the Greek text—so that the potential proof by extension to all planes is strictly speaking not necessary. Apparently, all of this was not so clear to the reviser we owe the received Greek text to.[40]

5.1.2. *Determiners of arbitrariness*

In the previous Section we have frequently met the main determiner of arbitrariness: the forms of the verb τυγχάνω “to happen”—which I shall almost always translate with the syntagm “(at) random”. The following table sets out the forms of this verb attested in the *Elements* and in the *Data*:[41]

[38] See Busard 1983, 312.335–337, and Busard 1984, c. 348.27–29, respectively.

[39] For instance by means of the phrase *quoniam protraham ex linea ab superficiem, quocumque modo producatur* in Gerard; see Busard 1984, c. 348.16–17.

[40] A final footnote for an adverbial form of πᾶς. In Book X there are 12 occurrences of πάντως “in every instance”, all in spurious material (the remaining occurrence is in the likewise spurious V.19por). For a detailed discussion of the textual problems of Book X, see Vitrac 1990–2001 III, 381–399; Rommevaux, Djebbar, Vitrac 2001, 250–276, 285, 291–292.

[41] There are 131 occurrences in the entire Euclidean corpus.

τυγχ–	III.25, 37, IV.5 (*quater*), XI.18, XII.4/5
aorist participle	I.5, 9, 11, 12, 23, 31, III.2 (*ter*), 13, 32, VI.9 (*bis*), 10, 11, 12, XI.2, 4, 7 (*ter*), 9, 14, 18, 21, 26, 35 (*bis*), **b**36, XII.17, **b**16, *alt*XI–XIII.3, *Data* 94, *alt*6
ἔτυχεν	II.1, 2 (*ter*), 3 (*ter*), 4 (*ter*), 7 (*ter*), 8 (*ter*), 12, 13, III.1, 5, 6, 8 (*bis*), IV.3, V.4 (*quater*), 7 (*quinquies*), 8 (*ter*), 11 (*quater*), 12 (*bis*), 13 (*quater*), 14, 16 (*bis*), 17 (*bis*), 22 (*quater*), 23, VI.1, XI.4, 11, *Data* 49 (*bis*), 51 (*bis*), 63, 77, 78
τύχοι	X.72
τυχόντως	XI.4[42]

Thus, two forms exhibit the majority of the occurrences:

- aorist participle that directly qualifies an object, this being very often a point (32 occurrences in the *Elements*);
- phrases like ὡς / ὃ / ἃ ἔτυχεν "(at) random", which play periphrastically the role of determinations of object (the last two in the list) or of action (59 occurrences).

No occurrences at all are provided by the arithmetic Books VII–IX, where generalizing qualifiers are instead greatly prevailing (Sect. 5.1.4). The only occurrence of a determiner of arbitrariness in Book X has the form of the incidental, obviously inauthentic clause εἰ τύχοι "if it happen" in X.72.[43] I have also shown elsewhere[44] that all the occurrences of ἄλλο ὃ ἔτυχεν / ἄλλα ἃ ἔτυχεν "another, random" / "others, random" in Book V are spurious. Let us discuss the rest of the sample.

In Book II, the phrase ὡς ἔτυχεν "at random" determines the operation of cutting an assigned straight line at a random point; it is the only phrase with this function in this book. After an isolated occurrence in the setting-out of II.1, this phrase sets a crucial constraint in II.2–4 and 7–8, repeated in enunciation, setting-out and conclusion; the occurrences in II.12 and 13 are instantiated citations of II.4 and 7, respectively. The phrase is employed only in the propositions in which it is required to cut the assigned straight line at one point only; in the other propositions, the assigned straight line is cut "in equal and unequal ‹segments›" (II.5 and 9) or it is "bisected, and some straight line is added to it" (II.6 and 10).[45] The determiner of arbitrariness in the setting-out of II.1 refers to the several points of section, and answers to the generalizing qualifier (see Sect. 5.1.4) "in as many segments as we please" of the enunciation. The transition from one determiner of generality to the other is justified by the shifting focus from segments (certainly a discrete sequence) to points (less certainly such a one).[46] All determiners in Book II are almost surely authentic, even if the indirect tradition is sharply divided: Gerard confirms all of them, Adelard I has none. I take it for certain that, as is customary with his translation, Adelard has curtailed the text—he does not translate any general conclusion in Book II, either.

[42] For this hapax see Sect. 5.1.1.
[43] At *EOO* III, 218.22, within the clause ἔστω, εἰ τύχοι, πρότερον μεῖζον τὸ ΑΒ τοῦ ΓΔ "let, if it happen, AB be first greater than ΓΔ", whose spurious character is patent (just compare I.6 or I.26)—the clause is also absent in the Theonine manuscripts and in Gerard, it is placed in anomalous position, and τύχοι is a hapax in the *Elements*. The case "AB less than ΓΔ", treated in X.72 by means of a potential proof (*EOO* III, 222.3–5), is useless there since the "greater than" case already provides the two irrational lines required in the enunciation (contrary to what happens in the twin proposition X.71). Thus, Heiberg rightly brackets this passage, which is absent both in **P** and in the indirect tradition—but then εἰ τύχοι is only attested in **P** and should be bracketed as well!
[44] In Acerbi 2003a, 201–205. I shall not resume that discussion here.
[45] Note the determiner of indefiniteness "some", and see Sect. 5.1.3.
[46] We shall return on this point at the end of this Section.

Elsewhere, ὡς ἔτυχεν determines the act of "drawing through" a straight line.[47] Two of these occurrences, in the enunciation and in the conclusion of III.8, are clearly out of place: they do not figure in the parallel enunciation of III.7, nor are they attested in the indirect tradition.[48] The 7 occurrences of the relative syntagm ἃ ἔτυχεν "random" in *Data* 49 (*bis*), 51 (*bis*), 63, 77, 78[49] refer to arbitrary "rectilinear figures"; the first four of these occurrences are almost certainly authentic. The only other occurrence of a determiner of arbitrariness in the *Data*, τις τυχοῦσα "some random ‹straight line›" in *Data* 94, is an instance of overdetermination (because of the co-occurrence with the determiner of indefiniteness τι "some") and for this reason of doubtful authenticity.[50]

Verb forms in the present stem τυγχ– are found only in almost certainly spurious portions of text:[51] we find them in III.25 (within a reference to the obvious), 37 (useless final addition in the form of a potential proof; the whole textual unit is absent in the indirect tradition), porism to IV.5 (*quater*), XI.18 (useless final addition in the form of a potential proof), XII.4/5 (a lemma that is absent in the indirect tradition).

The aorist participle is mainly employed to mark the arbitrariness of points taken at random, on straight lines or on circumferences, within constructive acts. In Book I, we read occurrences of τυχὸν σημεῖον "random point" only within constructive acts;[52] in VI.9, the same syntagm designates a point taken on a straight line. In Book III, the points are taken at random only on circumferences: this happens in III.2 and 13 (two points) and 32. In our paradigmatic proposition III.2, the determination of arbitrariness τυχόντα is supposed among the assignments, so that we find it repeated in enunciation, setting-out, and conclusion. Book XI offers the same record: XI.2 (two points), 4,[53] 7 (three occurrences, distributed as in III.2), 9, 14, 18, 21 (3 points), 26, 35.[54]

To conclude: the points qualified by forms of τυγχάνω are always taken on lines, never in the plane. This also happens in XII.17 (= 16 **b**), where four points are taken at random, two pairs on two parallel straight lines. The two Greek proofs in XII.17 and XII.16 **b** are fairly divergent, and in fact they result from a series of nested revisions and contaminations that stem from a shorter version, which also involves a Greek model of the version we read in the indirect tradition.[55] Still, the transformation, during this process of rewriting, from the instantiated step in redaction **b** to an instantiated citation (of XI.7) in the main text of the Greek tradition keeps the determiner of arbitrariness. The reason is that XI.7 does have such a determiner in the enunciation, and an instantiated step that applies a theorem may well be expected to preserve a typifying feature of the template, even if the instantiated step is not a fully-fledged instantiated citation.

When arbitrary points in a whole plane—or, more generally, on a surface, as in Apollonius, *Con.* I.6, or in Archimedes, *Con. sph.* 15—must be taken, these do not receive a determiner of

---

[47] In III.1, 5, 6, IV.3, XI.4 and 11. Only the next-to-last item in this list is certainly spurious, as we have seen in the previous Section. See Sect. 4.3 for the verb διάγω "to draw through".

[48] At *EOO* I, 182.24 and 188.25. See Busard 1984, c. 63.41, for Gerard's translation.

[49] The last three occurrences in the list are found in instantiated citations of proposition 49. In *Data* 49 and 51, there are exactly parallel occurrences in the enunciation and in the setting-out. Thus, there are just two independent occurrences in the *Data*.

[50] The Arabic tradition confirms only the four occurrences in *Data* 49 and 51: see Sidoli, Isahaya 2018, 110–111 and 114–115, respectively.

[51] These are all present participles, with the only exception of the occurrence in proposition III.37 and of one of the occurrences in IV.5por.

[52] In I.5, 9, 11, 12, 23, 31; two random points are taken in I.23 and the participle is in the plural.

[53] This is spurious; we have discussed it in Sect. 5.1.1.

[54] Two random points are taken on different straight lines, in the enunciation and in the setting-out—the general conclusion of proposition XI.35 is absent.

[55] See Vitrac 1990–2001 IV, 355–371.

arbitrariness, but a determiner of indefiniteness τι "some",[56] as in *El.* III.7–8, 36–37, IV.10,[57] III.9, XI.12.[58] A "random point" taken in the plane we find instead in the alternative proof of *Data* 30 and in proposition XI.38 *vulgo*, and this confirms the compilative character of these propositions.

A remarkable example of the resistance to taking "random" points in a plane is found in the Heronian proofs that serve as complements to I.1. The problem is to construct a scalene triangle on a given straight line *ab*; in the first case set out by Hero, the given line must be less than one of the two other sides of the triangle but greater than the third.[59] The construction starts as in I.1, tracing two equal circles *age*, *bgd* that intersect at *g*. In order to construct a triangle as required, it is enough to take a random point inside one of the two circles but outside the other and to join the straight lines from this point to the extremities of the given straight line. Let us read instead Hero's solution, in the Latin translation of the text of an-Nayrīzī:[60]

deinde signabo in arcu *ge* punctum qualitercumque contingat, quod sit punctum *z*, et coniungam *a* cum *z*. punctum quoque secundum signabo in linea que est inter punctum *z* et circumferentiam circuli *bgd*, quod sit punctum *h*, et coniungam *b* cum *h* et protraham ipsam lineam secundum rectitudinem usque ad punctum *t*. manifestum est ergo quod linea *ah* est longior linea *ab* et linea *ab* est longior linea *bh*.

Thus, the point that solves the problem is taken as the intersection of two straight lines that join the extremities of the given straight line and two points suitably taken on one of the circles. I am unable to explain such a contrived procedure otherwise than with the intention to avoid taking points "at random" in the plane.

Other geometric objects seldom receive determiners of arbitrariness. In Book VI (9–12), the participle τυχοῦσα qualifies the arbitrary angle contained by two straight lines. In XI.36 redaction **b**, a solid angle is qualified as "random"; the determiner is confirmed by the indirect tradition but has been eliminated in the main Greek tradition. Conversely, I have not found in the *Elements* examples of points actually taken at random but that do not receive a determiner of arbitrariness or of indefiniteness.[61]

I close this Section with a discussion of an interesting difference of formulation between Books I and II. Let us read to this purpose the construction of I.11 (*EOO* I, 32.9–11) and the enunciation of II.4 (*EOO* I, 124.18–21):

εἰλήφθω ἐπὶ τῆς ΑΓ τυχὸν σημεῖον τὸ Δ, καὶ κείσθω τῇ ΓΔ ἴση ἡ ΓΕ, καὶ συνεστάτω ἐπὶ τῆς ΔΕ τρίγωνον ἰσόπλευρον τὸ ΖΔΕ, καὶ ἐπεζεύχθω ἡ ΖΓ.	Let a random point, Δ, be taken on ΑΓ, and let a ‹straight line›, ΓΕ, be set equal to ΓΔ, and let an equilateral triangle, ΖΔΕ, be constructed on ΔΕ, and let a ‹straight line›, ΖΓ, be joined.
ἐὰν εὐθεῖα γραμμὴ τμηθῇ ὡς ἔτυχεν, τὸ ἀπὸ τῆς ὅλης τετράγωνον ἴσον ἐστὶ τοῖς τε ἀπὸ τῶν τμημάτων τετραγώνοις καὶ τῷ δὶς ὑπὸ τῶν τμημάτων περιεχομένῳ ὀρθογωνίῳ.	If a straight line be cut at random, the square on the whole ‹straight line› is equal both to the squares on the segments and to twice the rectangle contained by the segments.

[56] But see Pappus, *Coll.* III.64, for an exception.

[57] In all these propositions, a point outside a circle is taken; in III.9, the point is inside a circle.

[58] In XI.12, what is taken is an "elevated" point, hence located in space: the verb form is not εἰλήφθω "let it be taken", but the one of regular use in space: νενοήσθω "let it be conceived": see Sect. 4.2.

[59] To be clear: just one side is given; the scalene triangle is otherwise arbitrary. The non-trivial point of the construction is how to avoid hitting on an isosceles triangle—and, apparently, how to avoid taking a random point in the plane that contains the given straight line.

[60] At Tummers 1994, 40.18–24. As said, (large) portions of Hero's commentary on the *Elements* are only preserved in the analogous commentary of the Persian scholar an-Nayrīzī: see Acerbi, Vitrac 2014, 31–39.

[61] For Book I, I have also checked the indirect tradition.

In Book I, the determiner of arbitrariness refers to a point, in the form of the aorist participle τυχόν; in Book II, this kind of determiner modifies the verb that describes the operation of taking a section, by means of the periphrastic phrase ὡς ἔτυχεν. The ways of reference are strictly complementary,[62] even if in both cases a random point is taken on a straight line. The difference is explained by the fact that the focus of Book II is on the operation of taking a section and on its results, namely, the "segments"[63] on which the figures of interest in any proposition of this Book are constructed. True, the point of section is named in the setting-out, but this fact is in a sense incidental, since such a naming has just the function of qualifying the cut-operation as being performed κατὰ τὸ Γ "at Γ". Again, in both propositions the random point plays an "active" role in the subsequent construction—as an extremity of a straight line on which an equilateral triangle is constructed in I.11, as a point through which a suitable parallel straight line is drawn in II.4. However, the two constructions have different functions (see Sect. 4.3): the construction in I.11 is the standard auxiliary construction that is discharged during the proof; the construction in II.4 generates the geometric configuration intended in the enunciation. This difference persists if we compare all the relevant propositions of Book I (namely, those in which a point is taken at random within auxiliary constructions) and those of Book II.

5.1.3. *Determiners of indefiniteness*

Forms of the adjective τις "some" are currently used to strengthen the indefinite character of a statement or of a supposition, and thereby its genericity. The main text of the *Elements* exhibits 360 occurrences, which are distributed as in the following table:[64]

	I	II	III	IV	V	VI	VII	VIII	IX	X	XI	XII	XIII	tot.
# prop.	48	14	37	16	25	33	39	27	36	115	39	18	18	465
% # signs	7.6	3.3	6.9	3.4	4.9	7.6	5.7	5	5.1	26.1	9	8.3	6.9	100
forms of τις	14	13	71	3	11	9	80	7	39	39	26	46	2	360
% τις	3.9	3.6	19.7	0.8	3	2.5	22.2	1.9	10.8	10.8	7.2	12.7	0.6	100

Redaction **b** of Book XII contains 35 occurrences. The wild oscillations in the distribution show that the use of a determiner of indefiniteness is forced by specific mathematical requirements. The logical range of the determiner of indefiniteness are the determinations of objects whose role in a specific geometric configuration is in some respects, but not in all, left unconstrained. Thus, given objects cannot receive a determiner of indefiniteness. On the opposite extreme of the range, complete arbitrariness is marked by forms of ἔκκειμαι"to be set out" (cf. Sect. 4.2).[65] The only exceptions to the second rule can be found in propositions III.1 and XI.11, in which totally arbitrary straight lines are "drawn through" within a circle or in the plane and are provided with a double determiner, as in XI.11 (*EOO* IV, 32.9–10)—we shall return on these two propositions presently:

[62] By the way, this confirms the stylistic compartmentation of the Books of the *Elements*.
[63] That is, they are τμήματα (lit. "the cut ones"), *nomen rei actae* related to the verb τέμνω "to cut".
[64] The list includes the occurrences in definitions I.def.13, 14 (*bis*), 17, III.8, 9, V.def.18 (*bis*), VI.def.5, VII.def.14–18, X.def.4, XI.def.17.
[65] Geometric objects are marked by this verb in I.22*, IV.10*, VI.12, 23*, X.29, 48, 50, 52–53, 62, XI.23/24 XII.13 (straight line), IV.11 (triangle), XI.36 (solid angle), XIII.13–16, 18 (diameter), XIII.13, 16 (circle), XIII.14–15 (square), XIII.17 (two faces of a cube). Only sporadically is a τις introduced when a single straight line is at issue (with asterisk); in the other instances, there are at least two straight lines, one of which receives the determiner. As for arithmetic objects, the verb sets out a unit, in VIII.9 and in IX.32.

διήχθω γάρ τις ἐν τῷ ὑποκειμένῳ ἐπιπέδῳ εὐθεῖα ὡς ἔτυχεν ἡ ΒΓ.	In fact, let some straight line, ΒΓ, be drawn through at random in the underlying plane.

The indirect tradition only retains the second determiner, but this fact has no significance, as we shall see in a moment. All of this is slightly paradoxical: when there is complete freedom in assigning an object, a verb is employed with markedly particularizing connotations, like ἔκκειμαι. The difference between the two grades of indetermination, namely, a partial and a total indetermination, can be best appreciated by comparing VI.9 and VI.12 (*EOO* II, 104.6–13 and 108.17–24), of which we read the setting-out and the construction:

ἔστω ἡ δοθεῖσα εὐθεῖα ἡ ΑΒ. δεῖ δὴ τῆς ΑΒ τὸ προσταχθὲν μέρος ἀφελεῖν.	Let there be a given straight line, AB. Thus it is required to remove a prescribed part from AB.
ἐπιτετάχθω δὴ τὸ τρίτον, καὶ διήχθω τις ἀπὸ τοῦ Α εὐθεῖα ἡ ΑΓ γωνίαν περιέχουσα μετὰ τῆς ΑΒ τυχοῦσαν, καὶ εἰλήφθω τυχὸν σημεῖον ἐπὶ τῆς ΑΓ τὸ Δ, καὶ κείσθωσαν τῇ ΑΔ ἴσαι αἱ ΔΕ ΕΓ, καὶ ἐπεζεύχθω ἡ ΒΓ, καὶ διὰ τοῦ Δ παράλληλος αὐτῇ ἤχθω ἡ ΔΖ.	Thus let the third ‹part› be prescribed, and from A let some straight line, ΑΓ, be drawn through containing a random angle with AB, and let a random point, Δ, be taken on ΑΓ, and let ‹straight lines›, ΔΕ, ΕΓ, be set equal to ΑΔ, and let a ‹straight line›, ΒΓ, be joined, and through Δ let a ‹straight line›, ΔΖ, be drawn parallel to it.
ἔστωσαν αἱ δοθεῖσαι τρεῖς εὐθεῖαι αἱ Α Β Γ. δεῖ δὴ τῶν Α Β Γ τετάρτην ἀνάλογον προσευρεῖν.	Let there be three given straight lines, A, B, Γ. Thus it is required to find a fourth proportional of A, B, Γ.
ἐκκείσθωσαν δύο εὐθεῖαι αἱ ΔΕ ΔΖ γωνίαν περιέχουσαι [τυχοῦσαν] τὴν ὑπὸ ΕΔΖ, καὶ κείσθω τῇ μὲν Α ἴση ἡ ΔΗ τῇ δὲ Β ἴση ἡ ΗΕ καὶ ἔτι τῇ Γ ἴση ἡ ΔΘ, καὶ ἐπιζευχθείσης τῆς ΗΘ παράλληλος αὐτῇ ἤχθω διὰ τοῦ Ε ἡ ΕΖ.	Let two straight lines, ΔΕ, ΔΖ, be set out containing a [random][66] angle, ΕΔΖ, and let a ‹straight line›, ΔΗ, be set equal to A and HE equal to B and further ΔΘ equal to Γ, and ΗΘ being joined let a ‹straight line›, EZ, be drawn through E parallel to it.

Here as in VI.10–11, the determiners of arbitrariness strengthen the idea that the construction has some degrees of freedom. As to the difference between the two propositions, it is enough to note that, in VI.9, the straight line ἡ ΑΓ has a constrained extremity, namely, point τὸ Α, whereas in VI.12 the two straight lines are entirely arbitrary, so that they must be "set out": they cannot simply be "drawn through" while, for instance, containing a "random" angle. We must conclude that the bracketed determiner of arbitrariness in the setting-out of VI.12 is really Theon's addition, who aims at standardizing the texts of the propositions in the group VI.9–12.

The geometric objects determined by forms of τις are mainly straight lines or points. The entire range of variability is bookwise set out in the following table, which also lists the 26 occurrences in the *Data*:

point (19 + 4)	III (17), IV (1), XI (1), *Data* (4)
straight line (109 + 12)	I (14), II (13), III (52), IV (2), VI (5), X (8+6),[67] XI (8), XIII (1), *Data* (12)
part of a straight line (7)	XI (7)
arc (1)	III (1)

[66] The determiner is attested in the Theonine manuscripts only.
[67] The 6 occurrences qualify the ῥητή "expressible" straight line.

angle (1)	III (1)
segment of a circle (5)	XII (5)
rectangle (+ 2)	*Data* (2)
rectilinear figure (2)	X (1), XII (1)
plane (11)	XI (10), XII (1)
polyhedron (2)	XII (2)
pyramid (1)	XII (1)
sphere (5)	XII (5)
solid (22)	XII (21), XIII (1)
plane region (9)	XII (9)[68]
magnitude (35 + 8)	V (11), VI (3+1),[69] X (19), XII (1), *Data* (8)
number (126)	VII (80),[70] VIII (5), IX (39), X (2)
metalinguistic (5)	VIII (2),[71] X (3)[72]

A determiner of indefiniteness is often present in the enunciation; in this case, it is retained in the setting-out, which, however, may also carry additional determiners. In Book II, this happens in the entire string of propositions II.5–10; let us read an extract from the enunciation and the setting-out of II.6 (*EOO* I, 132.6–7 and 132.14–16):

ἐὰν εὐθεῖα γραμμὴ τμηθῇ δίχα προστεθῇ δέ τις αὐτῇ εὐθεῖα ἐπ’ εὐθείας [...]	If a straight line be bisected, and some straight line be added to it in a straight line [...]
εὐθεῖα γάρ τις ἡ ΑΒ τετμήσθω δίχα κατὰ τὸ Γ σημεῖον, προσκείσθω δέ τις αὐτῇ εὐθεῖα ἐπ’ εὐθείας ἡ ΒΔ.	In fact, let some straight line, AB, be bisected at point Γ, and let some straight line, ΒΔ, be added to it in a straight line.

Such additional occurrences are a priori suspect since they introduce a surplus of generality: the focus of the enunciation, in fact, is on the arbitrariness of the added straight line with respect to the main straight line, not on the arbitrariness of the main straight line. Unfortunately, the indirect tradition is of no help, for the absence of the definite article and the presence of a determiner of indefiniteness are rendered in Arabic by the same linguistic tool (the so-called "nunation"). A fortiori, the difference becomes invisible to a Latin translator from Arabic.

The distribution of the occurrences of the determiners of indefiniteness provides some useful indications as to their authenticity and pertinency. The point can be stated as follows. Greek mathematical style has an obvious tool to mark the indefinite relatum of a generic mathematical object: the absence of the article. So, what mathematical reasons might induce making indefiniteness explicit—a move that amounts to strengthening it? To see better what happens, the following table sets out the distribution of the occurrences of the determiners of indefiniteness by specific part of a proposition and by logico-deductive function within a proof.

[68] These, as long as all previous occurrences in Book XII, are found in the initializing clause of an argument by "exhaustion", to be read below in this Section.
[69] These are: an instantiated citation of V.7 applied to a parallelogram in VI.14, to a triangle in VI.2 and 15. Add to them the qualification of πηλικότης "numeric value" in the spurious VI.def.5.
[70] Sometimes, as in VII.4, a μέρος "part" of a number is so qualified.
[71] In the expression ὅσους ἄν τις ἐπιτάξῃ "as many as one prescribes".
[72] In the expression κατά τινα τῶν (προ)ειρημένων τρόπων "according to some of the said ways".

definition	I.13, 14 (*bis*), 17, III.8–9, V.18 (*bis*), [VI.5], VII.14–18, X.4, XI.17
enunciation	I.14, II.6, 10, III.1por (*bis*), 3 (*bis*), 7 (*bis*), 8 (*bis*), 9*, 18 (*bis*)*, 19, 32 (*bis*), 36–37, V.6*, 7por, VI.2*, VII.15 (*bis*), 16*, 17*, 18 (*bis*)*, 24, 27 (*bis*)*, 28*, 30 (*bis*)*, 31–32, 35, 37, VIII.13 (*bis*)*, IX.1*, 3–4*, 5, 7 (*bis*)*, 11, 16–17, 28*, 29*, 31, 36*, X.1, 13, 13/14, 16/17*, 18/19 (*ter*)*, 23por, XI.1 (*bis*), 8*, 16*, 18*, 19*, XII.7por
setting-out	I.13*, 14, II.5*, 6 (*bis*)*, 7 (*bis*)*, 9*, 10 (*bis*)*, III.3 (*bis*), 7 (*bis*), 8 (*bis*), 18, 19, 32 (*bis*), 36–37, V.7*, VII.15 (*bis*), 18, 23*, 24, 30, 31–32, 35, 37, IX.5, 7, 11, 16–17, 31, X.1, 13, 13/14, XI.1 (*bis*), 4*, 5*, 14*, 18*, XIII.3*
anaphora	III.19, V.7
ἐκκείσθω-clause	I.22, IV.10, VI.23, X.29, 48, 50 (*bis*),[73] 52–53, 62
construction	III.1, VI.9, IX.12, XI.4, 11, 12, XII.17
conclusion	I.14, II.6, 10, III.3 (*bis*), 7 (*bis*), 8 (*bis*), 9, 18 (*bis*), 19, 32 (*bis*), 36–37, V.6, VI.2, VII.31–32, IX.11,[74] 17,[75] X.16/17, XI.1**, 8, 16, 18, 19, XIII.17[76]
non-instantiated citation	III.9 (1por), VII.22 (VII.def.15), 31 (VII.def.14 *bis*), 32 (31), IX.7 (VII.def.14), 9 (3), 12 (VII.def.14), 13 (VII.31 *bis*), 14 (VII.30), 15 (VII.24), X.28/29I (IX.1), XI.11 (8), XII.2 (X.1)
instantiated citation	I.45, 47 (14), II.10 (I.29*), III.4 (3 *bis*), 10 (1por *bis*), 14 (3), 32 (19), 33 (32), 34 (32**), 35–36 (3), IV.2 (III.32**), 17 (III.37), V.20–21 (8*), VI.2, 14–15 (V.7), 32 (I.14), VII.1 (VII.def.15), 2 (1), 3 (VII.def.15), 19 (18**), 23 (VII.def.15), 24 (VII.def.15), 24 (23), 28 (VII.def.15 *bis*), 29 (VII.def.15), 31 (VII.def.14), IX.13 (11 and VII.31 *bis*), 20 (VII.31), 31 (VII.def.15), X.2 (1), 4 (X.def.1*), 5 (X.def.1*), 15 (X.def.1* *bis*), 16 (X.def.1* *bis*), XI.2 (1** *bis*)
pseudo-existential or its negation[77]	III.7, V.13, 18, VII.1§c,[78] 2§c (*bis*), 3§c (*ter*), 4, 22§c, 23§c, 24§c, 28§c (*bis*), 29§c, 31 (*bis*), 34§ (*bis*), 36§ (*bis*), IX.13, 31c, 34, 36§, X.2§, 3§, 4§, 16§ (*bis*), XII.2, 5, 10 (*bis*), 11, 12, 17 (*bis*)
exhaustion[79]	XII.2 (*nonies*), 5 (*septies*), 11 (*septies*), 12 (*septies*), 18 (*quinquies*)
existential	III.16, V.13, VII.21§, 33§, 39§, VIII.4 (*ter*), XI.1 (*sexties*)
ποιῶσί τινας-clause[80]	VII.def.17–18, VII.16–18, 27, 30, VIII.13, IX.1, 3–4, 5, 7, 9, 11, 14, 28–29, 36, X.28/29I
πλήν-clause[81]	I.14, 39–40, III.1, 18–19, XI.3

In the above table, an asterisk * marks the propositions in which enunciation and setting-out differ by one unit in the number of occurrences;[82] a double asterisk ** is assigned to citations or conclusions that carry less occurrences than the associated setting-outs; in the case of citations, the quoted proposition is indicated in parenthesis, the Book number being omitted if it is the same as the Book number of the host proposition; the number of the quoted proposition follows a string of propositions if it is the same for all of them. Here is the same table for the *Data*:

[73] One of the two occurrences qualifies a number, the other an "expressible" straight line.
[74] Instantiated conclusion.
[75] Instantiated conclusion.
[76] Within a partial conclusion, typical of this proposition.
[77] When a multiple occurrence is marked by a sign §, two of these occurrences are located in the initializing and in the closing clause of a reduction to the impossible, respectively (cf. Sect. 5.2.1). Otherwise, the determiner of indefiniteness τις occurs within sentences that formulate isolated pseudo-existential statements, which in their turn possibly (but not necessarily) initialize indirect arguments.
[78] When the occurrences are marked by a letter "c", one of them coincides with the associated citation, and for this reason it has already been listed among citations.
[79] This is a subcategory of the previous item, which I have separated for obvious reasons.
[80] These have already been listed among enunciations or citations.
[81] These special occurrences, within negative sentences that contain the clause ἄλλη τις πλήν "some other ‹straight line› with the exception of", are shortly presented below.
[82] The imbalance is more frequently in favour of the enunciation, as we shall see in a moment. The same for instantiated citations (richer in determiners) with respect to setting-outs (poorer).

enunciation	2, 5, 9, 22, 36*, 38, 73*, 78, 89*, 91 (*bis*), 92 (*bis*), 94
setting-out	2, 5, 9, 22, 38, 78, 91 (*bis*), 92 (*bis*), 94
instantiated citation	67[83]

The pattern of distribution is clear: most forms of τις figure in enunciations or in their retrievals: these are setting-outs, instantiated and non-instantiated citations, conclusions. In III.19, the determiner is even repeated in the anaphora.[84] The differences in the numbers of occurrences in the enunciation and in the setting-out make some regular pattern to emerge: the cases in which the enunciation has one occurrence more than the setting-out are nearly double of those in which the opposite is the case; Book XI has an equitable distribution; Books II and VII–IX show marked and opposite compartmentations.

The rest of the occurrences of the determiners of indefiniteness can be included in categories that display more or less marked existential connotations. I have called some such clauses "pseudo-existential" because their existential import is in a sense "concealed" by the formulation. Still, the homogenization I have introduced is harmless, as some examples will show. Let us read first a fully-fledged existential statement, in one of its occurrences in VIII.4 (*EOO* II, 282.21–284.3):

οἱ Ν Ξ Μ Ο ἄρα ἑξῆς ἀνάλογόν εἰσιν ἐν τοῖς τοῦ τε Α πρὸς τὸν Β καὶ τοῦ Γ πρὸς τὸν Δ καὶ ἔτι τοῦ Ε πρὸς τὸν Ζ λόγοις. λέγω δὴ ὅτι καὶ ἐλάχιστοι ἐν τοῖς ΑΒ ΓΔ ΕΖ λόγοις. εἰ γὰρ μή, ἔσονταί τινες τῶν Ν Ξ Μ Ο ἐλάσσονες ἀριθμοὶ ἑξῆς ἀνάλογον ἐν τοῖς ΑΒ ΓΔ ΕΖ λόγοις.

Therefore N, Ξ, M, O are successively in proportion in the ratios of both A to B and of Γ to Δ and further of E to Z. I now claim that they are also least in the ratios AB, ΓΔ, EZ. In fact, if not, there will be some numbers less than N, Ξ, M, O successively in proportion in the ratios AB, ΓΔ, EZ.

This example is representative insofar as most existential or pseudo-existential clauses are found in the arithmetic Books, and introduce indirect proofs. Thus, they are "local determinations" (cf. Sect. 4.2.1), as for instance the first pseudo-existential clause in VII.34 (*EOO* II, 256.21–23):

οἱ Α Β ἄρα τὸν Γ μετροῦσιν. λέγω δὴ ὅτι καὶ ἐλάχιστον. εἰ γὰρ μή, μετρήσουσί τινα ἀριθμὸν οἱ Α Β ἐλάσσονα ὄντα τοῦ Γ.

Therefore A, B measure Γ. I now claim that ‹Γ› is also the least. In fact, if not, A, B will measure some number that is less than Γ.

A supposition of existence of some measured or measuring numbers (as in the many references to VII.def.15) produces almost all occurrences of pseudo-existential clauses. The formulations of the two pseudo- / existential clauses just seen are strictly parallel, and I submit that the fully-fledged existential statements confirm the existential character of the pseudo-existential clauses, rather than the latter confirming a supposedly copulative character of the former.[85] The pseudo-existential formulation may be dictated by stylistic constraints, for instance avoiding a double circumstantial participle, as in the reconstructed conditional *εἰ γὰρ μή, ἔσται τις ἀριθμὸς ὑπὸ τῶν Α Β ἐλάχιστος

[83] This is a result that can be easily deduced from *El.* II.5; a proof of it is only attested as a scholium to this very step of the *Data* (scholium n. 133 in *EOO* VI, 296.2–297.8). The form of the reference in the text of the *Data* is an instantiated citation (within the anafora!); the enunciation contained in the scholium that provides the proof is of course an a posteriori reconstruction.

[84] The occurrence in V.7 is strictly speaking outside the anaphora, but it is a reference to the construction.

[85] Translation of the relevant portion of VIII.4 if "to be" has a copulative value: "In fact, if not, some numbers successively in proportion in the ratios AB, ΓΔ, EZ will be less than N, Ξ, M, O".

μετρούμενος ἐλάσσων ὢν τοῦ Γ "in fact, if not, there will be some least number measured by A B that is less than Γ".

A peculiar category of sentences marked for indefiniteness are the clauses typical of the method of exhaustion, an iterative argument that in the *Elements* is exclusive of Book XII and that is formulated as a reduction to the impossible (cf. Sect. 4.5.2). Let us read the clause that initializes the reduction in proposition XII.5 (*EOO* IV, 164.24–166.1):

εἰ γὰρ μή ἐστιν ὡς ἡ ΑΒΓ βάσις πρὸς τὴν ΔΕΖ βάσιν οὕτως ἡ ΑΒΓΗ πυραμὶς πρὸς τὴν ΔΕΖΘ πυραμίδα, ἔσται ὡς ἡ ΑΒΓ βάσις πρὸς τὴν ΔΕΖ βάσιν οὕτως ἡ ΑΒΓΗ πυραμὶς ἤτοι πρὸς ἔλασσόν τι τῆς ΔΕΖΘ πυραμίδος στερεὸν ἢ πρὸς μεῖζον.	In fact, if, as base ΑΒΓ is to base ΔΕΖ, so pyramid ΑΒΓΗ is not to pyramid ΔΕΖΘ, as base ΑΒΓ is to base ΔΕΖ, so pyramid ΑΒΓΗ will be either to some solid less than pyramid ΔΕΖΘ or to a greater one.

The standard, non-constructive assumption of existence of a fourth proportional is here at work;[86] its existential connotation is obvious even if only implicit. The same obvious-yet-implicit existential connotation is carried by the pseudo-existential clauses that describe the core of the method, as the procedural clause in XII.5 (*EOO* IV, 166.6–10):

καὶ πάλιν αἱ ἐκ τῆς διαιρέσεως γινόμεναι πυραμίδες ὁμοίως διῃρήσθωσαν, καὶ τοῦτο ἀεὶ γινέσθω ἕως οὗ λειφθῶσί τινες πυραμίδες ἀπὸ τῆς ΔΕΖΘ πυραμίδος αἵ εἰσιν ἐλάττονες τῆς ὑπεροχῆς ᾗ ὑπερέχει ἡ ΔΕΖΘ πυραμὶς τοῦ Χ στερεοῦ.	And again, let the pyramids resulting from the division be similarly divided, and let this come about continually until some pyramids have remained of pyramid ΔΕΖΘ that are less than the excess by which pyramid ΔΕΖΘ exceeds the solid X.

The negative sentence that follows immediately the "reference to the impossible" (see Sect. 5.2.1) at the very end of an exhaustion-reduction almost always retains the determiner included in the initializing sentence, as the table above shows.

In special instances, a reduction to the impossible is accompanied by a πλήν-clause (see again the table above), which usually contains a further τις and whose existential import is fairly explicit—as a matter of fact, this clause denies the existence of objects with a certain property, so that it can also be regarded as a uniqueness statement. Let us read the end of the proof of proposition III.18 (*EOO* I, 216.10–13) as an example (sorry for the very strained translation):

ὅπερ ἐστὶν ἀδύνατον· οὐκ ἄρα ἡ ΖΗ κάθετός ἐστιν ἐπὶ τὴν ΔΕ. ὁμοίως δὴ δείξομεν ὅτι οὐδ' ἄλλη τις πλὴν τῆς ΖΓ· ἡ ΖΓ ἄρα κάθετός ἐστιν ἐπὶ τὴν ΔΕ.	Which is really impossible; therefore it is not the case that ΖΗ is perpendicular to ΔΕ. Very similarly we shall prove that some other ‹straight line› with the exception of ΖΓ will not be either; therefore ΖΓ is perpendicular to ΔΕ.

Forms of τις in geometric constructive acts are extremely rare. If we exclude the ἐκκείσθω-clauses, only a handful of occurrences remain. The reason is that, in a construction, objects that are not completely determined are seldom generated. Four of the 6 occurrences of determiners of indefiniteness in constructive acts mark straight lines that are arbitrary or are drawn from a point

[86] On this assumption, see Becker 1932–33; Mueller 1981, 231–234. Recall that the proof of XII.18 is radically simplified by assuming the existence of a fourth proportional *in the form of a sphere*.

without further constraints:[87] these are in III.1, VI.9, XI.4, 11, of which we have read above the second and the fourth. Of a similar kind is the occurrence in XII.17 (*EOO* IV, 228.16–17):

τετμήσθωσαν αἱ σφαῖραι ἐπιπέδῳ τινὶ διὰ τοῦ κέντρου.	Let the spheres be cut with some plane through the centre.

Let us summarize our findings: the presence of the determiner of indefiniteness τις "some" is required whenever a mathematical entity, usually introduced within a supposition, is strictly linked with objects that are the mathematical subject of the proof, and which thereby impose constraints on the supposed mathematical entity. To clarify this by an example: there is no point in qualifying with "some" the circle that is the mathematical subject of any enunciation of Book III, whereas it is significant to refer to "some" tangent to it (III.16) or to "some" chord of it (III.3), because such entities are in a strong sense (that is, as a class) determined by the circle. Thus, the determiners of indefiniteness convey a *particularizing* connotation, devised to select, from an undifferentiated genus, mathematical objects by means of constraints that, however, do not identify them univocally.[88]

In number theory, the presence of determiners of indefiniteness is even more justified than in a geometric context, for numeric objects are, in an obvious sense, more markedly particular than geometric entities.[89] For this reason, about numeric objects statements can be made that carry a whole range of particularizing degrees: there is a sharp difference between statements that apply to "number" unconstrainedly or that apply to "number" as an arbitrary representative of a specific subclass. Let us clarify this with the example of VII.37 (*EOO* II, 264.20–22), where the subclass is that of the parts of an assigned number:[90]

ἐὰν ἀριθμὸς ὑπό τινος ἀριθμοῦ μετρῆται, ὁ μετρούμενος ὁμώνυμον μέρος ἕξει τῷ μετροῦντι.	If a number be measured by some number, the measured one will have a part homonymous to the measuring one.

A specific kind of clause in the arithmetic Books that includes determiners of indefiniteness is interesting in this perspective: these are phrases like ποιῶσί τινας "make some ‹numbers›". Reading an enunciation that contains such a clause makes it manifest that the presence or the absence of a determiner of indefiniteness in the setting-out bears no connection with the assignment of denotative letters. Here is VII.18 (*EOO* II, 224.22–226.3):

ἐὰν δύο ἀριθμοὶ ἀριθμόν τινα πολλαπλασιάσαντες ποιῶσί τινας, οἱ γενόμενοι ἐξ αὐτῶν τὸν αὐτὸν ἕξουσι λόγον τοῖς πολλαπλασιάσασιν.	If two numbers multiplying some number make some ‹numbers›, those resulting from them will have the same ratio as the multipliers.
δύο γὰρ ἀριθμοὶ οἱ Α Β ἀριθμόν τινα τὸν Γ πολλαπλασιάσαντες τοὺς Δ Ε ποιείτωσαν. λέγω ὅτι ἐστὶν ὡς ὁ Α πρὸς τὸν Β οὕτως ὁ Δ πρὸς τὸν Ε.	In fact, let two numbers, A, B, multiplying some number, Γ, make Δ, E. I claim that, as A is to B, so Δ is to E.

[87] For XI.12 see the previous Section, to which the occurrence in *Data* 67 must be added. Such straight lines are generated by using διάγω "to draw through"—this is the less constraining operation on straight lines and the only verb related to the operation of tracing a straight line that is not linked with any of the postulates (see Sect. 4.3).

[88] This property is not restricted to the use of such determiners in mathematics: see Humbert 1960, 26.

[89] And in fact, the three arithmetic Books exhibit more than one-third of the occurrences, while engaging a bit less than 16% of the *Elements*.

[90] Read also the constructive act in IX.12, *EOO* II, 362.21–22.

The second determiner in the enunciation falls, and is replaced by the letters; the first determiner cannot fall, because it names numbers arbitrarily chosen in a class. The persistence of determiners of generality in the presence of lettered designations is the main theme of the following Section.

5.1.4. *Generalizing qualifiers*

The generalizing qualifiers are used only in such propositions as deal with indefinite multiplicities of terms. It is no surprise, then, that almost two-thirds of the occurrences of generalizing qualifiers in the *Elements* are found in the arithmetic Books (45 out of 72). The generalizing qualifiers are (of course, they are all in the plural): ὁποσοιδηποτοῦν, ὁσοιδηποτοῦν, and ὁποσοιοῦν "as many as we please" (said of numbers and seldom of straight lines), or ὁποσαοῦν "as many as we please" (magnitudes).[91] The last qualifier is a lexical marker of Book V, where we do not find occurrences of qualifiers that carry the infix –δηποτ–. In the arithmetic Books, the above forms may be interchanged within the same proposition and even when reference is made to the same arithmetic sequence. The distribution of the forms is as in the table below. The occurrences in a given proposition are almost always by triads or by pairs, according to whether a general conclusion is present (in Book V) or not (in the arithmetic Books). The two propositions marked by an asterisk carry one single occurrence, in the enunciation or in the setting-out.[92]

	enunciation	setting-out	construction	citation	conclusion
ὁποσοιοῦν	VII.12, 14, 33, VIII.3, 4*, 6–7, IX.8–13, 21–23, 36	VII.12, 14, 33, VIII.1, 3, 6–7, 13, IX.8, 11, 13, 21, 23		VIII.2,[93] IX.28,[94] 32,[95] X.12[96]	
ὁποσαοῦν[97]	V.1 (*bis*), 12, 22	V.1 (*bis*), 12, 22, 23		V.12 (*bis*)[98]	V.1 (*bis*), 12, 22
ὁποσοιδηποτοῦν		IX.12, 35			
ὁσοιδηποτοῦν	II.1, VIII.1, 13, IX.17, 35, **b**XI.37	IX.9, 10, 17, 22, 32*, 36, **b**XI.37	VI.1 (*bis*), 33 (*bis*), XI.25 (*bis*), XII.13 (*bis*)[99]		II.1

The table shows that a generalizing qualifier that is included in the enunciation is retained in the setting-out. This phenomenon is prima facie surprising, since any indefinite multiplicity cannot but be set out by means of a finite, and well-defined, number of terms, and one wonders on what grounds can such a specific set of units be qualified "as many as we please". Let us see then how such qualifiers work, on the example of the geometric progressions of arbitrary ratio whose sum is determined in IX.35 (*EOO* II, 404.21–406.9):

[91] In the Euclidean minor works, adverbs are found like ὁπωσδηποτοῦν and ὁπωσοῦν "as we please" (both one occurrence, in *Opt.* 23 and 28 **A**), and qualifiers like ὁσδηποτοῦν "whatsoever" (*Phaen.*, introduction), but I shall not deal with them. Just two occurrences of ὁστισοῦν (form ὁτιοῦν) "any", in the enunciation and in the setting-out of VII.38.
[92] Gerard's translation confirms that the same happens in the indirect tradition.
[93] Non-instantiated citation of the enunciation of VIII.1.
[94] Non-instantiated citation of the enunciation of IX.21.
[95] Non-instantiated citation of the enunciation of IX.13.
[96] Instantiated citation of VIII.4.
[97] Add two occurrences, one in the enunciation and one in the construction, in *Data* 3.
[98] Non-instantiated citation of the enunciation of V.1.
[99] In these four propositions, the reference is to the arbitrariness of the equimultiples taken in the construction that preludes to an application of V.def.5.

ἐὰν ὦσιν ὁσοιδηποτοῦν ἀριθμοὶ ἑξῆς ἀνάλογον ἀφαιρεθῶσι δὲ ἀπό τε τοῦ δευτέρου καὶ τοῦ ἐσχάτου ἴσοι τῷ πρώτῳ, ἔσται ὡς ἡ τοῦ δευτέρου ὑπεροχὴ πρὸς τὸν πρῶτον οὕτως ἡ τοῦ ἐσχάτου ὑπεροχὴ πρὸς τοὺς πρὸ ἑαυτοῦ πάντας.	If there be as many numbers as we please successively in proportion and ‹numbers› equal to the first be subtracted both from the second and from the last, as the excess of the second to the first, so the excess of the last will be to all those before itself.
ἔστωσαν ὁποσοιδηποτοῦν ἀριθμοὶ ἑξῆς ἀνάλογον οἱ Α ΒΓ Δ ΕΖ ἀρχόμενοι ἀπὸ ἐλαχίστου τοῦ Α, καὶ ἀφῃρήσθω ἀπὸ τοῦ ΒΓ καὶ τοῦ ΕΖ τῷ Α ἴσος ἑκάτερος τῶν ΒΗ ΖΘ. λέγω ὅτι ἐστὶν ὡς ὁ ΗΓ πρὸς τὸν Α οὕτως ὁ ΕΘ πρὸς τοὺς Α ΒΓ Δ.	Let there be as many numbers as we please successively in proportion, Α, ΒΓ, Δ, ΕΖ, starting from a least ‹number› Α, and from ΒΓ and ΕΖ let each of ΒΗ, ΖΘ be subtracted equal to Α. I claim that, as ΗΓ is to Α, so ΕΘ is to Α, ΒΓ, Δ.

The ratio ὁ Α of the progression does not deserve a generalizing qualifier—a passing mention suffices. The multiplicity of the terms to be added, instead, receives the generalizing qualifier ὁ(πο)σοιδηποτοῦν "as many as we please", both in the enunciation and in the setting-out. The multiplicity of the terms is instantiated, in this case as elsewhere (see below), on the smallest number of terms that secure full generality. In our instance, the enunciation mentions "first", "second", and "last", but the proof requires one term more, in order for "those before ‹the last›" not to be the first and the second, which have an independent function. Therefore, four terms are set out—and the first and the third term, from which no number will be subtracted, are denoted by one letter only. The proof, even if it is carried out with four terms, is general simply because it does not use the fact that they are exactly four. Since the setting-out rigidly conforms to the enunciation, while introducing letters whenever possible,[100] the result is the bewildering practice of setting out four terms, denoted by letters, but still qualified as ὁποσοιδηποτοῦν "as many as we please".

There is, of course, no loss of generality in such a procedure, or in introducing the denotative letters: there is simply no difference if an object is designated "the excess of the second to the first" or ὁ ΗΓ, once it is said that ΗΓ = ΒΓ – ΒΗ, with ΒΗ = Α = "the first". Finally, theorems like IX.35 do have a strong particularizing connotation; the only free parameter in them is the number of the terms in the progression, which is accordingly the only item that receives a generalizing qualifier.

A further example is found in IX.36. The theorem provides a sufficient condition for a number to be perfect, that is, equal to the sum of its parts. The condition involves the sum of a suitable geometric progression of ratio 2 and first term a unit—2 is the only particular number about which something is proved in the arithmetic Books. The proof is general, but of course it can only be formulated on the basis of a specific progression; it is a fact that the proof works for the selected progression, namely, the progression with four terms: and in fact the associated perfect number is 496.[101] This notwithstanding, the multiplicity of the terms receives a generalizing qualifier both in the enunciation and in the setting-out of IX.36 (*EOO* II, 408.7–17):

ἐὰν ἀπὸ μονάδος ὁποσοιοῦν ἀριθμοὶ ἑξῆς ἐκτεθῶσιν ἐν τῇ διπλασίονι ἀναλογίᾳ ἕως οὗ ὁ σύμπας συντεθεὶς πρῶτος γένηται καὶ ὁ σύμπας ἐπὶ τὸν	If starting from a unit as many numbers as we please be successively set out in double proportion until the sum total become prime and the sum total

[100] The clause "starting from a least ‹number› A" is required in the setting-out but not in the enunciation, since in the enunciation the ordering is made explicit by the assignment of the ordinals.

[101] The Greek sources list four perfect numbers: 6, 28, 496, 8128; see for instance Nicomachus, *Ar.* I.16. He claims that the condition is also necessary for even perfect numbers, but he does not provide a proof, which we owe to Euler. Recall that, to Nicomachus, perfect numbers are by definition a subspecies of even numbers.

ἔσχατον πολλαπλασιασθεὶς ποιῇ τινα, ὁ γενόμενος τέλειος ἔσται.
ἀπὸ γὰρ μονάδος ἐκκείσθωσαν ὁσοιδηποτοῦν ἀριθμοὶ ἐν τῇ διπλασίονι ἀναλογίᾳ ἕως οὗ ὁ σύμπας συντεθεὶς πρῶτος γένηται, οἱ Α Β Γ Δ, καὶ τῷ σύμπαντι ἴσος ἔστω ὁ Ε, καὶ ὁ Ε τὸν Δ πολλαπλασιάσας τὸν ΖΗ ποιείτω. λέγω ὅτι ὁ ΖΗ τέλειός ἐστιν.

multiplied by the last make some ‹number›, the resulting ‹number› will be perfect.
In fact, starting from a unit let as many numbers as we please be set out in double proportion until the sum total become prime, A, B, Γ, Δ, and let E be equal to the sum total, and let E multiplying Δ make ZH. I claim that ZH is perfect.

Let me discuss the formulation of this setting-out. Note first that, in order for the proof to be general, it must be independent of the multiplicity of terms actually set out, for instance, five terms,[102] *even if* in this case the resulting number is not perfect. Still, some multiplicity must be chosen in the setting-out, and we might wonder whether the effectiveness of the proof would not be strengthened if it were carried out on a case that "does not work". Well, this problem was perceived, and the choice of four terms is deliberate. For IX.35 must be applied in the proof of IX.36, and this entails that the terms cannot be less than four, as we have just seen.[103] Six terms, which yield the perfect number 8128, are redundant once four terms suffice. It remains to choose between four (which do yield a perfect number) and five (which do not), and, by minimality, four must be selected. This said, a peculiarity of the setting-out allows carrying out a proof that is impeccable from the point of view of mathematical generality. For, as we have seen in Sect. 3.2, the denotative letters are abbreviations of designations—in our instance, the definite description implicit in the supposition underlined in the setting-out above. As a consequence, placing the "name" οἱ Α Β Γ Δ *after* the relative clause ἕως οὗ ὁ σύμπας συντεθεὶς πρῶτος γένηται "until the sum total become prime" entails including this clause in the definite description named by the letters—and this entails in turn that the sequence set out *must* give rise to a prime number. So the option "five terms" is excluded by the mere position of the denotative letters.

I pass now to give substance to my claim about the minimality of the terms that receive a generalizing qualifier.[104] The data are collected in the following tables; these set out, for the propositions listed at the beginning of this Section, the multiplicity of the terms with which they are instantiated. The first table includes all propositions that are instantiated with the smallest number of terms compatible with full generality, the second table includes all propositions that, seemingly, are not instantiated with the smallest number of terms; the last row on this table sets out the multiplicities in Gerard's translation.

II	V		VI		VII		VIII		IX									XI	XII	**b**XI
1	1	22	1	33	12	14	7	13	8	10	11	17	21	22	23	35	36	25	13	37
2	2	3	2	2	2+2	3	4	3	6	6	4	4	4	4	3	4	4	2	2	4

V	VII	VIII				IX			
12	33	1	3	4	6	9	12	13	32
3+3	3	4	4	2+2+2	5	6	4	4	3
3+3	3	4	4	2+2+2	5	4	4	4	3

[102] Three terms are not enough, as we shall see in a moment.
[103] Recall that the way a proposition is "applied" is by adherence to a template.
[104] The following problem arises in V.20, 21 and 23: the propositions are enunciated in full generality in Gerard, whereas Adelard and the Greek tradition expressly mention three pairs of magnitudes. A similar phenomenon occurs in VI.22 and in XI.37, while in V.22 the Greek text offers an enunciation in general terms, too. I have discussed the related textual evidence, thereby showing that minimality is kept in some formulations, and that these must be regarded as nearest to the original, in Acerbi 2003a, 205–208.

Comment on the first table.

The setting-outs instantiated with 2 terms are of course minimal. Larger multiplicities are forced by the mathematics implicit in the enunciation: this is the case in V.22 and in its numeric counterpart VII.12 (theorems "through an equal"), in VIII.13 (minimal multiplicity of terms in continuous proportion), IX.8 (seven terms in continuous proportion are required, a unit included), 21–22 (2 is not the best example of an even number), 23 (3 is the first odd number), XI.37 redaction **b** (minimal multiplicity of straight lines in a generic proportion). Forced by the formulation of the enunciation are the multiplicities in VIII.7, IX.17 (mention of "first", "second", and "last" term: so there must be at least another term in between), IX.10 (mention of the third term in the sequence and of οἱ δύο διαλείποντες πάντες "all those leaving out two"), 11 (mention of "lesser", "greater" and τινα τῶν ὑπαρχόντων ἐν τοῖς ἀνάλογον ἀριθμοῖς "some ‹number› among those featuring in the numbers in proportion": the plural entails that there are at least four terms), 35–36 (seen above).

Comment on the second table.

In V.12, 2 + 2 magnitudes would suffice, for V.1 is applied, which is instantiated with 2 magnitudes, and the numeric counterpart VII.12 is instantiated with 2 + 2 numbers (see the first table). In VII.33, two terms would suffice. In VIII.1, three terms are certainly enough. As for the other propositions, the minimality requirement is in fact satisfied. In VIII.6, it might seem that 3 terms suffice, and in fact two of the terms set out are not used in the proof. However, the setting-out of VIII.6 (*EOO* II, 228.9–11) refers to, or mentions, "first", "second" and "no other":

ἔστωσαν ὁποσοιοῦν ἀριθμοὶ ἑξῆς ἀνάλογον οἱ Α Β Γ Δ Ε, ὁ δὲ Α τὸν Β μὴ μετρείτω. λέγω ὅτι οὐδὲ ἄλλος οὐδεὶς οὐδένα μετρήσει.	Let there be as many numbers as we please successively in proportion, A, B, Γ, Δ, E, and let A not measure B. I claim that neither will any other measure any.

Thus, we must suppose that there are at least four terms: the three that are mentioned and the term measured by none of them. Moreover, the proof requires an additional term, since it also explains why two non-consecutive terms cannot be measured—and therefrom, by a potential proof, any farther terms in the sequence. Ergo: 5 terms are necessary. The attested proof falsifies my argument because it simply takes the first term as the "no other" term. Since ἄλλος "other" in the enunciation is exclusive, the Greek proof is not flawless in this respect. In VIII.3, three terms would suffice, but the explicitly iterative character of the construction suggests instantiating with 4 terms. In VIII.4, the construction with 2 + 2 terms is not general enough for an application to sequences with higher multiplicity, whereas the construction with 2 + 2 + 2 terms is. In IX.9, one of the 6 instantiated terms does not figure in the proof, which, however, expressly cites IX.8, where the 6 terms are required.[105] In IX.12 and 13, less than 4 terms would suffice, but the two proofs are explicitly iterative: exhibiting this feature requires one step—and hence one term—more. In IX.32, two terms would suffice, but 3 are required to apply IX.13 to the 3 arbitrary terms plus the dyad. I conclude that the only propositions in which minimality is not preserved are V.12, VII.33, and VIII.1, namely, a marginal sample (3 out of 31).

Minimality arguments also apply to the theorems that involve polygons (VI.20, XII.1, 6),[106] rectilinear figures (VI.18, 21, 22, 25, 28, and 29), or εἴδη "forms" (VI.31 and the porism to

[105] Recall that the 4 terms in Gerard fit minimality because his proof is different—see Busard 1984, cc. 215.48–216.7.
[106] XII.6 does not exist neither in the indirect tradition nor in **b**.

VI.19).[107] Every proof of these theorems is of course carried out with a particular number of sides, which is in fact the smallest number of sides compatible with full generality. In the case of polygons, a pentagon must be used,[108] and for two reasons. First, neither a triangle nor a quadrilateral are polygons: this neatly results from I.def.19, in which the three species of a rectilinear figure are listed: "polylateral ‹figures are› those contained by more than four straight lines".[109] Second, in VI.20 and XII.6, more than 4 sides are necessary in order for the proof to retain its validity for any number of sides. In fact, the pentagon is divided into triangles by drawing all diagonals from one of its vertices; of the three resulting triangles, the "innermost" triangle is not homologous to the other two, since only one of its sides coincides with a side of the pentagon. The proof proceeds by incomplete induction on the number of sides, and the transition from an "external" to the "innermost" triangle requires specific deductive steps. In a quadrilateral, instead, such an "innermost" triangle does not exist, so that the proof is not sufficiently general. Since, in the Greek text, both VI.20 and XII.6[110] are generalizations of proofs enunciated for triangles, the proof for quadrilaterals was very likely omitted as useless on account of deductive economy.[111] As for XII.1, in which the "innermost" triangle is of no use, it is enough to note that, in subsequent theorems, this proposition is applied to figures with arbitrarily many sides. Thus, proving XII.1 for a quadrilateral would have resulted in a loss of generality.

If a theorem is intended to be valid for triangles and quadrilaterals along with polygons, "rectilinear figures" are mentioned, and in fact the proofs of I.45 and VI.18 are instantiated with the minimal number of sides, namely, four. In II.14, VI.21, 28, and 29, the "rectilinear figures" are denoted by one single letter, so that minimality requires three sides, as all manuscripts have—Heiberg is wrong in using quadrilaterals.[112] VI.22 and 25 refer to pairs of independent figures: the manuscripts have for this reason a triangle and a quadrilateral, but the quadrilateral is denoted, in VI.22 by 3 letters, in VI.25 by just one: a pair of triangles would have worked as well.

The noun εἴδη "forms" in VI.19por and in VI.31 is synonymous with "rectilinear figures"; such "forms" are not even denoted by letters, but by a syntagm like τὸ εἶδος ἀπό "the form on …". Thus, the formulation does not set constraints on the number of sides:[113] accordingly, no associated figure is drawn in the manuscripts. In this case again, Heiberg over-interprets the text by adding "forms" shaped as quadrilaterals.

To conclude: explicitly deployed generality and the minimality requirements of the kind just discussed induce competing constraints in enunciations and in setting-outs; when a conflict arises, minimality prevails.[114]

Forms of ὁποιοσοῦν "whichever" mark the arbitrariness of a choice between two or more options.[115] The choice may select one item out of two: II.def.1, porism to VI.8; but also two items out of three: IX.15 and XI.20–22;[116] three items out of four: *Data* 83. In V.def.5, V.4, and *Data* 77 the

---

[107] The porism is firmly attested both in the direct and in the indirect tradition, apart from the obvious final interpolation (ἐπείπερ-clause in *EOO* II, 130.12–14).

[108] That the polygon in question is a pentagon *does not* result from the figure, but from the number of letters that in the text provide the name of the object called "polygon".

[109] At *EOO* I, 6.3–6. The term "polylateral" is a hapax in the Euclidean corpus, but this does not entail that the partition was not canonical: see *Def.* 39 and 64, at *HOO* IV, 38.9–11 and 46.8–10, respectively.

[110] But XII.6 is probably spurious, as we have just seen.

[111] This was not the opinion of the author of the first porism to VI.20, absent in the entire indirect tradition.

[112] And even a pentagon in VI.29.

[113] Of course, the porism does not have a figure.

[114] A further, and paradigmatic, example are V.2–3, 5–6 (cf. Sect. 4.5.1.4).

[115] Only 2 occurrences of ὁστισοῦν "any", both in VII.38 (*EOO* II, 266.16–18).

[116] But the occurrences in XI.21–22 are simply instantiated citations of XI.20.

number of possible options is unlimited, and one of them must be chosen.[117] As is to be expected, the indirect tradition presents a poorer record:[118] the qualifier is absent in II.def.1; it is attested in XI.20 but not in XI.21–22 because the instantiated citations in which it is contained are altogether absent.[119] The occurrences of δύο ὁποιοιοῦν "whichever two" in the enunciation and in the setting-out of IX.15 are confirmed by Gerard and by both families of the Arabic tradition.

The porism to VI.8 (*EOO* II, 102.23–104.2) raises a textual problem that allows me to explain how such issues must be handled:

ἐκ δὴ τούτου φανερὸν ὅτι ἐὰν ἐν ὀρθογωνίῳ τριγώνῳ ἀπὸ τῆς ὀρθῆς ἐπὶ τὴν βάσιν κάθετος ἀχθῇ, ἡ ἀχθεῖσα τῶν τῆς βάσεως τμημάτων μέση ἀνάλογόν ἐστιν, ὅπερ ἔδει δεῖξαι, [καὶ ἔτι τῆς βάσεως καὶ ἑνὸς ὁποιουοῦν τῶν τμημάτων ἡ πρὸς τῷ τμήματι πλευρὰ μέση ἀνάλογόν ἐστιν].

Thus it is manifest from this that, if in a right-angled triangle a ‹straight line› is drawn from the right ‹angle› perpendicular to the base, the drawn one is mean proportional of the segments of the base, which it was really required to prove. [And further, of the base and of one whichever of the segments, the side corresponding to the segment is mean proportional].[120]

Let us get a look at Heiberg's apparatus: the bracketed clause in the quote above is found in the main text in **PBp**, by the first hand but in erasure in **F**, by a second hand and in the margin in **V**.[121] This record forces Heiberg to keep the clause in the text, while declaring: "sine dubio interpolata est". Reasons supporting Heiberg's contention are: the clause follows the canonical QED phrase; the result it states is never used; it is missing in some good Theonine manuscripts. Heiberg does not assess the issue correctly, however. Obvious counterarguments are: the QED phrase can in fact be read only in **P**; plenty of perfectly well-attested propositions are never used in the *Elements*; strictly Lachmannian criteria (that is, the bracketed clause is found in both branches of a bipartite manuscript tradition) force us to neglect the testimony of **FV**. If this is no guarantee that the whole porism is original, it certainly dates back to early textual layers: therefore, it must be kept in its entirety. The indirect tradition confirms this assessment—there is a proliferation of determiners of generality, but the only such determiner attested in Greek retains its place:[122]

Adelard I

unde etiam manifestum est quia omnis trianguli rectanguli a cuius angulo recto perpendicularis ad basim exierit, erit perpendicularis proportio ad dividentia basis, unicuique etiam lateri ad totam basim atque ad dividentia singularia.

Gerard

unde ex hoc manifestum est quod perpendicularis protracta ab omni recto angulo cuiuslibet trianguli ad basim est proportionalis inter duas sectiones basis, et quod unumquodque duorum laterum trianguli est proportionale inter basim et unam sectionum eius quod ipsam sequitur.

[117] I neglect the occurrence in the spurious lemma X.27/28. *Data* 77 is almost certainly spurious, as it does not find a place in the description of the treatise we read in Pappus, *Coll.* VII.4.

[118] The occurrences in *Data* 77 and 83 are retained in the Arabic translation: see text and translation in Sidoli, Isahaya 2018, 164–165 and 174–175, respectively.

[119] Of the four occurrences in the Greek text of XI.20, only three are preserved in the indirect tradition (the fourth is contained in an obviously spurious passage). Gerard always translates δύο ὁποιαιοῦν with *omnes duo*, as in IX.15, Adelard once in the same way and twice with *duo quolibet/quilibet*.

[120] The English formulation is not idiomatic, but the alternative was a sentence in which "one whichever of the segments" has just been mentioned; the problem lies in the fact that the attested word order is idiomatic in Greek but not in English.

[121] The manuscript **F** has the hapax ὁποτερουοῦν.

[122] At Busard 1983, 173.209–212, and Busard 1984, cc. 144.53–145.4, respectively.

> vel aliter: unde ex hoc manifestum est quod perpendicularis producta ad basim a recto angulo cuiuslibet trianguli est media in proportione inter easdem sectiones basis. quod tamen in greco non invenitur, et unumquodque duorum laterum trianguli est medium in proportione unicuique basi et inter unamquamque duarum sectionum eius.

Gerard collates several sources, as is customary with him, and confirms that a curtailed version of the porism was circulating.

5.1.5. *The use of the article*

A subtle issue related to the generality of the *enunciations* is raised by the fact that they may apply to objects taken as generic representatives of a certain class, but designated in the singular and qualified by an article.[123] A priori, this might result in a loss of generality of the formulation. That this is not the case in some crucial instances we have already seen in Sect. 3.3. Let us discuss some other representative examples.

The first example comes from an interesting phenomenon that can be detected by comparing the *Elements* and the *Data*. In the *Data*, no passive participial form of δίδωμι "to give" modifying an indefinite noun has the article, in the *Elements*, every such form has an article.[124] All instances of this phenomenon in the *Elements* come from the enunciations of problems: the object that must be constructed carries no article, but the "givens" always have one, both in the enunciation and in the setting-out. Let us read again I.1 (*EOO* I, 10.14–18):

ἐπὶ τῆς δοθείσης εὐθείας πεπερασμένης τρίγωνον ἰσόπλευρον συστήσασθαι.	Construct an equilateral triangle on a given bounded straight line.
ἔστω ἡ δοθεῖσα εὐθεῖα πεπερασμένη ἡ AB. δεῖ δὴ ἐπὶ τῆς AB εὐθείας τρίγωνον ἰσόπλευρον συστήσασθαι.	Let there be a given bounded straight line, AB. Thus it is required to construct an equilateral triangle on straight line AB.

One might submit that we are here facing a defect of generality, since the given straight line is generic, and still it is designated by the articular noun phrase ἡ δοθεῖσα εὐθεῖα. We cannot simply use "neutralization" to restore genericity (see Sect. 3.3); the point is to explain the diverging practices in the *Elements* and in the *Data*. The problem can be given two complementary solutions: one of them is purely grammatical, the other is pragmatic.

The grammatical solution runs as follows:[125] when a noun is qualified by an attributive participle further determined by a complement in prepositional form, the participle must be preceded by the article: this we have seen in Sect. 3.3 and shall see again presently. The forms of the participle δοθείς, however, are (almost) never preceded in the *Elements* by such a prepositional complement.

[123] For classes of mathematical entities referred to in the plural, the article is mandatory in Greek; cf. again Galen, *Inst. Log.* XII.7, read in Sect. 5.1.
[124] For statistical data about these participles, see Acerbi 2011a, 127–128.
[125] See Federspiel 1995, 258–259.

Therefore, as a construct without the article attached to the forms of δοθείς is in principle possible (this is the evidence from the *Data*), the solution adopted as canonical in the *Elements* is the consequence of a stylistic choice that aims at standardizing the form of all attributive participles.

The second solution stems from an obvious fact: no proposition of the *Data* can be a problem, so that the status of the geometric "givens" in them is altogether different from the status of the "givens" in the *Elements*. What is at issue in each theorem of the *Data* is transferring the predicate "given" from some objects to others: the *Data* deal with mathematical objects qua given, not with mathematical objects an incidental trait of which is to be given, as is the case with the "givens" of a problem in the *Elements*. Still, such an incidental trait is crucial from the mathematical point of view, and singles out the "given" object from all the objects that figure in the enunciation of a problem in the *Elements*—whereas in a *Data*-theorem (almost) all objects are given. Now, a standard stylistic resource of Greek language in order to create a reference specifying modifier and to place emphasis on it consists in providing the modifier with an article and in preposing it to the noun (phrase) it is intended to modify:[126] the object of interest is "such-and-such an object, most notably, the given one". As a consequence, the articular noun phrase ἡ δοθεῖσα εὐθεῖα is made of the indefinite noun εὐθεῖα "a straight line" and of the preposed determiner ἡ δοθεῖσα "given", which carries the saliency of the syntagm.

Grammatical reasons may also force the presence of the article before the name of a generic object mentioned in the singular. Let us rule out first many trivial instances: if an enunciation is in conditional form, the anaphoric value of the article explains all the occurrences of articular noun phrases in the consequent of the conditional, since the entities there mentioned have normally been introduced in the antecedent of the same conditional—or are obviously connected with entities there mentioned. For instance, in XIII.9–12 reference is made to "*the* side of a regular **agon inscribed in a circle": the article is required by the fact that, whenever a regular polygon is inscribed in a given circle, its side is univocally determined.[127]

How to deal with the phenomenon of denominations in the singular preceded by an article is best explained by looking at Book X. Let us read the enunciations of X.54, 60, and 66—in which some property is predicated of the irrational line called "binomial"—and in parallel those of X.91, 97, and 103, which are the exact counterpart of propositions X.54, 60, and 66 for the irrational called "apotome":[128]

ἐὰν χωρίον περιέχηται ὑπὸ ῥητῆς καὶ τῆς ἐκ δύο ὀνομάτων πρώτης, ἡ τὸ χωρίον δυναμένη ἄλογός ἐστιν ἡ καλουμένη ἐκ δύο ὀνομάτων.	ἐὰν χωρίον περιέχηται ὑπὸ ῥητῆς καὶ ἀποτομῆς πρώτης, ἡ τὸ χωρίον δυναμένη ἀποτομή ἐστιν.
τὸ ἀπὸ τῆς ἐκ δύο ὀνομάτων παρὰ ῥητὴν παραβαλλόμενον πλάτος ποιεῖ τὴν ἐκ δύο ὀνομάτων πρώτην.	τὸ ἀπὸ ἀποτομῆς παρὰ ῥητὴν παραβαλλόμενον πλάτος ποιεῖ ἀποτομὴν πρώτην.
ἡ τῇ ἐκ δύο ὀνομάτων μήκει σύμμετρος καὶ αὐτὴ ἐκ δύο ὀνομάτων ἐστὶ καὶ τῇ τάξει ἡ αὐτή.	ἡ τῇ ἀποτομῇ μήκει σύμμετρος ἀποτομή ἐστι καὶ τῇ τάξει ἡ αὐτή.

[126] On this combination of articularity and of the principle of saliency of the liminal position in a complex noun phrase, see Bakker 2009a, with a synthesis of her findings at 288–290. Apollonius Dyscolus already equated articular definiteness with identifiability: *Synt*. I.43, in *GG* II.2, 38–39.

[127] In XIII.9, the sides take on the article because they are already mentioned in the antecedent of the conditional together with the associated polygons.

[128] Texts at *EOO* III, 158.18–21, 180.26–182.2, 200.4–6; 274.17–19, 304.11–12, 330.19–20 (in order, each column from top to bottom).

If a region be contained by an expressible and a first binomial, the ‹straight line› worth the region is irrational, the so-called binomial.	If a region be contained by an expressible and a first apotome, the ‹straight line› worth the region is an apotome.
The ‹square› on a binomial applied to an expressible makes a first binomial width.	The ‹square› on an apotome applied to an expressible makes a first apotome width.
A ‹straight line› commensurable in length with a binomial is itself also a binomial and the same in order.	A ‹straight line› commensurable in length with an apotome is itself an apotome and the same in order.

Comparing the formulations in the two columns, and further comparing them with the associated setting-outs (see just below), we conclude that in all these theorems the designations of the irrational lines are indefinite: the article in the designations of the "binomial" is required by the fact that, in Greek, ἐκ δύο ὀνομάτων "binomial" is a prepositional syntagm, which must receive an article if it has to acquire the declensional structure of a noun. The only exceptions to this rule may occur when the syntagm is certainly in the nominative; this can only be the case either when the syntagm is the nominal complement of the copula (X.66 above) or when it is the subject of a verb taken in absolute sense (setting-out of X.66 below). The ἀποτομή "apotome", instead, is designated by a noun, which does have a declensional structure: no article must be attached if the referent is indefinite, and in fact the article is absent in all enunciations just read. The article *in the dative* in the phrase in X.103 ἡ τῇ ἀποτομῇ μήκει σύμμετρος "a ‹straight line› commensurable in length with an apotome" is dictated by standardization of all statements of this kind in Book X, and, in a structural perspective, by adherence to the standard form of the transitivity statements (see Sect. 4.5.1.4).

Again, consider the article *in the nominative* in the expression ἡ τῇ ἐκ δύο ὀνομάτων μήκει σύμμετρος "a ‹straight line› commensurable in length with a binomial" of X.66 and in the parallel expression in X.103. This article is required by the fact that the attribute μήκει σύμμετρος of an understood "straight line" is further modified by a complement. The resulting expression ἡ τῇ ἐκ δύο ὀνομάτων μήκει σύμμετρος ‹εὐθεῖα› has all possible articles but is altogether indefinite: this is exactly the same phenomenon studied in Sect. 3.3 on the example of I.29.[129] As in I.29, the setting-out confirms the indefinite character of the expression, and accordingly we have to supply an understood εὐθεῖα in this case too. In fact, the setting-out of X.66 (*EOO* III, 200.7–9) reads:

ἔστω ἐκ δύο ὀνομάτων ἡ ΑΒ, καὶ τῇ ΑΒ μήκει σύμμετρος ἔστω ἡ ΓΔ. λέγω ὅτι ἡ ΓΔ ἐκ δύο ὀνομάτων ἐστὶ καὶ τῇ τάξει ἡ αὐτὴ τῇ ΑΒ.	Let there be a binomial, AB, and let a ‹straight line›, ΓΔ, be commensurable in length with AB. I claim that ΓΔ is a binomial and the same in order as AB.

Since the liminal ἔστω "let there be" has a presential value, ἐκ δύο ὀνομάτων is its (indefinite) grammatical subject and therefore does not receive an article.

As said, the article is also absent when the noun group is the nominal complement of the copula, as in the enunciations of X.54 and X.66. In X.54, the article ἡ qualifies the participle καλουμένη and not the subsequent syntagm ἐκ δύο ὀνομάτων: the expression is neutralized but the meaning is indefinite. Finally, the expression πλάτος ποιεῖ τὴν ἐκ δύο ὀνομάτων πρώτην in X.60 must be translated "makes a first binomial width" and not "makes a first binomial as width": the presence of the article proves in fact that τὴν ἐκ δύο ὀνομάτων πρώτην is attributive; the article would not be

[129] The first αὐτή in X. 66 and 103 does not have the article (= "itself"), the second does have it (= "the same").

required if the same expression were the direct complement of the verb because the adjective πρῶτος "first" does admit of a declension into cases: see for instance the occurrences in the final porism of Book X.[130]

The article is also required by grammar in XI.14 (*EOO* IV, 38.4–8):

πρὸς ἃ ἐπίπεδα ἡ αὐτὴ εὐθεῖα ὀρθή ἐστιν, παράλληλα ἔσται τὰ ἐπίπεδα.	Those planes will be parallel such that a same straight line is orthogonal to the planes.
εὐθεῖα γάρ τις ἡ AB πρὸς ἑκάτερον τῶν ΓΔ EZ ἐπιπέδων πρὸς ὀρθὰς ἔστω. λέγω ὅτι παράλληλά ἐστι τὰ ἐπίπεδα.	In fact, let some straight line, AB, be at right ‹angles› with each of the planes ΓΔ, EZ. I claim that the planes are parallel.

The first occurrence of ἐπίπεδα "planes" in the enunciation does not have an article,[131] the second occurrence is articular because the planes have been already mentioned. The article in ἡ αὐτὴ εὐθεῖα "a same straight line" is required by the meaning of αὐτή: the expression is in fact indefinite, as the setting-out confirms, strengthening this connotation with τις "some".

5.1.6. *Ordinals as variables*

In Stoic logic, syllogisms were often formulated using ordinals as generic terms that replace or denote the ἀξιώματα "statements". The syllogisms formulated in this way were called τρόποι "modes"; the λογότροποι "argumodes" are the syllogisms in which concrete terms are subsequently referred to by ordinals; let us read D.L. VII.76–77:[132]

λόγος δέ ἐστιν, ὡς οἱ περὶ τὸν Κρῖνίν φασι, τὸ συνεστηκὸς ἐκ λήμματος καὶ προσλήψεως καὶ ἐπιφορᾶς [...] τρόπος δέ ἐστιν οἱονεὶ σχῆμα λόγου, οἷον ὁ τοιοῦτος "εἰ τὸ πρῶτον, τὸ δεύτερον· ἀλλὰ μὴν τὸ πρῶτον· τὸ ἄρα δεύτερον". 77 Λογότροπος δέ ἐστι τὸ ἐξ ἀμφοτέρων σύνθετον, οἷον "εἰ ζῇ Πλάτων, ἀναπνεῖ Πλάτων· ἀλλὰ μὴν τὸ πρῶτον· τὸ ἄρα δεύτερον". παρεισήχθη δὲ ὁ λογότροπος ὑπὲρ τοῦ ἐν ταῖς μακροτέραις συντάξεσι τῶν λόγων μηκέτι τὴν πρόσληψιν μακρὰν οὖσαν καὶ τὴν ἐπιφορὰν λέγειν, ἀλλὰ συντόμως ἐπενεγκεῖν "τὸ δὲ πρῶτον· τὸ ἄρα δεύτερον".	An argument, as Crinis is reported to assert, is what is made of an assumption, a coassumption, and a conclusion [...] A mode is a sort of scheme of an argument, like such a one: "if the first, the second; but of course the first; therefore the second". 77 An argumode is made of both, like "if Plato lives, Plato breaths; but of course the first; therefore the second". The argumodes were introduced not to repeat, in the formulation of the longer arguments, the coassumption, if it is long, and the conclusion, but to conclude in a concise way: "and the first: therefore the second".

These ordinals are abbreviations—and therefore logical constants—and not variables: they are invariably preceded by a neuter article, showing that either ἀξίωμα "statement" must be understood, or a reference must be assumed to the particular statement contained in the statement that opens an argumode. The same must be supposed for the "modes". The reason is that the Stoic indemonstrable

[130] The text is at *EOO* III, 354.23–356.7.
[131] The relative ἃ has this function, in fact.
[132] Examples of "argumodes" are also in Sextus, *M* VIII.242, 306, and in the fragments of Chrysippus' *Logical investigations* in PHerc. 307, II.21–26; best discussion of the issue in Frede 1974, 136–148. For the "modes" of the first three indemonstrables see Sextus, *M* VIII.227, and Sect. 5.2.2.

syllogisms are particular arguments, as is clear from two facts: first, their general description contains indefinite references to logical objects, not to ordinals; second, a specific name—namely, "mode"—was invented for the forms that contain ordinals. In the case of complex arguments, the transition from a concrete syllogism to a mode is not univocal, and in fact Chrysippus wrote a treatise in one book περὶ τοῦ τάττεσθαι τὸν αὐτὸν λόγον ἐν πλείοσι τρόποις "About the same argument being possibly put in several modes" and a treatise in two books πρὸς τὰ ἀντειρημένα τῷ τὸν αὐτὸν λόγον ἐν συλλογιστικῷ καὶ ἀσυλλογίστῳ τετάχθαι "Against the objections to the same argument being put in a syllogistic and an asyllogistic ‹mode›" (D.L. VII.194)—see also the remarks by Galen at *Inst. Log.* XV.8.

It is likely that the modes were introduced in order to facilitate the analyses of syllogisms (see Sect. 2.4.1). The logical form of any particular syllogism was thereby made perspicuous, for the schematic structure makes a check of the syllogism's validity easier. However, a concrete syllogism was always taken as reference, as we may check in all the attested examples of analyses, and as is obviously the case when the λογότροποι are employed.

Seemingly analogous devices can be found in mathematical texts. Let us start with Diophantus' *Arithmetica*. The unknowns of a Diophantine problem are often denoted by ordinals,[133] either in the setting-out or in the immediately subsequent ὑπόστασις "supposition".[134] Such ordinals are consistently used throughout the proof. Since these ordinals are always preceded by an article, they are abbreviations and not variables. In the manuscripts, they are always written in full letters: δεύτερος "second" and not β′ "2nd".[135] The following table sets out the occurrences of πρῶτος as a designation in each Book of the *Arithmetica*:[136]

Book	I	II	III	IV	V	VI	tot.
# prop.	39	35	21	40	30	24	189
tot.	75	46	114	199	34	/	468

Ordinals are also systematically introduced to denote the terms of a ratio or of a proportion in Book V of the *Elements*, where the general theory of proportions among magnitudes is presented. Let us read the enunciation of proposition V.13 (*EOO* II, 38.18–22):

ἐὰν πρῶτον πρὸς δεύτερον τὸν αὐτὸν ἔχῃ λόγον καὶ τρίτον πρὸς τέταρτον τρίτον δὲ πρὸς τέταρτον μείζονα λόγον ἔχῃ ἢ πέμπτον πρὸς ἕκτον, καὶ πρῶτον πρὸς δεύτερον μείζονα λόγον ἕξει ἢ πέμπτον πρὸς ἕκτον.

If first to second have the same ratio as third to fourth and third to fourth have a greater ratio than fifth to sixth, first to second will also have a greater ratio than fifth to sixth.

The crucial difference between this kind of designation by means of ordinals and the Stoic ordinals or the Diophantine designations is that in this enunciation, as elsewhere in Book V, the

[133] The greatest such ordinal employed is "seventh".

[134] On the Diophantine style see Acerbi 2011e, 57–95. In a Diophantine "supposition", the numbers that serve as parameters of a problem are assigned particular values.

[135] Designations of the unknowns by means of ordinals are found in the following propositions; bracketed numbers refer to the propositions in which the ordinals are introduced in the setting-out and not in the "supposition": I.5, 6, 15, 16, 17, 18, 18*alt*, 19, 19*alt*, 20, (21), 22, 23, 24, 25, II.(3, 4, 8, 8*alt*, 9, 10, 15), 17, 18, (19, 28, 29, 30, 31, 32, 33, 34, 35), III.(1, 2, 3, 4, 5, 5*alt*, 6, 6*alt*, 7, 8, 9, 10, 11, 12, 13, 14, 15, 15*alt*, 16, 17, 18, 19, 20, 21), IV.(1), 7, (9, 10, 13), 15, (16, 17, 18, 19, 20, 21, 22, 23, 24, 25, 26, 27, 28, 28*alt*, 30, 31, 31*alt*, 32), 33, (33/34), 34, (34/35), 35, 36, 37, (38, 39, 40), V.(1, 2, 3, 4, 5, 6, 6/7, 7, 7/8, 8, 12, 13, 14, 15, 17, 24, 25, 28, 30).

[136] The absence of such designations in Book VI is motivated by the type of arithmetic problem there presented.

ordinals are not preceded by an article: they are introduced directly, and do not follow any explicit mention of the multiplicity of the magnitudes at issue (this would make the article necessary because of the anaphoric reference—see above, and again below). For this reason, the mathematical ordinals without an article of Book V are not abbreviations with a μέγεθος "magnitude" understood, but a way of denoting the terms in proportion that does not use to letters: they are *variables*, and the absence of the article further reflects the indefinite character of the enunciation. As a consequence, the function of the ordinals without an article in Book V is quite different from the function of the letters in geometric proofs: the ordinals are abstract mathematical objects, whereas the letters, as we have seen in Sect. 3.2, are names of linguistic objects. This is confirmed by the fact that, in the setting-out of V.13 (*EOO* II, 38.23–28), the ordinals are regularly instantiated by letters:

πρῶτον γὰρ τὸ Α πρὸς δεύτερον τὸ Β τὸν αὐτὸν ἐχέτω λόγον καὶ τρίτον τὸ Γ πρὸς τέταρτον τὸ Δ τρίτον δὲ τὸ Γ πρὸς τέταρτον τὸ Δ μείζονα λόγον ἐχέτω ἢ πέμπτον τὸ Ε πρὸς ἕκτον τὸ Ζ. λέγω ὅτι καὶ πρῶτον τὸ Α πρὸς δεύτερον τὸ Β μείζονα λόγον ἕξει ἤπερ πέμπτον τὸ Ε πρὸς ἕκτον τὸ Ζ.

In fact, let first, A, to second, B, have the same ratio as third, Γ, to fourth, Δ, and let third, Γ, to fourth, Δ, have a greater ratio than fifth, E, to sixth, Z. I claim that first, A, to second, B, will also have a greater ratio than fifth, E, to sixth, Z.

One might wonder whether, here and in the Stoic modes, we are not simply being deluded by a copyist who reads A but writes πρῶτον. For Book V this possibility must be excluded: as we have just seen, in the setting-out the ordinals are regularly assigned letters, and expressions are found there like πρῶτον τὸ Α, which would constitute intolerable repetitions. What is more, in V.13 the sixth term is arrived at, and the ordinal for 6 is not ζ, but ς. As for Stoic ordinals, at least one testimony refers to the abbreviations as *numeri*.[137]

Not all the occurrences of ordinals in Book V are non-articular.[138] The presence of the article can be forced by grammatical constraints: an article is added when the ordinals refer to already posited magnitudes, either because they are introduced as ordinals without an article—as in the enunciation of V.14 (*EOO* II, 42.6–10),

ἐὰν πρῶτον πρὸς δεύτερον τὸν αὐτὸν ἔχῃ λόγον καὶ τρίτον πρὸς τέταρτον τὸ δὲ πρῶτον τοῦ τρίτου μεῖζον ᾖ, καὶ τὸ δεύτερον τοῦ τετάρτου μεῖζον ἔσται, κἂν ἴσον, ἴσον, κἂν ἔλαττον, ἔλαττον.

If first to second have the same ratio as third to fourth and the first be greater than the third, the second will also be greater than the fourth, and if it be equal, equal, and if it be lesser, lesser.

or as specific multiplicities of magnitudes expressly mentioned at the beginning of the sentence, as in the enunciations of V.20–21 and in V.def.9 (*EOO* II, 4.8–10):[139]

ὅταν δὲ τρία μεγέθη ἀνάλογον ᾖ, τὸ πρῶτον πρὸς τὸ τρίτον διπλασίονα λόγον ἔχειν λέγεται ἤπερ πρὸς τὸ δεύτερον.

Whenever three magnitudes be in proportion, the first to the third is said to have a duplicate ratio than to the second.

[137] See Apuleius, *Int.*, 212.10–12.

[138] Let us take for instance δεύτερος "second" or τρίτος "third". The occurrences without an article are in V.def.5, 2 (*decies*), 3 (*sexties*), 4 (*ter*), 13 (*sexties*), 14(*ter*), 24 (*nonies*), those with an article in V.def.5.7(*bis*).9.10.17.18(*bis*), 3 (*ter*), 4 (*bis*), 14 (*bis*), 16, 18, 20 (*bis*), 21 (*bis*).

[139] The same occurs in VII.19.

Ordinals that denote magnitudes are also found in the *Data*, but they always carry an article, and are always preceded by the specification of the magnitudes at issue,[140] as in the enunciation of *Data* 74 (*EOO* VI, 142.6–10):

ἐὰν δύο παραλληλόγραμμα λόγον ἔχῃ δεδομένον ἤτοι ἐν ἴσαις γωνίαις ἢ ἀνίσοις μὲν δεδομέναις δέ, ἔσται ὡς ἡ τοῦ πρώτου πλευρὰ πρὸς τὴν τοῦ δευτέρου πλευρὰν οὕτως ἡ ἑτέρα τοῦ δευτέρου πλευρὰ πρὸς ἣν ἡ λοιπὴ τοῦ πρώτου λόγον ἔχει δεδομένον.	If two parallelograms, either in equal or unequal but given angles, have a given ratio, as the side of the first is to the side of the second, so the other side of the second will be to a ‹straight line› to which the remaining ‹side› of the first has a given ratio.

These ordinals are abbreviations of names of already posited objects.

In the sentences just read, the occurrences with an article are abbreviations, that is, logical constants with a well-defined referent. Conversely, when a theorem is directly enunciated with ordinals, the article is always absent, as we have seen. Of particular significance is the crucial definition of proportionality in V.def.5. At the beginning of the definition, generic μεγέθη "magnitudes" are mentioned, whose multiplicity is not specified even if they are obviously four. As a consequence, when the magnitudes are identified for the first time by ordinals, these carry no article, which they regularly take on at their second occurrence. Let us read V.def.5 (*EOO* II, 2.10–16):

ἐν τῷ αὐτῷ λόγῳ <u>μεγέθη</u> λέγεται εἶναι <u>πρῶτον</u> πρὸς δεύτερον καὶ τρίτον πρὸς τέταρτον, ὅταν τὰ <u>τοῦ πρώτου</u> καὶ τρίτου ἰσάκις πολλαπλάσια τῶν τοῦ δευτέρου καὶ τετάρτου ἰσάκις πολλαπλασίων καθ' ὁποιονοῦν πολλαπλασιασμὸν ἑκάτερον ἑκατέρου ἢ ἅμα ὑπερέχῃ ἢ ἅμα ἴσα ᾖ ἢ ἅμα ἐλλείπῃ ληφθέντα κατάλληλα.	<u>Magnitudes</u> are said to be in a same ratio, <u>first</u> to second and third to fourth, whenever the equimultiples of <u>the first</u> and third, according to whichever multiple, respectively, either exceed together or be equal together or fall short together of the equimultiples of the second and fourth, taken in corresponding order.

The enunciation with ordinals of V.def.5 determines an ordering of the four magnitudes in proportion, but does not specify which are the antecedents and which are the consequents of the two resulting ratios. In fact, the presence of both "exceed" and "fall short"—that is, of both "greater" and "less"—in the definition makes it symmetric with respect to the exchange of the pair of magnitudes termed "first" and "third" with those termed "second" and "fourth", taken in the same order.[141] For this reason, it is not necessary to prove that the ἀνάπαλιν "by inversion" property holds,[142] whereas for instance the ἐναλλάξ "by alternation" property requires a proof (V.16).[143] Moreover, since it is immaterial which of the ratios is taken first, the relation "being in a same ratio" is symmetric. As a consequence, V.def.5 automatically applies to all permutations of the magnitudes in proportion that preserve their original structure of ordered pairs. There are two such possibilities: by inversion 1234 → 2143 (ordered pair 13 interchanged with 24) and by symmetry 1234 → 3412 (12 interchanged with 34). Interchanging 14 with 23, namely, the middle terms with

[140] These can be general magnitudes of well-defined multiplicity, but also parallelograms, triangles, or straight lines.
[141] That the same order must be kept is specified by the final phrase ληφθέντα κατάλληλα. Of course, magnitudes can be interchanged only by pairs.
[142] That is, if $A:B::C:D$, then $B:A::D:C$. A porism to V.7 states that this is the case, but the porism is certainly spurious.
[143] That is, if $A:B::C:D$, then $A:C::B:D$ and $D:B::C:A$.

the extremes, gives the same result as inversion. The proportion "read from right to left" corresponds to the permutation 1234 → 4321, which is a combination of the two permutations above.

Thus, from the combinatorial point of view, there is a fundamental difference between the Stoic "modes" and the use of ordinals in V.def.5. In fact, the Stoic "modes" of presentation of an argument entail that the (validity of such an) argument is independent of the affirmative or negative character of the statements that compound it. However, the presentation by means of ordinals introduces a spurious asymmetry between the statements denoted by different ordinals. For instance, a "mode" of a third indemonstrable might be: "it is not the case that: both the first and the second; but of course the first: therefore not the second"—there is no symmetry between the two ordinals, for "the second" is negated and "the first" is not, even if conjunction is a symmetric connector. The descriptions, or "circumscriptions",[144] of the indemonstrables encompass instead several permutations within a single sentence (Sextus, *M* VIII.226; cf. D.L. VII.80 and Sect. 5.2.2):

τρίτος δέ ἐστι λόγος ἀναπόδεικτος ὁ ἐξ ἀποφατικοῦ συμπλοκῆς καὶ ἑνὸς τῶν ἐν τῇ συμπλοκῇ, τὸ ἀντικείμενον τοῦ λοιποῦ τῶν ἐν τῇ συμπλοκῇ ἔχων συμπέρασμα.	A third indemonstrable argument is the one ‹compounded› of a negation of a conjunction and by one of the items in the conjunction, having the contradictory of the remaining item in the conjunction as conclusion.

5.1.7. *The indefinite conditionals of Stoic logic*

The duality between instantiated statements and their indefinite counterparts was thematized in Stoic logic in a peculiar way. Chrysippus' school introduced a kind of non-simple sentence in view of formulating "general" statements.[145] These are conditionals with indefinite antecedent (they include an indefinite pronoun τις / τι "someone / -thing" and a predicate); the consequent of the conditional features a reference, realized by a demonstrative pronoun (ἐκεῖνος "that one")[146] anaphoric of the indefinite pronoun in the antecedent: for instance, εἴ τίς ἐστιν ἐνταῦθα, οὐκ ἔστιν ἐκεῖνος ἐν Ῥόδῳ "if someone is here, not: that one is in Rhodes".[147] This was the Stoic way of handling issues of universal quantification, that is, of formulating statements whose predicates apply to a class of objects. Such statements were called καθολικά "general";[148] they are in fact definitions.[149] Here is the exposition in Sextus, *M* XI.8:

τὸν γὰρ ὅρον φασὶν οἱ τεχνογράφοι ψιλῇ τῇ συντάξει διαφέρειν τοῦ καθολικοῦ, δυνάμει τὸν αὐτὸν ὄντα. καὶ εἰκότως· ὁ γὰρ εἰπὼν "ἄνθρωπός ἐστι ζῷον λογικὸν θνητόν" τῷ εἰπόντι "εἴ τί ἐστιν ἄνθρωπος, ἐκεῖνο ζῷόν ἐστι λογικὸν θνητόν" τῇ μὲν δυνάμει τὸ αὐτὸ λέγει, τῇ δὲ φωνῇ διάφορον.	Competent people say that a definition differs from a general ‹statement› as to syntax only, even if they are the same as to meaning. And reasonably so: for whoever says "a man is a mortal rational animal" says the same thing as to meaning, but a different thing as to wording, as whoever says "if something is a man, that one is a mortal rational animal".

[144] So Barnes 2007, 286–292, with a discussion.
[145] On indefinite conditionals see Crivelli 1994; Caston 1999, 192–204.
[146] In Greek mathematical style, the pronoun ἐκεῖνος always correlates with a proleptic relative.
[147] The example is at D.L. VII.82, and cf. also VII.75.
[148] At Sextus, *M* XI.8–11, and cf. Cicero, *De fato* 11–15, to be partly read just below.
[149] See also, most recently, Crivelli 2010, 409–415.

A Stoic indefinite conditional is true if and only if all its instances are true; the instances are said to "be subordinated" to the indefinite conditional.[150] Within arguments, an indefinite conditional of a kind different from the one just seen is supplemented by a definite coassumption, which contains a demonstrative or a noun that provides a specific reference to the indefinite pronoun in the antecedent of the conditional. Let us read an example in Sextus, *P* II.141:[151]

εἴ τίς σοι εἶπεν ὅτι πλουτήσει οὗτος, πλουτήσει οὗτος· οὑτοσὶ δὲ ὁ θεός – δείκνυμι δὲ καθ' ὑπόθεσιν τὸν Δία – εἶπέ σοι ὅτι πλουτήσει οὗτος· πλουτήσει ἄρα οὗτος.

If someone told you that this one will become rich, this one will become rich; and this god here—and I suppose I am pointing at Jupiter—told you that this one will become rich; therefore this one will become rich.

Generality is also at issue in the prescription of transforming some kinds of indefinite conditional into a negation of a conjunction, as results from Cicero, *De fato* 15–16:

15 hoc loco Chrysippus aestuans falli sperat Chaldaeos ceterosque diuinos, neque eos usuros esse coniunctionibus,[152] ut ita sua percepta pronuntient "si quis natus est oriente Canicula, is in mari non morietur" sed potius ita dicent: "non et natus est quis oriente Canicula et is in mari morietur". o licentiam iocularem: ne ipse incidat in Diodorum, docet Chaldaeos quo pacto eos exponere percepta oporteat. quaero enim, si Chaldaei ita loquentur ut negationes infinitarum coniunctionum potius quam infinita conexa ponant cur idem medici cur geometrae, cur reliqui facere non possint. medicus in primis quod erit ei perspectum in arte non ita proponet "si cui uenae sic mouentur, is habet febrim", sed potius illo modo "non et uenae sic ‹cui› mouentur, et is febrim non habet"; itemque geometres non ita dicet "in sphaera maximi orbes medii inter se diuiduntur", sed potius illo modo: "Non et sunt in sphaera maximi orbes, et ei non medii inter se diuiduntur". **16** quid est, quod non possit isto modo ex conexo transferri ad coniunctionum negationem? et quidem aliis modis easdem res efferre possumus. modo dixi "in sphaera maximi orbes medii inter se diuiduntur", possum dicere: "si in sphaera maximi orbes erunt", possum dicere "quia in sphaera maximi orbes erunt"; multa genera sunt enuntiandi nec ullum distortius quam hoc, quo Chrysippus sperat Chaldaeos contentos Stoicorum causa fore.

Contrary to what Cicero intimates with his geometric example (by the way, this is exactly Theodosius, *Sph.* I.11), the transformation recommended by the Stoics applies to subsistent states of affairs, which, however, cannot be linked by a Chrysippean συνάρτησις "connection": as a consequence, mathematical statements are ruled out (cf. Sect. 5.1.6).

The obvious difference between an indefinite conditional and a mathematical enunciation in conditional form is that the indefinite pronoun in the antecedent of a Stoic indefinite conditional refers to a *particular* entity.[153] The indefinite conditionals were in fact used to prevent general terms from being grammatical subjects in well-formed discourse, since in indefinite conditionals general terms can only figure as nominal complements of the copula.[154] So, despite the shared use of

[150] The Greek verb is ὑποτάττω; cf. Sect. 4.5.1.1 for the "inverse" technical term ἐπιτάττω "to subordinate".
[151] We find the same example in *M* VIII.308 and 313; note the repeated pronoun οὗτος "this one" (which has nothing to do with the indefinite character of the conditional in which it figures): the Stoics prescribed to avoid implicit references.
[152] To be corrected to *conexis*. The subsequent forms of *infinitus* mean of course "indefinite".
[153] Syntactically, the indefinite linguistic item in a Stoic indefinite conditional is a pronoun, whereas any determiner of indefiniteness in a mathematical sentence that formulates a statement is an adjective (see Sect. 5.1.3).
[154] See Caston 1999, 192–204; the aim of this linguistic device (admittedly quite contrived) is in fact to eliminate general terms from ontology, not only from discourse. Chrysippus, *apud* Alexander, *in APr.*, 402.16–18, adopts a similar strategy to give prominence to the existential import of singular statements; the syntactic shift he advocates is the same as the shift at work in the standard translation of the setting-out I have discussed in Sect. 3.1.

indefinite terms like τις as "quantifiers", the mathematical practice and the Stoic prescription appear to diverge irremediably.

However, some enunciations are better suited than others to fit the Stoic paradigm, especially if we free ourselves from the subject-predicate structure of Sextus' examples and, most importantly, if we accept that the indefinite pronoun be replaced by the indefinite designation of a mathematical species—after all, the gist of the Stoics' insight resides in the interplay between indefinite referent in the antecedent and anaphoric referent in the consequent. Let us read I.41 (*EOO* I, 96.5–8):

ἐὰν παραλληλόγραμμον τριγώνῳ βάσιν τε ἔχῃ τὴν αὐτὴν καὶ ἐν ταῖς αὐταῖς παραλλήλοις ᾖ, διπλάσιόν ἐστι τὸ παραλληλόγραμμον τοῦ τριγώνου.

If a parallelogram both have the same base as a triangle and be in the same parallels, the parallelogram is double of the triangle.

The anaphora is in this case realized by the article in the second occurrence of "parallelogram". A simple transformation produces a reasonably well-formed Stoic indefinite conditional:

*εἴ τί ἐστιν παραλληλόγραμμον τριγώνῳ βάσιν τε ἔχον τὴν αὐτὴν καὶ ἐν ταῖς αὐταῖς παραλλήλοις ὄν, ἐκεῖνο διπλάσιόν ἐστι τοῦ τριγώνου.

If something is a parallelogram both having the same base as a triangle and being in the same parallels, that one is double of the triangle.

Strongly anaphoric references in mathematical definitions are quite common, as the example of VII.def.7 (*EOO* II, 184.12–13) shows—I elaborate here on a suggestion of B. Wilck:

περισσὸς δὲ ὁ μὴ διαιρούμενος δίχα ἢ ὁ μονάδι διαφέρων ἀρτίου ἀριθμοῦ.

an odd ‹number› is that which cannot be divided into two ‹equals› or that which differs by a unit from an even number.

In both definitions of "odd", the article anaphorically refers to the grammatical subject (obviously ἀριθμός "number") of which the adjective περισσός "odd" that makes the whole *definiendum* is predicated. Since that grammatical subject is omitted, in principle this phrase could be read as follows: "somitem odd is that which cannot be divided into two ‹equals›", and finally "somitem is odd whenever that one cannot be divided into two ‹equals›", the coinage "somitem" getting rid of the disturbing presence of gender neuter "something" where the Greek displays a masculine gender context. The same applies to definitions such as V.def.2 ("multiple" for magnitudes) and VII.def.4–5 ("parts", and "multiple" for numbers), 9 ("even times odd"), 20 ("cube"): these are all definitions in which the grammatical subject of which the adjective that expresses the *definiendum* is predicated is omitted because it was expressly stated in the preceding definition.

VII.def.7 has other interesting features. First, it compiles two definitions; the first is exactly the negation of the definition of "even number" in VII.def.6, the second makes "odd" a definitional subordinate to "even". Second, the expression διαιρούμενος δίχα "divided into two ‹equals›" is exactly parallel to δίχα τεμνόμενος "bisected", widely used for straight lines. Third, the first definition of "odd" above and the definition of "even" in VII.def.6 exhibit a different word order: ὁ μὴ διαιρούμενος δίχα versus ὁ δίχα διαιρούμενος; the inversion is motivated both by the presence of the negation and by saliency, for what is salient in VII.def.6 is "can be divided into two ‹equals›", whereas in VII.def.7 it is "cannot be divided into two ‹equals›". Fourth, both definitions in VII.def.7 feature a participle, but διαιρούμενος carries a modal connotation of possibility, διαφέρων does not. And modals will be the subject of the next Section.

5.2. Modals

In mathematics, modals qualify states of affairs that obtain in specific configurations;[155] their linguistic expressions act on sentences according to the rules of ordinary speech. Modal operators normally have the syntactic function of predicates, whose subjects are declarative sentences: "such-and-such a state of affairs is impossible".[156]

Necessity qualifiers are scarce in mathematical texts, with the notable exception of a purely metalinguistic formula:[157] δεῖ "it is required" that introduces the two kinds of "determination" (see Sect. 4.2.1).[158] In the *Elements*, there are no occurrences of ἀναγκαῖος "necessary" or the like. The word is also exceedingly rare in the rest of the mathematical corpus; the only author who uses it is Archimedes: 9 occurrences in technical contexts and 1 in the preface of the *Stomachion*.[159] The necessity is predicated of the consequent of a conditional that introduces a reduction to the impossible, of a paraconditional (*bis*), of conclusions of inferences (*ter*), of a statement to be proved, of the reference to a known result, and of an immediate consequence of the geometric configuration.[160]

Other expressions that carry a connotation of necessity are δεικτέον "it must be proved"[161] and δέον (ἔστω) "let it be required". In the *Elements*, δεικτέον exhibits 16 occurrences, only in Book X,[162] plus 8 in the supplementary material, in this case always outside Book X.[163] The syntagm δέον (ἔστω) has a strong metadiscursive connotation; it occurs 8 times, in material that is either squarely inauthentic or stylistically and mathematically aberrant,[164] as for instance propositions IX.18–19. Outside the *Elements*, of some interest are the 4 occurrences of δέον (ἔστω) in the "gnomonic" propositions *Opt.* 18–21 **A**. Note also the 2 instances of θετέον "it must be posited", referring to assumptions, in the introduction to the *Phaenomena*.[165]

The predicates "true" and "false" are never used in the Euclidean corpus; they are sporadically met in other authors, but never within proofs.[166] In particular, Archimedes calls some statements "false" or "true", and qualifies the term συμπέρασμα "conclusion" as "true", while obviously referring to the "enunciation". The predicate ψευδής "false" was applied to fallacious proofs, as suggested by the title of the lost Euclidean *Pseudaria* and by a series of Aristotelian texts.[167] In general, explicit metadiscourse within proofs places emphasis on the fact that something has been proved, not on something being true because it has been proved (see Sect. 4.5.5).

---

[155] The Stoics held that modals are properties of statements; they are not operators that map statements into statements, as negation is. Cf. Alexander, *in APr.*, 177.25–178.4, D.L. VII.75, Boethius, *Int.* 2.11, 234.27–235.4; see also Sect. 5.3.3. On modals in Aristotle's *Int.*, see Weidemann 2005 and 2012, Aimar, unpublished typescript; on modals in Megaric and Stoic logic, see Bobzien 1999, 86–92 and 115–121.

[156] The verb δύναμαι "to be worth" and the δυνάμεις "powers" in the theory of irrational lines of Book X have nothing to do with modality.

[157] For the QED formula ὅπερ ἔδει δεῖξαι (ποιῆσαι) see Sect. 4.1.

[158] Recall that the "determination" that follows the setting-out of a problem is explicitly directive, and that a "real" determination states a necessary condition for the resolvability of a problem.

[159] The occurrence in Apollonius, prefatory epistle to *Con.* I (*AGE* I, 4.7), is irrelevant.

[160] At *AOO* I, 286.18, 304.6, 304.9, 328.8, 358.19, II, 56.9, 290.21, 304.7, 320.19, and II, 416.3–4, respectively.

[161] For its use in Archimedes as a substitute of λέγω see Sect. 4.2.1.

[162] In X.17–18 (both *bis*), 49, 53, 55–58, 60–63, 69–70. This verbal adjective introduces partial determinations, with the only exception of the last two occurrences, where it introduces the liminal determination.

[163] In XII.4/5, XIII.2/3, 13/14, 18/19, X.23por*alt*, 39*alt*, 117*alt*, XII.17*alt*.

[164] In III.33, IX.18–19, X.13/14, X.27*alt*, 29*alt*, 39*alt*.

[165] At *EOO* VIII, 2.8–14; for the propositions in *Optica*, see Acerbi 2011a, 133–134.

[166] In Archimedes there are 4 occurrences of ψεῦδος in the prefatory epistle to *Spir.*, 2 occurrences of ἀληθής in *Meth.* (*AOO* II, 4.29, 6.1, 6.11, 6.18, and 438.18–19, respectively). Apollonius exhibits 1 irrelevant occurrence in the prefatory epistle to *Con.* IV (*AGE* II, 4.10).

[167] A detailed discussion of the entire evidence is in Acerbi 2008c.

Since mathematical propositions establish truths that are necessary and time-independent, "false" can be upgraded to "necessarily false", that is, "impossible". Aristotle clearly states this, by using a well-known example of reduction to the impossible (*Metaph.* Δ.12, 1019b21–27):[168]

καὶ ἀδύνατα δὴ τὰ μὲν κατὰ τὴν ἀδυναμίαν ταύτην λέγεται, τὰ δὲ ἄλλον τρόπον, οἷον δυνατόν τε καὶ ἀδύνατον. ἀδύνατον μὲν οὗ τὸ ἐναντίον ἐξ ἀνάγκης ἀληθές – οἷον τὸ τὴν διάμετρον σύμμετρον εἶναι ἀδύνατον ὅτι ψεῦδος τὸ τοιοῦτον οὗ τὸ ἐναντίον οὐ μόνον ἀληθὲς ἀλλὰ καὶ ἀνάγκη· τὸ ἄρα σύμμετρον οὐ μόνον ψεῦδος ἀλλὰ καὶ ἐξ ἀνάγκης ψεῦδος.	And some items are called impossible in virtue of this kind of impossibility, while others are so in another sense, I mean possible and impossible. "Impossible" is that of which the contrary is of necessity true—for instance that the diagonal is commensurable is impossible, because such a ‹statement› is a falsity of which not only is the contrary true but also necessary; therefore that it is commensurable is not only false but of necessity false.

It is well known that the present indicative may carry a modal connotation of possibility. Examples of potential present in moods other than the indicative can be found in two principles: I.post.3 and X.def.1 (*EOO* I, 8.11–12, and III, 2.2–4)—see also Sects. 4.3 and 5.3.3:

καὶ παντὶ κέντρῳ καὶ διαστήματι κύκλον <u>γράφεσθαι</u>.	And that a circle <u>can be described</u> with any centre and radius.
σύμμετρα μεγέθη λέγεται τὰ τῷ αὐτῷ μέτρῳ <u>μετρούμενα</u> ἀσύμμετρα δὲ ὧν μηδὲν ἐνδέχεται κοινὸν μέτρον γενέσθαι.	Commensurable magnitudes are said those that <u>can be measured</u> with a same measure, incommensurable those of which it is not possible that any common measure come about.

Another example of potential present in a mathematical text[169] is VII.def.6 (*EOO* II, 184.11) ἄρτιος ἀριθμός ἐστιν ὁ δίχα <u>διαιρούμενος</u> "an even number is that which <u>can be divided</u> into two ‹equals›". Since Nicomachus rephrased the definition by making the potential connotation explicit,[170] the difference triggered a scholium in which the annotator shows himself unaware of the fact that the potential value of the present indicative can also be assumed by the present participle.[171]

As for making the modal connotation of (im)possibility explicit, three categories are witnessed to in the *Elements*:

a) Definitions such as V.def.4 (read in Sect. 2.2) or X.def.1 above (for "incommensurable").
b) Problems that carry an explicit emphasis on constructibility, exhibited by the presence of the modal phrase δυνατόν ἐστι "it is possible". In the *Elements*, only propositions IX.18–19 and XI.22 are of such a kind. Several occurrences can be found in the Archimedean corpus, and in the παράδοξα problems, ascribed to Erykinos, which Pappus presents in *Coll.* III.60–73 (see Sects. 4.2.1 and 4.3 for detail).
c) Phrases that initialize the main kind of indirect proof, a discussion of which I present in the following Section.

---

[168] The transmitted text is meaningless. I adopt Ross' athetesis, which is more economical than Jaeger's integration.
[169] But cf. Aristotle, *Top.* VI.4, 142b12. VII.def.7 (odd) is essentially the same definition; we have read it in Sect. 5.1.7.
[170] At *Ar.* I.7.2: ἔστι δὲ ἄρτιον μέν, ὃ <u>οἷόν τε</u> εἰς δύο ἴσα διαιρεθῆναι μονάδος μέσον μὴ παρεμπιπτούσης "an even ‹number› is that which <u>can</u> be divided into two equal ‹parts›, because a unit does not fall in the middle".
[171] See Riedlberger 2013, 130–131 for edition and translation of the scholium, 229–230 for comment; in particular, at 230 and n. 448 a preliminary discussion of the linguistic issue is provided; see also Acerbi 2017, 182, for an instance in an *Almagest* scholium.

5.2.1. *Reductions to the impossible*

In the *Elements*, there are 123 reductions to the impossible (*RI* henceforth, for both the singular and the plural), including the probably spurious propositions I.40 and III.12. Other *RI* can be found in the supplementary material: these are lemmas and disparate complements that Heiberg has kept in his main critical text (10 items),[172] alternative proofs (20),[173] redaction **b** of Book XII (12).[174] Add 3 items in the *Data* (propositions 28–30) and 10 in the minor works.[175] The following table sets out bookwise the distribution of the *RI* in the main text of the *Elements*:[176]

I (9)	4, 6–7, 14, 26 (*bis*), 27, 39–40
III (19)	1, 2, 4–8, 10–12, 13 (*bis*), 16 (*bis*), 18–19, 23–24, 27
IV (3)	4, 8, 13
V (1)	18
VI (3)	7 (*bis*), 26
VII (20)	1–2, 3 (*bis*), 20–24, 28 (*bis*), 29, 31, 33, 34 (*bis*), 35, 36 (*bis*), 39
VIII (4)	1, 4 (*bis*), (6)
IX (21)	10 (*bis*), 12, 13 (*sexties*), 14, 16–18, 19 (*bis*), 20, 30–31, 33–34, 36
X (21)	2–4, 13, 16 (*bis*), 26, 42 (*bis*), 43, (44), 45–47, 79–84, 111
XI (10)	1–3, 5, 7, 13–14, 19, 23 (*bis*)
XII (12)	2, 5, 10–12, 18 (each *bis*)

The "absurd" is reached in a variety of ways; the following list presents a complete typology:[177]

- contradiction with a previous result: I.26 (16), 27 (16), III.7 (7), 8 (8), 10 (5), 16 (I.17), 23 (I.16), 24 (10), IV.4, 8, 13 (each III.16), VI.7 (I.17), IX.13 (11, 12 + VII.def.12), X.42–43 (26), 44 (42), 45–46 (26), 47 (42), 79–80 (26), 81 (79), 82–83 (26), 84 (79), XI.2 (1), 14 (I.17), 19 (13), XII.2 (2), 5 (5), 11 (11), 12 (12), 18 (18);
- contradiction with a principle: I.4 (cn 7), VII.21, 23–24, 28 (each def. 13), 29 (def. 12), IX.13 (11, 12 + VII.def.12), 16, 17, 19, 31 (each VII.def.13), XI.1 (I.def.16), 3 (cn 9), 7 (cn 9);
- some magnitudes are both equal and unequal (ass.): straight lines III.5–6, 27, VI.26; angles I.7, 14, 26, III.1, 4, 13, 18, XI.5, 13, 23; triangles I.6, 39, 40;
- some magnitudes are both greater and less the one than the other (ass.): straight lines III.2, 16, III.11–12, 18; angles VI.7; regions or solids XII.2, 5, 10–12, 18; general magnitudes V.18;
- the same straight line is both internal and external (ass.) to a circle: III.13;

[172] These are VI.22/23, X.28/29II (*bis*), XIII.2/3 (*bis*), and XIII.18/19 (*quinquies*).
[173] In the alternative proofs of III.7–10, VII.31; in the propositions *vulgo* X.13, X.117 (*ter*), X.117*alt* (*bis*), XI.38; in the additional cases of III.11 (*bis*), X.9por, X.38, XI.23 (*quater*).
[174] These *RI* are analogous to those found in the corresponding propositions of the main redaction.
[175] In *Optica* 32 **A**; *Catoptrica* 2, 3, 21, 22 (*bis*), 27; *Phaenomena* 2 and 6*alt* **b**; *Sectio canonis* 3.
[176] Proposition X.44 is within brackets since the final reference to the impossible is there omitted; that this is simply an omission is shown by the presence of the standard initializing formula "in fact, if possible" (*EOO* III, 124.26). A likely final clause is attested in manuscript **F** (*EOO* III, 128.21 *app.*); the clause is also found in the indirect tradition, and includes an explicit reference to X.42. A similar accident of transmission must be called responsible for the omitted final clause in the first *RI* of XII.18. Proposition VIII.6 is in fact a fake *RI*, as we shall see below.
[177] Within brackets is the result that is contradicted, if it figures among the principles or if it is proved in a previous proposition; the Book of a proposition is the same as that at the beginning of a suitable string if the corresponding ordinal is omitted; the identity of the Arabic numeral means a previous case in the same proposition; "(ass.)" means that either of the conditions is assumed as a supposition.

- two points both coincide and do not coincide (ass.): X.42;[178]
- a straight line is both expressible and irrational: X.26, 111 (apotome);
- a number measures a number less than itself, or a unit: VII.1–3, 34, 36, VIII.1, 4, IX.20;
- a number both measures and does not measure another number (ass.): IX.12, 18, 19;
- a number is both even and odd (ass.): IX.30;
- an even number measures an odd number: IX.33;
- a number both is and is not a square (a cube) (ass.): IX.10;
- a number both is and is not a power of 2 (ass.): IX.34;
- two numbers both coincide and do not coincide (ass.): IX.36;
- some numbers are less than the least numbers in the same ratio (ass.): VII.20, 22;
- a number both is greater than the GCD of some numbers (ass.) and measures them: VII.33, 39;
- a number both is less than the LCM of some numbers (ass.) and is measured by them: VII.35, IX.14;
- infinite descent in numbers: VII.31;
- a magnitude measures a magnitude less than itself: X.2–4;
- two magnitudes both are and are not commensurable (ass.): X.13, 16.

This list makes it clear that the typology of number-theoretical "absurds" is more varied than the typology of geometric "absurds"; as a matter of fact, most geometric "absurds" belong to two main categories: contradiction with a proved result or with a principle; magnitudes satisfying two contradictory (in)equalities.

As a paradigmatic *RI*, let us go back to Sect. 1.1 and read again the relevant portion of our paradigmatic proposition III.2 (*EOO* I, 168.21–170.13):[179]

λέγω ὅτι ἡ ἀπὸ τοῦ Α ἐπὶ τὸ Β ἐπιζευγνυμένη εὐθεῖα ἐντὸς πεσεῖται τοῦ κύκλου.	I claim that the straight line joined from A to B will fall within the circle.
μὴ γάρ, ἀλλ' εἰ δυνατόν, πιπτέτω ἐκτὸς ὡς ἡ ΑΕΒ, καὶ εἰλήφθω τὸ κέντρον τοῦ ΑΒΓ κύκλου καὶ ἔστω τὸ Δ, καὶ ἐπεζεύχθωσαν αἱ ΔΑ ΔΒ, καὶ διήχθω ἡ ΔΖΕ.	In fact not, but if possible, let it fall outside as AEB, and let the centre of circle ABΓ be taken and let it be Δ, and let ‹straight lines›, ΔA, ΔB, be joined, and let a ‹straight line›, ΔZE, be drawn through.
ἐπεὶ οὖν ἴση ἐστὶν ἡ ΔΑ τῇ ΔΒ, ἴση ἄρα καὶ γωνία ἡ ὑπὸ ΔΑΕ τῇ ὑπὸ ΔΒΕ. καὶ ἐπεὶ τριγώνου τοῦ ΔΑΕ μία πλευρὰ προσεκβέβληται ἡ ΑΕΒ, μείζων ἄρα ἡ ὑπὸ ΔΕΒ γωνία τῆς ὑπὸ ΔΑΕ· ἴση δὲ ἡ ὑπὸ ΔΑΕ τῇ ὑπὸ ΔΒΕ· μείζων ἄρα ἡ ὑπὸ ΔΕΒ τῆς ὑπὸ ΔΒΕ· ὑπὸ δὲ τὴν μείζονα γωνίαν ἡ μείζων πλευρὰ ὑποτείνει· μείζων ἄρα ἡ ΔΒ τῆς ΔΕ· ἴση δὲ ἡ ΔΒ τῇ ΔΖ· μείζων ἄρα ἡ ΔΖ τῆς ΔΕ ἡ ἐλάττων τῆς μείζονος, *ὅπερ ἐστὶν ἀδύνατον*· οὐκ ἄρα ἡ ἀπὸ τοῦ Α ἐπὶ τὸ Β ἐπιζευγνυμένη εὐθεῖα ἐκτὸς πεσεῖται τοῦ κύκλου.	Then since ΔA is equal to ΔB, therefore angle ΔAE is also equal to ΔBE. And since one side AEB of a triangle ΔAE turns out to be produced, therefore angle ΔEB is greater than ΔAE; and ΔAE is equal to ΔBE; therefore ΔEB is greater than ΔBE; and the greater side extends under the greater angle; therefore ΔB is greater than ΔE; and ΔB is equal to ΔZ; therefore ΔZ is greater than ΔE, the less than the greater, *which is really impossible*; therefore it is not the case that the straight line joined from A to B will fall outside the circle.

A *RI* normally follows a liminal or partial determination (cf. Sect. 4.2.1) and is delimited by three canonical kinds of clause: the initializing clause (first underlined clause above), the reference

[178] The argument at *EOO* III, 122.3–10, is certainly spurious; see Vitrac 1990–2001 III, 216–218.
[179] For my translation of the preposed negative particle see Sect. 5.3.3.

to the impossible (in italics), and the negation of the reduction assumption (second underlined clause above). The inferences located between the first and the second clause cannot be distinguished from those normally occurring in a direct proof. Let us discuss the three kinds of clause just identified, in the order in which they appear in a *RI*.

1) *Initializing clause*. The initializing clause can be analyzed in a variety of ways. It is normally formulated as a conditional sentence (this is strictly speaking false, but see just below), whose antecedent states the "reduction assumption" or "reduction supposition", which is usually the negation of what must be proved and which usually coincides with the negation of the statement that immediately precedes the initializing conditional itself. The antecedent of the initializing conditional may be formulated in an abbreviated form I shall call "elliptical". In exceptional cases, for instance in very short *RI*, the conditional is absent. The distribution in the whole of the *Elements* of these three categories is as follows:[180]

elliptical antecedent	I.7, 26–27, 39, 40, III.1, 2, 4–8, 10–12, 13 (*bis*), 16 (*bis*), 18, 19, 23, VI.26, VII.20, 21, 34 (*bis*), 36 (*bis*), 39, VIII.1, 4, (6), IX.10 (*bis*), 12, 13 (*bis*), 14, 16–18, 19 (*bis*), 20, 30, 34, 36, X.2–4, 16, 26, 42 (*bis*), 43, (44), 45–47, 79–84, 111, XI.1, 3, 5, 7, 13, 14, 19, 23 (*bis*), XII.2, 5, 10, 11 (*bis*), 12, 18 // III.9, VII.31, X.28/29II (*bis*), X.117, X.117*alt* (*bis*), XI.23 (*ter*), 38, XII.2, 5, 10–12, 18, XIII.2/3
non-elliptical antecedent	I.4, 6, 14, 26 (*bis*), III.24, 27, IV.4, 8, 13, V.18, VI.7, VII.1, 2, 3 (*bis*), 22–24, 28 (*bis*), 29, 31, 33, 35, VIII.4, IX.13 (*quater*), 31, 33, X.2, 13, 16 (*bis*), XI.2, XII.2, 5, 10, 12, 18 // VI.22/23, X.9por, X.13, X.38, X.28/29II, X.117 (*bis*) , XII.2, 5, 10–12, 18
absent conditional	VI.7, IX.12, XI.19 // III.7–8, 10, 11 (*bis*), XI.23, XIII.2/3, XIII.18/19 (*quinquies*)

An elliptical antecedent may assume several forms. It may contain a negative particle, like εἰ γὰρ μή "in fact, if not" (61 in the Euclidean corpus, not only in *RI*), or a modal operator of possibility, like εἰ γὰρ δυνατόν "in fact, if possible" (62 occurrences), mainly when negative statements are reduced to the impossible, or μὴ γάρ, ἀλλ' εἰ δυνατόν "in fact not, but if possible",[181] when affirmative statements are reduced to the impossible. Strictly speaking, the initializing clause introduced by the last expression is not a conditional, but a severely abridged conjunction: the first conjunct, the only vestige of which is μὴ γάρ ("in fact, let so-and-so not be the case"), is a supposition with the verb in the imperative; the second conjunct is the conditional whose antecedent is εἰ δυνατόν. For our purposes, this complication is immaterial.

The distribution of the nexuses that introduce the initializing clause is as in the following table:

εἰ γὰρ δυνατόν	I.7, III.4–8, 10, 13 (*bis*), 16, 23, VII.20, VIII.(6), IX.10 (*bis*), 13–14, 16–18, 19 (*bis*), 20, 30, 36, X.4, 26, 42 (*bis*), 43, (44), 45–47, 79–84, 111, XI.1, 13, 23, XII.2, 5, 10–12, 18 // X.28/29II, X.117, X.117*alt* (*bis*), XI.23, XII.2, 5, 10–12, 18
εἰ γὰρ μή	I.27, 39, 40, III.18, VII.21, 34 (*bis*), 36 (*bis*), 39, VIII.1, 4, X.3, XI.3, 14, 23, XII.11 // VII.31, XI.23 (*bis*), XIII.2/3
μὴ γάρ, ἀλλ' εἰ δυνατόν	III.1, 2, 11–12, 16, 19, VI.26, XI.5, 7 // III.9, XI.38
μὴ γάρ	IX.12, XI.19
εἰ γάρ	IX.16
εἰ γὰρ οὔ	IX.34[182]
εἰ δυνατόν	I.26, X.2, 16 // X.28/29II, XIII.2/3

[180] The double slash separates the supplementary material.

[181] As said, the predicate δυνατόν "possible" can also be found in some problems, as for instance IX.18–19 and XI.22 (see the discussions in Sects. 4.2.1 and 4.3).

[182] The presence of the objective negative particle must be considered an anomaly, probably induced by the absence of a local determination.

The last clause in this list may occur (first 3 items) when a non-elliptical conditional whose consequent has the verb form in the indicative is followed by a conditional with proleptic consequent in the imperative. The function of the second conditional (which I take to be the real initializing clause) is to set out the object whose existence is asserted in the first conditional, as we may check in proposition X.16 (*EOO* III, 46.1–3):[183]

εἰ γὰρ μή ἐστιν ἀσύμμετρα τὰ ΓΑ ΑΒ, μετρήσει τι αὐτὰ μέγεθος. μετρείτω, εἰ δυνατόν, καὶ ἔστω τὸ Δ.	In fact, if ΓΑ, AB are not incommensurable, some magnitude will measure them. Let ‹some magnitude› measure ‹them›, if possible, and let it be Δ.

In general, the consequent of the initializing conditional has the verb in the imperative if it carries out a constructive act,[184] or if it contains the first instantiated occurrence of an object.[185] The consequents that do not carry out constructive acts[186] do not need an imperative, and we may find present or future indicative as their verb form. The optative mood, or unreal conditionals with the verb in the imperfect are exceedingly rare.[187] The occurrences of the verb mood in the consequent of the initializing conditionals are evenly distributed, both in the geometric and in the arithmetic Books, between indicative and imperative, as the following table shows:[188]

indicative	I.4, 6, 26 (*bis*), 27, III.24, 27, IV.4, 8, 13, V.18, VI.7, VII.1–2, 3 (*bis*), 21–24, 28 (*bis*), 29, 31, 33, 34 (*bis*), 35, 36 (*bis*), 39, VIII.4, IX.13 (*quinquies*), 31, 33, 34, X.2, 3, 13, 16 (*bis*), XI.2, 14, 23, XII.2, 5, 10–12, 18 // VI.22/23, X.9por, X.13, X.28/29II, X.38, X.117 (*bis*), XI.23 (*bis*), XII.2, 5, 10–12, 18
imperative	I.7, 14, 26, 39, 40, III.1, 2, 4–8, 10–12, 13 (*bis*), 16 (*bis*), 18, 19, 23, VI.26, VII.20, VIII.1, 4, (6), IX.10 (*bis*), 13–14, 16–18, 19 (*bis*), 20, 30, 36, X.2, 4, 16, 26, 42 (*bis*), 43, (44), 45–47, 79–84, 111, XI.1, 3, 5, 7, 13, 19, 23, XII.2, 5, 10–12, 18 // III.9, VII.31, X.28/29II (*bis*), X.117, X.117*alt* (*bis*), XI.23, 38, XII.2, 5, 10–12, 18, XIII.2/3

If the imperative is mandatory for constructive acts, in other cases the distinction between consequent in the indicative and in the imperative in the initializing clause is a stylistic choice, as is made clear by two strictly parallel passages in VIII.4 (*EOO* II, 280.13–16 and 282.24–284.4):

λέγω δὴ ὅτι καὶ ἐλάχιστοι. εἰ γὰρ μή εἰσιν οἱ Θ Η Κ Λ ἑξῆς ἀνάλογον ἐλάχιστοι ἔν τε τοῖς τοῦ Α πρὸς τὸν Β καὶ τοῦ Γ πρὸς τὸν Δ καὶ ἐν τῷ τοῦ Ε πρὸς τὸν Ζ λόγοις, ἔστωσαν οἱ Ν Ξ Μ Ο.	I now claim that they are also the least. In fact, if Θ, H, K, Λ are not least successively in proportion in the ratios of both A to B and of Γ to Δ and in that of E to Z, let them be N, Ξ, M, O.
λέγω δὴ ὅτι καὶ ἐλάχιστοι ἐν τοῖς ΑΒ ΓΔ ΕΖ λόγοις. εἰ γὰρ μή, ἔσονταί τινες τῶν Ν Ξ Μ Ο ἐλάσσονες ἀριθμοὶ ἑξῆς ἀνάλογον ἐν τοῖς ΑΒ ΓΔ ΕΖ λόγοις. ἔστωσαν οἱ Π Ρ Σ Τ.	I now claim that they are also the least in the ratios AB, ΓΔ, EZ. In fact, if not, there will be some numbers less than N, Ξ, M, O successively in proportion in the ratios AB, ΓΔ, EZ. Let them be Π, P, Σ, T.

[183] The proposition contains two *RI* that are almost identical, with two variations: the second *RI* does not have the formula εἰ δυνατόν, and the manuscripts show incertitudes as to the form of the verb ὑπόκειμαι in the reference to the impossible.
[184] The verb is in the passive perfect imperative, as seen in Sect. 4.3.
[185] Usually, this is a present imperative of the verb "to be", as in X.16 just read.
[186] Such are for instance the references, in the form of a postposed explanation, to previously proved propositions or to principles, as in propositions I.4, III.24, IV.4, 8, 13, VII.31. See also I.27 and XI.14, in both cases within instantiated citations of I.post.5.
[187] Additional case of X.38 and of X.117, respectively.
[188] In IX.13, a macro-*RI* contains five other *RI* in succession. The framing argument has the verb of the consequent in the imperative, the others have the verb of the consequent in the indicative.

The first formulation subsumes in the conditional both the statement of existence of the object and its instantiation; the second formulation employs two independent clauses. In principle, then, a conditional with imperative in the consequent can always be read as a shortening of a formulation with indicative in the consequent followed by an independent clause that includes an instantial imperative—of course, the opposite reading is also legitimate.

The consequent of the initializing clause may be formulated as a disjunction; this happens for instance in all proofs carried out according to the method of exhaustion.

However, initializing clauses that present more than one condition are normally formulated *in the consequent* with a principal clause determined by a participial construct; let us read an extract from VIII.1 (*EOO* II, 270.8–11) as an example:

λέγω ὅτι οἱ Α Β Γ Δ ἐλάχιστοί εἰσι τῶν τὸν αὐτὸν λόγον ἐχόντων αὐτοῖς.	I claim that A, B, Γ, Δ are the least among those having the same ratio as them.
εἰ γὰρ μή, ἔστωσαν ἐλάττονες τῶν Α Β Γ Δ οἱ Ε Ζ Η Θ ἐν τῷ αὐτῷ λόγῳ <u>ὄντες</u> αὐτοῖς.	In fact, if not, let thre be ‹numbers› less than A, B, Γ, Δ <u>that are</u> in the same ratio as them, E, Z, H, Θ.

or by a relative clause, as in VII.31 and 39. In the arithmetic Books, participial constructs that formulate reduction assumptions can also be found.[189]

2) *Reference to the impossible*. The *RI* is closed by the claim that an impossible state of affairs has been arrived at. This claim has the form of a relative clause attached to the sentence in which such an impossible state of affairs is formulated. The form is either ὅπερ (ἐστὶν) ἀδύνατον "which is really impossible" or ὅπερ (ἐστὶν) ἄτοπον "which is really absurd". In the *Elements*, the former clause is more frequent than the latter: 119 occurrences versus 29; in Archimedes we only find the formula ὅπερ (ἐστὶν) ἀδύνατον. In exceptional cases, any of these two clauses is replaced by an explicit reference to the fact that an assumption has been contradicted; such formulae are ὅπερ οὐχ ὑπόκειται "which has really not been supposed",[190] or ὅπερ οὐχ ὑπέκειτο "which had really not been supposed".[191] These expressions are obvious stylistic markers of the arithmetic Books—as they are of the spurious argument in X.42. Even in the presence of a canonical reference to the impossible, in propositions like X.16 (*bis*) the reference to the contradicted assumption is made explicit by ὑπέκει(ν)το "it had been supposed", whereas in VII.20, X.9por, and XI.23 (*bis*) the verb form is ὑπόκεινται "they have been supposed".

Some *RI* are not closed by a reference to the impossible: the four types just presented exhibit together 156 occurrences, but the *RI* in the *Elements* listed above are 165. The 9 *RI* without a canonical closing formula are VIII.6, X.44, XII.18 (1 argument out of 2), and, among the supplementary material, III.11 (1 out of 2), XI.23 (1 out of 4), XIII.2/3 (1 out of 2), XIII.18/19 (3 out of 5). The argument in VIII.6 is in fact a direct proof, as we shall see below. The closing formulae of X.44 and XII.18 have certainly disappeared as a consequence of some accident of transmission, as

[189] In VII.3 (*bis*), 21, 31, 33, 34 (*bis*), 35, 36 (*bis*), 39, VIII.1, 4 (*bis*), IX.13 (*ter*), 20. In VIII.4 the participle of the verb "to be" is understood.

[190] We find this formula and not οὐχ ὑπόκειται δέ "and this has not been supposed", which is typical of the kind of indirect proof I shall call "by contraposition", see Sect. 5.2.2.

[191] The first of these two formulae is found (within *RI* or other indirect arguments) in VII.2, IX.10, 13, 34, X.9/10, 42, X.13 *vulgo* and X.39*alt*; the second formula only in IX.10. In IX.13, the framing argument is identified by ἄτοπον, the 5 nested *RI* are identified by ἀδύνατον or (the third one) by the formula just seen.

we have seen above. The incomplete *RI* among the supplementary material are abbreviated or clumsy redactions that we may safely deem spurious.[192]

3) *Negation of the reduction assumption.* The reference to the impossible is followed by the negation of the reduction assumption, supposed in the antecedent of the initializing conditional. Such a negation is equivalent, modulo a double negation and if branchings are not required, to the statement that immediately precedes the initializing conditional itself; this statement is in turn formulated, as we have seen, as a "determination" (cf. Sect. 4.2.1). The negative particle in the negation of the reduction assumption is always preposed, as we shall see in Sect. 5.3.3. The negative particle may be missing (as for instance in I.4) if the negated statement is formulated by means of a sentence that already contains a negative particle or a denegative part of speech: the deductive step that contains the double negation is omitted, its affirmative counterpart is directly stated.

I shall now discuss some issues raised by the formulation of the reference to the impossible, passing then to a brief presentation of metamathematical divertissements with *RI* authored by Apollonius, Hero, and Menelaus, and, finally, to a discussion of the way Aristotle conceptualizes *RI*.

In the reference to the impossible, the qualifier "impossible" is adopted, instead of "false", which would have served the scope as well. This stylistic choice may have been dictated by a perception of greater persuasiveness conveyed by the modal connotation. This is confirmed by the frequency of the variant with ἄτοπον "absurd". It is not immediately clear, however, what state of affairs is declared to be impossible. In the final deductive step of a *RI*, the subsistence of a mathematical state of affairs (for instance, A is greater than B) is explicitly or implicitly contrasted with the subsistence of a mathematical state of affairs incompatible with it (say, A is equal to B). The impossibility qualifies in fact the *conjoined subsistence* of both states of affairs, and not the subsistence of a state of affairs incompatible with a state of affairs that has been assumed or previously proved. As a consequence, the referent of the relative ὅπερ that figures in the reference to the impossible is the conjunction of two statements, namely, the statements that formulate the two incompatible states of affairs. In this interpretation, the metalinguistic character of the reference to the impossible is particularly prominent—for it negates a *logical* fact—and its modal connotation is thereby fully justified, for it operates on (formulations of) statements, not on mathematical states of affairs. That this is the correct reading is confirmed by dozens of final strings of *RI*, among which I select the deductive steps in I.7 and VII.1 (*EOO* I, 26.3–6, and II, 190.12–13):

πάλιν ἐπεὶ ἴση ἐστὶν ἡ ΓΒ τῇ ΔΒ, ἴση ἐστὶ καὶ γωνία ἡ ὑπὸ ΓΔΒ γωνίᾳ τῇ ὑπὸ ΔΓΒ· ἐδείχθη δὲ αὐτῆς καὶ πολλῷ μείζων, ὅπερ ἐστὶν ἀδύνατον.	Again, since ΓΒ is equal to ΔΒ, an angle ΓΔΒ is also equal to an angle ΔΓΒ; and it was also proved that it is much greater than it, which is really impossible.
καὶ λοιπὴν ἄρα τὴν ΑΘ μονάδα μετρήσει ἀριθμὸς ὤν, ὅπερ ἐστὶν ἀδύνατον.	Therefore being a number it will also measure the unit ΑΘ as a remainder, which is really impossible.

Thus, the reference to the impossible marks the incompossibility of suitable states of affairs. A different role have modal clauses like εἰ γὰρ δυνατόν "in fact, if possible", which initialize the *RI*. The existence of variants such as εἰ γὰρ μή "in fact, if not", which stands for the contradictory of the statement to be proved, confirms that the modal connotation is not an essential feature of a *RI*,

[192] They also contain expressions like δείξομεν τὸ ἄτοπον "we shall prove the absurd" or πολλῷ γὰρ τὸ ἀδύνατον μεῖζον "for the impossible is much greater".

but a mere stylistic trait. Moreover, and contrary to what happens with the reference to the impossible, the initializing formula of a *RI* qualifies as possible, or simply supposes, the subsistence of a single state of affairs.[193]

The same phenomenon is made prominent by variants of the reference to the impossible that accentuate the metamathematical connotation and that clearly refer to a *single* statement. These are expressions like ὅπερ ἀδύνατον δέδεικται "which has really been proved impossible". Their occurrences in the *Elements* are set out in the following table; these are keyed on the verb form, on the preposed or postposed position of such a verb form, on referring to a result proved within the ongoing proof or elsewhere,[194] and on the adjective that formulates the reference to the impossible:

	δέδεικται	ἐδείχθη
ἄτοπον	/	IV.4, 8, 13 (all III.16), XI.2 (XI.1), XII.5
ἀδύνατον	**b** XII.2, 5, 10, 11	III.8, **X.84** (**X.79**), XII.2, 11–12, 18

All statements referred to in this table are proved by *RI*. This variant formulation clearly serves as a stylistic marker of all and only the *RI* of Book IV (all of which depend on III.16) and of those of Book XII in which the second *RI* that figures in the method of exhaustion refers to the first.[195] As a matter of fact,[196] the only occurrences of δείκνυμι "to prove" in the above list that are not spurious are those in Book XII, for all these references to something previously proved are internal to the ongoing proposition: it is this feature that dictates the form of the reference.

The indirect proofs in the *Elements*, and in particular the *RI*, triggered a series of scholarly interventions that eventually had a major mathematical import. First, the exclusive use of proofs by *RI* in Book IV of Apollonius' *Conica* brings a less pervasive but still obvious characteristic of Book III of the *Elements* to the extreme.[197] Second, in his commentary on the *Elements*, Hero transformed a number of indirect proofs into direct proofs; this and other interventions interfered with the textual transmission of the *Elements*.[198] Third, the initial string of propositions of Book I of Menelaus' *Spherica* can be read as a rewriting, in a different deductive order and without using indirect proofs, of the corresponding theorems of Book I of the *Elements*. The following table sets out the corresponding propositions:

Sphaerica	1	2	3	4	5	6	7	8	9	10	11	14–15	17
Elements	23	5	6	4, 8	20	21	19	24	18	16	32	26a	26b

[193] But note the counterexample in X.4.

[194] Instances of preposed verb forms are in boldface; the previous propositions that prove or state the results referred to are within brackets.

[195] This excludes XII.10 = 9 **b**.

[196] X.84 is the only item of a string of strictly analogous propositions that carries the verb from ἐδείχθη in the reference to the impossible; the same holds for III.8 in comparison with the twin theorem III.7. The Arabo-Latin tradition helps us clarify the issue. Gerard's translation has the verb "to prove" in the propositions that correspond to XII.2, 5, 10–11, 18, but neither in XI.2 nor in the constructions of Book IV. The version Adelard I has it only in XII.2. In both translations, the three short *RI* of Book IV are altogether absent. This is not surprising, since the form of the *RI* in Book IV suggests that they are *marginalia* that have found their way into the text. They are in fact made of a single conditional, whose consequent is identical to a part of the enunciation of III.16 (the only variation is the replacement of ἐκτός "outside" with ἐντός "inside", so as to produce a statement that contradicts the said enunciation). The conditional is framed as a postposed explanation and is immediately followed by "which was really proved impossible". Add that, in IV.13, a further metalinguistic element is added: the verb form in the consequent is συμβήσεται "it will be concluded" (*EOO* I, 310.12; the verb also occurs in VII.27, VIII.13, and X.115*altI*). Similar arguments hold for XI.2.

[197] The move did not escape Eutocius, see *AGE* II, 354.8–13. On Apollonius' scholarly approach to mathematics and to foundational issues, see Acerbi 2010b.

[198] For the Heronian commentary on the *Elements*, see Acerbi, Vitrac 2014, 31–39.

The several references of Aristotle to specific proofs by *RI* and to its general logical features show that, already in his times, this deductive method was currently applied in mathematics.[199] Aristotle employs the proof of incommensurability of the side and of the diagonal of a square as a paradigmatic example of *RI*; he also mentions a specific proof, in which, as a consequence of supposing commensurability, "the odd are equal to the even".[200] Aristotle's insistence on this example has certainly interfered with the text of the *Elements*.[201] On the proof-theoretical side, Aristotle asserts that a *RI* cannot be entirely reduced to a syllogistic formulation (*APr.* I.44, 50a29–38):

ὁμοίως δὲ καὶ ἐπὶ τῶν διὰ τοῦ ἀδυνάτου περαινομένων – οὐδὲ γὰρ τούτους οὐκ ἔστιν ἀναλύειν, ἀλλὰ τὴν μὲν εἰς τὸ ἀδύνατον ἀπαγωγὴν ἔστι (συλλογισμῷ γὰρ δείκνυται), θάτερον δ' οὐκ ἔστιν· ἐξ ὑποθέσεως γὰρ περαίνεται. διαφέρουσι δὲ τῶν προειρημένων ὅτι ἐν ἐκείνοις μὲν δεῖ προδιομολογήσασθαι, εἰ μέλλει συμφήσειν – οἷον ἂν δειχθῇ μία δύναμις τῶν ἐναντίων, καὶ ἐπιστήμην εἶναι τὴν αὐτήν· ἐνταῦθα δὲ καὶ μὴ προδιομολογησάμενοι συγχωροῦσι διὰ τὸ φανερὸν εἶναι τὸ ψεῦδος – οἷον τεθείσης τῆς διαμέτρου συμμέτρου τὸ τὰ περιττὰ ἴσα εἶναι τοῖς ἀρτίοις.

Similarly also for the ‹arguments› brought to conclusion through the impossible—for these cannot be analyzed either; rather, the reduction to the impossible can be (for this is proved by means of a syllogism), the rest cannot, for the conclusion is reached from a supposition. The latter are different from the ones just said in that in those cases one has to make an agreement in advance if the other person is to consent—for example, if it were proved that there is a single power of contraries, that the knowledge of them is also one and the same. But here people concede the point even without a prior agreement because the falsehood is manifest—such as, for example, the diagonal being posited commensurable, that the odd are equal to the even.

Here and in *APo.* I.23, Aristotle tries to clarify the structure of a *RI*, and of the more general class of proofs ἐξ ὑποθέσεως "from a supposition".[202] To him, a part of a *RI* can always be put in syllogistic form: this is the part that goes from the negation of that which is to be proved as far as what immediately precedes the statement of a contradiction or of a patent falsity (namely, in the typology here adopted, the part that goes from the "initializing clause" of a *RI* to the "reference to the impossible", this being excluded). What cannot be put in syllogistic form is the very final segment of the *RI*: this is recognizing the said contradiction and passing from this to the negation of the *RI* supposition, that is, to that which is to be proved (namely, the transition from the "reference to the impossible" to the "negation of the reduction assumption"). The validity of such a transition is always taken for granted; for this reason, in the text just read, Aristotle claims that "people concede the point even without a prior agreement because the falsehood is manifest".

A *RI* is a species of a syllogism from a supposition because such a syllogism proceeds by replacing the *demonstrandum* with a suppletive premise (called τὸ μεταλαμβανόμενον), from which the

[199] To Aristotle, direct proofs are better than indirect proofs, for the latter do not have explanatory power and therefore, while being valid arguments, do not produce scientific knowledge; see the in-depth discussion in Malink 2020.

[200] *APr.* I.23, 41a21–32. Other mentions are for instance in *APo.* I.33, 89a29–30, *Top.* I.15, 106a38–b1, and VIII.13, 163a11–13, *Ph.* IV.12, 221b24–25, *GA* II.6, 742b27, *Metaph.* Δ.12, 1019b21–27, and I.1, 1053a17–18, *EN* III.5, 1112a21–23. The next-to-last passage asserts that the side and the diagonal of a square must be measured with two different measures set as standards.

[201] A proof in Book X that conforms to the Aristotelian indications is transmitted by all manuscripts of the *Elements* (this is the so-called X.117 *vulgo*, which bears no connection with what precedes). The proof in Alexander's commentary (*in APr.*, 260.20–261.19) is similar but not identical to this proof: its presence suggests that Alexander did not read X.117 *vulgo* in his text of the *Elements*.

[202] Other relevant passages are in *APr.* I.23 and II.11–14, *Top.* I.18, 108b12–19, III.6, 119b35–120a2, VIII.2, 157b34–158a2. On syllogisms "from a supposition" see Striker 1979 and Crivelli 2011.

demonstrandum follows. The transition from the suppletive premise to the *demonstrandum* is licensed by a preliminary agreement, and cannot be formalized as a syllogism. In a *RI*, the premise of which Aristotle says that it is so obviously false that no agreement is necessary is the patent falsehood arrived at the end of the *RI*,[203] namely, a premise. In the general case of syllogisms from a supposition, it is not clear whether what Aristotle takes to be the object of an agreement is the suppletive supposition or the logical complex that corresponds to the transition from this to the *demonstrandum* (namely, an inference rule).[204]

In *APr.*, Aristotle uses twice (and repeatedly alludes to)[205] proofs "through the impossible" in showing that specific syllogistic moods are valid.[206] Let us read how such a proof works for the syllogistic mood *Baroco* (*APr.* I.5, 27a36–b1):[207]

πάλιν εἰ τῷ μὲν Ν παντὶ τὸ Μ τῷ δὲ Ξ τινὶ μὴ ὑπάρχει, ἀνάγκη τὸ Ν τινὶ τῷ Ξ μὴ ὑπάρχειν· εἰ γὰρ παντὶ ὑπάρχει, κατηγορεῖται δὲ καὶ τὸ Μ παντὸς τοῦ Ν, ἀνάγκη τὸ Μ παντὶ τῷ Ξ ὑπάρχειν· ὑπέκειτο δὲ τινὶ μὴ ὑπάρχειν.

Again, if M holds of all of N and not of some of Ξ, necessarily N does not hold of some of Ξ; in fact, if it holds of all, and M is also predicated of all of N, necessarily M holds of all of Ξ; and it had been supposed not to hold of some.

Aristotle also claims that every *RI* can be converted into a direct proof. This can be done by means of a form of argument contraposition similar to what the Stoics called "first *thema*",[208] which states that if two premises syllogize a conclusion, then the syllogism is also valid in which one of the premises and the conclusion are replaced by their contradictories and interchanged. Now, since

[203] In addition to the passage read above, which is quite explicit on the issue, see *APr.* I.23, 41a30–40, II.11, 61a21–25, II.14, 62b29–38, *Top.* VIII.2, 157b34–158a2.

[204] As said, Aristotle thought that direct proofs were better that indirect proofs (cf. *APo.* I.26); still, he regarded indirect proofs as perfectly sound from the logical point of view, and a remarkable dialectical tool (*Top.* VIII.2, 157b34–38).

[205] Relevant occurrences are in *APr.* I.5–7 are at I.5, 27a15, 28a7, I.6, 28a23 and 28a29, I.7, 29a32 and 29a35, 29b5.

[206] The best (terse and not interpretation-laden) overall account of Aristotle's assertoric syllogistic is Crivelli 2012. Aristotle's assertoric syllogistic has been translated into a number of systems of modern logic. The first fully-fledged proposal read syllogistic as an axiomatic propositional logic (that is, a syllogism is a non-simple statement): Łukasiewicz 1957, followed by Patzig 1968, who, however, ends in construing, following a formalization developed in Lorenzen 1965, Aristotle's syllogistic as a logic of relations. Ebbinghaus 1964 (which in fact depends again on Lorenzen's system), Smiley 1973, and Corcoran 1974 first realized that natural deduction not only models the system of syllogistic figures (that is, a syllogism is a deduction) but also the arguments by means of which Aristotle proves his results. A proof-theoretical analysis of these arguments, within the natural deduction model and with some important insights, is in von Plato 2016. See also the idiosyncratic approach in Thom 1981, who in fact combines features of the two main models. On the much more difficult issue of modal syllogistic see Malink 2013.

[207] Translation Crivelli 2012, 133, adapted to my conventions. The other mood proved to be valid "through the impossible" is *Bocardo* (*APr.* I.6, 28b17–21). Aristotle also states that other moods can be proved "through the impossible", and proves indirectly (without asserting that the proof is of such a kind) the rules of conversion of the syllogistic premises (*APr.* I.2, 25a14–26, and I.3, 25a27–36). Apart from "direct" arguments, Aristotle also proves syllogistic validity by ἔκθεσις (see Sect. 4.2), and sometimes presents this method as alternative to other arguments to the same effect, including the proofs "through the impossible" (Aristotle shows this for *Bocardo*). Actually, *RI* can be eliminated from Aristotle's syllogistic: Joray 2014 and Tennant 2014 develop different systems of ecthetic syllogistic; Dyckhoff 2019 puts forward a minimal set of (syllogistic) rules that allow disposing of *RI*. Syllogistic reductions in which *RI* is implicitly used by Aristotle are actually fairly complex systems of nested *RI*, as shown in Antonelli, von Plato, unpublished typescript.

[208] This is a rule for manipulating deductions. We read a description of it in Apuleius (*Int.*, 209.9–14), who apparently makes a confusion between a *RI* and the first *thema*: *est et altera probatio communis omnium etiam indemonstrabilium, quae dicitur "impossibile" appellaturque a Stoicis "prima constitutio" vel "primum expositum". quod sic definiunt: "Si ex duobus tertium quid colligitur, alterum eorum cum contrario illationis colligit contrarium reliqui"* "there is also another proof shared by all indemonstrables, which is said 'impossible" and is called 'first rule' or 'first position' by the Stoics. They define it as follows: 'If from two ‹premises› some third follows, from either of them with the contradictory of the conclusion the contradictory of the remaining one follows'". For the Stoic *themata*, see also Sect. 2.4.1.

the syllogistic part of a *RI* whose *demonstrandum* is *C* is $A, \neg C: \neg B$,[209] the rule allows converting it into $A, B: C$. This is nothing but a direct proof of *C*, and the transition from $\neg\neg B$ to *C*, which cannot be given a syllogistic form, is no longer necessary. There are two reasons why this scheme cannot work outside syllogistic.[210] First, *RI* such as the one in III.1: "Find the centre of a given circle", can only hinge upon a definition—here that of "circle" (cf. Sect. 5.1.1 for the peculiar form of this definition, and Sect 4.3 for the verb "to find")—because no properties of a circle have been proved so far.[211] But if we want to find *directly* the centre of a circle with a construction, the definition is useless since it presupposes the existence of the centre: and there are no other premises in the *Elements* from which the statement that such-and-such a point is the centre of a circle can be deduced directly. Second, try to convert, according to Aristotle's prescription, a *RI* in which the contradiction arrived at is, as is often the case and as is the case in III.1, of the kind "two angles are both equal and unequal" (this statement is $\neg B$ above): the contradictory of this statement is an empty tautology, useless as a premise of an inference. The only *RI* that possibly fit Aristotle's scheme are those in which what is contradicted is a previous theorem.

To convert a direct proof into a *RI* is instead immediate: it is the method used in VIII.6 and in Archimedes, *Con. sph.* 7–9, or Apollonius, *Con.* VI.1–2. It is enough to posit the contradictory of the statement to be proved as the reduction supposition, make it followed by a direct proof of the statement itself, claim at the end "and this has not been supposed, which is really absurd", and assert the statement itself. The trick is simply that the contradictory of the statement to be proved is supposed but not discharged in the direct proof.[212]

The sources suggest that the Stoics did not thematize *RI*; the Stoic school accorded instead prominence to a simpler argument, as we shall see presently.

5.2.2. *Arguments "for a contrapositive"*

The following table sets out the distribution, in the main text of the *Elements*, of the arguments "for a contrapositive" (*ArCo* henceforth, also in the plural; I use this denomination because the contrapositive of an established result is normally proved by means of an *ArCo*). In the last column, with an obvious correspondence principle, the propositions or principles are listed on which they hinge.

---

[209] This is just a notational shortcut of mine: Aristotle did not prescribe that the negative particle has to be preposed to a sentence in order for it to formulate a premise that is the contradictory of an assigned premise (see Sect. 5.3.3).

[210] *APr.* I.29, 45a23–b20, and II.11–14, and see Lear 1980, 49–52. Arguments have been framed whose purpose is to show that—within specific logical theories—it is possible to transform any indirect proof into a direct proof (Löwenheim 1946, a typescript translated and published posthumous by W.V.O. Quine; Gardies 1991), but the manipulations involved are non-constructive and in fact straightforwardly equivalent to assuming the double negation principle (whose $\neg\neg A \rightarrow A$ harm is a core step in any *RI* of an *affirmative* statement and is notoriously non-constructive). As a consequence, this logical metatheorem begs the question. The very instructive history of a (fallacious) proof to the same effect by H. Behmann, demolished by an amazing counterexample conceived by K. Gödel, is reconstructed in Mancosu 2002. See also a discussion of post-Kantian conceptions of indirect proofs in Mancosu 1996, sect. 4.3, critically assessed in Hodges, unpublished typescript. In a constructive framework, in which $\neg A$ is *defined* as $A \rightarrow \bot$ (which entails that $\neg\neg\neg A \rightarrow \neg A$ is constructive), only proofs by *RI* of negative statements can properly be termed "indirect": von Plato 2013, sect. 5.1; von Plato 2015, partly retrieved as von Plato 2017, sect. 1.3.

[211] The proof runs as follows: draw the perpendicular bisector of any chord and take the middle point of the segment of it cut off by the circle: this point is the centre, for a suitable construction uniformly carried out starting from any other point within the circle gives rise to a configuration in which two angles are both equal and unequal. As usual, the *Elements* neglects limiting positions of the point that in the proof is assumed to be the real centre while not being it; in these positions (the point lies on the perpendicular bisector), a contradiction is arrived at more straightforwardly.

[212] This is an "empty" *RI*, not a proof by *consequentia mirabilis* (for which see Acerbi 2019). The trick here described is called "vacuous discharge" in modern logic.

I	8, 19, 25 (*bis*), 29	7, 5 and 18, 4 and 24, I.post.5
III	16	16
V	9 (*bis*), 10 (*quater*)	8, 7 and 8
VII	2	1
VIII	7, 16 (*bis*), 17 (*bis*)	6, 14, 15
X	7, 8, 9 (*bis*)	6, 5, 9
XI	16	1 and XI.def.8 and XI.23

The two micro-arguments in XI.34[213] are interpolations; the first of them even corrects a previous interpolation. The two pairs VIII.14–15 and X.7–8, which complete a "logical square", are almost certainly inauthentic.[214] In the *Data*, propositions 25–27 are proved by *ArCo* (def. 4).

An *ArCo* is a short indirect argument that does not suppose the negation of the statement to be proved, nor contain modal operators or references to the impossible or to the absurd; the most standard form of *ArCo* is licensed by the deductive scheme traditionally called *modus tollens*: it is required to prove p: now, $\neg p \rightarrow q$ obtains because it was proved previously; but $\neg q$: therefore p. The borderline between an *ArCo* and a *RI* is not sharp: small linguistic changes—which in fact simply amount to adding the "initializing clause", in which $\neg p$ is assumed as a supposition ("in fact not, but let $\neg p$ be the case" in the schematic example just seen), and the "reference to the impossible" ("therefore $q \& \neg q$, which is impossible"), both discussed in the previous Section—allow transforming an *ArCo* into a *RI* (the vice versa is much more contrived). Aristotle was already explicit about this: he recognized that his own rule of syllogistic contraposition (seen at the end of Sect. 5.2.1) and a *RI* are different, yet the former can be easily transformed into the latter (*APr.* II.11, 61a21–25 and 61a32–33; I stick to the standard translation of ἀντιστροφή "conversion"):

ὅμοιον γάρ ἐστι τῇ ἀντιστροφῇ, πλὴν διαφέρει τοσοῦτον ὅτι ἀντιστρέφεται μὲν γεγενημένου συλλογισμοῦ καὶ εἰλημμένων ἀμφοῖν τῶν προτάσεων, ἀπάγεται δ' εἰς ἀδύνατον οὐ προομολογηθέντος τοῦ ἀντικειμένου πρότερον, ἀλλὰ φανεροῦ ὄντος ὅτι ἀληθές. […] ὅσα γὰρ ἀντιστροφὴν δέχεται, καὶ τὸν διὰ τοῦ ἀδυνάτου συλλογισμόν.

In fact, [the absurd] is similar to conversion, except that they differ insofar as one converts as soon as a syllogism turns out to result and both premises are assumed, whereas one reduces to impossible as soon as there is no preliminary agreement on the contradictory ‹statement›, but it is manifest that it is true. […] for what admits of conversion also ‹admits of› syllogism through the impossible.

But if this is true, the care with which the two kinds of indirect proof are distinguished in the *Elements* is significant.

ArCo are preferred to *RI* whenever the indirect argument can be formulated in a handful of steps: a comparison between *Data* 25–27 and *Data* 28–30, which are proved by *RI* as we have seen, nicely illustrates this fact. The only *ArCo* that are not of the type "two premises and a conclusion" are those in the strictly homologous propositions I.29 and XI.16. In a *RI*, instead, the mathematical content is almost always non-trivial: for instance, the *RI* in IX.36 lasts 34 lines in Heiberg's edition.

There is some latitude in the formulation of an *ArCo*. In I.19, 25, and V.10, the only propositions in which three alternatives are posited and two are proved not to be the case, the first clause is not

213 They are at *EOO* IV, 108.1–4 and 108.11–12.
214 On X.7–8, see Vitrac 1990–2001 III, 111.

a complete conditional but a simple statement that would have served as the consequent of such a conditional—in fact, it is formulated with ἄν + imperfect, marking unreality. The antecedent of the conditional is omitted because its negation is anticipated as a separate statement, which opens the *ArCo*. Let us read V.10 (*EOO* II, 32.8–15) as an example:

εἰ γὰρ μή, ἤτοι ἴσον ἐστὶ τὸ Α τῷ Β ἢ ἔλασσον. ἴσον μὲν οὖν οὐκ ἐστι τὸ Α τῷ Β – ἑκάτερον γὰρ ἂν τῶν Α Β πρὸς τὸ Γ τὸν αὐτὸν εἶχε λόγον· οὐκ ἔχει δέ· οὐκ ἄρα ἴσον ἐστὶ τὸ Α τῷ Β –· οὐδὲ μὴν ἔλασσόν ἐστι τὸ Α τοῦ Β – τὸ Α γὰρ ἂν πρὸς τὸ Γ ἐλάσσονα λόγον εἶχεν ἤπερ τὸ Β πρὸς τὸ Γ· οὐκ ἔχει δέ· οὐκ ἄρα ἔλασσόν ἐστι τὸ Α τοῦ Β –· ἐδείχθη δὲ οὐδὲ ἴσον· μεῖζον ἄρα ἐστὶ τὸ Α τοῦ Β.	In fact, if not, either A is equal to B or less. Then now, A is not equal to B—for each of A, B would have the same ratio to Γ; and it has not; therefore it is not the case that A is equal to B—; nor of course is A less than B—for A to Γ would have a lesser ratio than B to Γ; and it has not; therefore it is not the case that A is less than B—; and it was proved that it is not equal either; therefore A is greater than B.

ArCo that resemble very much a *modus tollens* are in I.8, I.29,[215] III.16, V.9, VII.2,[216] VIII.7, 16–17, X.7–9,[217] XI.16. These *ArCo* are arguments that comprise the following items: a conditional introduced by εἰ "if" and with the verb form in the present or the future indicative,[218] a coassumption that is the negation of the consequent of the conditional, a conclusion that is the negation of the antecedent of the conditional. Let us read the entire VIII.16 (*EOO* II, 314.17–316.5) as a remarkably well-formed example:

ἐὰν τετράγωνος ἀριθμὸς τετράγωνον ἀριθμὸν μὴ μετρῇ, οὐδὲ ἡ πλευρὰ τὴν πλευρὰν μετρήσει· κἂν ἡ πλευρὰ τὴν πλευρὰν μὴ μετρῇ, οὐδὲ ὁ τετράγωνος τὸν τετράγωνον μετρήσει.	If a square number do not measure a square number, the side will not measure the side either; and if the side do not measure the side, the square will not measure the square either.
ἔστωσαν τετράγωνοι ἀριθμοὶ οἱ Α Β, πλευραὶ δὲ αὐτῶν ἔστωσαν οἱ Γ Δ, καὶ μὴ μετρείτω ὁ Α τὸν Β. λέγω ὅτι οὐδὲ ὁ Γ τὸν Δ μετρεῖ.	Let there be square numbers, A, B, and let their sides be Γ, Δ, and let A do not measure B. I claim that Γ does not measure Δ either.
εἰ γὰρ μετρεῖ ὁ Γ τὸν Δ, μετρήσει καὶ ὁ Α τὸν Β· οὐ μετρεῖ δὲ ὁ Α τὸν Β· οὐδὲ ἄρα ὁ Γ τὸν Δ μετρήσει.	In fact, if Γ measures Δ, A will also measure B; and A does not measure B; therefore Γ will not measure Δ either.
μὴ μετρείτω δὴ πάλιν ὁ Γ τὸν Δ. λέγω ὅτι οὐδὲ ὁ Α τὸν Β μετρήσει.	Now again, let Γ do not measure Δ. I claim that A does not measure B either.
εἰ γὰρ μετρεῖ ὁ Α τὸν Β, μετρήσει καὶ ὁ Γ τὸν Δ· οὐ μετρεῖ δὲ ὁ Γ τὸν Δ· οὐδ' ἄρα ὁ Α τὸν Β μετρήσει, ὅπερ ἔδει δεῖξαι.	In fact, if A measures B, Γ will also measure Δ; and Γ does not measure Δ; therefore A will not measure B either, which it was really required to prove.

Such arguments are almost canonical examples of a Stoic second indemonstrable.[219] These are concrete arguments having a well-defined form, which the Stoics took as the base of their syllogistic. For these indemonstrables, it was "immediately clear that they validly deduce, namely, that

---

[215] In this case, the conditional is amplified to an argument featuring several steps.
[216] In the last two items, the antecedent is abridged to εἰ γὰρ / δὲ μή.
[217] The last six *ArCo* are very clean specimens, even if X.7–8 are interpolations.
[218] In V.9 the antecedent is elliptic and the consequent has ἄν + imperfect.
[219] I adopt the translation advocated in Bobzien 2020.

for them the conclusion is validly deduced from the premises" (Sextus, *M* VIII.228); for this reason, each of them was called ἀναπόδεικτος (συλλογισμός) "indemonstrable (syllogism)". Chrysippus recognized five species of such syllogisms; each species was identified by means of (*a*) a canonical concrete argument that serves as paradigmatic example; (*b*) a canonical description; (*c*) a template logical form, the τρόπος "mode", in which the sentences that formulate the statements featuring in (*a*) are replaced by ordinals (see Sect. 5.1.6). Let us read the canonical descriptions of four indemonstrables, which I homogenize in translation (cf. D.L. VII.80–81, Sextus, *M* VIII.224–226):[220]

A first indemonstrable is the one ‹compounded› of a conditional and of the antecendent in that conditional ‹as premises›, having the consequent as conclusion.
A third indemonstrable is the one ‹compounded› of a negation of a conjunction and of one of the conjuncts ‹as premises›, having the contradictory of the remaining conjunct as conclusion.
A fourth indemonstrable is the one ‹compounded› of a disjunction and of one of the disjuncts ‹as premises›, having the contradictory of the remaining disjunct as conclusion.
A fifth indemonstrable is the one ‹compounded› of a disjunction and of the contradictory of one of the disjuncts ‹as premises›, having the remaining disjunct as conclusion.

Let us read instead the original description, and the canonical example, of a second indemonstrable, as they are presented by Sextus (*M* VIII.225; cf. also D.L. VII.80):[221]

δεύτερος δ' ἐστὶν ἀναπόδεικτος ὁ ἐκ συνημμένου καὶ τοῦ ἀντικειμένου τῷ λήγοντι ἐν ἐκείνῳ τῷ συνημμένῳ, τὸ ἀντικείμενον τῷ ἡγουμένῳ ἔχων συμπέρασμα. [...] ὡς τὸ "εἰ ἡμέρα ἔστι, φῶς ἔστιν· οὐχὶ δέ γε φῶς ἔστιν· οὐκ ἄρα ἔστιν ἡμέρα".	A second indemonstrable is the one ‹compounded› of a conditional and of the contradictory of the consequent in that conditional ‹as premises›, having the contradictory of the antecendent as conclusion. [...] like "if it is day, there is light; and certainly not: there is light; therefore not: it is day".

A so-called "fifth indemonstrable with multiple disjuncts" formalizes the framing argument of an *ArCo* with more than two alternatives. Let us read again Sextus, who reports Chrysippus' contention that even a hunting dog applies Stoic logic (*P* I.69; see also the arguments, with explicit mention of the indemonstrable, in Cleomedes, *Cael.* I.5.23–28 and I.6.1–7):

ἤτοι τῇδε ἢ τῇδε ἢ τῇδε διῆλθε τὸ θηρίον· οὔτε δὲ τῇδε οὔτε τῇδε· τῇδε ἄρα.	The beast passed either here or here or here; and neither here nor here; therefore here.

The standard correlative used in mathematical arguments of this kind is οὐδέ ... οὐδέ ... "neither ... nor ..."; the Stoics seem to have recommended οὔτε ... οὔτε ... (cf. Sect. 5.3.4).

Of course, the Stoics would have objected to the identification of a mathematical *ArCo* with a second indemonstrable. For, even if *ArCo* are very good approximations of second indemonstrables, the scheme is not applied uniformly in the *Elements*: the negative particle may not be preposed, the first statement may not be a conditional, the coassumption may not be the negation

[220] The indemonstrable syllogisms from the third to the fifth go into themselves under application of the Stoic "first *thema*" read at the end of Sect. 5.2.1; a first and a second indemonstrable are instead transformed into one another; therefore, one of them is metalogically redundant.
[221] For my translation of the preposed negative particle see Sect. 5.3.3.

of the consequent of the conditional (if this is present), the required double negative particle may not be expressed (VIII.7). As seen above, the formulations we read in propositions I.19, 25, and V.10 are not orthodox.

When the coassumption of an *ArCo* refers to a supposition made in the setting-out of the ongoing proposition, the *ArCo* is closed by the following clause, which gives rise to a hybrid between a *RI* and an *ArCo*: ὅπερ οὐχ ὑπόκειται "which has really not been supposed", whereas we would have expected that this deductive unit were anticipated as a coassumption οὐχ ὑπόκειται δέ "and it has not been supposed", followed by the rest of the *ArCo*.[222] For this reason, I have included these deductions, with the exception of the first two listed in the previous footnote, among the *RI*. The two occurrences of the clause ὅπερ οὐχ ὑπόκειται in proposition IX.10 (but the first occurrence has the imperfect ὑπέκειτο) and the sole occurrence in X.42 are included in arguments to be regarded as *RI* both because of the length of the arguments and because, while of course not featuring the "reference to the impossible" clause that characterizes a *RI*, they do have the initializing clause and the "negation of the reduction assumption". The inverse phenomenon occurs in propositions IV.4, 8, 13, again to be classified as *RI*.

To see a hybrid of a similar kind and featuring a non-negated form of ὑπόκειμαι, let us read the first part of the proof of I.29 (*EOO* I, 72.6–16), which also contains a non-instantiated citation of V.post.5 (what is proved is that a straight line falling on parallel straight lines makes the alternate angles equal) and an interesting example of interaction of a preposed negation and of a privative part of speech (see Sect. 5.3.3):

εἰ γὰρ ἄνισός ἐστιν ἡ ὑπὸ ΑΗΘ τῇ ὑπὸ ΗΘΔ, μία αὐτῶν μείζων ἐστίν. ἔστω μείζων ἡ ὑπὸ ΑΗΘ· κοινὴ προσκείσθω ἡ ὑπὸ ΒΗΘ· αἱ ἄρα ὑπὸ ΑΗΘ ΒΗΘ τῶν ὑπὸ ΒΗΘ ΗΘΔ μείζονές εἰσιν· ἀλλὰ αἱ ὑπὸ ΑΗΘ ΒΗΘ δυσὶν ὀρθαῖς ἴσαι εἰσίν· αἱ ἄρα ὑπὸ ΒΗΘ ΗΘΔ δύο ὀρθῶν ἐλάσσονές εἰσιν· αἱ δὲ ἀπ' ἐλασσόνων ἢ δύο ὀρθῶν ἐκβαλλόμεναι εἰς ἄπειρον συμπίπτουσιν· αἱ ἄρα ΑΒ ΓΔ ἐκβαλλόμεναι εἰς ἄπειρον συμπεσοῦνται· οὐ συμπίπτουσι δὲ διὰ τὸ παραλλήλους αὐτὰς ὑποκεῖσθαι· οὐκ ἄρα ἄνισός ἐστιν ἡ ὑπὸ ΑΗΘ τῇ ὑπὸ ΗΘΔ· ἴση ἄρα.

In fact, if ΑΗΘ is unequal to ΗΘΔ, one of them is greater. Let ΑΗΘ be greater; let ΒΗΘ be added as common; therefore ΑΗΘ, ΒΗΘ are greater than ΒΗΘ, ΗΘΔ; but ΑΗΘ, ΒΗΘ are equal to two right ‹angles›; therefore ΒΗΘ, ΗΘΔ are less than two right ‹angles›; and ‹straight lines› from less than two right ‹angles› meet once unboundedly produced; therefore ΑΒ, ΓΔ once unboundedly produced will meet; and they do not meet because they have been supposed parallel; therefore it is not the case that ΑΗΘ is unequal to ΗΘΔ; therefore it is equal.

The above occurrence of the infinitive ὑποκεῖσθαι is include in a postposed explanation of the kind διὰ τό "because" + infinitive (Sect. 4.5.3); another, identical clause in a similar proof is found in XI.16. Both of these occurrences are likely to be authentic, even if this kind of clause is a priori suspect; certainly spurious is instead the residual occurrence of ὑποκεῖσθαι in lemma XII.4/5, within a different type of argument.

[222] This happens in VII.2, IX.13, IX.34, and, among the supplementary material, in lemma X.9/10, in the proof adjoined to X.39, and in the proposition that the *vulgata* and the Greco-Latin translation prepose to X.13. The Theonine manuscripts do have X.13 *vulgo* in the main text (but **B** seems to declare to have found it in another antigraph), **P** has it by collation, in a later hand. The same result, in the form of a canonical *RI*, is transmitted as a scholium, to X.17 in **P** or to X.18 in **V**; in **P** the scholium is in the first hand and carries the inscription "in others, this is placed as a theorem in order after the 12th" (see *sch.* X.125 in *EOO* V, 480.14–23 and *app.*). X.13 *vulgo* lacks the conclusion and is introduced by an aberrant clause: "lemma to the 13th from the reductio ad absurdum". Heiberg rightly regards it as an interpolation, see *EOO* III, 382.3–13. The Arabo-Latin tradition does not even have proposition X.13.

5.3. SENTENTIAL OPERATORS

Sentential operators form statements by combining or modifying statements; they can be binary or unary operators: subordinants and coordinants are examples of the former kind, negation is the most important specimen of the latter kind. Some binary operators may upgrade to polyadic operators. The present macro-Section surveys the way sentential operators are used in the demonstrative code, and does this in the order in which they appear in a mathematical proposition: Sect. 5.3.1 treats conditional clauses (they prominently figure in the enunciation); Sect. 5.3.2, "paraconditionals", namely, systems made of a causal subordinate and a principal clause (beginning of the proof); 5.3.4, disjunction; 5.3.5, conjunction; 5.3.6, syllogistic connectors. Negation is studied in Sect. 5.3.3. In these Sections, frequent parallels are drawn between mathematical practice and logical or grammatical doctrines. Ancient sources are in fact particularly generous: besides being a key doctrine of Stoic logic and a major theme in Galen's *Institutio logica*, sentential operators are presented, classified, and discussed in Dionysius Thrax's *Ars grammatica*, in the dedicated treatise *On connectors* of the great grammarian Apollonius Dyscolus (Galen's contemporary) as well as in scattered passages of his *Syntaxis*, and of course in the rich tradition of grammatical scholia.[223]

5.3.1. *Conditional*

If we exclude *RI* and *ArCo*, conditional statements are mainly found in the enunciations of theorems and in their citations within a proof. These statements are formulated by sentences featuring ἐάν "if" with aorist subjunctive or present indicative in the antecedent, present or future indicative in the consequent.[224] The passive aorist subjunctive is employed when an operation or a constructive act is undergone by the mathematical entity which is the grammatical subject of the antecedent of the conditional. In these cases, the setting-out contains a passive perfect imperative, as for the verb τέμνω "to cut" in the enunciation + setting out of II.2 (*EOO* I, 120.20–25)—the same phenomenon occurs in the entire string of propositions II.1–10:

ἐὰν εὐθεῖα γραμμὴ τμηθῇ ὡς ἔτυχεν, τὸ ὑπὸ τῆς ὅλης καὶ ἑκατέρου τῶν τμημάτων περιεχόμενον ὀρθογώνιον ἴσον ἐστὶ τῷ ἀπὸ τῆς ὅλης τετραγώνῳ.	If a straight line be cut at random, the rectangle contained by the whole ‹straight line› and by each of the segments is equal to the square on the whole.
εὐθεῖα γὰρ ἡ AB τετμήσθω ὡς ἔτυχεν κατὰ τὸ Γ σημεῖον.	In fact, let a straight line, AB, be cut at random at point Γ.

The conditionals of the mathematical enunciations differ from the canonical format of the conditional statements (συνημμένα litt. "connected") recommended in Stoic logic, for the canonical formulation of a Stoic conditional has εἰ "if" plus indicative in the antecedent (D.L. VII.71).[225]

[223] I shall refrain from citing scholia; they can be found in suitable volumes of *GG*.
[224] This is usually called the "present general" conditional; the future in the consequent strengthens the idea of necessity. On conditional sentences in ancient Greek see Wakker 1994. Dionysius Thrax (*GG* I.1, 91.2–92.1) defines conditionals as follows: συναπτικοὶ δέ εἰσιν ὅσοι ὕπαρξιν μὲν οὐ δηλοῦσι σημαίνουσι δὲ ἀκολουθία "connexive ‹connectors› are all those which do not make existence manifest but signify consequence". He lists εἰ, εἴπερ, εἰδή, εἰδήπερ.
[225] This is usually called the "simple" conditional. Aristotle did not thematize conditionals; he is also rather free in his formulations of conditions (for instance as a genitive absolute: see *APr*. I.32, 47a28–31). On the whole issue see the discussion in Ebrey 2015.

The mathematical practice is probably rooted in the requirement of attaining a maximum of generality: the antecedent with ἐάν refers to states of affairs whose actualization may occur at any time, and that necessitate the states of affairs in the consequent:[226] a more accurate translation of this subordinant is "whenever", which, however, I use for translating ὅταν. The particle ἄν marks in fact in the same way this connector, which is used, typically but not exclusively,[227] in definitions: ὅταν = ὅτε ἄν "exactly when", that is, "if and only if".[228]

A genitive absolute sometimes replaces the antecedent of the conditional sentence. This formulation is a bridge between the enunciation in conditional form and the enunciation in the form of a simple statement (cf. Sect. 4.1)[229] Here is the enunciation of I.32 (*EOO* I, 76.14–17):[230]

παντὸς τριγώνου μιᾶς τῶν πλευρῶν προσεκβληθείσης ἡ ἐκτὸς γωνία δυσὶ ταῖς ἐντὸς καὶ ἀπεναντίον ἴση ἐστίν, καὶ αἱ ἐντὸς τοῦ τριγώνου τρεῖς γωνίαι δυσὶν ὀρθαῖς ἴσαι εἰσίν.

An external angle of every triangle, one of its sides being produced, is equal to the two internal and opposite ‹angles›, and the three angles internal to the triangle are equal to two right ‹angles›.

Framing such an enunciation in conditional form would quite likely have made the quantifier to fall: *ἐὰν τριγώνου μία τῶν πλευρῶν προσεκβληθῇ, ἡ ἐκτὸς γωνία κτλ. "if one of the sides of a triangle be produced, the external angle etc.". It may be, thus, that the attested form has been retained to preserve the analogy with the enunciations of the theorems in the string I.16–20, all of which contain a quantifier and to which I.32 naturally belongs from the stylistic point of view.[231]

Conditionals with ἐάν plus subjunctive in the antecedent can also be found in definitions. A most interesting example of nested conditionals with an imperative of καλέων "to call" in the consequent is found in the definitions of the subspecies of two irrational lines in Book X. Let us read the six definitions of the subspecies of a binomial located between X.47 and 48 (*EOO* III, 136.2–19)—the ὀνόματα "names" of a binomial are its compounding straight lines; note the initial supposition formulated as a genitive absolute (underlined); the *variatio* between relative clause and conditional in the formulation of the first constraint on the "names" (italicized); the three nested conditionals correlated by μέν … δέ … δέ … (underlined):

ὑποκειμένης ῥητῆς καὶ τῆς ἐκ δύο ὀνομάτων διῃρημένης εἰς τὰ ὀνόματα *ἧς τὸ μεῖζον ὄνομα τοῦ ἐλάσσονος μεῖζον δύναται τῷ ἀπὸ συμμέτρου ἑαυτῇ μήκει*, ἐὰν μὲν τὸ μεῖζον ὄνομα σύμμετρον ᾖ μήκει τῇ ἐκκειμένῃ ῥητῇ, καλείσθω ἡ ὅλη ἐκ δύο ὀνομάτων πρώτη·

‹1› An expressible and a binomial divided into the names being supposed *whose greater name is worth more than the lesser by the ‹square› on a ‹straight line› commensurable in length with itself*, if the greater name be commensurable in length with the set out expressible, let the whole be called first binomial;

ἐὰν δὲ τὸ ἔλασσον ὄνομα σύμμετρον ᾖ μήκει τῇ ἐκκειμένῃ ῥητῇ, καλείσθω ἐκ δύο ὀνομάτων δευτέρα·

‹2› and if the lesser name be commensurable in length with the set out expressible, let ‹the whole› be called second binomial;

---

[226] See Kühner, Gerth 1898–1904, § 575.
[227] See III.20, where ὅταν replaces the canonical ἐάν.
[228] See e.g. Rijksbaron 2006, 83.
[229] We find it in I.16, 32, X.1, 71–72, 108–110.
[230] I regard παντὸς τριγώνου as not belonging to the genitive absolute phrase.
[231] In the Archimedean corpus, 24 problems out of 27, but no theorems, contain a genitive absolute in the enunciation. In Apollonius' *Conica*, 6 problems (out of 18) and 33 theorems (12 simple statements and 21 conditionals) out of 208 are enunciated in this way, even if most of the theorems so formulated are subcases of previous propositions.

ἐὰν δὲ μηδέτερον τῶν ὀνομάτων σύμμετρον ᾖ μήκει τῇ ἐκκειμένῃ ῥητῇ, καλείσθω ἐκ δύο ὀνομάτων τρίτη.	‹3› and if neither of the names be commensurable in length with the set out expressible, let ‹the whole› be called third binomial.
πάλιν δὴ ἐὰν τὸ μεῖζον ὄνομα τοῦ ἐλάσσονος μεῖζον δύνηται τῷ ἀπὸ ἀσυμμέτρου ἑαυτῇ μήκει, ἐὰν μὲν τὸ μεῖζον ὄνομα σύμμετρον ᾖ μήκει τῇ ἐκκειμένῃ ῥητῇ, καλείσθω ἐκ δύο ὀνομάτων τετάρτη·	‹4› Now again, *if the greater name be worth more than the lesser by the ‹square› on a ‹straight line› incommensurable in length with itself,* if the greater name be commensurable in length with the set out expressible, let ‹the whole› be called fourth binomial;
ἐὰν δὲ τὸ ἔλασσον, πέμπτη·	‹5› and if the lesser, fifth;
ἐὰν δὲ μηδέτερον, ἕκτη.	‹6› and if neither, sixth.

The definition is applied, curtailed and instantiated, in X.66, 103, 112–113, where all subspecies are identified in succession in some specific, and actually obtaining, geometric configuration. For this reason, in these propositions the conditional is set in the simple mode (that is, it is of the form εἰ "if" + indicative, so that the kind of negation changes from μηδέτερον to οὐδετέρα).[232] The conditional structure of the definition has in these propositions the rhetorical function of marking the division into cases. Let us read an extract from X.66 (*EOO* III, 202.17–22), in which I transcribe the last three identifications only:

καὶ εἰ μὲν ἡ ΑΕ σύμμετρός ἐστι τῇ ἐκκειμένῃ ῥητῇ, καὶ ἡ ΓΖ σύμμετρός ἐστιν αὐτῇ, καί ἐστιν ἑκατέρα τετάρτη· εἰ δὲ ἡ ΕΒ, καὶ ἡ ΖΔ, καὶ ἔσται ἑκατέρα πέμπτη· εἰ δὲ οὐδετέρα τῶν ΑΕ ΕΒ, καὶ τῶν ΓΖ ΖΔ οὐδετέρα σύμμετρός ἐστι τῇ ἐκκειμένῃ ῥητῇ, καὶ ἔσται ἑκατέρα ἕκτη.	And if AE is commensurable with the set out expressible, ΓZ is also commensurable with it, and each of them is a fourth; and if EB, also ZΔ, and each of them will be a fifth; and if neither of AE, EB, neither of ΓZ, ZΔ is commensurable with the set out expressible, and each of them will be a sixth.

Within the body of a proof, and if we exclude the non-instantiated citations of conditional enunciations (which are all of doubtful authenticity), the prevailing form of a conditional has εἰ "if" with present indicative in the antecedent (but the verb is very often understood), present or future indicative or imperative in the consequent. It is interesting that exactly this form (and not the unreal conditional clause, so frequently and unwarrantedly used in present-day mathematical discourse)[233] is the only one found in *RI* (cf. Sect. 5.2.1): this is well in keeping with the fact that the Stoic "suppositions" are not truth-apt (cf. Sect. 4.2) and confirms that a Greek *RI* is not committed to regarding the reduction assumption as false. Less frequent is the use of this kind of conditional to formulate an alternative between independent cases in a proof,[234] as in proposition X.66 read above or in the following extract from I.13 (*EOO* I, 36.9–11):[235]

[232] The gender changes from masculine to feminine, according to the transition from the "names" (masc.) to the "straight lines" (fem.) αἱ ΑΕ ΕΒ that represent the "names".
[233] This feature, and the widespread tendency to add determiners of necessity and of generality (like "always" or "never": parallel straight lines "never" meet), show that present-day mathematicians have a much less developed sense of mathematical generality than Greek mathematicians had: just open at random any modern textbook or advanced treatise to check my contention.
[234] The preferred form is an instantial imperative that introduces each case of the disjunction.
[235] The elliptic syntagm εἰ δὲ οὔ is employed to mark the second case of a dichotomy in propositions I.13, II.14, VI.28, VII.2, 4, 33, IX.19, XI.11, 20, 22.

εἰ μὲν οὖν ἴση ἐστὶν ἡ ὑπὸ ΓΒΑ τῇ ὑπὸ ΑΒΔ, δύο ὀρθαί εἰσιν. εἰ δὲ οὔ, ἤχθω ἀπὸ τοῦ Β σημείου τῇ ΓΔ εὐθείᾳ πρὸς ὀρθὰς ἡ ΒΕ.

Then if ΓΒΑ is equal to ΑΒΔ, they are two right ‹angles›. And if not, from point B let a ‹straight line›, BE, be drawn at right ‹angles› with straight line ΓΔ.

If one of the independent cases in a proof is dealt with by means of a potential proof, the conditional shifts back to the general mode, as we see in V.6 (*EOO* II, 20.24–22.2)[236]—the conditional immediately follows the conclusion of the argument potentially to be repeated:

ὥστε εἰ τὸ ΗΒ τῷ Ε ἴσον ἐστίν, καὶ τὸ ΘΔ ἴσον ἔσται τῷ Ζ. ὁμοίως δὴ δείξομεν ὅτι, κἂν πολλαπλάσιον ᾖ τὸ ΗΒ τοῦ Ε, τοσαυταπλάσιον ἔσται καὶ τὸ ΘΔ τοῦ Ζ.

So that, if HB is equal to E, ΘΔ will also be equal to Z. Very similarly we shall prove that, even if HB be multiple of E, ΘΔ will also be such a multiple of Z.

A peculiar use of the simple conditional is found in the applications of V.def.5, where this definition is checked as a conjunction of three conditionals, as in V.11 (*EOO* II, 34.13–18):

καὶ ἐπεί ἐστιν ὡς τὸ Α πρὸς τὸ Β οὕτως τὸ Γ πρὸς τὸ Δ καὶ εἴληπται τῶν μὲν Α Γ ἰσάκις πολλαπλάσια τὰ Η Θ τῶν δὲ Β, Δ ἄλλα, ἃ ἔτυχεν, ἰσάκις πολλαπλάσια τὰ Λ Μ, εἰ ἄρα ὑπερέχει τὸ Η τοῦ Λ, ὑπερέχει καὶ τὸ Θ τοῦ Μ, καὶ εἰ ἴσον ἐστίν, ἴσον, καὶ εἰ ἐλλείπει, ἐλλείπει.

And since, as A is to B, so Γ is to Δ, and of A, Γ turn out to be taken equimultiples H, Θ, of B, D other random equimultiples Λ, M, therefore if H exceeds Λ, Θ also exceeds M, and if it is equal, it is equal, and if it falls short, it falls short.

The strictly homologous enunciations of IX.18 and 19 (*EOO* II, 380.25–382.2 and 384.2–4) provide an interesting example of early corrections to a wording that was apparently regarded as aberrant with respect to the (canonical) conditional form:[237]

δύο ἀριθμῶν δοθέντων ἐπισκέψασθαι εἰ δυνατόν ἐστιν αὐτοῖς τρίτον ἀνάλογον προσευρεῖν.	Two numbers being given, investigate whether it is possible to find a third proportional to them.
τριῶν ἀριθμῶν δοθέντων ἐπισκέψασθαι πότε δυνατόν ἐστιν αὐτοῖς τέταρτον ἀνάλογον προσευρεῖν.	Three numbers being given, investigate when it is possible to find a fourth proportional to them.

The first enunciation has εἰ "if" where the second has πότε "when", but all Theonine manuscripts correct πότε to εἰ.[238]

A conditional is always a non-simple *statement*: the cases of a conditional with ἄρα in the consequent (a "hybrid" between a statement and an argument) are very rare: I have found them only in statements that initialize *RI*. A noteworthy cluster (which obviously serves as a stylistic marker) is in redaction **b** of Book XII;[239] let us read the occurrence in XII.10 (*EOO* IV, 404.26–30):

εἰ γὰρ μὴ ὁ ΑΒΓΔΚΛ κῶνος πρὸς τὸν ΕΖΗΘΜΝ τριπλασίονα λόγον ἔχει ἤπερ ἡ ΒΔ πρὸς τὴν ΖΘ, ἕξει

In fact, if cone ΑΒΓΔΚΛ to ΕΖΗΘΜΝ does not have a triplicate ratio than ΒΔ to ΖΘ, therefore

[236] Similar transitions, almost surely not authentic, are also found in III.25 and 37.
[237] See also Sect. 4.2.1 for a discussion of some peculiarities of these propositions.
[238] Of course, the determination of IX.19 has another πότε. The enclitic adverb ποτε "eventually" is found in parallel deductive steps in V.8 (*bis*), X.1 and X.1*alt*, X.3.
[239] In propositions **b** XII.5, 9–10, 17, but there is also a residue in XII.18 (*EOO* IV, 244.1–5) of the main redaction, which has escaped the homogenization.

ἄρα ὁ ΑΒΓΔΚΛ κῶνος ἤτοι πρὸς ἔλασσόν τι τοῦ ΕΖΗΘΜΝ κώνου στερεὸν τριπλασίονα λόγον ἤπερ ἡ ΒΔ πρὸς τὴν ΖΘ ἢ πρὸς τὸ μεῖζον.	cone ΑΒΓΔΚΛ either to some solid less than cone ΕΖΗΘΜΝ or to a greater one will have a triplicate ratio than ΒΔ to ΖΘ.

The phenomenon of hybridization between a non-simple statement and an inference becomes pervasive in the case of the logical object I shall discuss in the next Section, and which we have just seen at work in V.11.

5.3.2. *Paraconditional*

In Sect. 4.4 I have outlined the features of the initial segment of a proof; following M. Federspiel, I have called it "anaphora". This initial segment is characterized by a peculiar logico-syntactic form, recognized as such in ancient dialectic. In his *Dialectical Art*, the Stoic Crinis[240] introduced in fact a non-simple statement akin to the conditional; he called it παρασυνημμένον "paraconditional". Syntactically, it differs from a conditional only by the initial subordinant: ἐπεί "since" instead of εἰ "if". Semantically, it is stronger than a conditional:[241]

ἐπαγγέλλεται δ' ὁ σύνδεσμος ἀκολουθεῖν τε τὸ δεύτερον τῷ πρώτῳ καὶ τὸ πρῶτον ὑφεστάναι.	The connector announces both that the second is a consequence of the first and that the first subsists.

A paraconditional is true whenever the consequent follows (εἰς ἀκόλουθον λήγει) from the antecedent and the antecedent is true (D.L. VII.74)—this characterization is not truth-functional, as it depends on the notion of ἀκολουθία "consequence". Already Theophrastus studied this kind of statement and "made the reason for such a use clear in the first ‹Book› of his *First Analytics*".[242]

Note, in the definition above, the verb form ὑφεστάναι "to subsist". In Stoic metaphysics, it denotes the peculiar mode of existence of incorporeal entities like the λεκτά "sayables", a subclass of which are the ἀξιώματα "statements".[243] Still, we are not entitled to attach too much importance to this lexical peculiarity: in Dionysius Thrax (*GG* I.1, 92.2–3) we find an alternative definition, which employs the less connotated notion of ὕπαρξις "existence":[244]

παρασυναπτικοὶ δέ εἰσιν ὅσοι μεθ' ὑπάρξεως καὶ τάξιν δηλοῦσιν.	Paraconnexive ‹connectors› are all those which make an ordering with an existence manifest.

Prima facie, this is exactly what happens in mathematics: the antecedent of a paraconditional formulates a state of affairs that "subsists" since it has been supposed in the setting-out or in the construction; the consequent follows from the antecedent, either immediately or in virtue of previous results. There are problems, however, in establishing a link between Crinis' proposal and mathematical practice. First, the condition in the antecedent of a paraconditional has only been

[240] He is later than the second half of 2nd century BCE; we know next to nothing of him: *DPhA* II, 551.
[241] D.L. VII.71. For the ordinals, see Sect. 5.1.6; for the verb ἐπαγγέλλω "to announce", Sects. 5.3.4–5.
[242] Simplicius, *in Cael.*, 552.31–553.5 (= fr. 112C Fortenbaugh). Simplicius asserts that οἱ νεώτεροι παρασυναπτικὸν καλοῦσι "more recent authors call it 'paraconnexive'" and that ἐν οἷς τὸ ἡγούμενον οὐ μόνον ἀληθές ἐστιν, ἀλλὰ καὶ ἐναργὲς καὶ ἀναμφίλεκτον "in them, the antecedent is not only true, but also manifest and unquestionable".
[243] See for instance Brunschwig 2003, sect. 3, and also Sect. 4.2 above. Bronowski 2019 mainly offers a thick verbal fog.
[244] On the two terms see Goldschmidt 1972. Dionysius Thrax lists ἐπεί, ἐπείπερ, ἐπειδή, ἐπειδήπερ as exemples (for the latter three see Sect. 4.5.3).

previously *supposed*: therefore, it is neither true nor false. So, the semantic feature that distinguishes a paraconditional from a conditional does not apply exactly to those conditions that invariably constitute the antecedent of a paraconditional. An easy reply to this objection lies in the fact that the above semantic feature distinguishes a *true* paraconditional from a *true* conditional and in recalling our discussion in Sect. 4.4: simply stated, a proof need not begin with a *true* paraconditional.

The second problem is more serious: by Stoic standards, ἄρα cannot figure in a sentence that formulates a statement, be it simple or non-simple, because ἄρα characterizes the conclusion of an *argument*; thus, such a syntactic unit as a paraconditional with ἄρα introducing the consequent formulates a hybrid between a statement and an argument.[245] Well, it is a fact that the majority of mathematical paraconditionals do have ἄρα in the consequent (an "apodotic" ἄρα);[246] this is, moreover, the only conspicuous case in which ἄρα does not mark the conclusion of an inference. To clarify the issue, let us compare the paraconditionals that open the proofs of I.43 and 44 (*EOO* I, 100.12–14 and 102.20–22)—the propositions they refer to are I.34 and I.29, respectively:

ἐπεὶ γὰρ παραλληλόγραμμόν ἐστι τὸ ΑΒΓΔ διάμετρος δὲ αὐτοῦ ἡ ΑΓ, ἴσον ἐστὶ τὸ ΑΒΓ τρίγωνον τῷ ΑΓΔ τριγώνῳ.	In fact, since there is a parallelogram, ΑΒΓΔ, and a diagonal of it, ΑΓ, triangle ΑΒΓ is equal to triangle ΑΓΔ.
καὶ ἐπεὶ εἰς παραλλήλους τὰς ΑΘ ΕΖ εὐθεῖα ἐνέπεσεν ἡ ΘΖ, αἱ ἄρα ὑπὸ ΑΘΖ ΘΖΕ γωνίαι δυσὶν ὀρθαῖς εἰσιν ἴσαι.	And since a straight line, ΘΖ, fell on parallel ‹straight lines›, ΑΘ, ΕΖ, therefore the angles ΑΘΖ, ΘΖΕ are equal to two right ‹angles›.

The first sentence above is something Crinis would have accepted as a well-formed paraconditional,[247] the second formulates a hybrid between a statement and an argument—I shall insist on calling each of them "paraconditional", while specifying, if needed, whether I mean the paraconditional in "pure" form or the paraconditional in "hybrid" form.

To assess this unwelcome discrepancy, I have carried out a detailed survey of all paraconditionals found in some Books of the *Elements*; the aim was to test possible correlations between a number of characteristics of the textual unit introduced by ἐπεί. These are:

- Form of the subordinant nexus: καὶ ἐπεί, ἐπεὶ οὖν, ἐπεὶ καί, ἐπεὶ γάρ, ἐπεί, or πάλιν ἐπεί.
- Nature of the sentence that formulates the statement: pure paraconditional, hybrid with ἄρα, or hybrid with δή.
- Type of the assumption in the antecedent: reference to the setting-out, to the construction, to a previous proposition, or to a result previously proved in the ongoing proof.
- Mood of the verb in the consequent: indicative or imperative; they correspond to a geometric statement or to a constructive act, respectively.
- Position within the proof: liminal or non-liminal.
- Presence or absence of a reference to a previous proposition.[248]
- Position of the antecedent: preposed or postposed.
- Antecedent of the paraconditional in the form of a simple statement or of a conjunction.

245 Of course, this would be a debased argument: it is a μονολήμματος "one-assumption" argument (and these were proscribed until Antipater: see Bobzien 1999, 155, for a discussion); it is not a system of principal clauses.
246 For comparison, there are hundreds of "paraconditional" clauses in the Aristotelian corpus, but only 8 of them contain an apodotic ἄρα.
247 The antecedent is made of two sentences, conjoined by δέ; see below and Sect. 5.3.5.
248 But this may be difficult to assess, see Sect. 4.5.4.

I have analysed in this way the "anaphorae" of Books I, II, IV, V, VI, VII, XII of the *Elements*. The results are set out in the following series of tables:[249]

Book I

	pure	hybrid with ἄρα	hybrid with δή
καὶ ἐπεί	1c, 3c, **13c**, 17c, 18c+, 32c§, 33a§, 42c, 45c+, **47c**	5a+§, 20d+§*, 21p*, 24d+§*, 30c, 31c, 33d+, 34d+, 36a§, 44c,[250] 44c+§[251], 45c, 45c+§, 45d+§	26c+§, 33a+§, 47c+, 47c+§,[252] 48c+§
ἐπεὶ οὖν	2c, 7c, 20c, 43d+§[253]	5d+§, 14c, 23c§	5a§*, 6c§, 16c§, 24a§, 26c§
ἐπεὶ καί	18c+***		
ἐπεὶ γάρ	**18a****, 22c, **24a****, 34a§, 35a, 43a§[254]	15a, 28a§, 46c+	9c§, 10c§, 11c§, 12c§, 34a+§
ἐπεί	48c		
πάλιν ἐπεί	1c+, 2c+, 7c+, **13c+**, **21p+**, 22c+, 24c+, 32c+§, 34a+§, 43a+§[255]	15a+, 21+*, 28a+§, 30c+	

Book II

	pure	hybrid with ἄρα
καὶ ἐπεί	4c§, **5d**, 9c, 9c+, 10c+, 10c+	4dp+§, 8cp+§, 8c+§, 8c+, 9c+, 9dp+§, 10c, 10d+
ἐπεὶ οὖν	**6a**, **7p**	8cp§, 14c§*
ἐπεὶ καί	4c+***, 5a+***	
ἐπεὶ γάρ		4c+§, 11c§*, 12a*, 13a*
ἐπεί		
πάλιν ἐπεί	9d+, 10d+	8cp+§, 9dp+§, 10cp+§

Book IV

	pure	hybrid with ἄρα	hybrid with δή
καὶ ἐπεί	10d+, 12c+	3c§, 3d+§, 5c§ (*bis*), 6c§,[256] 7cp+§, 7cd+§, 8ad§, 9ad+§,[257] 10c§, 10cd+§, 12c§, 12d+ (*bis*), 13ac+§, 14ac+§, 15c+*	4c§*, 9a§, 12a+§, 13a§, 13c+§*
ἐπεὶ οὖν	1c+, **10d+**	2c§, 7c§, 10cd+§, 11cd§	
ἐπεὶ καί	10c+***		
ἐπεὶ γάρ	**11d+**, 15c, **15d+**	6c+, 7cd+§, 12d+§	
ἐπεί			
πάλιν ἐπεί	15c+		

[249] The signs that accompany the numbers of the propositions have the following meaning: species of the assumption in the antecedent: a = *a*ssumption in the setting-out, c = in the *c*onstruction, d = result proved in the ongoing *d*emonstration, p = result proved in a previous *p*roposition. The occurrences with an imperative in the consequent are in boldface. Moreover, + = non-liminal position in the proof; * = citation (possibly instantiated) of an enunciation; ** = paraconditional located in the construction: *** = postposed antecedent; § = conjoined antecedent. I have also bracketed the occurrences Heiberg deemed spurious. The associated footnotes list all variant readings in the reference manuscripts.

[250] om. ἄρα **P**.

[251] om. ἄρα **B**.

[252] om. δή **BFVbp**.

[253] add. ἄρα **b**.

[254] add. ἄρα **F**.

[255] add. ἄρα **P**.

[256] om. ἄρα **Bp**.

[257] om. ἄρα **p**.

Book V

	pure	hybrid with ἄρα
καὶ ἐπεί	18c+*, 19a+*,[258] [19d+], 23a+, 24d+*	1c+, 3ac+§, 4ac§, 4ac+§, 5c, 5ad+§, 8c+, 11ac§, 12ac§, 13ac+§, 15c+§, 16c*§, 17c, 17a+§, 22ac§, 23ap*§, 23ap+*§, 25d+*, 25c+§, [25p+*]
ἐπεὶ οὖν		3d+§*, 6d+, 7ac§, 8c+, 22d+*, 23d+*§, 24ad+§, [25ac§]
ἐπεὶ καί		
ἐπεὶ γάρ	**8a****, **13a**§**, 19a	1a, 2a, 3a, 14a§, 15a, 20a*§, 21a§, 24a
ἐπεί		6§
πάλιν ἐπεί	17c+§*[259]	8c+§, 11ac+§, 16c+, 17c+, 23a+

Book VI

	pure	hybrid with ἄρα	hybrid with δή
καὶ ἐπεί	1c, 4a+,[260] 20a, 20p§,[261] 22/23a+, 24a+, **28p+**, 29c+, **30c+**	1d+§*, 3c, 3c+*, 4ad§, 4c+*, 5cd+§, 7ac, 10c+*, 13c+*, 16ac§, 17ac§, 18cd+, 19cp+§*, 20p+§, 20d+§, 20dp+§*, 22ap§, 22cd§, 23ap+§, 24d+, 25cp§*, 28c, 29c, 31d+, 31d+*, 32ad+§*	
ἐπεὶ οὖν	11c*, 31c,[262] 33c	1d+, 4d+, 9c*, 12c*, 14ac§*, 15ac§*, 19a, 20dp+§*, 22c§, 23d+, 26c, 27cp+§, 28c	5d+§
ἐπεὶ καί	27c+***, 28c+***, 29p+***		
ἐπεὶ γάρ	21a, 24c, 27c,[263] 32§a*	8ap§, 8ad§, 14ap§	
ἐπεί	13c	2p§, 3d§, 15ap§, 16c§, 17ac§c, 29c§	
πάλιν ἐπεί	3c+, 4a+, 21a+, 24c+	4d+, 10c+*, 23ap+§	

Book VII (no consequent in the imperative)

	pure	hybrid with ἄρα	hybrid with δή
καὶ ἐπεί	3c+, 4c+, 10c+	2c+§, 3c+§, 5c+,[264] 6c+, 7c, 7c+,[265] 8ac+§, 9c+§, 15c+§, 20c+§, 21c+, 24ac+§,[266] 28c+, 29c+, 31c+§, 33c+ (*bis*), 34c+ (*bis*), 35ac§	
ἐπεὶ οὖν		1c+§, 2c+§, 3c+, 21p+*, 27ad+§, 27ad+*, 28c+, 30c+	19ac§
ἐπεὶ καί			
ἐπεὶ γάρ	31a*	3a+*, 5a, 6a, 9a, 10a, 12a, 13a, 14a, 15a, 16a, 17a, 18a,[267] 26a, 27a§, 38ap§	
ἐπεί		11a, 19a+, 22c+, 23ac+§, 25ac+§, 29ac+§, 36c+, 36c+§, 36c+, 37cp§	
πάλιν ἐπεί		8ac+§, 14a+, 16+, 27a+§	19ac+§

[258] add. ἄρα **Bp** m. 2 **V** m. rec. **P**.
[259] add. ἄρα **BpF**.
[260] add. ἄρα **BVp** m. 2 **F**.
[261] add. ἄρα **F**.
[262] add. ἄρα **V**.
[263] add. ἄρα **Bp**.
[264] om. ἄρα **BVFp**.
[265] om. ἄρα **F**.
[266] om. ἄρα **Vφ**.
[267] om. ἄρα **φ** del. **V**.

Book XII (no consequent in the imperative)

	pure	hybrid with ἄρα
καὶ ἐπεί	1a, 3c+* (*ter*), 3c+, 17c+, 17c+§, 17c+ (*bis*)	3c+§, 3p+*, 4a+, 4/5a+*, 6c+, 7cd+§, 8a, 8d+§, 9ad§, 12a+, 12d+§, 13c*,[268] 14c+*, 15a+, 16c+§, 17c+*, 17c+, 17c+§[269], 17c+*, 17c+§ (*ter*)
ἐπεὶ οὖν	6c*, 14a*	11d+§,12d+§*, 13c+§
ἐπεὶ καί		
ἐπεὶ γάρ	2/3d+	4a§
ἐπεί		3c§, 7a§, 9d+§, 15a+§
πάλιν ἐπεί	7a+§	12c+§, 12d+§ (*bis*)

These raw data call for the following remarks, which will be subsequently assessed.

First. The pure paraconditionals are a strict minority. The data do not show significant correlation with any of the above parameters,[270] apart from features that are obvious or not supported by large enough statistical samples: the paraconditionals with postposed antecedent are all pure; the same for the paraconditionals with an imperative in the consequent.

Second. The occurrences of ἐπεί not accompanied by a particle are infrequent;[271] the higher frequency of isolated ἐπεί in Book VII may serve as a stylistic marker. In our sample, there are only 17 variant readings concerning the ἐπεί-syntagm; most of them add a particle to a solitary ἐπεί. The presence of a particle marks, as a scope particle, the role of a given paraconditional in the proof. We thus find paraconditionals that initialize a proof or a new case in the proof (these are accompanied by γάρ); paraconditionals with a progressive function (accompanied by καί or οὖν), both at the beginning and in the body of a proof; paraconditionals with a reiterative function (πάλιν), all within the body of a proof. As a matter of fact, solitary ἐπεί are mainly found, as is natural, in postposed antecedents, or in peculiar constructs such as the genitive absolute phrase τῶν γὰρ αὐτῶν κατασκευασθέντων, ἐπεί … "in fact, the same constructions being performed, since …",[272] where the scope particle γάρ must be placed in a liminal position and is thus disconnected from ἐπεί .

Third. Paraconditionals in whose consequent constructive acts are performed are scarce.[273]

Fourth. An interesting variant of semi-hybrid paraconditional contains the mildly resultative particle δή in the consequent. Most occurrences of this kind in Book I are within citations of the enunciations of I.4 and I.8.[274] These are simple alignment steps, statements that obviously derive from the assigned conditions and whose inferential character is thereby extremely weak[275]: for this reason we find δή in the consequent (cf. Sect. 5.3.6). I have called these steps "alignment steps" because they reformulate the conjunction of two assumptions so as to "align" them (see Sect. 4.5.1.3) with the canonical formulation contained in the antecedent of the cited enunciation. Let us read a sequence that refers to I.4 and one that refers to I.8;[276] they are included in the immediately subsequent propositions I.5 and I.9 (*EOO* I, 20.15–17 and 30.18–20), respectively:

[268] om. ἄρα **P**.
[269] om. ἄρα m. 1 **P**.
[270] Only a correlation with the number of assumptions conjoined in the antecedent seems to be detectable.
[271] In the main text of the *Elements*, there are only 51 occurrences of ἐπεί not accompanied by a particle.
[272] Occurrences in III.3, VI.2, 3, VI.16–17, VII.19, XI.34, XII.9, 15.
[273] Examples are provided in Sect. 4.4.
[274] The only exception is the first occurrence in I.47, included in a reference to I.14. The obvious fact is that the sum of two right angles makes two right angles, and this is enough to satisfy the assumption of I.14. Recall that I.4 and I.8 are the SAS and the SSS criterion of congruence of triangles, respectively.
[275] But not all steps of this kind carry a particle δή; read for instance VII.5 (*EOO* II, 202.6–8): καὶ ἐπεὶ ἴσος ἐστὶν ὁ μὲν ΒΗ τῷ Α ὁ δὲ ΕΘ τῷ Δ, καὶ οἱ ΒΗ ΕΘ ἄρα τοῖς Α Δ ἴσοι "and since BH is equal to A and EΘ to Δ, therefore BH, EΘ are also equal to A, Δ".
[276] For the form of the citation of these theorems, see Sect. 4.5.4.

ἐπεὶ οὖν ἴση ἐστὶν ἡ μὲν ΑΖ τῇ ΑΗ ἡ δὲ ΑΒ τῇ ΑΓ, δύο δὴ αἱ ΖΑ ΑΓ δυσὶ ταῖς ΗΑ ΑΒ ἴσαι εἰσὶν ἑκατέρα ἑκατέρᾳ.	Then since AZ is equal to AH and AB to AΓ, thus two ‹sides›, ZA, AΓ, are equal to two ‹sides›, HA, AB, respectively.
ἐπεὶ γὰρ ἴση ἐστὶν ἡ ΑΔ τῇ ΑΕ κοινὴ δὲ ἡ ΑΖ, δύο δὴ αἱ ΔΑ, ΑΖ δυσὶ ταῖς ΕΑ ΑΖ ἴσαι εἰσὶν ἑκατέρα ἑκατέρᾳ.	In fact, since AΔ is equal to AE and AZ is common, thus two ‹sides›, ΔA, AZ, are equal to two ‹sides›, EA, AZ, respectively.

The same happens in VII.19 (*EOO* II, 226.25–228.1), where the "obvious" statement that performs the "alignment" reads as follows:[277]

ἐπεὶ οὖν ὁ Α τὸν Γ πολλαπλασιάσας τὸν Η πεποίηκεν τὸν δὲ Δ πολλαπλασιάσας τὸν Ε πεποίηκεν, ἀριθμὸς δὴ ὁ Α δύο ἀριθμοὺς τοὺς Γ Δ πολλαπλασιάσας τοὺς Η Ε πεποίηκεν.	Then since A multiplying Γ turns out to make H and multiplying Δ turns out to make E, thus a number, A, multiplying two numbers, Γ, Δ, turns out to make H, E.

Fifth. The absence of a paraconditional at the very beginning of a proof can be induced by the facts that the proof requires a division into cases or that it is framed as a *RI*.

Sixth. A look at Heiberg's apparatus reported in the footnotes above shows that the variant readings that concern ἄρα in the consequent are sporadic. Therefore, we are not entitled to suppose that the presence of ἄρα is connected with later revisions or with accretions that occurred in the process of copying. Stated in a different way: if one or more campaigns of insertion of apodotic ἄρα in paraconditionals have taken place, they must have occurred at a pre-traditional stage of the transmission of the *Elements*.

Seventh. Paraconditionals in geometric proofs naturally draw the statement located in their antecedent from the construction more frequently than paraconditionals in proofs related to proportion theory (Book V) or to number theory (Books VII–IX).

Eighth. The presence of the specific syntagm ἐπεὶ γάρ does not correlate with the absence of γάρ in a previous setting-out or in a previous construction. Just looking at Book I, a paraconditional introduced by ἐπεὶ γάρ follows setting-out and construction, neither of them introduced by γάρ, in propositions 9–11, 22, 46; it follows setting-out and construction, the latter introduced by γάρ, in proposition 12; it immediately follows the setting-out in propositions 15, 28, 34 (*bis*), 35, 43; it initializes the construction in propositions 18 and 24.

How shall we assess these results? The paraconditional is a key logico-syntactic item in the Greek demonstrative code, because it is the only non-simple statement with inferential import used systematically in a proof. It would be surprising that the Stoics did not pay attention to it. Moreover, a paraconditional might be read as a first indemonstrable (= a *modus ponens*) in which the coassumption is subsumed in the antecedent of the conditional that figures in the first indemonstrable (with shift of the subordinant from εἰ to ἐπεί): this reading might justify the presence of apodotic ἄρα. Crinis' prescription would simply have resolved the ambiguity inherent in mathematical practice. However, I find this argument rather weak: the data set out above corroborate the hypothesis

[277] Note the use of the perfect stem, typical in this kind of formula.

that the mathematical practice and the Stoic doctrine as far as the paraconditional is concerned are not only independent, but divergent.

This view is confirmed by the fact that, whenever the antecedent of a paraconditional is in conjoined form, a striking variability is apparent in the use of coordinants within the antecedent, and among them we find particles that Stoic doctrine does not categorize as conjunctive but as coassumptive.[278] What happens in Book VI is set out in the following table:[279]

correlative form	proposition
nothing ... ἀλλά ...	3, 17, 23 (*bis*), 29
nothing ... ἀλλὰ μέν ... δέ ...	2, 14, 15
nothing ... ἀλλὰ μήν ...	8, 20
nothing ... δέ ...	4, 5, 14, 15, 16, 17, 19, 20, 22 (*bis*), 25, 27, 32
nothing ... δὲ καί ...	20
nothing ... καί ...	8, 20, 32
nothing ... καὶ μέν ... δέ ...	16, 22
nothing ... καὶ ἔτι ...	20
μέν ... ἀλλά ...	5
μέν ... δέ ...	1, 4, 7, 18, 23, 24
μέν ... δέ ... δέ ...	1

I end this Section with a table that displays the distribution of all paraconditionals in the main text of the *Elements*; I have categorized the occurrences according to the form of the ἐπεί-syntagm, when this is accompanied by a particle.[280] The density of paraconditionals depends on the length and on the articulation of a proof, that is, on the number of signs (and not on the number of propositions) in a Book:[281]

	I	II	III	IV	V	VI	VII	VIII	IX	X	XI	XII	XIII	tot.
# prop.	48	14	37	16	25	33	39	27	36	115	39	18	18	465
# signs	58	25.3	52.6	26	37.7	58	44	38.3	39	199.7	68	63.7	52.7	763.4
% # signs	7.6	3.3	6.9	3.4	4.9	7.6	5.7	5	5.1	26.1	9	8.3	6.9	100
καὶ ἐπεί	29	14	39	24	25	35	23	38	31	149	64	31	55	557
ἐπεὶ οὖν	12	4	21	6	8	17	9	2	5	65	5	5	5	164
ἐπεὶ καί	1	2	/	1	/	3	/	/	1	/	1	/	4	13
ἐπεὶ γάρ	14	4	2	6	11	7	16	13	14	64	5	2	10	168
πάλιν ἐπεί	14	5	12	1	6	7	5	17	4	31	3	4	8	117
tot.	70	29	74	38	50	69	53	70	55	309	78	42	82	1019

A markedly negative correlation with the number of signs is evident in the case of Book XII; this can be explained by the length of the theorems there contained. To check this, look at the correlation between the number of propositions and the number of signs in each Book.

[278] Cf. Sect. 4.5.5 and Federspiel 2008b, 543–545.
[279] I include with hesitation the correlatives μέν ... δέ ... governed by the same verb.
[280] Recall that there are 51 occurrences of solitary ἐπεί.
[281] Add three occurrences of διὰ τὰ αὐτὰ δὴ ἐπεί "for the very same ‹reasons›, since" in XI.17, 31, XIII.16.

5.3.3. *Negation*

The propositions of the *Elements* enunciated in negative form[282] are set out in the following table:[283]

statement with negation of the predicate	denegative
I.7, III.4–6, 10, 13, 16,[284] 23, VIII.6, 16–17, IX.16–17, X.7, 9,[285] 26, 111, XI.1, 13	VIII.6,[286] IX.10, 13–14, X.115

The terminology I have used conforms to the Stoic classification of negative statements, witnessed to in D.L. VII.69–70:[287]

ἐν δὲ τοῖς ἁπλοῖς ἀξιώμασίν ἐστι τὸ ἀποφατικὸν καὶ τὸ ἀρνητικὸν καὶ τὸ στερητικὸν [...] ἀποφατικὸν μὲν οἷον "οὐχὶ ἡμέρα ἐστίν"· εἶδος δὲ τούτου τὸ ὑπεραποφατικόν. ὑπεραποφατικὸν δ' ἐστὶν ἀποφατικὸν ἀποφατικοῦ, οἷον "οὐχὶ ἡμέρα ἔστι"· τίθησι δὲ τὸ "ἡμέρα ἐστίν". **70** ἀρνητικὸν δέ ἐστι τὸ συνεστὸς ἐξ ἀρνητικοῦ μορίου καὶ κατηγορήματος, οἷον "οὐδεὶς περιπατεῖ". στερητικὸν δέ ἐστι τὸ συνεστὸς ἐκ στερητικοῦ μορίου καὶ ἀξιώματος κατὰ δύναμιν, οἷον "ἀφιλάνθρωπός ἐστιν οὗτος".

Among simple statements there is the negative, the denegative, and the privative [...] A negative ‹is the one made of a negation and of a statement›, like "not: it is day"; the supernegative is a species of this. A supernegative is a negative of a negative, like "not: ‹not:› it is day". It posits "it is day". **70** A denegative is the one made of a denegative part and of a predicate, like "no one walks". A privative is the one made of a privative part and of a potential statement, like "this one is asocial".

With the exception of X.115, the proofs of all the propositions listed above are *RI*.[288] As we have seen in Sect. 4.2, some propositions enunciated in negative form do not have a setting-out.[289] Thus, the enunciation is immediately followed by the conditional that initializes a *RI*; the consequent of this conditional normally contains an instantiation of the "impossible" configuration, which in its turn is immediately followed by the construction, as in X.26 (*EOO* III, 74.8–13):

μέσον μέσου οὐχ ὑπερέχει ῥητῷ.
εἰ γὰρ δυνατόν, μέσον τὸ ΑΒ μέσου τοῦ ΑΓ ὑπερεχέτω ῥητῷ τῷ ΔΒ, καὶ ἐκκείσθω ῥητὴ ἡ ΕΖ, καὶ τῷ ΑΒ ἴσον παρὰ τὴν ΕΖ παραβεβλήσθω παραλληλόγραμμον ὀρθογώνιον τὸ ΖΘ πλάτος ποιοῦν τὴν ΕΘ, τῷ δὲ ΑΓ ἴσον ἀφῃρήσθω τὸ ΖΗ.

A medial does not exceed a medial by an expressible.
In fact, if possible, let a medial, AB, exceed a medial, ΑΓ, by an expressible, ΔΒ, and let an expressible, EZ, be set out, and let a rectangular parallelogram, ZΘ, equal to AB be applied to EZ making a width EΘ, and let a ‹region›, ZH, be removed equal to ΑΓ.

Let us return to the list set out at the beginning of this Section. The enunciation of the impossibility theorem IX.13 (*EOO* II, 366.14–18) is a denegative statement:

[282] See Moorhouse 1959 for negatives in Greek prose (but the sample of post-classical writers is very limited); Cavini 1985 for an in-depth logical analysis of Aristotle's and the Stoics' doctrines of negation.
[283] Of course, conditional enunciations must have a negative consequent.
[284] Only the second statement of the three in the enunciation.
[285] Only the second part of the enunciation.
[286] The enunciation contains both a negative particle and denegative pronouns.
[287] Add to this passage the long discussion in Alexander, *in APr.*, 401.16–405.16. Diogenes' text requires a couple of integrations, which I only provide in translation.
[288] As we have seen in Sect. 4.3, negations have a prominent role in the formulation of statements of uniqueness.
[289] These are I.7, III.10, 13, 23, X.26, XI.1, 13.

ἐὰν ἀπὸ μονάδος ὁποσοιοῦν ἀριθμοὶ ἑξῆς ἀνάλογον ὦσιν ὁ δὲ μετὰ τὴν μονάδα πρῶτος ᾖ, ὁ μέγιστος ὑπ᾽ οὐδενὸς ἄλλου μετρηθήσεται παρὲξ τῶν ὑπαρχόντων ἐν τοῖς ἀνάλογον ἀριθμοῖς.

If from a unit as many numbers as we please be successively in proportion and the one after the unit be prime, the greatest one will not be measured by any other except those featuring among the numbers in proportion.

Privative statements are fairly common; in them, the nominal complement of the copula is formed by prefixing a privative particle to an adjective. For instance, the predicate ἄνισος "unequal" is found about 200 times in the Euclidean corpus and 75 times in the main text of the *Elements*. However, theorems showing that two mathematical objects are unequal are exceedingly rare. An example can be found in proposition 35 of *Optica* redaction **A** (*EOO* VII, 64.23–27):

ἐὰν δὲ ἡ ἀπὸ τοῦ ὄμματος πρὸς τὸ κέντρον τοῦ κύκλου προσπίπτουσα μήτε πρὸς ὀρθὰς ᾖ τῷ ἐπιπέδῳ τοῦ κύκλου μήτε τῇ ἐκ τοῦ κέντρου ἴση μήτε ἴσας γωνίας περιέχουσα, αἱ διάμετροι ἄνισοι φανήσονται πρὸς ἃς ποιεῖ ἀνίσους γωνίας.

If the ‹straight line› falling from the eye on the centre of the circle neither be at right ‹angles› with the plane of the circle nor be equal to the radius nor such as to contain equal angles, the diameters with which it makes unequal angles will appear unequal.

A crucial privative notion in the *Elements* is the relation of incommensurability, defined in X.def.1 (*EOO* III, 2.2–4):

σύμμετρα μεγέθη λέγεται τὰ τῷ αὐτῷ μέτρῳ μετρούμενα ἀσύμμετρα δὲ ὧν μηδὲν ἐνδέχεται κοινὸν μέτρον γενέσθαι.

Commensurable magnitudes are said those that can be measured with a same measure, incommensurable those of which it is not possible that any common measure come about.

The *definiens* of this definition also contains the denegative part μηδέν with denegative particle μή—as it must be in a subordinate clause—and the explicit modal connotation conveyed by the operator ἐνδέχεται "it is possible".[290] These features emphasize the operational character of the test procedure for incommensurability (X.2). An enunciation that formulates an incommensurability result is that of X.16 (*EOO* III, 44.17–20):

ἐὰν δύο μεγέθη ἀσύμμετρα συντεθῇ, καὶ τὸ ὅλον ἑκατέρῳ αὐτῶν ἀσύμμετρον ἔσται· κἂν τὸ ὅλον ἑνὶ αὐτῶν ἀσύμμετρον ᾖ, καὶ τὰ ἐξ ἀρχῆς μεγέθη ἀσύμμετρα ἔσται.

If two incommensurable magnitudes be compounded, the whole will also be incommensurable with each of them; and if the whole be incommensurable with one of them, the original magnitudes will also be incommensurable.

In the common formulation of a negative statement, the negative particle is an independent part of speech and acts on the predicate, as in the enunciation of III.10 (*EOO* I, 192.16–17) :

κύκλος κύκλον οὐ τέμνει κατὰ πλείονα σημεῖα ἢ δύο.

A circle does not cut a circle at more than two points.

[290] I take the definition of commensurability to carry an implicit modal connotation; see Sect. 5.2 for a discussion.

When the verb form of the principal clause is negated, the negative particle is οὐ [οὐκ(χ) before a vowel]; no significant occurrences can be found of the form οὐχί, recommended by the dialectical tradition because it makes fallacies of division impossible.[291] The subjective negative particle μή is employed within subordinate clauses,[292] with participial forms, and with nominal complements of the copula in the form of complex syntagms. Let us read one example of each species.

Negation of the subordinate clause, in the enunciation of VIII.16 (*EOO* II, 314.17–21)—the principal clause features an objective negative particle in the form οὐδέ:

ἐὰν τετράγωνος ἀριθμὸς τετράγωνον ἀριθμὸν μὴ μετρῇ, οὐδὲ ἡ πλευρὰ τὴν πλευρὰν μετρήσει· κἂν ἡ πλευρὰ τὴν πλευρὰν μὴ μετρῇ, οὐδὲ ὁ τετράγωνος τὸν τετράγωνον μετρήσει.	If a square number do not measure a square number, the side will not measure the side, either; and if the side do not measure the side, the square will not measure the square either.

Participial form, in the enunciation of XI.9 (*EOO* IV, 28.4–6):

αἱ τῇ αὐτῇ εὐθείᾳ παράλληλοι καὶ μὴ οὖσαι αὐτῇ ἐν τῷ αὐτῷ ἐπιπέδῳ καὶ ἀλλήλαις εἰσὶ παράλληλοι.	‹Straight lines› parallel to a same straight line and not being in a same plane as it are also parallel to one another.

Nominal complement of the copula, included in a denegative pronoun, in the supposition of the *RI* in IX.36 (*EOO* II, 410.17–19):

εἰ γὰρ δυνατόν, μετρείτω τις τὸν ΖΗ ὁ Ο, καὶ ὁ Ο μηδενὶ τῶν Α Β Γ Δ Ε ΘΚ Λ Μ ἔστω ὁ αὐτός.	In fact, if possible, let some ‹number› O measure ZH, and let O not be the identical to any of A, B, Γ, Δ, E, ΘK, Λ, M.

In negative enunciations, the negative particle is always preposed to the predicate, which is normally located in a non-liminar position in the enunciation. Within a proof, however, there are interesting examples of negative particles preposed to a whole sentence that formulates a statement: these particles are thereby well separated from the verb form that expresses the predicate. The occurrences of such separated negative particles are all and only within inferences by *RI* or by *ArCo*; most of these occurrences introduce the conclusion, when this is in a negative form. Let us read the supposition and the conclusion of the first *RI* in I.26 (*EOO* I, 62.19–20 and 64.6–7):[293]

εἰ γὰρ ἄνισός ἐστιν ἡ ΑΒ τῇ ΔΕ, μία αὐτῶν μείζων ἐστίν.	In fact, if AB is unequal to ΔE, one of them is greater.
οὐκ ἄρα ἄνισός ἐστιν ἡ ΑΒ τῇ ΔΕ.	Therefore it is not the case that AB is unequal to ΔE.

[291] This form occurs only twice, in the introductions to *Optica* **B** (certainly spurious) and to the *Sectio canonis*. There are no occurrences in Archimedes and in Apollonius.
[292] This normally happens in the antecedent of a conditional, most notably in reduction suppositions.
[293] Here as elsewhere, I translate the occurrences of preposed οὐκ ἄρα in mathematical texts with "therefore it is not the case that". It appears, however, that the Stoics admitted of the speech act of stating a negation, but did not admit of the speech act of negating a statement—see Cavini 1985, 50 and 61. For this reason, I translate the same syntagm in logical texts with "therefore not:" instead of "therefore it is not the case that".

Now, the obviously emphatic preposed position of the negative particle is attested as a well-defined stylistic trait in ordinary Greek prose.[294] Are we entitled to overload what happens in mathematical texts with logical connotations? I think we are, and for the following reasons:

1) The preposed negative particle can only be found in the deductive configurations mentioned above. The negative conclusions of such inferences are invariably introduced[295] by οὐκ ἄρα "therefore it is not the case that", or by οὐδὲ ἄρα "therefore it is not the case that ... either",[296] if the reduction supposition was in conjunctive form. Sometimes,[297] the conclusion of a *RI* is directly provided in non-instantiated form, and coincides with the conclusion of the entire theorem. This gives rise to one of the extremely rare instances in which the conclusion of a theorem is not identical to the enunciation, even if this discrepancy only affects the position of the negative particle.[298] Here is the conclusion of III.10 (*EOO* I, 194.16–17), whose enunciation we have read above:

οὐκ ἄρα κύκλος κύκλον τέμνει κατὰ πλείονα σημεῖα ἢ δύο.	Therefore it is not the case that a circle cuts a circle at more than two points.

If the conclusion is instantiated, as in IX.13 (*EOO* II, 372.5–7), the negative particle is extracted from the denegative part and preposed to the whole sentence that formulates the statement:

οὐκ ἄρα ὁ μέγιστος ὁ Δ ὑπὸ ἑτέρου ἀριθμοῦ μετρηθήσεται παρὲξ τῶν Α Β Γ.	Therefore it is not the case that the greatest ‹number› Δ will be measured by another number except A, B, Γ.

2) In some proofs by *ArCo*, the coassumption comprises three terms, always in this order: οὐκ–verb form–δέ "and ... not". A double example can be found in V.10 (*EOO* II, 32.8–15):

εἰ γὰρ μή, ἤτοι ἴσον ἐστὶ τὸ Α τῷ Β ἢ ἔλασσον. ἴσον μὲν οὖν οὔκ ἐστι τὸ Α τῷ Β – ἑκάτερον γὰρ ἂν τῶν Α Β πρὸς τὸ Γ τὸν αὐτὸν εἶχε λόγον· οὐκ ἔχει δέ· οὐκ ἄρα ἴσον ἐστὶ τὸ Α τῷ Β –· οὐδὲ μὴν ἔλασσόν ἐστι τὸ Α τοῦ Β – τὸ Α γὰρ ἂν πρὸς τὸ Γ ἐλάσσονα λόγον εἶχεν ἤπερ τὸ Β πρὸς τὸ Γ· οὐκ ἔχει δέ· οὐκ ἄρα ἔλασσόν ἐστι τὸ Α τοῦ Β –· ἐδείχθη δὲ οὐδὲ ἴσον· μεῖζον ἄρα ἐστὶ τὸ Α τοῦ Β.	In fact, if not, either A is equal to B or less. Then now, A is not equal to B—for each of A, B would have the same ratio to Γ; and it has not; therefore it is not the case that A is equal to B—; nor of course is A less than B—for A to Γ would have a lesser ratio than B to Γ; and it has not; therefore it is not the case that A is less than B—; and it was proved that it is not equal either; therefore A is greater than B.

The presence of a preposed negative particle in these very short clauses forces the displacement of δέ from its canonical position as the second word in a sentence. In classical authors, analogous clauses assign the last position to the negative particle.[299] Therefore, the preposed negative particle is a stylistic marker of mathematical texts.

[294] For the position of the negative particle, see the thorough analysis, featuring statistical surveys, in Moorhouse 1959, sects. IV–VI and the Appendix.
[295] There are 157 occurrences in the whole of the *Elements*.
[296] There are 23 occurrences, in V.8, VIII.16, 17 (*bis*), X.10, 30, 51–52, 85 (*bis*), 87 (*ter*), 88 (*bis*), 89 (*bis*), 90 (*quater*), XI.34 (*bis*), plus 1 occurrence of οὐδὲ ἄρα without elision in VIII.16.
[297] For instance in I.7, III.10, 23, X.42–47, XI.13.
[298] The case of X.42–47 is only seemingly more complex. Of course, account is here taken of the fact that an enunciation and the associated conclusion always differ by an ἄρα.
[299] See Moorhouse 1959, 76.

3) There are instances of a double negative particle, the one preposed to the sentence that formulates a statement, the other to the predicate.[300] The two positions are equivalent from the mathematical point of view because they may cancel out one another; the inferential character of this operation is often made explicit by a further deductive step, as in IX.31 (*EOO* II, 400.18–20):

οὐκ ἄρα ὁ Α πρὸς τὸν Γ πρῶτος οὔκ ἐστιν· οἱ Α Γ ἄρα πρῶτοι πρὸς ἀλλήλους εἰσίν	Therefore it is not the case that A is not prime to Γ; therefore A, Γ are prime to one another.

A further, interesting, example is in I.6 (*EOO* I, 24.6–7), in which the preposed negative particle neutralizes the negation implicit in a privative part of speech:

οὐκ ἄρα ἄνισός ἐστιν ἡ ΑΒ τῇ ΑΓ· ἴση ἄρα.	Therefore it is not the case that AB is unequal to ΑΓ; therefore it is equal.

However, the "double negation rule" is often obliterated. The obliteration of the deductive step that allows to cancel out the double negative particle is systematic whenever an indirect proof validates a negative enunciation, as in X.26. Let us read the supposition that opens the *RI* of proposition III.10 (*EOO* I, 192.18–19), whose enunciation and conclusion we have read above:

εἰ γὰρ δυνατόν, κύκλος ὁ ΑΒΓ κύκλον τὸν ΔΕΖ τεμνέτω κατὰ πλείονα σημεῖα ἢ δύο τὰ Β Η Ζ Θ.	In fact, if possible, let a circle, ΑΒΓ, cut a circle, ΔΕΖ, at more than two points, Β, Η, Ζ, Θ.

The fact that the enunciation to be validated is negative can be recovered by the opening clause εἰ γὰρ δυνατόν "in fact, if possible".

We may safely conclude that preposing the negative particle stresses the fact that inferences by *ArCo* or by *RI* falsify what has been supposed as a whole: this is the entire statement naturally associated with the supposition, not simply the predicate included in such a statement. This view is confirmed by the fact that the verb form is kept in its canonical position within the sentence, clearly separated from the negative particle.

The ancient theories of negation aim at eliminating the ambiguities of scope induced by the position of the negative particle in ordinary speech. The Aristotelian doctrine denies the existential import of negative sentences; this entails that the negative particle may simply be preposed to the main verb form of a sentence. The Stoics, and in particular Chrysippus, held the contrary view as to the existential import of negative statements. In this case, however, keeping the negative particle preposed to the main verb form of a sentence may give rise to ambiguities of scope if the grammatical subject of the sentence is non-existent.[301] To the Stoics, then, the negative particle must assume a formally unambiguous position in a sentence that formulates a statement; it is obvious that preposing it to the whole sentence is the only possible solution. In this way, the negative particle has scope over the whole sentence; a statement formulated by a sentence in which the negative particle

[300] The only exceptions are the occurrences of οὐκ ἄρα οὐ(κ) in VI.26 and VII.35. These are accidents induced by the form of the negated sentences: the first has the verb form preposed, the second is very short. Still, the two negative particles are separated by ἄρα.

[301] See the discussion in Cavini 1985, 36–45 and 67–84, and also Crivelli 1989. Sources for Aristotle are *Cat.* 10, 13b27–35, *Metaph.* Δ.7, 1017a13–22, and I.3, 1054b18–21; for the Stoics (but they are not expressly mentioned) Alexander, *in APr.*, 401.16–405.16.

has scope only over the predicate—that is, it is such that normally it is preposed to it—is not a negative statement but an affirmative statement. This entails that the negation is not a connector, but an operator that maps statements into statements. The negation of a statement generates its contradictory, that is, a statement whose truth value is the opposite of the truth value of the original statement. The exposition in Sextus, *M* VIII.89–90, is brilliant:[302]

φασὶ γὰρ "ἀντικείμενά ἐστιν ὧν τὸ ἕτερον τοῦ ἑτέρου ἀποφάσει πλεονάζει", οἷον "ἡμέρα ἔστιν· οὐχ ἡμέρα ἔστιν" – τοῦ γὰρ "ἡμέρα ἔστιν" ἀξιώματος τὸ "οὐχ ἡμέρα ἔστιν" ἀποφάσει πλεονάζει τῇ οὐχί, καὶ διὰ τοῦτ' ἀντικείμενόν ἐστιν ἐκείνῳ – ἀλλ' εἰ τοῦτ' ἔστι τὸ ἀντικείμενον, ἔσται καὶ τὰ τοιαῦτα ἀντικείμενα, τό τε "ἡμέρα ἔστι καὶ φῶς ἔστιν" καὶ τὸ "ἡμέρα ἔστιν καὶ οὐχὶ φῶς ἔστιν" – τοῦ γὰρ "ἡμέρα ἔστιν καὶ φῶς ἔστιν" ἀξιώματος ἀποφάσει πλεονάζει τὸ "ἡμέρα ἔστιν καὶ οὐχὶ φῶς ἔστιν" –· οὐχὶ δέ γε κατ' αὐτοὺς ταῦτα ἀντικείμενά ἐστιν· οὐκ ἄρα ἀντικείμενά ἐστι τῷ τὸ ἕτερον τοῦ ἑτέρου ἀποφάσει πλεονάζειν. **90** ναί, φασίν, ἀλλὰ σὺν τούτῳ ἀντικείμενά ἐστι, σὺν τῷ τὴν ἀπόφασιν προτετάχθαι τοῦ ἑτέρου – τότε γὰρ καὶ κυριεύει τοῦ ὅλου ἀξιώματος, ἐπὶ δὲ τοῦ "ἡμέρα ἔστιν καὶ οὐχὶ φῶς ἔστιν", μέρος οὖσα τοῦ παντός, οὐ κυριεύει πρὸς τὸ ἀποφατικὸν ποιῆσαι τὸ πᾶν. ἐχρῆν οὖν, ἐροῦμεν, προσκεῖσθαι τῇ ἐννοίᾳ τῶν ἀντικειμένων ὅτι τότε ἀντικείμενά ἐστιν, ὅταν μὴ ψιλῶς τὸ ἕτερον τοῦ ἑτέρου ἀποφάσει πλεονάζῃ, ἀλλ' ὅταν ἡ ἀπόφασις προτάττηται τοῦ ἀξιώματος.

For they say "‹statements› are contradictory such that the one exceeds the other by a negation", for instance "it is day; not: it is day"—for "not: it is day" exceeds the statement "it is day" by a negation "not", and for this ‹reason› it is contradictory of that one—but if this is "contradictory", the following ones will also be contradictory, both "it is day and there is light" and "it is day and not: there is light"—for "it is day and not: there is light" exceeds the statement "it is day and there is light" by a negation—; but of course not: these are contradictory according to them; therefore not: they are contradictory insofar as the one exceeds the other by a negation. **90** Fo sho!, they say, still they do are contradictories with this, namely, with the negation being preposed to one of them—for then it also governs the whole statement, whereas in "it is day and not: there is light", while being a part of the whole, does not govern it in such a way as to make the whole a negative statement. Then it was mandatory, we shall say, to add, to the notion of the contradictories, that they are indeed contradictory, not whenever the one simply exceed the other by a negation, but whenever the negation be preposed to the statement.

Note (*a*) the technical term προτάττω "to prepose", which designates the position of the negative particle; (*b*) the Sextan argument, refuted by the fictitious Stoic opponent, in which Sextus parodies some Stoic prescriptions about the form of an argument, among which the type and the position of the negative particle. As said, the use of the particle οὐχί was devised to forestall fallacies of division that might originate from using οὐκ, as expounded in PPar. 2.[303]

The possibility that a preposed negative particle interacts with the negation implicit in a privative part of speech is recognized as a genuine Stoic doctrine in Plutarch, *Comm. not.* 39, 1080C:[304]

[302] Add *M* VIII.225–227 and other scattered examples in Sextus, for instance *P* II.231 and 241, in which a paradox arises from the ambiguity of scope of the negative particle—the useless expunctions proposed by some scholars, as for instance Mates 1961, 31 n. 27, make the paradox unintelligible.
[303] Edition and commentary in Cavini 1985.
[304] Note the indefinite conditional expressing a general statement—cf. Sect. 5.1.7. The two statements can be both false because the Chrysippean conditional is not truth-functional, whereas conjuction is: see Sects. 5.2.1 and 5.3.5.

αὐτοὶ τὰ τοιαῦτα ἀξιώματα ψευδῆ λέγοντες εἶναι· "εἴ τινα μή ἐστιν ἴσα ἀλλήλοις, ἐκεῖνα ἄνισά ἐστιν ἀλλήλοις" καί "οὐκ ἔστι μὲν ἴσα ταῦτ' ἀλλήλοις οὐκ ἄνισα δ' ἐστὶ ταῦτ' ἀλλήλοις".

They claim that the following statements are false: "if some items are not equal to one another, those are unequal to one another" and "not: these are equal to one another and not: these are unequal to one another".

The context is the Democritean paradox of the cone. If the first conditional is false, a negation of the verb form "to be" cannot be made to react with the adjectival complement of the copula "equal": this shows that the negative particle in this sentence is wrongly located on Stoic standards. In the second sentence, the negation has scope over the entire subsequent conjunction,[305] whose second conjunct is negated in its turn. Therefore, if the conjunction is false, its negation is true, and we get this true conjunction by simply eliminating the first negation. As a consequence, the two predicates "equal" and "not unequal" are equivalent.

Arguments ascribed to such early logicians as Diodorus Cronus and Arcesilaus[306] contain a negative particle preposed to, and separated from, the verb form.[307] If Sextus did not modify the position of the negative particle,[308] such arguments attest to the use, in dialectical contexts and before the Stoic normalization, of a preposed negative particle in the coassumption of an *ArCo*. Let us read the argument ascribed to Arcesilaus (*M* VII.157):

εἰ τῶν συγκατατιθεμένων ἐστὶν ὁ σοφός, τῶν δοξαστικῶν ἔσται ὁ σοφός· οὐχὶ δέ γε τῶν δοξαστικῶν ἐστιν ὁ σοφός [...]· οὐκ ἄρα τῶν συγκατατιθεμένων ἐστὶν ὁ σοφός.

If the sage is among those who give their assent, the sage is among thoses who entertain beliefs; and certainly not: the sage is among thoses who entertain beliefs [...]; therefore not: the sage is among those who give their assent.

Earlier still, a passing remark by Aristotle suggests that the dialecticians were used to formulating negative conclusions by preposing the negative particle; moreover, this practice has an affirmative pendant in the presence of the operator of necessity ἀνάγκη "necessarily" as an "assertion sign" (*Rh.* II.24, 1400b34–1401a6; see also *SE* 15, 174b8–11):

ἐπεὶ δ' ἐνδέχεται τὸν μὲν εἶναι συλλογισμὸν τὸν δὲ μὴ εἶναι μὲν φαίνεσθαι δέ, ἀνάγκη καὶ ἐνθύμημα τὸ μὲν εἶναι τὸ δὲ μὴ εἶναι ἐνθύμημα φαίνεσθαι δέ, ἐπείπερ τὸ ἐνθύμημα συλλογισμός τις. τόποι δ' εἰσὶ τῶν φαινομένων ἐνθυμημάτων εἷς μὲν ὁ παρὰ τὴν λέξιν, καὶ τούτου ἓν μὲν μέρος, ὥσπερ ἐν τοῖς διαλεκτικοῖς τὸ μὴ συλλογισάμενον συμπερασματικῶς τὸ τελευταῖον εἰπεῖν, "οὐκ ἄρα τὸ καὶ τό", "ἀνάγκη ἄρα τὸ καὶ τό", ἐν τοῖς ἐνθυμήμασι τὸ συνεστραμμένως καὶ ἀντικειμένως εἰπεῖν φαίνεται ἐνθύμημα – ἡ γὰρ τοιαύτη λέξις χώρα ἐστὶν ἐνθυμήματος.

Since it is possible that this is a syllogism whereas that is not but it seems to be, necessarily, too, this is an enthymeme whereas that is not an enthymeme but it seems to be, since an enthymeme is really a sort of syllogism. And one of the divisions of the apparent enthymemes is the one depending on wording, and one subdivision of this is, as in dialectic wording the last step conclusionwise, even if it does not result from a syllogism: "therefore not: so-and-so", "therefore necessarily: so-and-so", ‹so› for enthymemes wording concisely and antithetically seems an enthymeme—for such a wording is typical of an enthymeme.

---

305 For the correlative μέν ... δέ ... see Sect. 5.3.5.

306 Both end 4th – beginning 3rd century BCE.

307 We read these arguments in Sextus, *M* X.85–101 and VII.157.

308 This is certainly the case for the coassumptive nexus of particles δέ γε.

We may thus safely surmise that the mathematical practice in the use of negation conforms to a well-established dialectical context. I have also checked the largest corpora in prose earlier than the 3rd century CE, in search of οὐκ ἄρα preposed to a conclusion: there are 549 occurrences, 476 of which concentrated in Aristotle (64), Plato (116), Sextus Empiricus and Diogenes Laertius (296). These are logically connotated authors, and the last two transmit abundant Stoic material. Even with this limitation, and even if many occurrences in Aristotle are in short clauses, I conclude that preposing the negative particle in the conclusion of an inference was a shared and widespread practice; mathematical style appropriated it to mark the inferences whose dialectical character is more prominent, namely, indirect proofs. No conclusions can be drawn as to whether mathematical practice and Stoic doctrines were purposely devised so as to be the one in conformity with the other.

5.3.4. *Disjunction*

Greek mathematics formulates the exclusive disjunction[309] in a typical way. This formulation is characterized by the disyllabic, weakly intensive form ἤτοι "either" of the standard disjunctive particle ἤ "or".[310] In the *Elements*, the disyllabic form occurs 92 times;[311] it is always preposed to the disjuncts and followed by at least another particle ἤ, placed among the disjuncts. The enunciation of I.13 (*EOO* I, 36.2–4) is a first example:

ἐὰν εὐθεῖα ἐπ᾽ εὐθεῖαν σταθεῖσα γωνίας ποιῇ, ἤτοι δύο ὀρθὰς ἢ δυσὶν ὀρθαῖς ἴσας ποιήσει.

If a straight line set up on a straight line make angles, it will make ‹them› either two right ‹angles› or equal to two right ‹angles›.

This disjunction only involves the consequent of the conditional enunciation. This disjunction is exclusive because the two possibilities are taken to be incompatible: as a consequence, the proof is composed of two independent deductions, according to whether the two angles are equal (and hence they are two right angles) or not. The consequent of the above conditional is weakly brachylogical: a well-formed enunciation according to Stoic standards would ban any cross-reference:

*εἰ εὐθεῖα ἐπ᾽ εὐθεῖαν σταθεῖσα γωνίας ποιῇ, ἤτοι ἡ εὐθεῖα ποιήσει δύο ὀρθὰς ἢ ἡ εὐθεῖα δυσὶν ὀρθαῖς ἴσας ποιήσει γωνίας.

If a straight line set up on a straight line makes angles, either the straight line will make two right ‹angles› or the straight line will make angles equal to two right ‹angles›.

Exclusive disjunctions are seldom found in the enunciation of propositions; another example is in the enunciation of VII.32 (*EOO* II, 252.8–9):

ἅπας ἀριθμὸς ἤτοι πρῶτός ἐστιν ἢ ὑπὸ πρώτου τινὸς ἀριθμοῦ μετρεῖται.

Every number either is prime or is measured by some prime number.

---

[309] This means that only one of the disjuncts is the case. If the disjuncts taken together exhaust all possible cases, and hence at least one of the disjuncts is the case, the disjunction is said to be exhaustive. We must recall that only true statements normally figure in mathematics. In particular, this happens if an exclusive disjunction is also exhaustive: all cases are taken into account and at least one of them must be the case. Such semantic features must be kept distinct from the syntactic issues that pertain to the actual formulation of a disjunction. This and the following Section are a reworking of Acerbi 2008b, which in its turn was an extract from the first redaction of the present book.

[310] Most of the occurrences of ἤ in the *Elements* are in comparative phrases, with a meaning analogous to Latin *quam*.

[311] Heiberg's critical apparatus records no variants that pertain to missing or added ἤτοι.

The very short proof of this theorem hinges upon an unstated exclusive and exhaustive disjunction (obtained by combining VII.def.12 and VII.def.14), namely, that every number is either prime or composite, and reads (*EOO* II, 252.12–14):

εἰ μὲν οὖν πρῶτός ἐστιν ὁ Α, γεγονὸς ἂν εἴη τὸ ἐπιταχθέν. εἰ δὲ σύνθετος, μετρήσει τις αὐτὸν πρῶτος ἀριθμός.

Then if A is prime, that which was prescribed would have come to be. And if it is composite, some prime number will measure it.

The proof itself consists of two independent branches ("If … If …" just read), and this shows that the enunciation above is in fact the disjunction of two independent enunciations, depending on whether the number at issue is prime or composite. We shall see below other ways in which several independent results can be enunciated within one and the same statement.

Very often, exclusive disjunctions introduce proofs by *RI*[312] or by *ArCo*.[313] In these cases, the disjunctions are usually dichotomic, and state that two of the three terms of a trichotomy will be proved not to be the case. For instance, the initial statement of a proof by *ArCo* that involves a trichotomy is a conditional that features a disjunction in the consequent, as in proposition I.19 (*EOO* I, 46.23–24)—what is supposed not to be the case is that ΑΓ is greater than AB:

εἰ γὰρ μή, <u>ἤτοι</u> ἴση ἐστὶν ἡ ΑΓ τῇ ΑΒ <u>ἢ</u> ἐλάσσων.

In fact, if not, <u>either</u> ΑΓ is equal to AB <u>or</u> less.

The same happens in a proof by *RI* typical of the method of exhaustion, as the one in proposition XII.10 (*EOO* IV, 186.17–19), where it must be proved that a cylinder is triple of the cone that has the same base as it and equal height:[314]

εἰ γὰρ μή ἐστιν ὁ κύλινδρος τοῦ κώνου τριπλασίων, ἔσται ὁ κύλινδρος τοῦ κώνου <u>ἤτοι</u> μείζων <u>ἢ</u> τριπλασίων <u>ἢ</u> ἐλάσσων ἢ τριπλασίων.

In fact, if the cylinder is not triple of the cone, the cylinder will be <u>either</u> greater than triple of the cone <u>or</u> less than triple.

In direct proofs, trichotomic disjunctions are found whenever a suitable distinction of cases is required, as in III.25 (*EOO* I, 228.3–4),

ἡ ὑπὸ ΑΒΔ γωνία ἄρα τῆς ὑπὸ ΒΑΔ <u>ἤτοι</u> μείζων ἐστὶν <u>ἢ</u> ἴση <u>ἢ</u> ἐλάττων.

Therefore angle ΑΒΔ is <u>either</u> greater than ΒΑΔ <u>or</u> equal <u>or</u> less.

or in III.24 and 33. Sporadic occurrences of quadrichotomic disjunctions are found either in proposition X.71 (*EOO* III, 212.16–19), where several species of irrational lines are listed,

ῥητοῦ καὶ μέσου συντιθεμένου τέσσαρες ἄλογοι γίγνονται <u>ἤτοι</u> ἐκ δύο ὀνομάτων <u>ἢ</u> ἐκ δύο μέσων πρώτη <u>ἢ</u> μείζων <u>ἢ</u> ῥητὸν καὶ μέσον δυναμένη.

An expressible and a medial ‹region› being compounded, there result four irrationals, <u>either</u> a binomial <u>or</u> a first bimedial <u>or</u> a major <u>or</u> a ‹straight line› worth an expressible and a medial ‹region›.

---

[312] All *RI* typical of the method of exhaustion contain disjunctions.

[313] The best examples are in I.19, V.10, XI.16. See Sects. 5.2.1 and 5.2.2.

[314] Note the presence of both disjunctive and comparative ἤ.

or in IX.19, within a passage, however, which is clearly corrupt (see also below).[315]

Disjunctions that are formulated solely with an interposed ἤ are never exclusive: such a syntactic distinction between exclusive and inclusive disjunctions is rigidly adhered to in the main text of the *Elements*.[316] A beautiful example, worth a discussion, is provided by the expression "part or parts". Let us recall VII.def.3–4 (*EOO* II, 184.6–8):[317]

μέρος ἐστὶν ἀριθμὸς ἀριθμοῦ ὁ ἐλάσσων τοῦ μείζονος, ὅταν καταμετρῇ τὸν μείζονα· μέρη δέ, ὅταν μὴ καταμετρῇ.	A number is a part of a number, the less of the greater, whenever it measure the greater out;[318] and parts, whenever it do not measure ‹it› out.

In modern terminology, a μέρος "part" of a number is any of its divisors (or "factors"), whereas μέρη "parts" (which is not the plural of "part", but a new technical term) of a given number is any number less than the given number and that is not a divisor of it.[319] Let us now read the enunciation of proposition VII.4 (*EOO* II, 198.15–16):

ἅπας ἀριθμὸς παντὸς ἀριθμοῦ ὁ ἐλάσσων τοῦ μείζονος <u>ἤτοι</u> μέρος ἐστὶν <u>ἢ</u> μέρη.	Every number is <u>either</u> part <u>or</u> parts of every number, the less of the greater.

The disjunction is exclusive because only one of the disjuncts can be the case for a given number: the definitions of "part" and "parts" are themselves exclusive and in addition exhaustive.[320] The two notions of "part" and "parts" are brought together again in VII.9–13 and only there. Well, let us read the enunciation of VII.9 (*EOO* II, 210.6–10):[321]

ἐὰν ἀριθμὸς ἀριθμοῦ μέρος ᾖ καὶ ἕτερος ἑτέρου τὸ αὐτὸ μέρος ᾖ, καὶ ἐναλλάξ, ὃ μέρος ἐστὶν <u>ἢ</u> μέρη ὁ πρῶτος τοῦ τρίτου, τὸ αὐτὸ μέρος ἔσται <u>ἢ</u> τὰ αὐτὰ μέρη καὶ ὁ δεύτερος τοῦ τετάρτου.	If a number be part of a number and another be the same part of another, by alternation too, what part <u>or</u> parts is the first of the third, the same part <u>or</u> the same parts will also be the second of the fourth.

One might wonder why "part or parts" is disjoined inclusively in this enunciation, if the two notions are incompatible. The reason is that both possibilities envisaged in the consequent of the conditional may occur, given the single condition assumed in the antecedent. This condition cannot

[315] Heiberg, who in his apparatus comments at length on the corruption, is here wrong in choosing the (incorrect) reading of **P**, the main testimony of his edition. The Theonine manuscripts present a more correct, and shorter, proof. The copyist of **P** inserts in the text, though marked off by *obeli*, a scholium lifted from his exemplar and containing a correction; this text is at *EOO* II, 386.1 *app.* On this proposition see Vitrac, forthcoming.

[316] Inclusive disjunctions are found in I.28, III.def.1.11, VII.def.19.20.21, and VII.9–13. That definitions prominently figure in the list is no surprise: when alternative definitions are proposed of the same object, they are most naturally connected by an inclusive disjunction.

[317] Recall that the definitions are not numbered in the manuscripts. As in so many other cases, an independent definitory item is a cluster of definitions; the several items in the cluster are coordinated by δέ. In this case, the cluster comprises VII.def.3–5.

[318] The prefix κατα– indicates that the operation of measuring has been carried out without a remainder.

[319] VII.def.3 defines what counts for a number to be a part of another number. A unit, which is not a "number" in Greek sense, is a part of any number by VII.def.2. The "parts" in VII.def.4 are not simply sums of more than one "part": for instance, the "parts" of a prime number could be sums of more than one "part" only in the rather debased sense of being sums of units.

[320] The enunciation has a structure that is exactly parallel to the enunciation of VII.32 read above.

[321] The ordinals are here abbreviations that refer to the four numbers just introduced, as the presence of the article confirms. The occurrences in V.def.5, to be read below, must be interpreted as variables. See Sect. 5.1.6 on this issue.

rule out one or the other of two arithmetically possible states of affairs, even if these are mutually incompatible. The goal of the theorem is only to prove that the cases "part" and "parts" in the two correlated clauses in the consequent cannot intermingle: there is no need to insist on the exclusive character of the partition. The same formulaic expression contained in the enunciation above is repeated several times, either instantiated or in its general formulation,[322] as far as VII.13. Residual puzzlement may subsist about the adequacy of the inclusive disjunction in the instantiated occurrences, as in the proof of VII.13 (*EOO* II, 218.5–10):

ἐπεὶ γάρ ἐστιν ὡς ὁ Α πρὸς τὸν Β οὕτως ὁ Γ πρὸς τὸν Δ, ὃ ἄρα μέρος ἐστὶν ὁ Α τοῦ Β ἢ μέρη, τὸ αὐτὸ μέρος ἐστὶ καὶ ὁ Γ τοῦ Δ ἢ τὰ αὐτὰ μέρη· ἐναλλὰξ ἄρα, ὃ μέρος ἐστὶν ὁ Α τοῦ Γ ἢ μέρη, τὸ αὐτὸ μέρος ἐστὶ καὶ ὁ Β τοῦ Δ ἢ τὰ αὐτὰ μέρη· ἔστιν ἄρα ὡς ὁ Α πρὸς τὸν Γ, οὕτως ὁ Β πρὸς τὸν Δ.

In fact, since, as A is to B, so Γ is to Δ, therefore, what part is A of B or parts, the same part is also Γ of Δ or the same parts; therefore, by alternation, what part is A of Γ or parts, the same part is also B of Δ or the same parts; therefore, as A is to Γ, so B is to Δ.

The problem is that the letters correspond to actual numbers. However, this problem is easily neutralized if we note that the instantiated numbers are generic. Therefore, the instantiated occurrences must also leave it undecided which relation between "part" and "parts" will actually be the case, even if they cannot be the case together. All theorems VII.9–13 leave both possibilities open and prove both cases by one and the same proof. The proofs of the two cases are not given in succession as independent proofs nor are they subordinated to one another; they are in a sense carried out the one superimposed on the other. A sign of this is the peculiar ordering of the terms, which seems to suggest that only the disjunct "part" is instantiated: placing the phrase "or parts" at the end of the clause makes the phrase look like a gloss to a statement whose primary meaning can be extracted from the "part" case only.

A linguistic problem of this kind is unique in the *Elements* and could not be easily dealt with in a compact and mathematically meaningful form. The retained solution hinges upon a very concise and peculiar formulaic expression, which contains a proleptic relative in the singular that also works for the "parts", a fact that would make the insertion of an ἤτοι preposed to the relative pronoun ὅ highly problematic. In other words, in VII.9–13 the "part or parts" formula does not state an alternative as in VII.4, but simply lists, in a correct disjunctive form, the possible configurations. Therefore, a weak form of disjunction is needed: this is rightly kept distinct from the exclusive disjunction and is analogous to the disjunction formulated by means of the correlative τε … τε … "both … and …", which we shall see in the next Section.

The explanation just proposed can be exemplified by comparing the enunciation of VII.9 read above with the enunciation of V.6 (*EOO* II, 20.2–5):

ἐὰν δύο μεγέθη δύο μεγεθῶν ἰσάκις ᾖ πολλαπλάσια καὶ ἀφαιρεθέντα τινὰ τῶν αὐτῶν ἰσάκις ᾖ πολλαπλάσια, καὶ τὰ λοιπὰ τοῖς αὐτοῖς ἤτοι ἴσα ἐστὶν ἢ ἰσάκις αὐτῶν πολλαπλάσια.

If two magnitudes be equimultiple of two magnitudes and some removed ‹magnitudes› be equimultiple of the same, the remainders are also either equal to the same or equimultiple of them.

[322] The general formulation occurs just once, in the enunciation of VII.10, which is analogous to the enunciation of VII.9.

The consequent of this conditional is a single disjunction, and not a principal clause and a relative subordinate set in strict parallel, as instead in VII.9. Only one of the stated alternatives may occur once a well-defined multiple of the magnitudes at issue is taken: therefore, the disjunction must be exclusive. True, the result holds for any such multiple, but they are taken into account severally in the proposition, and not as a whole.

An argument similar to the one expounded above for VII.9 explains the inclusive disjunction in VII.def.21 (*EOO* II, 188.5–7),[323] where we might have expected to find an exclusive disjunction:

ἀριθμοὶ ἀνάλογόν εἰσιν, ὅταν ὁ πρῶτος τοῦ δευτέρου καὶ ὁ τρίτος τοῦ τετάρτου ἰσάκις ἢ πολλαπλάσιος <u>ἢ</u> τὸ αὐτὸ μέρος <u>ἢ</u> τὰ αὐτὰ μέρη ὦσιν.	Numbers in proportion are whenever the first of the second and the third of the fourth be equimultiple <u>or</u> the same part <u>or</u> be the same parts.

At any rate, this definition has several problems and its value as an example is limited. First, not all possible cases of "first greater than second" are envisaged: only multiples are taken into account and clearly they do not suffice to encompass all ratios. The partition of the "less than" case, instead, is exhaustive; this partition would have sufficed to shape a complete definition, since the terms can always be arranged so that the first term is less than the second. Second, when the definition is employed, the disjunct "equimultiple" is never referred to, as we have seen above with VII.13. Third, one family of the Arabic tradition of the *Elements* lacks the "equimultiple" disjunct even in the definition.[324] Fourth, two manuscripts of the *Elements* suggest that a syntactic problem was perceived, since we find in them a further ἤ preposed to ἰσάκις "equi–",[325] a move that would partially set the definition in line with the canonical formulation of exclusive disjunction. There are thus several reasons to suspect that the mention of the equimultiple is spurious. However, it would be naive to think that the mere expunction of the syntagm that concerns the equimultiple would produce a text where the first of the interposed ἤ automatically becomes a preposed connector.

The only occurrence of a preposed monosyllabic ἤ is found in the celebrated definition 5 of Book V (*EOO* II, 2.10–16):

ἐν τῷ αὐτῷ λόγῳ μεγέθη λέγεται εἶναι πρῶτον πρὸς δεύτερον καὶ τρίτον πρὸς τέταρτον, ὅταν τὰ τοῦ πρώτου καὶ τρίτου ἰσάκις πολλαπλάσια τῶν τοῦ δευτέρου καὶ τετάρτου ἰσάκις πολλαπλασίων καθ' ὁποιονοῦν πολλαπλασιασμὸν ἑκάτερον ἑκατέρου <u>ἢ</u> ἅμα ὑπερέχῃ <u>ἢ</u> ἅμα ἴσα ᾖ <u>ἢ</u> ἅμα ἐλλείπῃ ληφθέντα κατάλληλα.	Magnitudes are said to be in a same ratio, first to second and third to fourth, whenever the equimultiples of the first and third, according to whichever multiple, respectively, <u>either</u> exceed together <u>or</u> be equal together <u>or</u> fall short together of the equimultiples of the second and fourth, taken in corresponding order.

Despite the preposed connector, the disjunction is not exclusive, and in fact we do not find the disyllabic form ἤτοι. The disjunction is not exclusive because the set of multiples is unbounded, and any of the three disjuncts is actually the case for some choice of the multiples.[326] Had a pair of multiples been chosen from the very outset, of course the disjunction would have been exclusive,

[323] The weak character of the disjunction in VII.def.21 has already been pointed out in Vitrac 1990–2001 II, 264 n. 96.
[324] But this might well be an a posteriori correction, devised to make the formulation of the definition fit the formulation of the applications.
[325] *EOO* II, 188.6 *app.* One of the copyists overwrites the particle. At any rate, it is unclear whether the additional connector is a scribal innovation, the original reading of the antigraph, or a variant of collation.
[326] As a matter of fact, there are infinitely many choices for each of the disjuncts.

but in V.def.5 the whole clause refers to the entire set of multiples.[327] Therefore, the disjunction is inclusive and exhaustive. I am unable to explain why a monosyllabic ἤ is preposed, unless by supposing that the formulation of this definition is earlier than the (authorial) regularizing campaign of the formulation of disjunctions in the *Elements*.[328]

Sometimes, an inclusive disjunction is not expressed by a sentence disjoined by means of a connector, but by a pronominal indefinite adjective such as ὁπότερος "either", which is the most frequent item of this kind.[329] We find it at work in the following, very short, extract from proposition I.24 (*EOO* I, 58.9–10):

καὶ κείσθω <u>ὁποτέρᾳ</u> τῶν ΑΓ ΔΖ ἴση ἡ ΔΗ.	And let a ‹straight line›, ΔΗ, be set equal to <u>either</u> of ΑΓ, ΔΖ.

The term ΔΗ can be made equal to any of the other two, and even to both (since they are equal), but not in the sense that they have to be added: if this had to be the case, no adjective would have been present, for the absence of determinations normally entails that the sum of the listed objects is to be taken. Of intensive forms as ὁποτεροσοῦν, where the idea of genericity is strengthened by the suffix –οῦν, there are 10 occurrences in the Euclidean corpus,[330] but none in the *Elements*. The adjective ἕτερος "one of them",[331] when preceded by an article, has an exclusive meaning; compare the beginning of I.6 (*EOO* I, 22.25–26):

εἰ γὰρ ἄνισός ἐστιν ἡ ΑΒ τῇ ΑΓ, <u>ἡ ἑτέρα</u> αὐτῶν μείζων ἐστίν.	In fact, if AB is unequal to ΑΓ, <u>one of them</u> is greater.

When ἕτερος is not preceded by an article,[332] it has the more usual meaning "another" (the choice is between two alternatives), as in the enunciation of VII.9 read above.

Indirect proofs with an initial trichotomy clearly show that the negation of an exclusive disjunction is not the conjunction of the negation of each disjunct.[333] The text of I.19 (*EOO* I, 46.26–48.3), whose introductory sentence we have read above, makes this point clear:

οὐδὲ μὴν ἐλάσσων ἐστὶν ἡ ΑΓ τῆς ΑΒ – ἐλάσσων γὰρ ἂν ἦν καὶ γωνία ἡ ὑπὸ ΑΒΓ τῆς ὑπὸ ΑΓΒ· οὐκ ἔστι δέ· οὐκ ἄρα ἐλάσσων ἐστὶν ἡ ΑΓ τῆς ΑΒ –· ἐδείχθη δὲ ὅτι οὐδὲ ἴση ἐστίν· μείζων ἄρα ἐστὶν ἡ ΑΓ τῆς ΑΒ.	Nor of course is ΑΓ less than AB—for angle ΑΒΓ would also have been less than ΑΓΒ; and it is not; therefore it is not the case[334] that ΑΓ is less than AB—; and it was proved that it is not equal either; therefore ΑΓ is greater than AB.

[327] Recall the completely opposite state of affairs seen above in V.6.
[328] See Sect. 5.1.6 and Federspiel 2003, 336–349, on the formulation of V.def.5. A preposed monosyllabic ἤ was present in **P** in the quadrichotomy that opens IX.19. Heiberg is wrong in correcting it to ἤτοι (*EOO* II, 384.8 *app.*). However, the beginning of this proposition (whose proof is the only invalid proof in the *Elements*) is clearly corrupt, and no conclusion can be drawn from such an occurrence (see Vitrac, forthcoming).
[329] There are 25 occurrences in the whole of the *Elements*; 2 in the *Data*, propositions 37–38.
[330] There are 1 occurrence in proposition 42 of *Optica* redaction **A**, 9 occurrences in the *Phaenomena*, scattered among the two redactions.
[331] There are 184 occurrences in the Euclidean corpus. The adjective is absent from Books IV–VI and VIII.
[332] This produces the overwhelming majority of occurrences.
[333] Recall that this kind of duality works only if the disjunction is inclusive. In Stoic logic, this kind of duality is excluded from the very outset, since, as we shall see in the next Section, conjunction is truth-functional and disjunction is not.
[334] We have here an instance of preposed negative particle, which has scope over the whole subsequent sentence: see the discussion in Sect. 5.3.3.

Apparently, thus, οὐδέ … οὐδέ … "neither … nor …" replaces a (fictitious) conjunction of negations such as *καὶ οὐ … καὶ οὐ …*. A form of Stoic "fifth indemonstrable with multiple disjuncts" formalizes instead the kind of deductive pattern at work here. A mode of it can be "either the first or the second or the third; but neither the second nor the third: therefore the first".[335] The examples found in ancient sources allow us to assume that the canonical form of the coassumption was either οὔτε … οὔτε … or οὐδέ … οὐδέ … The form with οὐδέ is preferred in mathematical texts, the form with οὔτε seems to have been prescribed by the Stoics.[336]

A final remark. A very refined dialectical training was often provided in the standard rhetorical curricula of late antiquity: as a consequence, the logical operators were handled with such a degree of awareness as to make their appropriate setting out in fairly complex statements an expressive goal. On the other hand, in geometric texts, such complex enunciations of problems or theorems were currently elaborated as encompass with a single statement several geometric configurations that apply to different mathematical objects, or obtained the one from the other by small variations of the assignments of the problem or of the theorem. Typical examples of such long enunciations can be found in Apollonius' *Conica* and in Theodosius' *Sphaerica*. The practice of formulating complex statements became a fashionable one among the commentators of late antiquity, who often found, in the treatises they were commenting on, fully-fledged proofs lacking a general enunciation.[337] The outcome was the production of enunciations that must be more properly regarded as pertaining to the domain of rhetorical bravura than as syntactic lesser evils induced by mathematical constraints. In Book VII of the *Collectio*, Pappus engages himself several times in this exercise: he embraces under a single statement the contents of entire treatises or of large portions of them. Let us read a problem first; this is the general enunciation that encompasses the entire *Determinate section* of Apollonius (*Coll.* VII.9):

ἑξῆς δὲ τούτοις ἀναδέδοται τῆς διωρισμένης τομῆς βιβλία β΄, ὧν ὁμοίως τοῖς πρότερον μίαν πρότασιν πάρεστιν λέγειν, διεζευγμένην δὲ ταύτην· τὴν δοθεῖσαν ἄπειρον εὐθεῖαν ἑνὶ σημείῳ τεμεῖν ὥστε τῶν ἀπολαμβανομένων εὐθειῶν πρὸς τοῖς ἐπ' αὐτῆς δοθεῖσι σημείοις ἤτοι τὸ ἀπὸ μιᾶς τετράγωνον ἢ τὸ ὑπὸ δύο ἀπολαμβανομένων περιεχόμενον ὀρθογώνιον δοθέντα λόγον ἔχειν ἤτοι πρὸς τὸ ἀπὸ μιᾶς τετράγωνον ἢ πρὸς τὸ ὑπὸ μιᾶς ἀπολαμβανομένης καὶ τῆς ἔξω δοθείσης ἢ πρὸς τὸ ὑπὸ δύο ἀπολαμβανομένων περιεχόμενον ὀρθογώνιον, ἐφ' ὁπότερ' ἂν χρῇ τῶν δοθέντων σημείων. καὶ ταύτης ἅτε δὶς διεζευγμένης καὶ περισκελεῖς διορισμοὺς ἐχούσης διὰ πλειόνων ἡ δεῖξις γέγονεν ἐξ ἀνάγκης.

Next after these the 2 books of the *Determinate section* have been handed down, for which, as for those above, it is possible to state a single enunciation, although a disjoined one, namely, this: cut a given unbounded straight line by one point so as to have, of the straight lines cut off to given points on it, either the square on one or the rectangle contained by two cut off ‹straight lines› a given ratio either to the ‹square› on one or to that by one cut off ‹straight line› and one given besides, or to the rectangle contained by two cut off ‹straight lines›, extending to whichever one uses of the given points. And this ‹enunciation› being twice disjoined and having intricate diorisms, of necessity the proof comes about through many ‹cases›.

---

[335] See for instance Sextus, *P* I.69, and Cleomedes, *Caelestia* I.5.24–26 and I.6.2–4. On Stoic τρόποι "modes" of indemonstrable arguments see Sects. 5.1.6 and 5.2.2.

[336] See Sextus, *AM* VIII.434, *P* I.69 and II.150, and Philoponus, *in APr.*, 246.3–4.

[337] The case of Ptolemy's *Almagest*, where no theorem is enunciated in non-instantiated form, and of Theon providing all such enunciations in his *Commentary*, is typical.

The only two occurrences of the technical term διεζευγμένος "disjoined" in the *Collectio* can be read in this passage. Note Pappus' appropriate remark about the enunciation being "twice disjoined": he wants to point out that the conditions are to be combined in all possible ways in order to find all the cases and subcases solved by Apollonius.[338]

In conditional form, let us read the general enunciation Pappus frames to encompass several loci of Apollonius' *Plane loci* (*Coll.* VII.23):

ἐὰν δύο εὐθεῖαι ἀχθῶσιν ἤτοι ἀπὸ ἑνὸς δεδομένου σημείου ἢ ἀπὸ δύο καὶ ἤτοι ἐπ' εὐθείας ἢ παράλληλοι ἢ δεδομένην περιέχουσαι γωνίαν καὶ ἤτοι λόγον ἔχουσαι πρὸς ἀλλήλας ἢ χωρίον περιέχουσαι δεδομένον ἅπτηται δὲ τὸ τῆς μιᾶς πέρας ἐπιπέδου τόπου θέσει δεδομένου, ἅψεται καὶ τὸ τῆς ἑτέρας πέρας ἐπιπέδου τόπου θέσει δεδομένου ὁτὲ μὲν τοῦ ὁμογενοῦς ὁτὲ δὲ τοῦ ἑτέρου, καὶ ὁτὲ μὲν ὁμοίως κειμένου πρὸς τὴν εὐθεῖαν ὁτὲ δὲ ἐναντίως. ταῦτα δὲ γίνεται παρὰ τὰς διαφορὰς τῶν ὑποκειμένων.	If two straight lines be drawn either from one given point or from two and either being in a straight line or parallel or containing a given angle and either having a ‹given› ratio to one another or containing a given area and the extremity of one touches a plane locus given in position, the extremity of the other will also touch a plane locus given in position, sometimes of the same kind sometimes of the other, and sometimes similarly situated with respect to the straight line, sometimes oppositely. This comes to be in accordance with the differences of what has been supposed.

The correlative ὁτέ "sometimes" employed by Pappus is found 19 times in the Euclidean corpus, 3 of which in the *Elements*—all of them in a porism to IV.5 that is almost certainly spurious—and none in the *Data*. No occurrences are found in the extant works of Apollonius or in Theodosius' *Sphaerica*; just one pair of correlated instances is attested in the Archimedean corpus. The absence of ὁτέ in strictly formal contexts marks it as a typical product of late departures, for rhetorical purposes, from the minimal lexical equipment typical of Hellenistic mathematics.

5.3.5. *Conjunction*

Since the use of conjunction in Greek mathematics is tricky, this section must begin with a caveat. The connectors introduced in Stoic logic are sentence-forming operators: they connect clauses. However, the connective particles that identify the standard connectors may have other syntactic functions: they can be employed inside subsentential clauses, for instance to conjoin nouns or adjectives. The two functions should in principle be kept distinct, but the borderline between them is seldom sharp, since in ancient Greek the formation of highly elliptical expressions is a matter of course.[339] Doubts may arise, then, as to whether certain single-looking clauses are in fact the result of connecting several, very elliptical, clauses.[340] As a consequence, certain dubious cases must be discussed in detail before being regarded as significant from the statistical point of view. On the other extremity of the variation spectrum, a conjunction, which is a non-simple sentence, must be

[338] Similar combinatorial concerns are implicit in the comprehensive enunciations that Pappus concocts for Apollonius' *Tangencies* and for some of the Euclidean *Porisms* (*Coll.* VII.11 and 16).
[339] Cf. the discussion in Barnes 2007, 168–173.
[340] This is one of the reasons why the Stoics prescribed to withdraw any form of ellipsis in logically well-formed statements, for instance by requiring that cross-references be avoided in non-simple statements.

carefully distinguished from the system of sentences that formulates an inference: some particles, most notably δέ, often stand for our punctuation signs, thereby linking independent principal clauses in an inference,[341] but may also serve as alternative connectors in a conjunction.

The use of conjunctive connectors is particularly unconstrained in the Greek mathematical corpus. Conjunctions are most frequently expressed by καί "and" interposed between the conjuncts but, contrary to what happens with disjunction, preposed καί is almost never found. One sporadic instance is found in X.70 (*EOO* III, 212.9–12):

ὥστε καὶ τὸ συγκείμενον ἐκ τῶν ἀπὸ τῶν ΓΖ ΖΔ τετραγώνων μέσον ἐστὶ καὶ τὸ ὑπὸ τῶν ΓΖ ΖΔ μέσον καὶ ἔτι ἀσύμμετρον τὸ συγκείμενον ἐκ τῶν ἀπὸ τῶν ΓΖ ΖΔ τετραγώνων τῷ ὑπὸ τῶν ΓΖ ΖΔ.

So that both the ‹region› compounded of the squares on ΓΖ, ΖΔ is medial and the ‹rectangle contained› by ΓΖ, ΖΔ is medial and further the ‹region› compounded of the squares on ΓΖ, ΖΔ is incommensurable with the ‹rectangle contained› by ΓΖ, ΖΔ.

Proposition X.70 is one in a series of theorems where the sentence just read, which characterizes two among the irrational lines introduced in Book X, is repeated with almost identical wording. Thanks to this, it is easy to check that the preposed καί in X.70 is only a local variant. In fact, the reference form has beyond doubt τε postposed to the article as the first correlative.[342]

As the previous example shows, the preferred form of the correlative conjunction is τε … καί … "both … and …". We also find this form when two or more items in a list are to be considered in some sense a unitary whole and cannot be read as the conjunction of two elliptical sentences. This happens either when non-homologous parts, of which some geometric figure is composed, are referred to, as in I.def.18 (*EOO* I, 4.19–6.1),

ἡμικύκλιον δέ ἐστι τὸ περιεχόμενον σχῆμα ὑπό τε τῆς διαμέτρου καὶ τῆς ἀπολαμβανομένης ὑπ' αὐτῆς περιφερείας.

A semicircle is a figure contained both by the diameter and by the circumference cut off by it.

or in contexts in which a sum of mathematical objects is intended, as in the enunciation of proposition II.3 (*EOO* I, 122.19–23),

ἐὰν εὐθεῖα γραμμὴ τμηθῇ ὡς ἔτυχεν, τὸ ὑπὸ τῆς ὅλης καὶ ἑνὸς τῶν τμημάτων περιεχόμενον ὀρθογώνιον ἴσον ἐστὶ τῷ τε ὑπὸ τῶν τμημάτων περιεχομένῳ ὀρθογωνίῳ καὶ τῷ ἀπὸ τοῦ προειρημένου τμήματος τετραγώνῳ.

If a straight line be cut at random, the rectangle contained by the whole ‹straight line› and one of the segments is equal both to the rectangle contained by the segments and to the square on the said segment.

[341] Other coordinating conjunctions, for instance ἀλλά "but", have both in mathematics and in Stoic logic the function of introducing coassumptions. The Stoics seem to have taken ἀλλά as the standard particle with this function, whereas in mathematics it is employed in a minority of cases, the preferred particle being in fact δέ "and": see Sect. 4.5.5.

[342] We find τε … τε … καὶ ἔτι … "both … and … and further" (see just below) in X.78 and 84 (enunciation), τε … καί … καὶ ἔτι … in X.84 (twice, both in the proof and in the conclusion) and X.107, τε … τε … ἔτι τε … in X.96, τε … καί … καί … in X.102. A parallel syntactic structure in the first part of Book X (props. 41, 47 and 65) gives only the form τε … καί … καὶ ἔτι … Granted, the first καί in X.70 might be adverbial and qualify ὥστε. If this were the case, however, we should expect that a correlative connects the first conjunct to the others, as in all other occurrences of the expression. Since this does not happen, our καί really replaces here the canonical τε. The presence of τε in the exactly corresponding proposition X.107 confirms this hypothesis.

or to indicate the formation of a single, composite predicate out of two predicates, as in the very concise enunciation of IV.12 (*EOO* I, 302.5–6):

περὶ τὸν δοθέντα κύκλον πεντάγωνον ἰσόπλευρόν τε καὶ ἰσογώνιον περιγράψαι.

About a given circle circumscribe a pentagon both equilateral and equiangular.

Contrary to what happens in Attic prose writers of the classical age, who made scarce use of juxtaposed τε καί,[343] this nexus is rather frequent in the ancient mathematical corpus, and in Euclidean writings more than elsewhere.[344] As in the last example just seen, this contiguous correlative is almost invariably found when two appositive adjectives identify, taken in conjunction, a well-defined mathematical object as a species of a wider genus: besides ἰσόπλευρόν τε καὶ ἰσογώνιον, which makes all the occurrences in Book IV, we also find for instance ὅμοιός τε καὶ ὁμοίως κείμενον "both similar and similarly placed",[345] ἴσος τε καὶ ὅμοιος "both equal and similar",[346] ἴση τε καὶ παράλληλος "both equal and parallel".[347] In all these instances, the absence of the article makes the two connectors τε and καί necessarily contiguous: otherwise τε, which must be in second position within the syntagm that constitutes the first conjunct, should have followed the article preposed to the first predicate.

As we have seen on the example of X.70, when more than two terms are listed, the last item is usually introduced by καὶ ἔτι "and further". A rather compact example can be found in proposition XIII.17 (*EOO* IV, 316.24–318.1), in fact a variant of the compound predicate of IV.12:

λέγω ὅτι τὸ ΥΒΧΓΦ πεντάγωνον ἰσόπλευρόν τε καὶ ἐν ἑνὶ ἐπιπέδῳ καὶ ἔτι ἰσογώνιόν ἐστιν.

I claim that pentagon ΥΒΧΓΦ is both equilateral and in one plane and further equiangular.

The occurrences of καὶ ἔτι can be independent of the presence of previous τε … καί …: see the setting-out of I.22 (*EOO* I, 52.21–24), where καὶ ἔτι comes in third position after a simple correlative μέν … δέ …:[348]

ἔστωσαν αἱ δοθεῖσαι τρεῖς εὐθεῖαι αἱ Α Β Γ, ὧν αἱ δύο τῆς λοιπῆς μείζονες ἔστωσαν πάντῃ μεταλαμβανόμεναι, αἱ μὲν Α Β τῆς Γ αἱ δὲ Α Γ τῆς Β καὶ ἔτι αἱ Β Γ τῆς Α.

Let there be three given straight lines, A, B, Γ, of which let two, however permuted, be greater than the remaining one, A, B of Γ and A, Γ of B and further B, Γ of A.

We must regard the above clause as a true conjunction of sentences, although stated in a very elliptical form, and not as a list of objects.

---

[343] See Denniston 1954, 512–513. For the same nexus in post-classical prose, see Blomqvist 1974, who draws the same conclusion: non-Attic style (that is, early Ionian prose writes, technical Hellenistic treatises, asianic style) more frequently uses juxtaposed τε καί.

[344] There are 107 occurrences in the whole of *Elements*, but 42 in the *Phaenomena*. Add to these the 15 occurrences of τέ ἐστι καί, all in the main text of the *Elements*. In the first four Books of Apollonius' *Conica*, we find 11 occurrences of contiguous τε καί, and 32 in the entire Archimedean corpus. Diophantus' *Arithmetica* exhibits 23 occurrences.

[345] These are most of the occurrences in Books VI and XI.

[346] There are 5 occurrences, one in XI.def.13 and two in each of XII.3 and 8.

[347] There are 11 occurrences in I.33, 36 and 45; 4 occurrences in XIII.16.

[348] The expression καὶ ἔτι occurs 81 times in the whole of the *Elements*, but only 8 times in the rest of Euclidean corpus (once in the *Data* and 7 times in the *Phaenomena*).

The following table sets out the variability of the correlative expressions that contain καὶ ἔτι and their occurrences in the whole of the *Elements*:

καί … δέ … καὶ ἔτι … καὶ ἔτι	VIII.2
καί … καί … καὶ ἔτι	VIII.19
καί … καὶ ἔτι	XI.22, 23, 27, 33, XII.17
μέν … δέ … καὶ ἔτι	I.22, 23, 26, 34, II.6, V.22, VI.4, 5 (*bis*), 12, IX.15, X.25, XI.22, 23, 25 (*bis*), XI.33
μέν … δέ … δέ … καὶ ἔτι	II.1, III.31 (*bis*)
μέν … δέ … καί … καὶ ἔτι	X.32/33
τε … καί … καὶ ἔτι	II.1 (*bis*), VI.18, VIII.4 (*sexties*), 19 (*ter*), X.35 (*bis*), 41, 47, 65, 70, 84 (*bis*), 107 (*bis*)
τε καί … καὶ ἔτι	XIII.17
τε … τε … καὶ ἔτι	X.78, 84
nothing … καί … καὶ ἔτι	I.17, VI.5, VIII.4, 13 (*bis*), 19, X.53, 70, 112, 113 (*bis*), XI.30, **b** XII.3
nothing … δέ … καὶ ἔτι	VIII.15
nothing … δέ … δέ … καὶ ἔτι	VI.24
nothing … ἤ … καὶ ἔτι	VIII.19
nothing … καὶ ἔτι … καὶ ἔτι	X.112
nothing … καί … καί … καὶ ἔτι	XI.4
nothing … καὶ ἔτι	IV.15por, VI.8 (*quater*), 20

A great analytic precision is achieved by an appropriate use of τε … καί … in conjoining sentences. This is the case when several results are to be combined in the enunciation of the same theorem. The enunciations of I.28 and 29 (*EOO* I, 68.13–17 and 70.20–24; see also I.34), which are transcribed one after the other below, are a small précis of logic:

ἐὰν εἰς δύο εὐθείας εὐθεῖα ἐμπίπτουσα τὴν ἐκτὸς γωνίαν τῇ ἐντὸς καὶ ἀπεναντίον καὶ ἐπὶ τὰ αὐτὰ μέρη ἴσην ποιῇ ἢ τὰς ἐντὸς καὶ ἐπὶ τὰ αὐτὰ μέρη δυσὶν ὀρθαῖς ἴσας, παράλληλοι ἔσονται ἀλλήλαις αἱ εὐθεῖαι.	If a straight line falling on two straight lines make an external angle equal to the internal and opposite and on the same side one or those internal and on the same side equal to two right ‹angles›, the straight lines will be parallel to one another.
ἡ εἰς τὰς παραλλήλους εὐθείας εὐθεῖα ἐμπίπτουσα τάς τε ἐναλλὰξ γωνίας ἴσας ἀλλήλαις ποιεῖ καὶ τὴν ἐκτὸς τῇ ἐντὸς καὶ ἀπεναντίον ἴσην καὶ τὰς ἐντὸς καὶ ἐπὶ τὰ αὐτὰ μέρη δυσὶν ὀρθαῖς ἴσας.	A straight line falling on parallel straight lines makes both the alternate angles equal to one another and an external one equal to the internal and opposite one and those internal and on the same side equal to two right ‹angles›.

The enunciation of I.28 is characterized by an inclusive disjunction: the two conditions on the angles are in fact equivalent to one another and to the condition formulated in I.27. Accordingly, a preposed ἤτοι is absent. As a consequence, the proof is composed of two independent proofs, each of which applies one of the suppositions made explicit in the antecedent of the conditional. The transition between the two proofs is canonically marked by the connector πάλιν "again". The whole proof amounts to validating two independent conditionals, which are combined in the enunciation.[349] The change of grammatical subject from the assumptions to the *demonstrandum* entails that the form of the enunciation must be a conditional (cf. Sect. 4.1).

349 Since the disjunction in the enunciation is inclusive, duality with respect to conjunction holds.

In I.29, instead, the same grammatical subject can be used throughout the enunciation; accordingly, the enunciation has not the form of a conditional. The three conditions on the angles that in I.27–28 occur in the antecedent are here located in the consequent, and for this reason they are conjoined and not disjoined. Proposition I.29 is in fact the converse of the pair I.27–28. The interchange of antecedent and consequent in these enunciations gives rise to a duality between conjunction and inclusive disjunction; the proofs of the three conditions are given in succession.

The structure of the consequent of I.29 has several interesting features: τε is placed after the very first element of the first conjunct, namely, the article; two of the four καί present in the text mark the "external" ternary conjunction; the remaining two καί (broken underlining) mark the two nested binary conjunctions. Maybe the Stoics would have preposed some καί, or maybe the nested conjunctions might have been stated in the correlative form τε … καί … in order to forestall possible ambiguities. In our text, instead, the same goal is reached by a clever disposition of the terms: the relational operator ἴσος "equal" is postpositive and canonically placed outside the terms in relation (Sect. 4.5.1.3); as a consequence, the conjoined syntagms that express (the second of the terms of) the pair in relation are nested between the article that follows immediately an "external" conjunctive particle and the relational operator. No ambiguities may arise.

A second example comes from the enunciation of II.10 (*EOO* II, 146.15–22); this is composed of two nested conjunctions, and the postposed τε forestalls possible ambiguities:

ἐὰν εὐθεῖα γραμμὴ τμηθῇ δίχα προστεθῇ δέ τις αὐτῇ εὐθεῖα ἐπ' εὐθείας, τὸ ἀπὸ τῆς ὅλης σὺν τῇ προσκειμένῃ καὶ τὸ ἀπὸ τῆς προσκειμένης τὰ συναμφότερα τετράγωνα διπλάσιά ἐστι τοῦ τε ἀπὸ τῆς ἡμισείας καὶ τοῦ ἀπὸ τῆς συγκειμένης ἔκ τε τῆς ἡμισείας καὶ τῆς προσκειμένης ὡς ἀπὸ μιᾶς ἀναγραφέντος τετραγώνου.

If a straight line be bisected and some straight line be added to it in a straight line, the ‹square› on the whole ‹straight line› plus the added one and that on the added one, both squares together, are double both of the ‹square› on the half and of the square described on the ‹straight line› compounded of both the half and the added one as on one ‹straight line›.

It is not easy to grasp at first reading what such an elliptical enunciation says. The first correlative τε … καί …, namely, the most "external" correlative, refers to the sum of two squares. In its turn, the second of these squares is identified by a complex expression: the square is described on a straight line that is the sum (the "composition" of the text) of two other straight lines: the first of these straight lines is half the given straight line, the second is the straight line added to the given one. This sum is also formulated by means of a correlative τε … καί … To understand what is associated with what the different prepositions ἀπό and ἐκ already serve a first, partial purpose. Second, note the different placement of the two prepositions with respect to the sums, and hence to the correlatives. The outermost sum has the structure συναμφότερα τετράγωνα διπλάσιά ἐστι τοῦ[sum of squares, which includes ἀπό], whereas the innermost sum has the structure εὐθεῖα συγκειμένη ἐκ[sum of straight lines]. The outermost sum is governed by an article, the innermost sum by a preposition; still, whereas the article τοῦ is repeated, the preposition ἐκ is not repeated and has scope over both genitives conjoined by τε … καί … This is the rule in similar expressions (cf. I.def.18 above):[350] a typical example with preposition ὑπό is τὸ περιεχόμενον ὀρθογώνιον ὑπό

[350] See also, in III.8 (*EOO* I, 184.5–6), ἡ μεταξὺ τοῦ τε σημείου καὶ τῆς διαμέτρου "the ‹straight line› between both the point and the diameter", but this is the only joint occurrence of μεταξύ and τε.

τε τῆς ΑΒ καὶ τῆς ΓΔ "the rectangle contained both by AB and by ΓΔ"; in all cases, the preposition must be repeated in translation, while in Greek it is understood.[351] The preposition ἀπό, instead, identifies the squares of which the sum is taken: it is repeated since only by its repetition each of the two mathematical objects can be identified as a square. Accordingly, the postpositive τε is placed after the first article, which is repeated after the καί correlated to it, since otherwise the case in which the prepositional noun τὸ ἀπὸ τῆς "the ‹square described› on" is inflected would remain undetermined. As a final touch, the second τε makes it clear that the last of the four feminine genitives in the conjunction, namely, the only genitive not preceded by a preposition, is ruled by ἐκ and not by ἀπό, settling an ambiguity which syntax alone could not decide. In sum, postpositive τε is pretty useful when it is employed in brachylogic conjunctions and in lists, where ambiguities of scope might arise.

The correlative τε ... τε ... "both ... and ..." is seldom used. Interesting examples can be found in the conditional enunciations of propositions III.13 (*EOO* I, 198.18–20), III.26–27, and VI.33, of which we read the first:

κύκλος κύκλου οὐκ ἐφάπτεται κατὰ πλείονα σημεῖα ἢ καθ' ἕν, ἐάν <u>τε</u> ἐντὸς ἐάν <u>τε</u> ἐκτὸς ἐφάπτηται.

A circle is not tangent to a circle at more points than one, <u>both</u> if it is tangent inside <u>and</u> if it is tangent outside.

This enunciation is a conditional clause but its form is not canonical: the elliptical double antecedent is postpositive. Usually, when two theorems are conflated into one, the two enunciations are complete conditional sentences conjoined by means of καί: this typically happens when a theorem and its converse are given under one and the same enunciation. If, as in I.28 above, a single conditional is formulated, the alternative conditions are disjoined, and not conjoined. In III.13, however, the conditions detailed in the two antecedents partially overlap (for the two circles are tangent in both cases), and the conjunction τε simply makes the division into cases necessary to the proof manifest. Accordingly, the proof comprises two rigidly separated sub-proofs. In III.26–27 and VI.33 the proofs are not even divided into cases, since the two conditions in the double postpositive antecedent are in fact one a consequence of the other. The division into cases is here induced by the choice of subsuming several categories of geometric objects under a single term, even if they are regarded as intrinsically different. In III.27 and in VI.33 the proofs are explicitly given as one subordinated to the other, while in III.26 both conditions are applied to achieve a single proof.[352]

From the above discussion it must be concluded that sometimes, when highly elliptical expressions are at issue, the distinction between conjunction and disjunction cannot be clear-cut, not even in mathematical texts.[353] The use of ἐάν τε ... ἐάν τε ..., namely, of a conjunctive form without καί, shows that the problems involved in formulating such a hybrid enunciation were perceived by the author of the text. The same construction can be found, in a treatise that we may well regard as technical, in a passage of Sextus' *Against the grammarians* (*M* I.173):

---

[351] Cf. Denniston 1954, 518–519.

[352] In the case of III.26, then, the formulation of the enunciation cannot be regarded as entirely correct.

[353] The oscillation between conjunctive and disjunctive meaning in very elliptical contexts as the ones just presented is a well-known syntactic phenomenon; see Kühner, Gerth 1898–1904, II.ii, § 539. The same remark was already made by Apollonius Dyscolus, *Conj.*, *GG* II.1.1, 219.12–220.22.

οὐδὲν γὰρ βλαπτόμεθα, ἐάν τε σὺν τῷ ι γράφωμεν τὴν δοτικὴν πτῶσιν ἐάν τε μή, καὶ ἐάν τε διὰ τοῦ σ τὸ σμιλίον καὶ τὴν Σμύρναν ἐάν τε διὰ τοῦ ζ, καὶ ἐπὶ τοῦ Ἀριστίων ὀνόματος ἐάν τε τῇ προηγουμένῃ συλλαβῇ τὸ σ προσμερίζωμεν ἐάν τε τῇ ἐπιφερομένῃ τοῦτο συντάττωμεν.	For we shall not be injured, both if we write the dative with ι and if not, and both if we write σμιλίον and Σμύρναν with σ and if with ζ, and for the name Ἀριστίων, both if we assign the σ to the preceding syllable and if we range it with the subsequent one.

We find only 5 occurrences of εἴτε … εἴτε … "either if … or if …" in the *Elements*, all in Book X.[354] In the occurrence inside the third item of the "first definitions" (read in Sect. 5.3.1), the formulation is particularly elliptical, and gives rise to a formulation that is again intermediate between a disjunction and a conjunction. What is more, here εἴτε … εἴτε … connects two terms in one of those disjoined partitions for which the possibility of coexistence of the terms is excluded by definition. In propositions X.14 (*bis*), 67, 104, unabridged conditionals are conjoined; the antecedents carry incompatible conditions, and the meaning of the construct is clearly conjunctive. In X.14, the fact that the two occurrences are instantiated confirms this hypothesis, whereas the enunciation, canonically given as a conjunction of two conditionals introduced by ἐάν, shows that one single statement conflates two distinct mathematical propositions under the same proof.

When two or more objects are compared with other sets of objects, ambiguities may arise as to whether the objects must be taken as a sum or severally, and in the latter case whether the comparison is between objects that hold the same place in the sequence or not. The first possibility just envisaged, namely, that the objects listed are to be added together, is normally dealt with by (1) simple conjunction (often in the correlative form τε … καί …, as in II.3 and 10 read above), or by (2) the absence of any qualifier attached to a mention of the objects in the plural (a practice that is the opposite of modern usage),[355] or by (3) the use of the adjective συναμφότερος "both together".[356] All these phenomena can be detected in the enunciation and setting-out + determination of proposition II.7 (*EOO* I, 134.22–136.3):

ἐὰν εὐθεῖα γραμμὴ τμηθῇ ὡς ἔτυχεν, τὸ ἀπὸ τῆς ὅλης καὶ τὸ ἀφ' ἑνὸς τῶν τμημάτων τὰ συναμφότερα τετράγωνα ἴσα ἐστὶ τῷ τε δὶς ὑπὸ τῆς ὅλης καὶ τοῦ εἰρημένου τμήματος περιεχομένῳ ὀρθογωνίῳ καὶ τῷ ἀπὸ τοῦ λοιποῦ τμήματος τετραγώνῳ.	If a straight line be cut at random, the ‹square› on the whole and that on one of the segments, both squares together, are equal both to twice the rectangle contained by the whole and[357] the said segment and to the square on the remaining segment.
εὐθεῖα γάρ τις ἡ AB τετμήσθω ὡς ἔτυχεν κατὰ τὸ Γ σημεῖον. λέγω ὅτι τὰ ἀπὸ τῶν AB ΒΓ τετράγωνα ἴσα ἐστὶ τῷ τε δὶς ὑπὸ τῶν AB ΒΓ περιεχομένῳ ὀρθογωνίῳ καὶ τῷ ἀπὸ τῆς ΓΑ τετραγώνῳ.	In fact, let some straight line, AB, be cut at random at point Γ. I claim that the squares on AB, ΒΓ are equal both to twice the rectangle contained by AB, ΒΓ and to the square on ΓΑ.

[354] In the introduction to the *Phaenomena*, εἴτε correlates with ἤ, and what results is a true disjunction.
[355] In a strict minority of cases, the sum is explicitly formulated by means of the prepositions μετά (as in II.5) or σύν (as in II.10); see Sect. 1.5 for the occurrences of σύν in the whole of the *Elements*.
[356] There are 53 occurrences in the *Elements*, 85 in the *Data*, 1 in the *Phaenomena*. The occurrences are found in clusters: both in the *Elements* and in the *Data*, only 18 propositions are involved in each treatise. See Acerbi 2012a.
[357] This conjunction disappears in the setting-out, where it is replaced by the simple mention of the denotative letters. No ambiguities arise with the "external" conjunction τε … καί … because the καί "and" to which this footnote is attached is followed by a genitive, whereas the καί in the external conjunction is followed by a dative.

The two squares to be added, which in the enunciation are listed in conjunction and whose sum is expressly marked by the strengthening phrase that contains συναμφότερα, are simply put in the plural in the setting-out: contrary to modern usage, the absence of any qualifier entails that they are not to be taken severally.

If the objects are instead to be taken severally, Greek mathematical language mainly makes an appropriate use of the adjective ἑκάτερος "each of" to settle possible ambiguities. The 695 occurrences of this adjective in the Euclidean corpus (538 in the whole of the *Elements*) can be divided into two categories: isolated occurrences and coupled occurrences. An instance of the isolated occurrences can be read in I.1 (*EOO* I, 12.8–10):

ἴση ἐστὶν ἡ ΒΓ τῇ ΒΑ· ἐδείχθη δὲ καὶ ἡ ΓΑ τῇ ΑΒ ἴση· ἑκατέρα ἄρα τῶν ΓΑ ΓΒ τῇ ΑΒ ἐστὶν ἴση.

ΒΓ is equal to ΒΑ; and ΓΑ was also proved equal to ΑΒ; therefore each of ΓΑ, ΓΒ is equal to ΑΒ.

Writing αἱ ΓΑ, ΓΒ τῇ ΑΒ εἰσιν ἴσαι means instead that the sum of ΓΑ, ΓΒ is equal to ΑΒ. Coupled occurrences are typically found in applications of the congruence criteria for triangles, as in proposition I.12 (*EOO* I, 34.18–20):

ἐπεὶ γὰρ ἴση ἐστὶν ἡ ΗΘ τῇ ΘΕ κοινὴ δὲ ἡ ΘΓ, δύο δὴ αἱ ΗΘ ΘΓ δύο ταῖς ΕΘ ΘΓ ἴσαι εἰσὶν ἑκατέρα ἑκατέρᾳ.

In fact, since ΗΘ is equal to ΘΕ and ΘΓ is common, thus two ‹sides›, ΗΘ, ΘΓ, are equal to two ‹sides›, ΕΘ, ΘΓ, respectively.

This means (1) that the straight lines in each of the two pairs are to be taken severally, and (2) that their ordering must be retained, so that the equal straight lines are those that occupy the same place in each pair (see Sect. 4.5.4).

Emphatic καί is usually translated with the adverb "also", but the equivalence is not so straightforward. Emphatic καί is preposed to whole relations or clauses, but it is not really a scope particle: it is an "inclusive focus particle".[358] The position of καί is in this instance analogous to the position of the emphatic negative particle, and should not be mistaken for the preposed καί that features in a conjunction. This usage, typical of Greek ordinary prose, may cause problems in translation, since in this case English tends to emphasize the main verb.

Emphatic καί is frequently found in the inferences by transitivity of relations. As observed in Sect. 4.5.1.3, in mathematical relations such as equality, proportionality, etc., the relational predicate lies outside the pair of terms set in relation. Now, a preposed καί stresses the fact that, in the conclusion of an inference, a relation is being established, analogous to the relation that figures in the premises, between two terms that in the premises lay unrelated. However, the emphatic particle cannot interfere with the ordered-pair structure of the terms: as a consequence, it must be preposed to them, even if not necessarily to the relational operator. Such a placement is in fact overdetermined, since the mere fact that what is emphasized is a conclusion of an inference would make the καί be preposed. The first example comes from a double application of transitivity to proportions in VI.14 (*EOO* II, 112.25–114.5):

[358] See Bakker 1993, 288 and 297–298. I shall translate "also", but in the first example below a most fitting rendering is the adverb "actually".

ἐπεὶ γάρ ἐστιν ὡς ἡ ΔΒ πρὸς τὴν ΒΕ οὕτως ἡ ΗΒ πρὸς τὴν ΒΖ ἀλλ' ὡς μὲν ἡ ΔΒ πρὸς τὴν ΒΕ οὕτως τὸ ΑΒ παραλληλόγραμμον πρὸς τὸ ΖΕ παραλληλόγραμμον ὡς δὲ ἡ ΗΒ πρὸς τὴν ΒΖ οὕτως τὸ ΒΓ παραλληλόγραμμον πρὸς τὸ ΖΕ παραλληλόγραμμον, καὶ ὡς ἄρα τὸ ΑΒ πρὸς τὸ ΖΕ οὕτως τὸ ΒΓ πρὸς τὸ ΖΕ.	In fact, since, as ΔB is to BE, so HB is to BZ but, as ΔB is to BE, so parallelogram AB is to parallelogram ZE and, as HB is to BZ, so parallelogram BΓ is to parallelogram ZE, therefore, as AB is to ZE, so BΓ is also to ZE.

This sentence has the logical form of a hybrid paraconditional (see Sect. 5.3.2); the conjunction in the coassumption (which is canonically introduced by ἀλλά) is formulated in the weakest possible form, by means of the correlative μέν … δέ … The ordering of the words in the nexus καὶ ὡς ἄρα is explained as follows: ὡς is always placed first in the canonical formula of proportionality; as a consequence, ἄρα, which is normally postpositive and holds the second place, must follow it; emphatic καί must be preposed to the whole formula. The following example comes from an application of transitivity of equality, in II.8 (*EOO* I, 140.1–6):

πάλιν ἐπεὶ ἴση ἐστὶν ἡ ΓΒ τῇ ΒΔ ἀλλὰ ἡ μὲν ΒΔ τῇ ΒΚ, τουτέστι τῇ ΓΗ, ἴση ἡ δὲ ΓΒ τῇ ΗΚ, τουτέστι τῇ ΗΠ, ἐστιν ἴση, καὶ ἡ ΓΗ ἄρα τῇ ΗΠ ἴση ἐστίν. καὶ ἐπεὶ ἴση ἐστὶν ἡ μὲν ΓΗ τῇ ΗΠ ἡ δὲ ΠΡ τῇ ΡΟ, ἴσον ἐστὶ καὶ τὸ μὲν ΑΗ τῷ ΜΠ τὸ δὲ ΠΛ τῷ ΡΖ.	Again, since ΓB is equal to BΔ but BΔ is equal to BK, that is to ΓH, and ΓB is equal to HK, that is to HΠ, therefore ΓH is also equal to HΠ. And since ΓH is equal to HΠ and ΠP to PO, AH is also equal to MΠ and ΠΛ to PZ.

The first paraconditional is hybrid, the second paraconditional is pure. The second καί is better read as a coordinant "and", even if a reading as a preposed adverbial "also" would not be incorrect. The first and the third καί occur in the two possible emphatic positions: the first καί is preposed to the whole relational expression, the third καί to the ordered pair only. In the latter case the inversion with respect to the operator ἴσον ἐστί is motivated by the presence of two ordered pairs: καί, preposed to both pairs set in correlation by μέν … δέ …, is distributed between them. The same requirement of distributivity forces the preposed position of the relational operator itself in the second paraconditional. In the first paraconditional, on the contrary, the presence of the two identifications of objects introduced by τουτέστι "that is" entails that the relational operator has to be postpositive and repeated in both disjuncts.

Another problem arises in translation as a consequence of the position of emphatic καί. The determination of a theorem is introduced by the canonical formula λέγω ὅτι "I claim that", which specifies what it is required to prove. It may happen that this clause asserts a condition that is similar to another condition already stated among the assumptions of the proposition, as for instance occurs in I.6 (*EOO* I, 22.22–24):

ἔστω τρίγωνον τὸ ΑΒΓ ἴσην ἔχον τὴν ὑπὸ ΑΒΓ γωνίαν τῇ ὑπὸ ΑΓΒ γωνίᾳ. λέγω ὅτι καὶ πλευρὰ ἡ ΑΒ πλευρᾷ τῇ ΑΓ ἐστιν ἴση.	Let there be a triangle, ABΓ, having angle ABΓ equal to angle AΓB. I claim that side AB is also equal to side AΓ.

In this case, the emphatic καί specifies that the sides are also equal to one another, very much as this was supposed for the angles: καί acts inside the that-clause.

A different translation must be adopted in the following case. When several statements are proved in a single theorem, normally the formula "I claim that" is repeated for any of the statements;

an emphatic καί is preposed to recall that an additional result is proved, as in I.34 (*EOO* I, 80.23–82.2), whose enunciation is

τῶν παραλληλογράμμων χωρίων αἱ ἀπεναντίον πλευραί τε καὶ γωνίαι ἴσαι ἀλλήλαις εἰσὶ καὶ ἡ διάμετρος αὐτὰ δίχα τέμνει.	Both the opposite sides and angles of parallelogrammic regions are equal to one another and the diagonal bisects them.

The fact that each diagonal bisects the parallelogram is proved after the first result stated in the enunciation of I.34 has been fully demonstrated; the partial retrieval of the enunciation of I.34 (*EOO* I, 84.3) reads as follows:[359]

λέγω δὴ ὅτι <u>καὶ</u> ἡ διάμετρος αὐτὰ δίχα τέμνει.	I now <u>also</u> claim that the diagonal bisects them.

Apart from the insertion of the particle δή, which marks a mild transition, the word order is the same as in I.6 read just above. Here, however, the emphatic καί must be taken as acting on the that-clause (if not on the whole sentence), and not inside it.

In what follows and until the end of this Section, I shall assess the evidence that concerns disjunction and conjunction in logical and grammatical writings.

The emphatic role of prepositive ἤτοι characterizes the exclusive disjunction as a connective structure that is both strongly marked from the logical point of view and hardly recognized as a prominent and widely used syntactic structure in ordinary prose style.[360] The fact that in ordinary language the meaning of ἤ can be decided only in context shows that the choice of the mathematicians of carefully distinguishing between exclusive and inclusive disjunction is intentional, and parallel to the Stoic prescription which we shall see presently.[361] This suggests that such a prescription was elaborated against the background of mathematical practice. It is instead highly implausible that the *Elements* were thoroughly revised in accordance with the Stoic prescription.

In fact, the systematic use of preposed ἤτοι meets one of the Chrysippean prescriptions for exclusive disjunction. Let us get a look at the evidence. Diogenes Laertius' account contains two unrelated characterizations of disjunction.[362] The first characterization is purely syntactic and looks very much like a definition; the second characterization is semantic and states an essential property of a disjunction (D.L. VII.72):

διεζευγμένον δέ ἐστιν ὃ ὑπὸ τοῦ "ἤτοι" διαζευκτικοῦ συνδέσμου διέζευκται, οἷον "ἤτοι ἡμέρα ἐστὶν ἢ νύξ ἐστιν". ἐπαγγέλλεται δ ὁ σύνδεσμος οὗτος τὸ ἕτερον τῶν ἀξιωμάτων ψεῦδος εἶναι.	A disjunction is what turns out to be disjoined by the disjunctive connector ἤτοι, such as "either it is day or it is night". This connector announces that one of the statements is false.

[359] This follows a fully-fledged partial conclusion stated without letters, which is unusual in the *Elements*: the partial conclusion is normally given in instantiated form (cf. Sect. 4.3).
[360] It is even disputable whether one can speak of ἤ meaning *aut* or *vel* before an adequate formalization of the notion of exclusiveness is proposed.
[361] Apollonius Dyscolus (*Conj.*, *GG* II.1.1, 220.23–221.15) regards the presence of ἤ *between* the disjuncts as the marking feature of a disjunction (either inclusive or exclusive), and never relates the preposed particle ἤτοι as singling out the exclusive disjunction.
[362] See the end of this Section for further detail on this multiple definition.

A point to be stressed in the first characterization, which can be assumed to be the standard syntactic definition in Stoic logic, is that the disjunctive particle referred to is the prepositive ἤτοι and not ἤ. This confirms that the presence of ἤτοι was regarded as essential, for it singles out the disjunction and confirms that the preposed position of a suitable connector was a crucial feature of Stoic non-simple statements. The second characterization in Diogenes Laertius would make disjunction a truth-functional connector, but this turns out not to be the case.[363] We find, for instance, more detail in Aulus Gellius (*NA* XIV.8.12–14),

Est item aliud, quod Graeci διεζευγμένον ἀξίωμα, nos, "disiunctum" dicimus. Id huiusmodi est: "aut malum est voluptas aut bonum aut neque bonum neque malum est". Omnia autem, quae disiunguntur, pugnantia esse inter sese oportet, eorumque opposita, quae ἀντικείμενα Graeci dicunt, ea quoque ipsa inter se adversa esse. Ex omnibus, quae disiunguntur, unum esse verum debet, falsa cetera. Quod si aut nihil omnium verum aut omnia plerave, quam unum, vera erunt aut quae disiuncta sunt, non pugnabunt aut quae opposita eorum sunt, contraria inter sese non erunt, tunc id disiunctum mendacium est et appellatur παραδιεζευγμένον, sicuti hoc est, in quo, quae opposita, non sunt contraria: "aut curris aut ambulas aut stas".

and in Sextus Empiricus, where the truth-conditions for a disjunction are definitely not truth-functional (*P* II.162 and II.191):

ἐν τῷ διεζευγμένῳ τὸ μὲν ἀληθές ἐστι τὸ δὲ ψεῦδος μετὰ μάχης τελείας, ὅπερ ἐπαγγέλλεται τὸ διεζευγμένον.	In the disjunction one ‹disjunct› is true, the other false with complete conflict, which is really what the disjunction announces.
τὸ γὰρ ὑγιὲς διεζευγμένον ἐπαγγέλλεται ἓν τῶν ἐν αὐτῷ ὑγιὲς εἶναι, τὸ δὲ λοιπὸν ἢ τὰ λοιπὰ ψεῦδος ἢ ψευδῆ μετὰ μάχης.	For a sound disjunction announces that one ‹disjunct› in it is sound, while the remaining one(s) is / are false with conflict.

It is clear from these definitions that the disjunction is exclusive: at most one disjunct can be true. That also exhaustiveness (at least one disjunct is true)[364] had to be an essential feature of the Chrysippean disjunction is made clear by the example reported by Sextus (*M* VIII.434 and *P* II.150) of an argument that is invalid "by deficiency". The argument, a fifth indemonstrable, is invalid exactly because a (trichotomic) disjunction is not stated by setting out all its disjuncts.[365] This nicely fits with Galen's claim that only a complete conflict between the disjuncts makes both the fourth and the fifth indemonstrable work (*Inst. Log.* V.3); for otherwise only the fourth will do (cf. Sects. 5.1.6 and 5.2.2). Galen (*Inst. Log.* IV.2 and V.2) even seems to identify the dichotomy complete / incomplete conflict with the exhaustive / non exhaustive alternative: the notion of conflict straightforwardly entails the mutual incompatibility of the disjuncts, and hence exclusiveness.

There is more in the Sextan accounts of disjunction: since they hinge upon the concept of μάχη "conflict" among the disjuncts,[366] disjunction is not truth-functional, even if the condition that one

---

[363] This was remarked as early as Casari 1958.
[364] Exhaustivity is currently formulated in ancient sources as the requirement that the disjuncts cannot be destroyed together, that is, that they cannot be all false.
[365] In all accounts of Stoic syllogisms involving multiple disjunctions, the several disjuncts are always partitioned as one vs. the others (by the way, passages like *P* II.191 just read show that disjunction is a polyadic connector). We might wonder whether different partitions might be admitted, but it seems that the very definition of multiple disjunction rules out these possibilities. However, this is a minor problem in mathematical arguments, where the relevant disjunctions are never more than trichotomic.
[366] As seen above (underlined syntagms), Aulus Gellius translates it with the verb *pugnare*.

and only one of the disjuncts is true remains necessary. Providing a satisfactory definition of the crucial Chrysippean notion of "conflict" is a notorious exegetic problem, and one is led to suspect that the attempts at formalizing it simply take a wrong route.[367] Surprisingly enough, the notion has been mainly studied in relation with the Chrysippean definition of a conditional, and evidence from disjunction has been brought to bear on the subject only very recently.

The mathematical exclusive disjunction seems to represent well the Stoic notion of conflict, although Chrysippus probably wanted it transferred to an "empirical" level too. First, statements such as "either A is greater than B or A is equal to B or A is less than B" may well be taken as representative of complete conflict. Second, any mathematical statement has the virtue of being time-independent. In addition, mathematics only deals with formal or analytic statements and mathematical necessity coincides with mathematical truth. These features shortcut a series of problems. First and foremost, the crucial notion of (complete) conflict is readily seen to reduce in mathematics to the usual truth-functional characterization in terms of (exhaustive and) exclusive alternatives. Second, disjunction and conjunction are defined by the Stoics on different grounds: disjunction is not truth-functional while conjunction is, and the Stoic requirements about disjunction make it more restrictive a connector than a mere truth-functional disjunction would be. This fact entails problems as far as duality between disjunction and conjunction is concerned, even if we have seen that duality is not essential to make indirect proofs work. In this respect, the underlying logic at work in mathematical reasoning is quite a simplified version of Stoic logic.

However, we must not forget that the incompatibility of the disjuncts in "either A is greater than B or A is equal to B or A is less than B" is always embodied in some kind of mathematical object: A and B are angles, or straight lines, or more complex figures. On the one hand, thus, there is some structural point to their being incompatible, namely, the very definition of relations such as "being greater than" or "being equal to". On the other hand, these definitions must build on some intuitive grasp on the essential features of the mathematical objects at issue—and usually these features are not borne out by the very definitions of these objects. What counts for a triangle to be contained in (and hence to be less than) another triangle cannot be read from the definition of a triangle, and theorems such as the congruence criteria for triangles simply make the meaning of the notion of equality more articulated. In this perspective, then, we might surmise that Chrysippus and the later Stoics actually took incompatible statements in mathematics as a model for developing the notion of conflict. These statements are analytic in an obvious sense, but they also refer to an underlying substratum of mathematical objects, whose ontological status Chrysippus seems to have considered less abstract than usually done by ancient thinkers.[368]

Two forms of παραδιεζευγμένον "paradisjunction" were introduced after Chrysippus. The transmitted definitions are grounded on semantic properties only,[369] thereby impoverishing the primary notion. In one of the paradisjunctions, at most one of the disjuncts is true (but there can be none); in the other, at least one of the disjuncts is true, but it is not said that it is the only one: these variants correspond to dropping one or the other of the conditions that single out an exhaustive and exclusive disjunction. Therefore, the truth-condition of the second kind of paradisjunction makes it correspond to our inclusive disjunction. The first kind of paradisjunction is also mentioned by Galen (*Inst. Log.* V.1), who calls them παραπλήσια διεζευγμένοις "quasi-disjunctions" and asserts that

[367] See Castagnoli 2004 on this aspect. A complete survey of the interpretations of μάχη, as well as of the complementary notion of συνάρτησις "connexion", is found in this paper.
[368] See for instance Robertson 2004.
[369] Cf. Aulus Gellius, *NA* XIV.8.14. We have read this passage above.

they announce incomplete conflict. He also points out that the quasi-disjunction is usually taken to coincide with the negation of a conjunction—thereby reducing it to a truth-functional definition—but maintains that this is not the right characterization.[370]

The paradisjunction and the quasi-disjunction try to catch relevant features of ordinary language, such as inclusiveness, but appear to be clumsy attempts that deeply misunderstand one of the main motivations of the Chrysippean approach, which apparently intended to give logical dignity to the exclusive and exhaustive disjunction only: para- and quasi-disjunctions are in fact indistinguishable on syntactic grounds from the exclusive and exhaustive disjunction. Maybe the paradisjunctions were defined on truth-functional grounds only, and Galen and others with him wanted to force them into the jargon of conflict.[371]

The paradisjunction, which announces neither conflict nor consequence and seems to be reported by Galen in truth-functional terms only, is dealt with at length by Apollonius Dyscolus. The ancient Greek grammarians paid in fact particular attention, in their classifications of linguistic objects, to the σύνδεσμοι "connectors". Some interesting features characterize Apollonius Dyscolus' account of disjunctions at the beginning of his *On connectors* (references are to the pages of Schneider's edition in *GG* II.1.1):[372]

* Apollonius does not use the notion of conflict to define disjunction, asserting that a disjunction "announces the existence of only one ‹disjunct›, and the refutation of the remaining one or of the remaining ones" (216.14–16). On the other hand, in the context of a digression that concerns the difference made by the Stoics between conflicting and contradictory statements (218.20–219.11), he provides a definition of conflict: "conflicting is what cannot be accepted in the same circumstances" (218.22–23). He introduces instead a distinction between statements that are disjoined κατὰ φύσιν "by nature" and πρὸς καιρόν "incidentally" (216.17–217.15). Incidentally disjoined statements give rise to a disjunction that is certainly non exhaustive and possibly non exclusive; Apollonius observes that the same statements that are disjoined by such an incidental disjunction may also be connected by a conjunction or by a conditional.

* He asserts in fact that, for statements that entail by themselves what kind of logical relationship must obtain between them (217.16–218.7), the three basic connectors are mutually incompatible: what conditionals with connection (συνάρτησις) and conjunctive connectors announce is in conflict with what disjunctive connectors announce; conversely, there is no logical consequence (ἀκολουθία) between the conjuncts or the disjuncts (218.7–19).

* Only one form of paradisjunction is admitted, namely, inclusive disjunction (219.12–220.22). Paradisjunction cannot be established between statements that are in conflict; the disjunctive connector in an inclusive disjunction may be replaced by καί. Apollonius also sets a parallel between paradisjunction and paraconditional (220.13–22), on the grounds that both subsume two basic connectors in their announcements (conjunction and disjunction in the case of a paradisjunction, conjunction and conditional in the case of a paraconditional).

---

[370] Of course, this would entail neglecting the conflict that is inherent in the definition of any disjunction; note, moreover, that Galen always defines the several disjunctions he introduces in modal terms.

[371] Evidence from Galen's *Institutio logica* should not be taken as referring to Stoic doctrines, as Bobzien 2004 has shown. Her interpretation of the Galenic hypothetical statements in terms of relations between nonempty classes, in particular the reading of disjunction in terms of "classes of things which with regard to some feature N is completely dividable into *non-empty* subclasses, and where the division is based on non-contingent properties of the things" (ibid., 70–71), establishes a non-trivial connection between the doctrines expounded in the *Institutio logica* and the subject-matter on which disjunction operates in mathematics.

[372] The issue of the relationships between Galen's and Apollonius' accounts is partly dealt with in Barnes 2005.

* In the *Syntaxis* (I.12, *GG* II.2, 14.6), Apollonius mentions ἤτοι as possibly conjunctive. The same remark is made, concerning inclusive paradisjunction, at *Conj*., 220.10–13; in this passage, Apollonius expressly refuses assimilation, pointing out that a conjunction only posits that both conjuncts are the case, and not either of them.[373]

* The connector ἤ can be either prepositive and postpositive, whereas ἤτοι is prepositive only. The marking feature of disjunction (both inclusive and exclusive) is the presence of ἤ between the disjuncts, "as if it were to keep one statement away from the other" (220.23–221.15).

* In a final recapitulation of the differents kinds of the disjunctive connector ἤ (222.24–223.4), Apollonius introduces a double conditional ἐὰν τοῦτο, οὐ τοῦτο· εἰ τοῦτο μή, τοῦτο "if this, not this; if not this, this" in order to explain what ἤ announces in the case of proper disjunctive meaning, whereas the dissertive (διασαφητικόν; it is discussed at 221.16–222.23)[374] meaning of ἤ announces τοῦτο, οὐ τοῦτο "this, not this". Apollonius adds that this is the reason why no external connector is admitted in this proposition, rather the intensive adverb μᾶλλον.[375]

* Prepositive ἤ or ἤτοι introduces either a disjunction or a paradisjunction, while interpositive ἤ is ambiguous between disjunction and dissertive connector (223.4–22).

In ordinary language, conjunction is more a routine construction than disjunction is;[376] the variety of accepted conjunctive particles and structures makes it almost impossible to single out exclusively one of them for use in mathematical language. Conjunction, although strictly formulaic in its essence, is nevertheless intended to cover a rich supply of states of affairs; for instance, conjunction of predicates is the exclusive way of singling out an object as belonging to some subclass within a wider genus of objects. Such a variety was exploited in mathematics, as we have seen, to convey shades of meaning that allow a remarkable elasticity in the expression of subtle mathematical distinctions. Among the correlative forms, τε … καί provides some emphasis, if any was needed, by singling out both conjuncts. I am led to conclude that the preposed καί of Stoic conjunction is an independent elaboration.

One might wonder why the Stoic doctrine prescribed to prepose a καί when the equivalent resource of postpositive τε was already at work in ordinary language, and it was specifically employed in mathematical practice to forestall possible ambiguities. Several reasons might be envisaged. First, a postpositive τε, since it is an enclitic particle, provides less emphasis than a prepositive καί. Second, maybe the Stoics intended to mark formally correct language by requiring for it a formulation that found very infrequent use in ordinary prose. A third reason I would suggest looks at the most striking common feature of Stoic connectors. All non-simple statements are typified by

---

[373] In Dionysius Thrax, ἤτοι is listed, with ἤ and ἠέ, among conjunctive and disjunctive connectors (*GG* I.1, 88.3–91.1); the definition of disjunctive connectors is διαζευκτικοὶ δέ εἰσιν ὅσοι τὴν μὲν φράσιν ἐπισυνδέουσιν ἀπὸ δὲ πράγματος εἰς πρᾶγμα διιστᾶσιν "disjunctive ‹connectors› are all those which connect the expression but separate an item to [*sic*] an item" (ibid., 90.1–91.1).

[374] This connector is also of Stoic origin, as we may gather from D.L. VII.72–73; Apollonius calls it "a further distinction of disjoining".

[375] Bobzien 1999, 111, suggests that the dissertive connector bears to disjunction the same relationships as paraconditional to conditional, and proposes that the truth conditions of the dissertive connector are identical to "both either the first or the second and the second".

[376] According to Galen, *Inst. Log.* IV.4, conjunction is the weakest connector. Dionysius Thrax (*GG* I.1, 88.3–89.2); defines conjunctions as follows: συμπλεκτικοὶ μὲν οὖν εἰσιν ὅσοι τὴν ἑρμηνείαν ἐπ' ἄπειρον ἐκφερομένην συνδέουσιν "conjunctive ‹connectors› are all those which connect an expression that is carried unboundedly". He lists among others μέν, δέ, τε, καί, ἀλλά, ἤτοι, ἄν.

a connector preposed to their formulation as sentences; negation, while not being a connector, was in canonical position only if preposed to an entire clause.[377] Moreover, the Stoics conceived of three, to some extent complementary, ways of presenting a non-simple statement. The first way, usually clarified by an example, is strictly syntactic and simply specifies which is the involved connective particle. The third is a semantic characterization, which establishes the truth conditions of the statement. The second way is something like a mixture of the other two, and it is usually presented by the clause (cf. the definition of disjunction in Sextus Empiricus read above) "the connector announces that…", which strongly emphasizes the preposed position.[378] By way of conclusion of this long survey, I surmise that the Chrysippean characterization of disjunction, both its being exclusive and its having a preposed ἤτοι, owes much to contemporary mathematical practice. I suspect instead that the rule of preposing καί in conjunctions was suggested to him by an *esprit de système* of sorts, with the goal of making the structures of all non-simple statements similar.[379]

5.3.6. *Syllogistic connectors*

As seen, the Stoic school and the ancient Greek grammarians paid particular attention, in their classifications of linguistic objects, to the σύνδεσμοι "connectors". Dionysius Thrax (*GG* I.1, 95.2–100.1) lists συλλογιστικοί and παραπληρωματικοὶ σύνδεσμοι "syllogistic" and "completive connectors";[380] among syllogistic connectors feature ἄρα "therefore" and, by conflating coassumptive and conclusive particles, ἀλλά "but" and ἀλλαμήν "but of course"; among completive connectors are δή "thus" and οὖν "then". Apollonius Dyscolus (*Conj*., *GG* II.1.1, 252.1–2) regards ἄρα as the main syllogistic particle; he also points out the difference between his own terminology and the Stoics' (συλλογιστικοὶ σύνδεσμοι vs. ἐπιφορικοί "conclusive" ibid., 251.27–29). Apollonius also clarifies the function of argumentative hiatus typical of δή by likening it to a παραγραφή "marginal sign".[381] Finally, οὐκοῦν "and then", frequently used by Aristotle and in Euclid's optical treatises,[382] is categorized by Apollonius as ἐπιλογιστικός "inferential" ibid., 257.18.

The connector ἄρα, invariably in second position within a clause,[383] announces that the sentence in which it is inserted formulates the conclusion of an inference. It is by far the most frequent such connector—4042 occurrences in the whole of the *Elements*—but it does not have this exclusive function: as we have seen in Sect. 5.3.2, it also introduces the consequent of a paraconditional. A

[377] Interferences of ordinary linguistic practice and copying mistakes must be at the root of a preposed καί lacking in many instances of conjunction, even mentioned for technical purposes, reported in our sources on Stoic logic. On the other hand, a preposed ἤτοι is never absent in disjunctions, and a couple of exceptions in the ancient literature are careless integrations of lacunae in the text, in particular those at Apollonius Dyscolus, *Conj*., 216.8 and 217.2. See also the remarks in Frede 1974, 94.

[378] See D.L. VII.71–72 and Sextus, *P* II.148, 162, 189, 191. The role of the second way of presentation is not limited to what is suggested here.

[379] A different view on this is argued in Brunschwig 1978.

[380] On completive connectors in Apollonius Dyscolus see Dalimier 1999.

[381] *Conj*., *GG* II.1.1, 253.15, and cf. 251.19–23 and *Synt*., *GG* II.2, 379.8 and 380.16.

[382] There are 21 occurrences in *Optica* redaction **A**, 58 in *Optica* **B**, 23 in *Catoptrica*, 5 in the *Sectio canonis*. No occurrences in Apollonius, just 2 in Archimedes, *Sph. cyl*. I.38 and *Quadr*. 16, in *AOO* I, 140.5, and II, 296.17, respectively. As for Aristotle, there are 21 occurrences in arguments related to the syllogistic figures, in *APr*., 26a24, 26b9, 36b38, 38a23, 42a9, 42a13, 45a29, 50b36, 51a4, 51a14, 59b21, 59b29, 60a1, 60b11, 61b12, 61b27, 62a25, 62b16, 63a9, 63a26, 63a41; 3 in *APo*., 79a39, 79b30, 82b20; 9 in *Ph*., 232a33, 233b25, 234a27, 236b11, 237a21, 237a31, 240b23, 248b2, 266b12; 1 in *Cael*., 275a16; 1 in *GC* 332b20; 1 in *Sens*. 448b6; 2 in *Metaph*., 1047b18 and 1092b35.

[383] On the position of ἄρα see Sect. 1.1.

further type of non-canonical ἄρα introduces a sentence that formulates a statement that immediately follows from a constructive act; this is in I.44 (*EOO* I, 104.1–8):

ἐκβεβλήσθωσαν καὶ συμπιπτέτωσαν κατὰ τὸ K, καὶ διὰ τοῦ K σημείου ὁποτέρᾳ τῶν EA ZΘ παράλληλος ἤχθω ἡ KΛ, καὶ ἐκβεβλήσθωσαν αἱ ΘA HB ἐπὶ τὰ Λ M σημεῖα. παραλληλόγραμμον ἄρα ἐστὶ τὸ ΘΛKZ διάμετρος δὲ αὐτοῦ ἡ ΘK περὶ δὲ τὴν ΘK παραλληλόγραμμα μὲν τὰ AH ME τὰ δὲ λεγόμενα παραπληρώματα τὰ ΛB BZ.

Let them be produced and let them meet at K, and through point K let a ‹straight line›, KΛ, be drawn parallel to either of EA, ZΘ, and let ΘA, HB be produced as far as points Λ, M. Therefore there is a parallelogram, ΘΛKZ, and a diagonal of it, ΘK, and about ΘK parallelograms, AH, ME, and the so-called complements, ΛB, BZ.

The frequency of ὥστε "so that" is particularly uneven in the *Elements*: just a few occurrences in Books III and VII–VIII, 134 out of 290 in Book X—of which it is an obvious stylistic marker—and similar frequencies in Books II, IV, and IX. Moreover, in Book X the classification of the irrationals by addition (X.36–72) exhibits 74 occurrences, that of the irrationals by removal (X.73–110) only 14. The following table sets out the distribution of the occurrences in the *Elements*:[384]

	I	II	III	IV	V	VI	VII	VIII	IX	X	XI	XII	XIII	tot.
# prop.	48	14	37	16	25	33	39	27	36	115	39	18	18	465
% # signs	7.6	3.3	6.9	3.4	4.9	7.6	5.7	5	5.1	26.1	9	8.3	6.9	100
ὥστε	12	12	4	14	8	24	6	4	22	134	16	16	18	290
% ὥστε	4.1	4.1	1.4	4.8	2.8	8.3	2	1.4	7.6	46.2	5.6	5.6	6.2	100

This connector is employed in a variety of ways. It figures as a subordinant in subinferential consecutive clauses, ὥστε "so as" + infinitive, where it specifies a relevant property of the object at issue, as in the enunciation of II.11 (*EOO* I, 152.5–8):

τὴν δοθεῖσαν εὐθεῖαν τεμεῖν ὥστε τὸ ὑπὸ τῆς ὅλης καὶ τοῦ ἑτέρου τῶν τμημάτων περιεχόμενον ὀρθογώνιον ἴσον εἶναι τῷ ἀπὸ τοῦ λοιποῦ τμήματος τετραγώνῳ.

Cut a given straight line so as to be the rectangle contained by the whole and by one of the segments equal to the square on the remaining segment.

or in constructs that are frequently used in Book X—here we read them in propositions X.48 and X.63 (*EOO* III, 136.22–26 and 192.5–6):

ἐκκείσθωσαν δύο ἀριθμοὶ οἱ AΓ ΓB ὥστε τὸν συγκείμενον ἐξ αὐτῶν τὸν AB πρὸς μὲν τὸν BΓ λόγον ἔχειν ὃν τετράγωνος ἀριθμὸς πρὸς τετράγωνον ἀριθμὸν πρὸς δὲ τὸν ΓA λόγον μὴ ἔχειν ὃν τετράγωνος ἀριθμὸς πρὸς τετράγωνον ἀριθμόν.

Let two numbers, AΓ, ΓB, be set out so as to have the ‹number› compounded of them, AB, to BΓ a ratio that a square number ‹has› to a square number and so as not to have to ΓA a ratio that a square number ‹has› to a square number.

ἔστω μείζων ἡ AB διῃρημένη κατὰ τὸ Γ ὥστε μείζονα εἶναι τὴν AΓ τῆς ΓB.

Let there be a major, AB, that turns out to be divided at Γ so as to be AΓ greater than ΓB.

[384] Add 14 occurrences in the alternative proofs and 1 in recension **b**. In the rest of the Euclidean corpus the frequency is only approximately proportional to the size and to the deductive density of a treatise (31 occurrences in the *Data*). A number of items are found in the *Phaenomena* (21, distributed between the two redactions), less in recension **A** of the *Optica* than in recension **B** (16 occurrences vs. 39).

In the construction of X.48, ὥστε is constructed ἀπὸ κοινοῦ; the two consecutive clauses it governs are correlated by μέν … δέ ….

However, ὥστε is mainly an inferential connective that marks the conclusion of a deduction; it has exactly the same function as ἄρα.[385] A very simple example (an inference by transitivity) is in proposition IX.17 (*EOO* II, 380.18–19):

μετρεῖ ἄρα καὶ ὁ Γ τὸν Δ· ἀλλ' ὁ Α τὸν Γ ἐμέτρει· ὥστε ὁ Α καὶ τὸν Δ μετρεῖ.

Therefore Γ also measures Δ; but A measured Γ; so that A also measures Δ.

Still, a detailed survey shows that ὥστε is typical of "marginal", and quite neatly defined, deductive contexts. In particular, ὥστε is found when the conclusion of an inference is followed by another conclusion (two consecutive ἄρα are very infrequent), as in IX.4 (*EOO* III, 346.22–14):

τῶν Α Β ἄρα δύο μέσοι ἀνάλογον ἐμπίπτουσιν ἀριθμοί· ὥστε καὶ τῶν Δ Γ δύο μέσοι ἀνάλογον ἐμπεσοῦνται ἀριθμοί.

Therefore between A, B two numbers fall as mean proportionals; so that between Δ, Γ two numbers will also fall as mean proportionals.

Here as often elsewhere, the omitted coassumption would simply consist in a direct reference to a previous theorem, in this case VIII.8. The conclusions introduced by ὥστε are sometimes so evident—even trivially evident—as not to require the presence of a coassumption; let us read an extract from X.81 (*EOO* III, 242.17–18):

λοιπὸν ἄρα τὸ ΕΛ ἴσον ἐστὶ τῷ ἀπὸ τῆς ΑΒ· ὥστε ἡ ΑΒ δύναται τὸ ΕΛ.

Therefore ΕΛ as a remainder is equal to the ‹square› on AB; so that AB is worth ΕΛ.

Thus, the second of two consecutive sentences marked as conclusions is often introduced by ὥστε. Within this genus, it often happens that the first of the two conclusions is the consequent of a paraconditional (see Sect. 5.3.2); it does not matter whether the paraconditional is pure or hybrid, as we can see in the two consecutive occurrences in X.60 (*EOO* III, 184.12–18):

καὶ ἐπεὶ σύμμετρόν ἐστι τὸ ἀπὸ τῆς ΑΓ τῷ ἀπὸ τῆς ΓΒ, σύμμετρόν ἐστι καὶ τὸ ΔΘ τῷ ΚΛ· ὥστε καὶ ἡ ΔΚ τῇ ΚΜ σύμμετρός ἐστιν. καὶ ἐπεὶ μείζονά ἐστι τὰ ἀπὸ τῶν ΑΓ ΓΒ τοῦ δὶς ὑπὸ τῶν ΑΓ ΓΒ, μεῖζον ἄρα καὶ τὸ ΔΛ τοῦ ΜΖ· ὥστε καὶ ἡ ΔΜ τῆς ΜΗ μείζων ἐστίν.

And since the ‹square› on ΑΓ is commensurable with that on ΓΒ, ΔΘ is also commensurable with ΚΛ; so that ΔΚ is also commensurable with ΚΜ. And since those on ΑΓ, ΓΒ are greater than twice the ‹rectangle contained› by ΑΓ, ΓΒ, therefore ΔΛ is also greater than ΜΖ; so that ΔΜ is also greater than ΜΗ.

There is just one instance of ὥστε serving to introduce the general conclusion of an entire theorem; it is found in X.66 (*EOO* III, 202.23–204.2):

ὥστε ἡ τῇ ἐκ δύο ὀνομάτων μήκει σύμμετρος ἐκ δύο ὀνομάτων ἐστὶ καὶ τῇ τάξει ἡ αὐτή, ὅπερ ἔδει δεῖξαι.

So that a ‹straight line› commensurable in length with a binomial is also a binomial and the same in order, which it was really required to prove.

[385] Therefore, in this function ὥστε must be preceded by an upper point, not by a comma. Of course, ὥστε as a sentential adverb keeps the same position as the position it has as a conjunction: it is the very first word in a sentence.

Two adjacent ὥστε are exceedingly rare; here is X.55 (*EOO* III, 166.16–18):

αἱ ΒΑ ΑΗ ΗΕ ἄρα ῥηταί εἰσι δυνάμει μόνον σύμμετροι· ὥστε μέσον ἐστὶν ἑκάτερον τῶν ΑΘ ΗΚ· ὥστε καὶ ἑκάτερον τῶν ΣΝ ΝΠ μέσον ἐστίν.

Therefore BA, AH, HE are expressibles commensurable in power only; so that each of AΘ, HK is medial; so that each of ΣN, NΠ is also medial.

Sometimes, ὥστε introduces a conclusion that immediately follows an analogical proof or a potential proof, as in I.36 (*EOO* I, 88.4–6):

διὰ τὰ αὐτὰ δὴ καὶ τὸ ΕΖΗΘ τῷ αὐτῷ τῷ ΕΒΓΘ ἐστιν ἴσον· ὥστε καὶ τὸ ΑΒΓΔ παραλληλόγραμμον τῷ ΕΖΗΘ ἐστιν ἴσον.

For the very same ‹reasons› EZHΘ is also equal to the same EBΓΘ; so that parallelogram ABΓΔ is also equal to EZHΘ.

Other "marginal" inference patterns with ὥστε have it preceded by a postposed explanation (see Sect. 4.5.3), as in IV.10 (*EOO* I, 296.18–21),

καὶ ἡ ὑπὸ ΒΔΑ ἄρα ἴση ἐστὶ τῇ ὑπὸ ΒΓΔ· ἀλλὰ ἡ ὑπὸ ΒΔΑ τῇ ὑπὸ ΓΒΔ ἐστιν ἴση – ἐπεὶ καὶ πλευρὰ ἡ ΑΔ τῇ ΑΒ ἐστιν ἴση –· ὥστε καὶ ἡ ὑπὸ ΔΒΑ τῇ ὑπὸ ΒΓΔ ἐστιν ἴση.

Therefore BΔA is also equal to BΓΔ; but BΔA is equal to ΓBΔ—since a side AΔ is also equal to AB —; so that ΔBA is also equal to BΓΔ.

or include the deductive chains in which the second of two conclusions entails a shift of relational operator—for instance, when a quantity is moved to the other side of an equality, as in the following extract from II.13 (*EOO* I, 158.26–30):

τὰ ἄρα ἀπὸ τῶν ΓΒ ΒΑ ἴσα ἐστὶ τῷ τε ἀπὸ τῆς ΑΓ καὶ τῷ δὶς ὑπὸ τῶν ΓΒ ΒΔ· ὥστε μόνον τὸ ἀπὸ τῆς ΑΓ ἔλαττόν ἐστι τῶν ἀπὸ τῶν ΓΒ ΒΑ τετραγώνων τῷ δὶς ὑπὸ τῶν ΓΒ ΒΔ περιεχομένῳ ὀρθογωνίῳ.

Therefore those on ΓB, BA are equal both to that on AΓ and to twice that by ΓB, BΔ; so that that on AΓ alone is less than the squares on ΓB, BA by twice the rectangle contained by ΓB, BΔ.

To sum up, ὥστε is preferably (but not exclusively) employed in deductions that are not canonical inferences with two premises and one conclusion. However, the example of proposition X.109 (*EOO* III, 346.11–20) makes it clear that the difference between a fully-fledged inferential structure and an enthymeme, in arguments whose conclusion is introduced by ὥστε, may be dictated by stylistic choices or even be the consequence of later revisions—the two deductions below are identical: the first of them makes the coassumption explicit, the second of them does not:[386]

εἰ μὲν οὖν ἡ ΘΖ τῆς ΖΚ μεῖζον δύναται τῷ ἀπὸ συμμέτρου ἑαυτῇ καί ἐστιν ἡ προσαρμόζουσα ἡ ΖΚ σύμμετρος τῇ ἐκκειμένῃ ῥητῇ μήκει τῇ ΖΗ, ἀποτομὴ δευτέρα ἐστὶν ἡ ΚΘ· ῥητὴ δὲ ἡ

Then if ΘZ is worth more than ZK by the ‹square› on a ‹straight line› commensurable with itself and the fitting ‹straight line› ZK is commensurable in length with the set out expressible ZH, KΘ is a second apotome; and ZH

[386] In the second deduction, ἑαυτῇ "with itself" and the identification of objects "that is EΓ" are also omitted.

ZH· ὥστε ἡ τὸ ΛΘ, τουτέστι τὸ ΕΓ, δυναμένη μέσης ἀποτομὴ πρώτη ἐστίν.
εἰ δὲ ἡ ΘΖ τῆς ΖΚ μεῖζον δύναται τῷ ἀπὸ ἀσυμμέτρου καί ἐστιν ἡ προσαρμόζουσα ἡ ΖΚ σύμμετρος τῇ ἐκκειμένῃ ῥητῇ μήκει τῇ ΖΗ, ἀποτομὴ πέμπτη ἐστὶν ἡ ΚΘ· ὥστε ἡ τὸ ΕΓ δυναμένη μετὰ ῥητοῦ μέσον τὸ ὅλον ποιοῦσά ἐστιν.

is expressible; so that the ‹straight line› worth ΛΘ, that is ΕΓ, is a first apotome of a medial.
And if ΘΖ is worth more than ΖΚ by the ‹square› on a ‹straight line› incommensurable and the fitting ‹straight line› ΖΚ is commensurable in length with the set out expressible, ΚΘ is a fifth apotome: so that the ‹straight line› worth ΕΓ is a ‹straight line› that produces with an expressible ‹region› a medial whole.

The following tables set out the distribution, in the main text of the *Elements*, of the clauses introduced by ὥστε, according to the typology just expounded. An asterisk * marks the occurrences that fall under two headings; the number of the associated proposition is in this case given twice; the numbers of the occurrences for every type are provided in brackets after each heading. The occurrences in Book X are set out in the second table:

Books I–IX and XI–XIII

consecutive (36)	I.7, 23, 44, II.11 (*quater*), III.4, 35, IV.10–12, V.13, 17–18, VI.10, 15, 19 (*bis*), 20, 23, 32 (*bis*), VIII.5, IX.19, 21–22, XI.23, 27, XII.17, XIII.13, 15–17, 18 (*bis*)
inferential (20)	I.4, 41*, IV.10*, V.8, 11, VI.20por (*bis*), 23, VIII.25, IX.1, 15[387], 17, 34, XI.30*, 33, XII.3 (*bis*), 10, 17, 17/18
avoids double ἄρα (76)	I.3–4, 8, 48, II.4 (*bis*), 9 (*bis*), 10 (*bis*), 12*, 13*, III.8, 12*, IV.8–10, 14–15, V.6, 12, 14, VI.3 (*bis*), 7 (*bis*), 20 (*ter*), 21, 22/23, 27*, VII.4, 9, VII.36–37, VIII.15, 20, IX.4, 10–12, 13 (*bis*), 17–19, 29–30, XI.4–5, 8 (*bis*), 11, 18, 23/24*, 28–29, 36, XII.2, 2/3, 4/5, 10, 12 (*bis*), 15, 17, XIII.3–4, 8, 10 (*bis*), 15 (*bis*), 16, 17 (*bis*)
avoids double concl. (8)	I.4*, VII.4, XI.23, 29, XII.2*, 10*, XIII.3, 7
follows analogical (9)	I.35–36, IV.7, 12, VI.31–32, VII.10, IX.24, 26
follows potential (3)	IV.5 (*bis*), VI.7
follows postposed (9)	I.4*, 41*, IV.10*, IX.33, XI.28*, 30*, XII.2* (*bis*), 3*, 10*
change of rel. op. (5)	II.12*, 13*, III.12*, VI.27*, 23/24*

Book X

consecutive (49)	12, 13/14, 28/29I (*bis*), 28/29II, 29 (*ter*), 30 (*bis*), 31 (*bis*), 32 (*ter*), 33, 34–35, 42, 43 (*bis*), 44 (*bis*), 45 (*bis*), 46 (*bis*), 47 (*bis*), 48–53, 53/54, 54, 55 (*bis*), 58–60, 62–64, 84, 88–89, 113
inferential (15)	4, 9/10 (*ter*), 17, 23/24, 34, 53, 66, 72, 72/73, 93, 109, 110, 112
avoids double ἄρα (56)	17*, 17 (*bis*), 18 (*ter*), 21, 24, 33 (*bis*), 35 (*bis*), 36, 38 (*bis*), 39 (*bis*), 40, 41/42, 44 (*bis*), 47–49, 51–52, 54 (*bis*), 55 (*bis*), 59/60, 60 (*bis*), 61 (*bis*), 62 (*bis*), 63–65, 71 (*sexties*), 78, 81, 91, 93 (*bis*), 99, 112 (*bis*), 113 (*bis*)
avoids double concl. (9)	4, 28/29II, 41, 55 (*bis*), 60–62, 109
follows analogical (0)	/
follows potential (4)	56, 69–70, 106
follows postposed (0)	/
change of rel. op. (1)	17*

[387] The clause is erroneously printed twice: see *EOO* II, 376.5–7.

The connectors οὖν "then" and δή "thus", "now" are in charge of weaker forms of transition than those marked by ἄρα or ὥστε. The connector οὖν "then" (238 occurrences in the main text of the *Elements*) is typically liminal, with a prominent role as a scope particle: it canonically opens the entire proof, or partial proofs, or at least fresh deductive chains. The most frequent location of οὖν is to accompany the connector that introduces the paraconditional that initializes a proof, in the nexus ἐπεὶ οὖν "then since": as seen in Sect. 5.3.2, this nexus exhibits 164 occurrences in the main text of the *Elements*. The liminal character of οὖν is again prominent in syntagms such as εἰ οὖν "then if", which introduces a division into cases, and in similar nexuses we shall see presently. There are (a few) examples of simple sentences or of suppositions introduced by οὖν: in these cases, the particle marks the liminal character of such linguistic units.

It is not immediately obvious what we have to mean by "liminal" in this context. In fact, a Greek mathematical proposition admits of some elasticity in identifying the borders, *normally* quite well-defined, between its own specific parts. Such a phenomenon crucially, and more frequently, affects the beginning of the proof (cf. Sect. 4.3). We have thus to provide supplementary criteria for determining where a deductive chain starts. Thus, "liminal" is what follows suppositions formulated in the imperative, such as the several steps that complete the construction, even if the construction contains some inferential steps, as in IV.11 (read in Sect. 4.3), or even if the construction has a complex structure, like that of the constructions in VI.27–28. The same criterion must be applied to the inferences that follow the "fictitious" construction or the dichotomy typical of a *RI*. In cases such as IX.1 or X.25–26, the initial paraconditional must be regarded as liminal by logic, even if strictly speaking it is preceded by steps that are a consequence of the construction, as in the following extract from IX.1 (*EOO* II, 340.8–12):

ὁ γὰρ Α ἑαυτὸν πολλαπλασιάσας τὸν Δ ποιείτω· ὁ Δ ἄρα τετράγωνός ἐστιν. ἐπεὶ οὖν ὁ Α ἑαυτὸν μὲν πολλαπλασιάσας τὸν Δ πεποίηκεν τὸν δὲ Β πολλαπλασιάσας τὸν Γ πεποίηκεν, ἔστιν ἄρα ὡς ὁ Α πρὸς τὸν Β οὕτως ὁ Δ πρὸς τὸν Γ.

In fact, let A multiplying itself make Δ; therefore Δ is a square. Then since A multiplying itself turns out to make Δ and multiplying B turns out to make Γ, therefore, as A is to B, so Δ is to Γ.

Conversely, the linguistic units preceded by units of the same sentential type in a given specific part of a proposition are certainly "non-liminal". In particular, no paraconditional can be liminal that is preceded by at least another paraconditional within a proof. This is the only large category of paraconditional that provides us with occurrences of οὖν within a proof.

Most of such non-liminal paraconditionals introduced by οὖν fall under at least one of three subcategories, all of which mark a more or less strong intermission of the deductive flow: these paraconditionals introduce independent deductive sequences, which possibly wind up a complex of arguments and constructive acts—this means that the connector does not have an inferential role, but a metadeductive function, serving in fact as a scope particle.

The three subcategories are:

1) Paraconditionals located at the very end of a proof, which recapitulate several deductions and possibly contain an instantiated citation, as in X.99 (*EOO* III, 316.26–318.4):

ἐπεὶ οὖν δύο εὐθεῖαι ἄνισοί εἰσιν αἱ ΓΜ ΜΖ καὶ τῷ τετάρτῳ μέρει τοῦ ἀπὸ τῆς ΖΜ ἴσον παρὰ τὴν ΓΜ παραβέβληται ἐλλεῖπον εἴδει τετραγώνῳ καὶ εἰς σύμμετρα αὐτὴν διαιρεῖ, ἡ ΓΜ ἄρα τῆς ΜΖ μεῖζον δύναται τῷ ἀπὸ συμμέτρου ἑαυτῇ· καὶ οὐδετέρα τῶν ΓΜ ΜΖ σύμμετρός ἐστι μήκει τῇ ἐκκειμένῃ ῥητῇ τῇ ΓΔ· ἡ ἄρα ΓΖ ἀποτομή ἐστι τρίτη.

Then since there are two unequal straight lines, ΓΜ, ΜΖ, and a ‹rectangle› equal to the fourth part of the ‹square› on ΖΜ turns out to be applied on ΓΜ falling short of a square form and divides it in commensurable ‹segments›, therefore ΓΜ is worth more than ΜΖ by the ‹square› on a ‹straight line› commensurable with itself; and neither of ΓΜ, ΜΖ is commensurable in length with the set out expressible ΓΔ; therefore ΓΖ is a third apotome.

2) Paraconditionals that re-initialize the deductive flow after the hiatus represented by an analogical proof or a potential proof, as in VIII.11 (*EOO* II, 304.4–8)—this paraconditional is immediately followed by another analogical proof:

διὰ τὰ αὐτὰ δὴ καὶ ὁ Δ ἑαυτὸν πολλαπλασιάσας τὸν Β πεποίηκεν. ἐπεὶ οὖν ὁ Γ ἑκάτερον τῶν Γ Δ πολλαπλασιάσας ἑκάτερον τῶν Α Ε πεποίηκεν, ἔστιν ἄρα ὡς ὁ Γ πρὸς τὸν Δ οὕτως ὁ Α πρὸς τὸν Ε.

For the very same ‹reasons› Δ multiplying itself also turns out to make Β. Then since Γ multiplying Γ, Δ turns out to make Α, Ε, respectively, therefore, as Γ is to Δ, so Α is to Ε.

3) Paraconditionals that suspend the deductive flow to introduce a particularly explicit instantiated citation that serves as local micro-recapitulation, as in XI.23 (*EOO* IV, 66.1–4)—citation of I.25; this step is immediately followed by a potential proof:

ἐπεὶ οὖν δύο αἱ ΑΒ ΒΓ δυσὶ ταῖς ΟΞ ΞΠ ἴσαι εἰσὶν καὶ βάσις ἡ ΑΓ βάσεως τῆς ΟΠ μείζων ἐστίν, γωνία ἄρα ἡ ὑπὸ ΑΒΓ γωνίας τῆς ὑπὸ ΟΞΠ μείζων ἐστίν.

Then since two ‹sides›, ΑΒ, ΒΓ, are equal to two ‹sides›, ΟΞ, ΞΠ, and base ΑΓ is greater than base ΟΠ, therefore angle ΑΒΓ is greater than angle ΟΞΠ.

Add to these the species in which μὲν οὖν "then now" introduces a sentence that formulates a simple statement,[388] in the form of a coassumption as in V.10 or in the form of a conclusion, as in XIII.18 or in IX.14 (*EOO* II, 372.25–374.3):

οἱ Β Γ Δ ἄρα ἕνα τῶν Ε Ζ μετρήσουσιν. τὸν μὲν οὖν Ε οὐ μετρήσουσιν – ὁ γὰρ Ε πρῶτός ἐστι καὶ οὐδενὶ τῶν Β Γ Δ ὁ αὐτός –· τὸν Ζ ἄρα μετροῦσιν ἐλάσσονα ὄντα τοῦ Α.

Therefore Β, Γ, Δ will measure one of Ε, Ζ. Then now, they do not measure Ε—for Ε is prime and the same as none of Β, Γ, Δ—; therefore they measure Ζ that is less than Α.

In the same way as ὥστε, the particle οὖν is an obvious stylistic marker of Book X, where we find almost half of its occurrences. The distribution in the main text of the *Elements* is set out in the following table:[389]

---

[388] What makes μὲν οὖν a relevant nexus in our present perspective is that there is no subsequent δέ that responds to μέν.
[389] Add the 23 occurrences in the alternative proofs and the 9 occurrences in recension **b**. The rest of the Euclidean corpus is less generous: 28 occurrences out of 94 propositions in the *Data* (plus 10 in the alternative proofs); the particle is a stylistic marker of *Optica* **A**, with 48 occurrences vs. 24 of recension **B**.

	I	II	III	IV	V	VI	VII	VIII	IX	X	XI	XII	XIII	tot.
# prop.	48	14	37	16	25	33	39	27	36	115	39	18	18	465
% # signs	7.6	3.3	6.9	3.4	4.9	7.6	5.7	5	5.1	26.1	9	8.3	6.9	100
οὖν	15	5	22	7	10	19	14	5	13	104	11	7	6	238
% οὖν	6.3	2.1	9.2	2.9	4.2	8	5.9	2.1	5.5	43.7	4.6	2.9	2.5	100

In the subsequent tables, I set out the distribution, in the main text of the *Elements*, of the occurrences of οὖν according to the typology expounded above. Occurrences marked by * are within recapitulative paraconditionals; those marked by + follow an analogical proof or a potential proof; those marked by § are in the form of an instantiated citation of a theorem or of a definition; the numbers of the occurrences for every type are provided in brackets after each heading. The occurrences in Book X are set out in the second table.

Books I–IX and XI–XIII

liminal parac. (74)	I.2, 5–7, 14, 16, 19–20, 23–26, II.6–8, 14, III.2, 4, 6, 7–9, 10–12, 13, 14, 18–20, 22–23, 25, 33–34, IV.1–2, 7, 11, V.7–8, 25, VI.9, 11–12, 14–15, 19, 22, 26–28, 31, 33, VII.1, 2, 3, 19, 21, 28, 30, VIII.25, IX.1, 6–7, 32, XI.12, 23/24, XII.6, 11, 14, XIII.4, 6, 13/14
supposition (6)	VI.28, VII.3, VIII.21, XI.34, XII.2, 17
simple statem. (4)	V.10 (*bis*), IX.14, XIII.18
εἰ μὲν οὖν (14)	I.13, II.14, III.35, IV.1, VI.28, VII.2, 4, 32–33, IX.19, XI.11, 20, 22, 34
ὅτι μὲν οὖν (6)	VIII.6, IX.9 (*bis*), 32–34
λέγω οὖν (1)	VIII.18
ἤτοι οὖν … ἤ (1)	IX.19
ἐὰν οὖν (1)	XI.23/24
non-liminal parac. (27)	I.5*, 43+, III.33§, 35§, IV.10§, 10, V.3*§, 6, 22+§, 23§–24§, VI.1*, 4*–5+, 20, 23*§, VII.27§ (*bis*), VIII.11+, IX.8, XI.15+§, 23§, 35+, XII.12§–13, XIII.11+, 11§ (*bis*)

Book X

liminal parac. (46)	2, 3 (*bis*), 4 (*bis*), 5–6, 9, 12, 15 (*bis*), 16 (*bis*), 21/22, 23, 25–26, 28/29II, 29, 43, 46, 59/60, 64, 83, 86, 87 (*bis*), 88–89, 90 (*bis*), 91, 92 (*bis*), 93 (*bis*), 94 (*ter*), 96, 102–103, 108, 111–113
supposition (27)	def.3, 1, 4, 48, 50–54, 87–88, 90, 92 (*bis*), 93 (*bis*), 94 (*ter*), 95 (*bis*), 96 (*bis*), 99–102
simple statem. (0)	/
εἰ μὲν οὖν (6)	3, 66, 103, 109, 112–113
λέγω οὖν (1)	28/29II
δεικτέον οὖν (1)	18
ἤτοι οὖν … ἤ (1)	71
εἴτε οὖν … εἴτε (3)	14, 67, 104
non-liminal parac. (19)	38, 63 (*bis*), 64, 91–92*, 94*, 96–97*§, 98, 98*§, 99*§–101*§, 105, 111 (*bis*), 111/112, 112*§

Besides stylistic choices, the high frequency of οὖν in Book X is induced by the articulated structure of many of the propositions there contained. In fact, the propositions in this Book freely alternate constructive acts and argumentative chains, even beyond the constraints imposed by their enunciation, which in itself often dictates a multi-layered proof. The "proof" of a proposition of Book X, thus, often proceeds by stops and fresh starts, and any such start is canonically marked by οὖν. This fact also explains the number of constructional suppositions introduced by this particle in Book X, whereas this kind of supposition is absolutely scarce in the rest of the *Elements*; in three such instances (VI.28, XII.2, 17) the propositions involved have quite complex a deductive structure as well. The phenomenon just outlined is particularly conspicuous in the final segment of Book X: 50 occurrences of οὖν out of the 104 in Book X are distributed in the string X.91–113, with a peak of 24 occurrences in the hexad X.91–96; the corresponding hexad in the classification of the irrationals by addition, X.54–59, only exhibits 2 occurrences.

As usual, in the *Data* the language is less varied; almost all non-liminal paraconditionals are recapitulative or contain particularly explicit instantiated citations—the meaning of the signs is the same as in the previous table:

liminal parac. (13)	1, 3–4, 40–41, 56, 66, 68, 71, 74, 80, 91–92
εἰ μὲν οὖν (1)	69
non-liminal parac. (14)	32§, 35*§–38*§, 43*§, 47§–48, 62*§, 75*§, 77*+, 84§–85§, 89*§

The connector δή "thus", "now" exhibits 703 occurrences in the main text of the *Elements*. Its value is emphatic and weakly resultative: it amounts to stating more an obvious fact than a conclusion of a deduction.[390] The connector can often be found marking the transition to a new stage of a proof, especially if a further mathematical fact that figures in the enunciation is proved. This transition may take the form of a further determination, introduced by λέγω δή "I now claim" (96 occurrences), or by its variant δεικτέον δή "now it must be proved" typical of Book X,[391] or by πάλιν δή "now, again" (9 occurrences) and ἀλλὰ δή "but now" (43 occurrences). The particle simply has an emphatic function in formulaic expressions like διὰ τὰ αὐτὰ δή "for the very same ‹reasons›" (131 occurrences) or ὁμοίως δή "very similarly" (110 occurrences), which introduce analogical proofs or potential proofs, respectively. In these syntagms, the particle amounts to a zero-grade stylistic marker, since Greek language requires a particle at the beginning of such a kind of argumentative break.[392] The particle has a mildly resultative value in the canonical clause δεῖ δή "thus it is required" (71 occurrences), which introduces the determination in those problems in which a determination is present.[393]

The connector δή acquires its strongest resultative value when a claim is made to the effect that some objects fit the conditions of a previous theorem. Among such theorems, the congruence criteria of triangles proved in Book I have a prominent role. In this case, the references are usually introduced by an ordinal followed by the connector: δύο δή "thus two".[394] As noted in Sect. 5.3.2, the claim that the conditions of the congruence criteria are fulfilled rests upon an obvious inference. Of the same kind are the 3 instantiated citations of I.14 in I.45, 47, and VI.32, introduced by πρὸς δή τινι εὐθείᾳ " thus … to some straight line". An analogous connotation of stating something obvious have the 8 occurrences[395] of clauses introduced by ἴσον δή ἐστι "thus … is equal"; these immediately follow a construction and infer a conclusion that is evident on the grounds of the geometric configuration just generated.[396] Let us read II.2 (*EOO* I, 122.4–7):[397]

[390] See also van der Pas 2014 on this connector.
[391] There are 11 occurrences, in X.49, 53, 55–56, 57 (*bis*), 58, 60–63.
[392] Likewise, we put a full stop before the break, and sometimes we feel that making a paragraph is required.
[393] For there are problems in which the determination is absent, as we have seen in Sect. 4.2.1. See again that Section for the two types of "determination" in which δεῖ δή occurs.
[394] There are 59 occurrences, to which add 1 citation of VI.6 in XII.1 and 1 of XI.10 in XI.24.
[395] In I.41, II.1–3, XI.30, 31 (*bis*), 32.
[396] Add the clause with existential import that opens the proof of XI.1.
[397] The straight line ἡ ΓΖ simply divides the square τὸ ΑΕ in two regions τὰ ΑΖ ΓΕ, so that τὸ ΑΕ is obviously equal to the sum of τὸ ΑΖ and τὸ ΓΕ: this much states the final clause.

ἀναγεγράφθω γὰρ ἀπὸ τῆς ΑΒ τετράγωνον τὸ ΑΔΕΒ, καὶ ἤχθω διὰ τοῦ Γ ὁποτέρᾳ τῶν ΑΔ ΒΕ παράλληλος ἡ ΓΖ. ἴσον δή ἐστι τὸ ΑΕ τοῖς ΑΖ ΓΕ.	In fact, let a square, ΑΔΕΒ, be described on ΑΒ, and through Γ let a ‹straight line›, ΓΖ, be drawn parallel to either of ΑΔ, ΒΕ. Thus ΑΕ is equal to ΑΖ, ΓΕ.

Making what is obvious explicit is also the function of a small constellation of statements;[398] let us read those in propositions VII.22, XI.13, XII.5 (*EOO* II, 234.22–236.4, and IV, 36.12–16 and 166.2–6) within their context:

ἐπεὶ ὁ Γ τὸν Α μετρεῖ κατὰ τὰς ἐν τῷ Δ μονάδας, ὁ Γ ἄρα τὸν Δ πολλαπλασιάσας τὸν Α πεποίηκεν. διὰ τὰ αὐτὰ δὴ καὶ ὁ Γ τὸν Ε πολλαπλασιάσας τὸν Β πεποίηκεν. ἀριθμὸς δὴ ὁ Γ δύο ἀριθμοὺς τοὺς Δ Ε πολλαπλασιάσας τοὺς Α Β πεποίηκεν.	Since Γ measures Α according to the units in Δ, therefore Γ multiplying Δ turns out to make Α. For the very same reasons Γ multiplying Ε turns out to make Β. Thus a number, Γ, multiplying two numbers, Δ, Ε, turns out to make Α, Β.
ἀπὸ τοῦ αὐτοῦ σημείου τοῦ Α τῷ ὑποκειμένῳ ἐπιπέδῳ δύο εὐθεῖαι αἱ ΑΒ ΑΓ πρὸς ὀρθὰς ἀνεστάτωσαν ἐπὶ τὰ αὐτὰ μέρη, καὶ διήχθω τὸ διὰ τῶν ΒΑ ΑΓ ἐπίπεδον. τομὴν δὴ ποιήσει διὰ τοῦ Α ἐν τῷ ὑποκειμένῳ ἐπιπέδῳ εὐθεῖαν.	From a same point Α let two straight lines, ΑΒ, ΑΓ, be erected on the same side at right ‹angles› with the underlying plane, and let the plane through ΒΑ, ΑΓ be drawn through. Thus it will make a rectilinear section through Α in the underlying plane.
καὶ διῃρήσθω ἡ ΔΕΖΘ πυραμὶς εἴς τε δύο πυραμίδας ἴσας ἀλλήλαις καὶ ὁμοίας τῇ ὅλῃ καὶ εἰς δύο πρίσματα ἴσα. τὰ δὴ δύο πρίσματα μείζονά ἐστιν ἢ τὸ ἥμισυ τῆς ὅλης πυραμίδος.	And let pyramid ΔΕΖΘ be divided both in two pyramids equal to one another and similar to the whole and in two equal prisms. Thus the two prisms are greater than the half of the whole pyramid.

The referents of the sentences marked by δή in the last two passages are propositions XI.3 and XII.3, respectively.

Something obvious is also stated in syntagms like ἔσται δὴ ἴσον τὸ πλῆθος "thus the multiplicity … will be equal",[399] whose context we read on the example of VII.20 (*EOO* II, 230.13–16):

διῃρήσθω ὁ μὲν ΓΔ εἰς τὰ τοῦ Α μέρη τὰ ΓΗ ΗΔ ὁ δὲ ΕΖ εἰς τὰ τοῦ Β μέρη τὰ ΕΘ ΘΖ. ἔσται δὴ ἴσον τὸ πλῆθος τῶν ΓΗ ΗΔ τῷ πλήθει τῶν ΕΘ ΘΖ.	Let ΓΔ be divided into the parts ΓΗ, ΗΔ of Α and ΕΖ into the parts ΕΘ, ΘΖ of Β. Thus the multiplicity of ΓΗ, ΗΔ will be equal to the multiplicity of ΕΘ, ΘΖ.

Seven scattered occurrences of the same kind include statements of absurd conclusions as those that contradict theorems I.16 (in I.26–27) or I.17 (in III.16, VI.7, XI.14), or the two "Archimedean" statements in V.8—these are certainly spurious (*EOO* II, 28.18–20):

τὸ δὴ ἔλαττον τῶν ΑΕ ΕΒ πολλαπλασιαζόμενον ἔσται ποτὲ τοῦ Δ μεῖζον.	Thus the lesser of ΑΕ, ΕΒ, once multiples are taken, will eventually be greater than Δ.

[398] There are 18 occurrences, in VII.3, 18, 19, 22, X.42, 109, XI.3, 5, 7, 13–14, XII.2, 5, 10 (*bis*), 17 (*ter*).
[399] There are 10 occurrences, in V.1, 3, 15, VII.5–6, 8–10, 15, 20.

We must finally add to the same genus the 13 occurrences[400] in exhaustive and exclusive disjunctions—dichotomic or trichotomic—of which we read the examples in III.33, VII.4, and X.109 (*EOO* I, 250.9–10, II, 200.5, and III, 346.8–10):

ἡ δὴ πρὸς τῷ Γ γωνία ἤτοι ὀξεῖά ἐστιν ἢ ὀρθὴ ἢ ἀμβλεῖα.	Thus the angle at Γ is either acute or a right ‹angle› or obtuse.
ὁ δὴ ΒΓ τὸν Α ἤτοι μετρεῖ ἢ οὐ μετρεῖ.	Thus ΒΓ either measures or does not measure A.
ἤτοι δὴ ἡ ΘΖ τῆς ΖΚ μεῖζον δύναται τῷ ἀπὸ συμμέτρου ἑαυτῇ ἢ τῷ ἀπὸ ἀσυμμέτρου.	Thus ΘΖ is worth more than ΖΚ either by the ‹square› on a ‹straight line› commensurable with itself or by that on an incommensurable.

To the first horn of the dichotomy in X.110 corresponds in the proof a hapax like the nexus εἰ μὲν δή "thus now, if".

Akin to these are the 22 occurrences, in the introductory phrase of porisms, of ἐκ δὴ τούτου φανερὸν ὅτι "thus it is manifest from this that";[401] and again: 14 similar expressions φανερὸν δή (ἐστιν) ὅτι "thus it is manifest that" within proofs;[402] 3 syntagms διὰ δὴ τοῦτο "for this very ‹reason›" in IX.30, X.42, and XI.38; 1 phrase κατὰ τὰ αὐτὰ δή "according to the very same ‹reasons›" in X.44; 6 clauses that initialize the core procedure of the method of exhaustion.[403]

More interesting are the 44 occurrences in suppositions, that is, in sentences whose verb is in the imperative.[404] They identify prescriptions, as those featuring ἐπιτετάχθω(σαν) "let it be prescribed" in VI.9 and VIII.2, or non-liminal constructions such as the construction in proposition XI.23 (*EOO* IV, 66.13–17):

ἀνεστάτω δὴ ἀπὸ τοῦ Ξ σημείου τῷ τοῦ ΛΜΝ κύκλου ἐπιπέδῳ πρὸς ὀρθὰς ἡ ΞΡ, καὶ ᾧ μεῖζόν ἐστι τὸ ἀπὸ τῆς ΑΒ τετράγωνον τοῦ ἀπὸ τῆς ΛΞ, ἐκείνῳ ἴσον ἔστω τὸ ἀπὸ τῆς ΞΡ, καὶ ἐπεζεύχθωσαν αἱ ΡΛ ΡΜ ΡΝ.	Thus from point Ξ let a ‹straight line›, ΞΡ, be erected at right ‹angles› with the plane of circle ΛΜΝ, and by how much the square on AB is greater than that on ΛΞ, to that much let the ‹square› on ΞΡ be equal, and let ‹straight lines›, ΡΛ, ΡΜ, ΡΝ, be joined.

Specific clauses introduced by ὁσάκις "how many times" are the correlate of a constructive act;[405] here is the occurrence in VIII.8 (*EOO* II, 292.21–23):

ὁσάκις δὴ ὁ Η τὸν Ε μετρεῖ, τοσαυτάκις καὶ ἑκάτερος τῶν Θ Κ ἑκάτερον τῶν Μ Ν μετρείτω.	Thus, how many times H measures E, so many times let Θ, K also measure M, N, respectively.

This sentence occurs 26 times; it admits of variants, for instance when καί "and" replaces δή. Note also the 4 occurrences in applications of V.def.5, within VI.1, 33, XI.25, XII.13; they are introduced by a genitive absolute, as in XI.25 (*EOO* IV, 76.12–13):[406]

[400] In III.33, IV.5, VII.3–4, 36, IX.18 (*bis*), 20, X.4, 103, 108–110.
[401] These are the porisms to I.15, II.4, III.1, 16, 31, IV.15, V.7, 19, VI.8, 19, VII.2, VIII.2, X.3–4, 6, 23, XI.33, 35, XII.7–8, XIII.16–17.
[402] In VIII.9, 15, IX.3, 13, 15, X.13/14, 28/29II, 42, 44, 54–55, 57–59; 1 variant δῆλον δή in X.44.
[403] In XII.2, 10 (*bis*), 11–12, 16; it is the same clause with participial *incipit* read in Sect. 4.5.2.
[404] In III.20, 35, IV.11, 13–14, V.9–10, 17, VI.9, 28, VII.3–4, 19, 28, 34, 36, VIII.2, 4, 16, 18–19, IX.19, 36, X.3–4, 16, 18, 38, 49–50, 53–54, 91, XI.23, 31 (*bis*), 34 (*ter*), XII.10, 12, XIII.11, 17–18.
[405] There are 7 occurrences, in VII.21, 24, VIII.8, 20 (*bis*), 21 (*bis*).
[406] Other 4 genitive absolutes marked by δή, this time within constructive acts, are found in I.4, 8, VII.4, 31.

τεσσάρων δὴ ὄντων μεγεθῶν, δύο μὲν βάσεων τῶν ΑΖ ΖΘ δύο δὲ στερεῶν τῶν ΑΥ ΥΘ [...]

Thus, there being four magnitudes, two bases, ΑΖ, ΖΘ, and two solids, ΑΥ, ΥΘ [...]

The 4 conditionals with ἐὰν δή in XII.16–17, XIII.13, 15 are also noteworthy. The last two occurrences describe the generation of a sphere by rotation of a semicircle, according to XI.def.14;[407] the first two occurrences are characterized by personal verb forms (see Sect. 4.5.2) and describe a specific constructive act in a very synthetic way, as in XII.16 (*EOO* IV, 228.2–6):

ἐὰν δὴ τῇ ΛΔ εὐθείᾳ ἴσας κατὰ τὸ συνεχὲς ἐναρμόσωμεν εἰς τὸν ΑΒΓΔ κύκλον, ἐγγραφήσεται εἰς τὸν ΑΒΓΔ κύκλον πολύγωνον ἰσόπλευρόν τε καὶ ἀρτιόπλευρον μὴ ψαῦον τοῦ ἐλάσσονος κύκλου τοῦ ΕΖΗΘ.

Thus if we continuously fit in circle ΑΒΓΔ ‹straight lines› equal to straight line ΛΔ, a polygon both equilateral and even-sided will be inscribed in circle ΑΒΓΔ not touching the lesser circle ΕΖΗΘ.

The occurrences of δή show that this particle is a negative stylistic marker of Book X and a positive marker for Books VII–VIII, XI, and even more for Book IV. As a matter of fact, this Book abounds in analogical and potential proofs, besides comprising only problems:

	I	II	III	IV	V	VI	VII	VIII	IX	X	XI	XII	XIII	tot.
# prop.	48	14	37	16	25	33	39	27	36	115	39	18	18	465
% # signs	7.6	3.3	6.9	3.4	4.9	7.6	5.7	5	5.1	26.1	9	8.3	6.9	100
δή	57	11	53	61	27	43	53	45	41	125	88	51	48	703
% δή	8.1	1.6	7.5	8.7	3.8	6.1	7.5	6.4	5.8	17.8	12.5	7.3	6.8	100

[407] Of course, since the definition is checked in one direction, ἐάν "if" replaces ὅταν "whenever" of the definition.

APPENDICES

F. Acerbi, *The Logical Syntax of Greek Mathematics*, Sources and Studies in the History of Mathematics and Physical Sciences,
https://doi.org/10.1007/978-3-030-76959-8

APPENDIX A. PROBLEMS IN THE GREEK MATHEMATICAL CORPUS

Euclid, *Elements*

	# prop.	problems
I	48	1–3, 9–12, 22–23, 31, 42, 44–46
II	14	11, 14
III	37	1, 17, 25, 30, 33–34
IV	16	1–16
V	25	/
VI	33	9–13, 18, 25, 28–30
VII	39	2–3, 33–34, 36, 39
VIII	27	2, 4
IX	36	18–19
X	115	3–4, 10, 27–35, 48–53, 85–90
XI	39	11–12, 23, 26–27
XII	18	16–17
XIII	18	13–18 (*bis* each)
tot.	465	95(101)

Problems III.1 and VI.11–13 are enunciated using the verb form (προσ)εὑρεῖν "find". The same for all problems in the arithmetic Books (VII–IX), IX.18–19 excepted, where the initializing clause is ἐπισκέψασθαι εἰ / πότε δυνατόν ἐστι "investigate whether / when it is possible". The problems in Book X have a markedly existential connotation; they all use the verb form εὑρεῖν, too. Problems IV.10, X.27–35, 48–53, 85–90 do not present "givens" on which the construction is carried out; XIII.13–18 do present a "given", but this is used in a peculiar way. Problems XIII.13–18 also have a twofold enunciation: this comprises two problems and a theorem enunciated in problematic form (that is, whose verb is in the infinitive).

Euclid, *Optica* A

Propositions (18–21), 37–38, 44–47, 48, (49). Very peculiar format and wording; their existential character is explicit in the problems from proposition 37 on. The propositions 18–21 are *Data*-style theorems.

Euclid, *Catoptrica*

Propositions (13–15), 29. Problems 13–15 display a peculiar wording and make their own existential import explicit. Problem 29 is enunciated using the verb form εὑρεῖν: the construction is not univocally determined by the givens of the problem.

Archimedes, *De sphaera et cylindro*

Propositions I.2–5, 6 (*ter*), II.1, 3–7. With the exception of I.5, problems in Book I are enunciated using δυνατόν ἐστι "it is possible" + infinitive: the construction is not univocally determined by the givens of the problem. Problems II.1 and 6 are enunciated using the verb form εὑρεῖν, even if the construction is univocally determined by the givens of the problem. The first and the third axiom in *Sph. cyl.* I are the only principles in Greek mathematics that overtly state existence.

Archimedes, *De conoidibus et sphaeroidibus*

Propositions 7–9, 19–20. Problems 7–9 are enunciated using δυνατόν ἐστιν εὑρεῖν: thus, the construction is not univocally determined by the givens of the problem. Problems 19–20 are enunciated using δυνατόν ἐστι + infinitive: again, the construction is not univocally determined by the givens of the problem.

Archimedes, *De lineis spiralibus*

Propositions 3–9, 21–23. All problems are enunciated using δυνατόν ἐστι + infinitive; in problems 3–4 and 6–9 the infinitive is (ποτι)λαβεῖν; in all problems, the construction is not univocally determined by the givens of the problem.

Archimedes, *De planorum aequilibriis*

Propositions II.6. It is enunciated using δυνατόν ἐστι + infinitive: the construction is not univocally determined by the givens of the problem.

Apollonius, *Conica*

Propositions I.52–60, II.4, II.44–47, 49–51, 53 (18 items out of 226 propositions in *Con.* I–IV). The constructions of problems I.52–60 are not univocally determined by the givens of the problem; only 52, 54, 56, 59 use the verb form εὑρεῖν. Problems 53, 55, 57, 58 (further cases of the problems immediately preceding each of them) do not have a general enunciation. Problems II.44–47 use the verb form εὑρεῖν; in all cases except II.44 the construction is univocally determined by the givens of the problem. The other problems have standard enunciations. Proposition II.4 is spurious.

Theodosius, *Sphaerica*

Propositions I.2, 18–21, II.14–15. Standard construction problems.

Hero, *Metrica*

The entire treatise contains problems, with just a few exceptions (lemmas to subsequent problems). Systematic use of the verb form εὑρεῖν; these problems do not require to construct an object but to calculate the area or the volume of a plane or solid figure, respectively, or to cut it according to assigned constraints.

Diophantus, *Arithmetica*

The entire treatise without exceptions contains determinate or indeterminate numeric problems: systematic use of the verb form εὑρεῖν, along with specific verb forms such as διελεῖν "divide", etc.

Serenus, *De sectione cylindri*

Propositions 7, 21–24, 26, 27, 28. Problems 21–24 use the verb form εὑρεῖν. Problems 27–28 are enunciated using δυνατόν ἐστι + infinitive: in all such problems, the construction is not univocally determined by the givens; in 27–28, Serenus makes this explicit by using the adverb ἀπειραχῶς "in unboundedly many ways".

Serenus, *De sectione coni*

Propositions 9, 12, 14, 15, 22, 25, 27. Standard construction problems.

Pappus, *Collectio*

Coll. II.2, 4, 5, 7–8, 12, 14–15 (all problems in Book II are arithmetic); III.23–27 (anthology of duplications of a cube), 28, 31–33, 36–43 (geometric constructions of means), 47–56 (arithmetic constructions of means; peculiar wording), 58–74 [Erykinos-style problems; most of them are enunciated using (ἀ)δυνατόν ἐστι + infinitive: the construction is not univocally determined by the givens of the problem], 75–77 (preliminary to what follows), 86–95 (construction of regular polyhedra), 97–100 (a duplication of a cube); IV.15 (proved as a *Data*-theorem), 41 (construction of Nicomedes' curve; use of δυνατόν ἐστι + infinitive), 43–44 (duplication of a cube using Nicomedes' curve), 45 (construction of the quadratrix), 57–71 (trisection of an angle), 72–76 (use of δυνατόν ἐστι + infinitive), 80; V.10 (use of δυνατόν ἐστι + infinitive), 15, 51 (use of δυνατόν ἐστι + infinitive and of the adverb ἀπειραχῶς), 69; VI.103 (a problem of the same kind as Euclid, *Optica* 37 ff.); VII.43–44, 64, 123–124, 128–129, 142–143, 146, 167–168, 170–174, 182–183, 223, 232, 247, 274–275, 294 (*Data*-problem: a *unicum*); VIII.12–13, 15, 17–18 (use of εὑρεῖν), 26 (a duplication of a cube), 27–29 (use of εὑρεῖν), 31, 32–39 (use of εὑρεῖν), 40–41 (use of εὑρεῖν), 42–44, 47–48, 49–51. The *Collectio* was edited by a literary executor; Pappus frequently writes quite informally.

Eutocius, *in De sphaera et cylindro*

At *AOO* III, 50–54, 56–106 (anthology of duplications of a cube), 130–176 (solutions of a problem left out by Archimedes), 278–284 (use of δυνατόν ἐστι + infinitive).

Appendix B. Theorems of the *Data* that have a Synthetic Counterpart and Extant Sources on Greek Analysis and Synthesis

Theorems of the *Data* that have a synthetic counterpart

Data	*Ele-ments*	sta-tus	elsewhere	The *Data* theorem employs *Data*	The *Data* theorem employs *Elements*
def.3	VI.def.1	def			
def.6	I.def. 15 post.3	def post	*Coll.* VII.24(1)		
3	cn.2	cn		def.1	
4	cn.3	cn		def.1	
5	V.19	t		def.2; 1, 2, 4	V.19
6	V.18	t		def.2; 1–3	V.18, 19
8, 9	V.22	t		def.2; 1, 2; 8	V.22
26	post.1	post		def.4	
27	I.3	p		def.4	
28	I.31	p		def.4	I.30
29	I.11, 23	p		def.4	
30	I.12	p		def.4	I.16
34	VI.2	t		def.2; 1, 25, 26, 30	I.12, VI.2
39	I.8, 22	t, p		def.1, 3, 6; 1, 25–27	post.3, VI.def.1; I.3, 8
40	VI.4	t		def.1, 3; **, 25, 26, 29, 39	I.23, 32, VI.4
41	VI.6	t		def.1, 2, 3; 2, 26, 27, 29, 39	I.23, VI.6
42	VI.5	t		def.1, 2, 3; 2, 39	I.22, V.7, 22, VI.5
44	VI.7	t		def.1, 3; 3, 4, 8, 40, 43	post.2; I.12, 13, 32
47	VI.20	t		def.3; 4, 8, 41	post.1
49	VI.18, 20	p, t		6, 8, 47, 48	post.1
50	VI.18, 20	p, t		def.2; 8	VI.11, 19por
57	I.44	p		def.1, 2, 3; 1–4, 8, 40, 52, **	post.2, 4; I.32, 35, 46, VI.1
58	VI.28	p		def.1, 3; 2, 4, 7, 52, 55	I.10, 34, 36, 43, VI.18, 24, 26
59	VI.29	p		def.1, 3; 2–4, 7, 52, 55	I.10, 34, 36, 43, VI.18, 24, 26
64	I.12	t		def.2, 3; 3, 4, 8, 40	I.12, 13, 32, 41, II.12, VI.1
65	I.13	t		def.2, 3; 3, 4, 8, 40	I.12, 32, 41, II.13, VI.1
66		t	*Coll.* VII.214	def.2, 3; 3, 4, 8, 40	I.12, 32, 41, VI.1
69	VI.20	t		def.3; 3, 4, 8, 40, 68	I.23, 29, 31, 32, 35
70	VI.20	t		def.2, 3; 3, 4, 8, 40	I.14, 23, 29, 31, 32, 35, 45, VI.1, 14
73	VI.14	t		def.2, 3; 3, 4, 40, 56	I.14, 23, 29, 31, 32, 35, 45, V.11, VI.1, 14
74	VI.14	t		def.2, 3; 3, 4, 8	I.14, 23, 29, 31, 32, 35, 45, V.11, VI.1, 14
75	VI.15	t		def.2; 74	I.31, 41
87	III.29	t	*Alm.* I.10	def.1, 3, 5; 2–4, 40	post.2, 3; I.32, III.1, 31, 34
88	III.28	t	*Opt.* 37–38 **A** *Coll.* VII.24(2) *Alm.* I.10	def.3, 5; 1, 3, 43, 87	post.2, 3; III.1, 31
89	III.26, 34	t, p		def.6; 2, 25, 26, 29	post.2; III.1, 20
90	III.17, 37	p, t		def.8; 25, 26	post.2; III.1, 18, 31
91	III.36	t		def.1; 52, 90	III.17, 36
92	III.35	t		def.1; 25, 26, 52	post.3; III.1, 35
93		t	*Alm.* I.10	def.1, 2; 1, 52, 87	post.2; I.15, 32, III.21, V.11, 12, VI.3, 4, 16

Extant sources on Greek analysis and synthesis

	Euclid	Archimedes	Apollonius	Diocles
Description of the corpus				
Description of treatises				
Definition of analysis and synthesis				
Lemmas on treatises				
Geometric problems solved by analysis and synthesis		*Sph. cyl.* II.1, 3–7	*Cutting off of a ratio* *Con.* II.44, 46–47, 49–51	*apud* Eutocius, *AOO* III, 160–176
Locus theorems				*On burning mirrors* 4–5
Data-style theorems	*Data* 1–94, *Optics* 18–21			
Theorematic analyses		*Sph. cyl.* II.8 *al.*	*Con.* III.24–26	
Validation of proofs by chains of givens				
Validation of algorithms by chains of givens (always within problems)				
Validation of procedures by chains of givens (always within problems)				

	Dionysodorus	Hero	Ptolemy	Diophantus
Description of the corpus				
Description of treatises				
Definition of analysis and synthesis		*ante El.* XIII.1–5 *al.* *ante El.* II.2–10 *al.*		
Lemmas on treatises				
Geometric problems solved by analysis and synthesis	*apud* Eutocius, *AOO* III, 152–160			
Locus theorems				
Data-style theorems				
Theorematic analyses		*El.* XIII.1–5 *al.* *El.* II.2–10 *al.*		
Validation of proofs by chains of givens		*Metrica*, III.10–16, 19	*Alm.* I.10, 13, II.3, 7, III.5, V.19, VIII.5, 6, XI.10, XIII.3 *Analemma*, *POO* II, 203–210	*Ar.* V.10
Validation of algorithms by chains of givens (always within problems)		*Metrica*, passim *Dioptra* 13–14, 25–30		
Validation of procedures by chains of givens (always within problems)		*Metrica*, passim		*De pol. num.*

	Pappus	Theon of Alexandria	Eutocius
Description of the corpus	*Coll.* VII.3		
Description of treatises	*Coll.* VII.4–42		
Definition of analysis and synthesis	*Coll.* VII.1–2		
Lemmas on treatises	*Coll.* VII.43–321, complement to what follows		
Geometric problems solved by analysis and synthesis	*Coll.* III.70, 86–95, IV.60–61, 65–66, 68, 72–76, 80, VII.64, 128–129, 142–143, 146, 167–168, 170–174, 182–183, 223, 232, 274–275, 294, VIII.17–18, 42–44, 47–48		
Locus theorems	*Coll.* IV.51–52, 67, 78–79, VII.37–40, 276, 312–318, VIII.12, 14		*AGE* II, 180–184
Data-style theorems	*Coll.* III.4, 9–13, IV.10–15, VII.192, 305, VIII.30		
Theorematic analyses	*Coll.* IV.6–7, 17–18, VI.16–20, 21–22, 26, 125, VII.73, 191, 225–226, 231, 321		
Validation of proofs by chains of givens			
Validation of algorithms by chains of givens (always within problems)	*in Alm. V.5, 13, 17, 19, VI.5*	*in Alm. I.10, 12, 13, II.3, 5, 9, 11, III.4, IV.2, 9*	
Validation of procedures by chains of givens (always within problems)			

APPENDIX C. ONOMASTICON

The list gives the name, the date (certain or presumed), and the mathematical works of the main Greek mathematicians. After the name, references to the corresponding biographical entries in a number of standard reference works are provided. The bibliographical references are to the critical editions of the surviving texts or to collections of fragments, to important mentions of the lost works in ancient writings (cited by page only), or to secondary literature I regard as fundamental.

Anthemius of Tralles (*DSB* I, 169–170; *RE* I.2, 2368–2369): d. 534
On Surprising Mechanisms, in a fragmentary state: theory of conic sections applied to optics (*MGM*, 78–87; *CG*, 349–359).

Apollonius of Perge (*DSB* I, 179–193; *RE* II.1, 151–160): early 2nd BCE
Conics: 8 books (7 extant), from V to VII in Arabic translation only (*AGE* I and II, 2–96; Decorps-Foulquier *et al.* 2008–2010; Toomer 1990; Zeuthen 1886; Decorps-Foulquier 2000; Fried, Unguru 2001). *Cutting off of a Ratio* in 2 books, only in Arabic translation (Rashed, Bellosta 2010; Pappus, *Coll.* VII.5–6). Lost works, traces of which can be found in Arabic authors (Hogendijk 1986; Pappus, *Coll.* VII.7–8, 9–10, 11–12, 21–26, 27–29, and 43–67, 69–119, 158–184, 120–157, 185–192; Jones 1986, 510–546): *Cutting off of an Area* in 2 books, *Determinate Section* in 2 books, *Tangencies* in 2 books, *Plane Loci* in 2 books, *Vergings* in 2 books. A treatise on the comparison of the dodecahedron and the icosahedron (Hypsicles in *EOO* V, 2–8). Investigations into: astronomy (*Alm.* XII.1), unordered irrationals (Pappus, *in X Elem.* I.1, II.1; Proclus, *iE*, 74; *EOO* V, 414), the ratio between a circumference and its diameter, numeric systems devised to express large numbers (Pappus, *Coll.* II), the cylindrical helix and more generally homeomeric lines, and foundational themes (*AGE* II, 101–139; Acerbi 2010a, 2010b).

Archimedes of Syracuse (*DSB* I, 213–231; *RE* II.1, 507–539): d. 212 BCE
On the Sphere and the Cylinder I and II (in origin two treatises), *Measurement of the Circle*, *On Conoids and Spheroids* (*AOO* I); *On Spirals*, *On the Equilibrium of Planes* in 2 books, *The Sand-reckoner*, *Quadrature of the Parabola*, *On Floating Bodies* in 2 books, *Stomachion*, *Method* (*AOO* II). *On Mutually Tangent Circles*, in Arabic only (*AOO* IV). Best overall account Dijksterhuis 1987; relative chronology of the treatises in Knorr 1978.

Archytas of Tarentum (*DSB* I, 231–233; *RE* II.1, 600–602; *DPhA* I, 339–342; DK 47; Timpanaro Cardini 1958–64 II, 272–385; Huffman 2005): late 5th BCE
A work on harmonic theory (various titles: *Harmonics*, *On Mathematics*, *On Music*). Solution, by intersection of surfaces, of the problem of duplication of a cube (Eutocius in *AOO* III, 84–88). A theorem on numeric mean proportionals (Boethius, *Inst. mus.* III.11).

Aristaeus (*DSB* I, 245–246; *RE* Suppl. III, 157–158): 4th BCE
Solid Loci in 5 books, lost (Pappus, *Coll.* III.21, VII.3, 29–31, 33–35; Jones 1986, 574–587). A treatise *On the Comparison of the Five Figures*, cited by Hypsicles and likewise lost (*EOO* V, 6).

Aristarchus of Samos (*DSB* I, 246–250; *RE* II.1, 873–876; *DPhA* I, 356–357): early 3rd BCE
On the Sizes and Distances of the Sun and Moon (Heath 1913; Berggren, Sidoli 2007a).

Autolycus of Pitane (*DSB* I, 338–339; *RE* II.2, 2602–2604): late 4th BCE
On a Moving Sphere, *On Risings and Settings*: geometry of a sphere and descriptive astronomy (Mogenet 1950; Aujac 1979).

Carpus of Antiochia (*RE* X.1, 2008–2009; *DPhA* II, 228–230): 1st
Investigations into mechanics (Theon, *iA*, 524), into astronomy in an *Astronomical Treatise* (Proclus, *in Rmp.* II, 218), and into foundational themes (Pappus, *Coll.* VIII.3; Proclus, *iE*, 125, 241–243). Quadrature of the circle by means of higher-order curves (Simplicius, *in Ph.*, 60 = *in Cat.*, 192).

Cleomedes (*DSB* III, 318–320; *RE* XI.1, 679–694; *DPhA* II, 436–439): 2nd
Cyclic Theory: astronomical handbook inspired by Stoic doctrines (Todd 1990; Goulet 1980).

Conon of Samos (*DSB* III, 391; *RE* XI.2, 1338–1340): 3rd BCE
Investigations into the plane spiral (Pappus, *Coll.* IV.30) and into the mutual intersections of conics (Apollonius in *AGE* II, 2). A treatise in 7 books *De astrologia* (Probus, *ad Ecl.* III.40).

Damianus of Larissa (*RE* IV.2, 2054–2055; *DPhA* II, 594–597): early 6th
Optical Hypotheses: a popular compendium of optics and catoptrics (Schöne 1897; Acerbi 2007, 2589–2628).

Demetrius of Alexandria (*RE* IV.2, 2849): before 1st
Linear Investigations: a study of higher-order curves (Pappus, *Coll.* IV.58).

Dinostratus (*DSB* IV, 103–105; *RE* IV.2, 2396–1398; *DPhA* II, 619; Lasserre 1987, 127–129): early 4th BCE
Quadrature of the circle by means of higher-order curves (Pappus, *Coll.* IV.45–50).

Diocles (*DSB* IV, 105; *RE* V.1, 813–814): 2nd BCE
On Burning Mirrors, a rearranged epitome in Arabic translation, partly attested in Greek as extracts: focal properties of parabola and circumference (Toomer 1976; *CG*, 98–141; Eutocius in *AOO* III, 66–70, 160–176; Acerbi 2011b).

Diodorus (*RE* V.1, 710–712, *DPhA* II, 782–783): 1st BCE
Investigations into astronomy and gnomonics, mentioned by several sources (Acerbi 2007, 2495–2496). Attempts at proving the fifth postulate (Tummers 1994, 31 and 55).

Dionysodorus (*DSB* IV, 108–110; *RE* V.1, 1005–1006; *DPhA* II, 875): 2nd BCE
Solution to a problem left open by Archimedes in *Sph. cyl.* II.4 (Eutocius in *AOO* III, 152–160). A treatise *On the Torus* (Hero, *Metr.* II.13). Design of a conic sundial (Vitruvius, *Arch.* IX.8.1).

Diophantus of Alexandria (*DSB* IV, 110–119; *RE* V.1, 1051–1073): 1st–3rd
Arithmetics in 13 books, 6 of which transmitted in Greek (*DOO* I, 2–448), 4 in Arabic version only (Sesiano 1982; *DA*). *On Polygonal Numbers*, incomplete (*DOO* I, 450–480; Acerbi 2011e). Lost

works: *Porisms* (*DOO* I, 316, 320, 358), *Moriastica* (scholium to Iamblichus, *in Nic.*, 127 Pistelli, and *DOO* II, 72).

Domninus (*DSB* IV, 159–160; *RE* V.1, 1521–1525): 5th
Pocketbook of Arithmetics: a compendium of arithmetic (Riedlberger 2013). A short text on how to remove a ratio from a ratio is wrongly ascribed to him (Acerbi, Riedlberger 2014).

Dositheus (*DSB* IV, 171–172; *RE* V.2, 1607–1608): 3rd BCE
Investigations into catoptrics (Diocles in Toomer 1976, 34; *CG*, 98). Weather forecast on calendric grounds: Dositheus is repeatedly mentioned in the *parapegmata* ending Geminus' *Introduction* and in Ptolemy's *Phases*. A writing on the 8-year intercalary cycle introduced by Eudoxus (Censorinus, *De die natali* 18.5).

Eratosthenes of Cyrene (*DSB* IV, 388–393; *RE* VI.1, 358–389; *DPhA* III, 188–236): active 240–195 BCE
A lost treatise *On Means* possibly included in the analytic corpus (Pappus, *Coll.* VII.3, 22, 29). Contributions to the theory of numeric means (Theon of Smyrna, *Exp.*, 106–111 and 113–119). Design of a device to find arbitrarily many mean proportionals between two given lines (Eutocius in *AOO* III, 88–96).

Euclid (*DSB* IV, 414–459; *RE* VI.1, 1003–1052; *DPhA* III, 252–272): 3rd BCE
Elements in 13 books (*EOO* I–IV; Vitrac 1990–2001; Acerbi 2007), *Data* (*EOO* VI; Acerbi 2007, 439–554; Sidoli, Isahaya 2018), *Optics* redactions **A** and **B**, *Catoptrics* (*EOO* VII; Acerbi 2007, 584–641; Kheirandish 1998), *Phenomena* redactions **a** and **b**, *Sectio canonis*: on the location of notes on the string of a monochord (*EOO* VIII; Berggren, Thomas 1996; Acerbi 2007, 677–702). An elementary treatise of mechanics, only in Arabic version (Woepcke 1851; Acerbi 2007, 2455–2484). Lost writings: *Porisms* in 3 books (Pappus, *Coll.* VII.13–20 and 193–232; Simson 1776; Hogendijk 1987; Jones 1986, 547–572; Acerbi 2007, 733–744), *Pseudaria*, on fallacious proofs (Acerbi 2008), *On the Division of Figures* (Archibald 1915; Acerbi 2007, 2383–2454), *Loci on a Surface* in 2 books (Pappus, *Coll.* VII.312–318; Jones 1986, 591–595; Acerbi 2007, 745–754), *Conics* in 4 books (Pappus, *Coll.* VII.30).

Eudoxus of Cnidos (*DSB* IV, 465–467; *RE* VI.1, 930–950; *DPhA* III, 293–302; Lasserre 1966): 4th BCE
Theory of means, proportion theory, method of exhaustion, mathematical and descriptive astronomy. We have titles such as *Phenomena*, *Mirror*, *On Speeds*, *Disappearances of the Sun*, *On the 8-year Cycle*, *Astronomy*, *On the Arachne*, *Circuit of the Earth*.

Eutocius of Ascalon (*DSB* IV, 488–491; *RE* VI.1, 1518; *DPhA* III, 392–396): early 6th
Commentary on three Archimedean treatises: *On the Sphere and the Cylinder* I–II, *Measurement of the Circle*, *On the Equilibrium of Planes* (*AOO* III, 2–318). Edition of, and commentary on, Apollonius' *Conics* I–IV (*AGE* II, 168–360). Assessment of his editorial and exegetic techniques in Acerbi 2012c.

Geminus (*DSB* V, 344–347; *RE* VII.2, 1026–1050; *DPhA* III, 472–477): 1st BCE
Introduction to the Phenomena: a treatise of descriptive astronomy (Manitius 1898; Aujac 1975; Evans, Berggren 2006). *Theory of Mathematics*, historico-philosophically oriented mathematical compendium, lost (Tittel 1895; Acerbi 2010a).

Hero of Alexandria (*DSB* VI, 310–315; *RE* VIII.1, 992–1080; *DPhA* Suppl., 87–103): 1st–3rd
Pneumatics in 2 books, *Automaton Construction* (*HOO* I), *Mechanics* (only in Arabic) and *Catoptrics* (*HOO* II; the last also in Jones 2001), *Metrics* in 3 books, *Dioptra* (*HOO* III; the first also in Acerbi, Vitrac 2014). *Artillery Construction* and *Cheiroballistra*: artillery manuals (Wescher 1867; Marsden 1971). Surely spurious but traditionally included in the Heronian corpus (some of them are Heiberg's philological artefacts): *Definitions*, *Geometry* (*HOO* IV), *Geodesy*, *Stereometry* I–II, *On Measurements* (*HOO* V), *Handbook of Agriculture*.

Hipparchus of Nicaea (*DSB* XV, Suppl. 1, 207–224; *RE* VIII.2, 1666–1681): active 147–127 BCE
Commentary of Aratus' and Eudoxus' Phenomena (Manitius 1894). Lost treatises on astronomy, number theory, optics, geography: *On Chords*, *On the Monthly Motion in Latitude of the Moon*, *On the Variation of Equinoctial and Solsticial Points*, *On the Length of the Year*, *On Intercalary Months and Days*, *On the Rising Times of the Twelve Zodiacal Signs*, *Against Eratosthenes' Geography*, *On the Partition of Numbers*, *On the Art of Algebra*, alias *The Rules* (Acerbi 2003b), *On Objects Moving Downwards because of Weight*.

Hippocrates of Chios (*DSB* VI, 410–418; *RE* VIII.2, 1780–1801; *DPhA* III, 762–770; Timpanaro Cardini 1958–64 II, 28–73): 5th BCE
A compilation of *Elements* (Proclus, *iE*, 66). A writing on the quadrature of lunes (Simplicius, *in Ph.*, 54–69). Reduction of the problem of duplication of a cube to finding a mean proportional of two given segments (Proclus, *iE*, 213; Eutocius in *AOO* III, 88).

Hypsicles (*DSB* VI, 616–617; *RE* IX.1, 427–433): 2nd BCE
The so-called Book XIV of the *Elements* (*EOO* V, 2–36; Vitrac, Djebbar 2011 and 2012). *On Rising Times* of the zodiacal signs (De Falco, Krause 1966). Lost works on number theory (*DOO* I, 470–472; Acerbi 2011e, 196–197).

Iamblichus of Chalcis (*DSB* VII, 1; *RE* IX.1, 645–651; *DPhA* III, 824–836): 3rd
Popular writings of number theory inspired by Neopythagorean doctrines: *On Nicomachus' Introduction to Arithmetics* (not a commentary but a rewriting; Pistelli 1894; Vinel 2014), *On Common Mathematical Science* (Festa 1891). Spuriously ascribed to him is the arithmological compilation *Theology of Arithmetics* (De Falco 1922).

Marinus of Neapolis (*RE* XIV.2, 1759–1767; *DPhA* IV, 282–284): late 5th
Prolegomena to Euclid's *Data* (*EOO* VI, 234–256; Acerbi 2007, 2485–2524). Specific comments on Ptolemaic treatises (Tihon 1976).

Menaechmus (*DSB* IX, 268–277; *RE* XV.1, 700–701; *DPhA* IV, 401–407; Lasserre 1987, 117–124): late 5th BCE
Investigations into the foundations of geometry (Proclus, *iE*, 72–73, 75–78, 181, 253–254) and into conic sections (Eutocius in *AOO* III, 78–84).

Menelaus of Alexandria (*DSB* IX, 296–302; *RE* XV.1, 834–835; *DPhA* IV, 456–464): late 1st
Spherics in 3 books, only in a number of Arabic revisions: geometry of the surface of the sphere, with explicit astronomical applications (Halley 1758; Krause 1936; Rashed, Papadopoulos 2017; Acerbi 2015). *Geometric Elements*, only a few fragments in Arabic authors (Hogendijk 2000), *Method for the Determination of the Magnitude of Each of the Mixed Bodies*, only in Arabic version (Würschmidt 1925). Maybe a work on planetary theory (Jones 1999, 69–80).

Nicomachus of Gerasa (*DSB* X, 112–114; *RE* XVII.1, 463–464; *DPhA* IV, 686–694): 2nd
A popular writing on number theory: *Introduction to Arithmetics* (Hoche 1866; D'Ooge, Robbins, Karpinski 1926). *Handbook of Harmonics* (*MSG*, 235–265). A lost numerological writing of the genre *Theology of Arithmetics* (Photius, *Bibliotheca*, codex 187).

Nicomedes (*DSB* X, 114–116; *RE* XVII.1, 500–504): 2nd BCE
Invention of the conchoids and their application to the duplication of a cube (Eutocius in *AOO* III, 98–106, *Coll.* IV.39–44, Simplicius, *in Ph.*, 60 = *in Cat.*, 192). Use of the quadratrix to square the circle (Pappus, *Coll.* IV.45–50).

Pappus of Alexandria (*DSB* X, 293–304; *RE* XVIII.3, 1084–1106; *DPhA* V, 147–149): 4th
Collection in 8 books, the first being lost, the last also transmitted in Arabic (Hultsch 1876–78; Jones 1986; Sefrin-Weis 2010; Jackson 1972). A commentary on Ptolemy's *Almagest*, of which only Books V–VI out of 13 survive (*iA* I). A commentary on *Elements* X in 2 books, only in Arabic translation (Junge, Thomson 1930). A commentary on Diodorus' *Analemma* (Pappus, *Coll.* IV.40).

Philo of Tyana (*RE* XX.1, 55)
Investigations into higher-order curves (Pappus, *Coll.* IV.58).

Posidonius of Apamea (*DSB* XI, 103–106; *RE* XXII.1, 559–826; *DPhA* V, 1481–1501; Theiler 1982; Edelstein-Kidd 1988–99): early 1st BCE
Investigations into foundational themes (Proclus, *iE*, 80–81, 143–144, 170–171, 168, 176, 214–218; Acerbi 2010b).

Proclus (*DSB* XI, 160–162; *RE* XXIII.1; 186–247; *DPhA* V, 1546–1674): d. 485
Commentary on the First Book of Euclid's Elements (*iE*; Morrow 1970). *Outline of Astronomical Models*, a popular writing on Ptolemaic astronomy (Manitius 1909).

Ptolemy (*DSB* XI, 186–206; *RE* XXIII.2, 1788–1859; *DPhA* V, 1718–1735): active 127–160
Almagest in 13 books (*POO* I; Manitius 1963; Toomer 1984; Kunitzsch 1974). Minor astronomical works: *Phases of the Fixed Stars*, *Planetary Models* in 2 books, a primer to the *Handy Tables*, *Analemma*, *Planisphere* (*POO* II), *Canobic Inscription* (*POO* II; Jones 2005), *Handy Tables* (Tihon, Mercier 2011–). The *Planetary Models* is transmitted in Arabic in complete form (Goldstein 1967; Morelon 1993). The *Planisphere* in *POO* II is a Latin translation from Arabic (the Arabic text is edited in Sidoli, Berggren, 2007). *Geography* in 8 books (Stückelberger, Graßhoff 2006; Berggren, Jones 2000). *Optics*, only in Arabo-Latin translation (Lejeune 1989). *Harmonics* (Düring 1930; Barker 2000). *Astrological Outcomes*, or *Quadripartite*, a treatise of horoscopic astrology (*POO* III.1). *On the Criterion of Truth*, epistemology (*POO* III.2).

Serenus of Antinoe (*DSB* XII, 313–315; *RE* IIA.2, 1677–1678; *DPhA* VI, 212–214): 3rd
On the Section of a Cylinder, which aims at recovering the main properties of the ellipse. *On the Section of a Cone*, by a plane passing through its apex (Heiberg 1896). A commentary on Apollonius' *Conics*, lost (Heiberg 1896, 52).

Sporus of Nicaea (*DSB* XII, 579–580; *RE* IIIA.2, 1879–1883; *DPhA* VI, 554–555): 3rd
Investigations into the use of curves for squaring the circle (Pappus, *Coll.* IV.45–50). Solution to the problem of duplication of a cube (Eutocius in *AOO* III, 76–78). A commentary on Aratus' *Phenomena* (Maass 1898, 562).

Theodosius of Bithynia (*DSB* XIII, 319–321; *RE* VA.2, 1930–1935): 2nd BCE
Spherics in 3 books: geometry of the sphere, with implicit astronomical applications (Heiberg 1927; Czinczenheim 2000). *On Habitations*: on phenomena pertaining to the rotation of the celestial sphere, in function of the observer's latitude; *On Days and Nights* in 2 books: on the relative lengths of day and night in function of the Sun's position on the ecliptic (Fecht 1927). A commentary on Archimedes' *Method*, works in astronomy and gnomonics (*Suda* Θ 142; Vitruvius, *Arch.* IX.8.1).

Theon of Alexandria (*DSB* XIII, 321–325; *RE* VA.2, 2075–2080; *DPhA* VI, 1008–1015): 4th
Commentaries on Ptolemy's *Almagest* and *Handy Tables* (*iA* II–III, only Books I–IV; Tihon 1978; Mogenet, Tihon 1985, Tihon 1991 and 1999).

Theon of Smyrna (*DSB* XIII, 325–326; *RE* VA.2, 2067–2075; *DPhA* VI, 1016–1028): 2nd
Exposition of the Things Useful for Reading Plato: arithmetic and astronomical compendium inspired by Middle-Platonic doctrines (Hiller 1878).

Zenodorus (*DSB* XIV, 603–605; *RE* XA, 18): 2nd BCE
On Isoperimetric Figures: extracts survive in Book I of Theon's commentary on the *Almagest* (*iA*, 355–379), in Pappus' *Collection* (V.3–19, 38–40), and in the anonymous *Prolegomena to the Almagest* (Acerbi, Vinel, Vitrac 2010; Acerbi 2022).

BIBLIOGRAPHY

F. Acerbi, *The Logical Syntax of Greek Mathematics*, Sources and Studies in the History of Mathematics and Physical Sciences,
https://doi.org/10.1007/978-3-030-76959-8

AGE = *Apollonii Pergaei quae Graece exstant cum commentariis antiquis*, ed. J.L. Heiberg, 2 vol., Leipzig, B.G. Teubner 1891–93.

AOO = *Archimedis opera omnia cum commentariis Eutocii*, ed. J.L. Heiberg, 3 vol., Leipzig, B.G. Teubner 1910–15.

CAG = *Commentaria in Aristotelem Graeca*, 23 vol. in 51 tomes, Berlin, G. Reimer 1883–1909.

CG = *Les catoptriciens grecs I. Les miroirs ardents*, ed. R. Rashed, Paris, Les Belles Lettres 2000.

DA = *Diophante. Les Arithmétiques*, ed. R. Rashed, 2 vol., Paris, Les Belles Lettres 1984.

DOO = *Diophanti Alexandrini opera omnia cum Graeciis commentariis*, ed. P. Tannery, 2 vol., Leipzig, B.G. Teubner 1893–95.

DPhA = *Dictionnaire des philosophes antiques*, ed. R. Goulet, 7 vol., Paris, CNRS Éditions 1994–2018.

DSB = *Dictionary of Scientific Biography*, ed. Ch.C. Gillispie, 17 vol., New York, Ch. Scribner's Sons 1970–1981.

EOO = *Euclidis opera omnia*, ed. J.L. Heiberg, H. Menge, 8 vol., Leipzig, B.G. Teubner 1883–1916.

GG = *Grammatici Graeci*, ed. G. Uhlig, R. Schneider, A. Lentz, A. Hilgard, 4 vol. in 9 tomes, Leipzig, B.G. Teubner 1867–1910.

HOO = *Heronis Alexandrini opera quae supersunt omnia*, ed. L. Nix, W. Schmidt, H. Schöne, J.L. Heiberg, 5 vol., Leipzig, B.G. Teubner 1899–1914.

iA = *Commentaires de Pappus et de Théon d'Alexandrie sur l'Almageste*, ed. A. Rome, 3 vol., Città del Vaticano, Biblioteca Apostolica Vaticana 1931–43.

iE = *Procli Diadochi in primum Euclidis Elementorum librum commentarii*, ed. G. Friedlein, Leipzig, B.G. Teubner 1873.

MGM = *Mathematici Graeci Minores*, ed. J.L. Heiberg, København, Blanco Lunos 1927.

POO = *Claudii Ptolemaei opera quae exstant omnia*, ed. J.L. Heiberg, 2 vol. in 3 tomes, Leipzig, B.G. Teubner 1898–1907.

RE = *Paulys Real-Encyclopädie der classischen Altertumswissenschaft*, ed. G. Wissowa *et al.*, 1st series, 24 vol. in 43 tomes, Stuttgart, A. Druckenmüller 1894–1963, 2nd series, 10 vol. in 19 tomes, Stuttgart, A. Druckenmüller 1914–1972.

F. Acerbi, "Drowning by Multiples. Remarks on the Fifth Book of Euclid's *Elements*, with Special Emphasis on Prop. 8", *Archive for History of Exact Sciences* 57 (2003a), 175–242.

F. Acerbi, "On the Shoulders of Hipparchus. A Reappraisal of Ancient Greek Combinatorics", *Archive for History of Exact Sciences* 57 (2003b), 465–502.

F. Acerbi, *Euclide, Tutte le Opere*, Milano, Bompiani 2007.

F. Acerbi, "In What Proof Would a Geometer Use the ποδιαία?", *Classical Quarterly* 58 (2008a), 120–126.

F. Acerbi, "Conjunction and Disjunction in Euclid's *Elements*", *Histoire, Épistémologie, Langage* 30 (2008b), 21–47.

F. Acerbi, "Euclid's *Pseudaria*", *Archive for History of Exact Sciences* 62 (2008c), 511–551.

F. Acerbi, "The Meaning of πλασματικόν in Diophantus' *Arithmetica*", *Archive for History of Exact Sciences* 63 (2009a), 5–31.

F. Acerbi, "Transitivity Cannot Explain Perfect Syllogisms", *Rhizai* 6 (2009b), 23–42.

F. Acerbi, "Homeomeric Lines in Greek Mathematics", *Science in Context* 23 (2010a), 1–37.

F. Acerbi, "Two Approaches to Foundations in Greek Mathematics: Apollonius and Geminus", *Science in Context* 23 (2010b), 151–186.

F. Acerbi, "The Language of the 'Givens': its Forms and its Use as a Deductive Tool in Greek Mathematics", *Archive for History of Exact Sciences* 65 (2011a), 119–153.

F. Acerbi, "The Geometry of Burning Mirrors in Greek Antiquity. Analysis, Heuristic, Projections, Lemmatic Fragmentation", *Archive for History of Exact Sciences* 65 (2011b), 471–497.

F. Acerbi, "Perché una dimostrazione geometrica greca è generale", in G. Micheli, F. Franco Repellini (eds.), *La scienza antica e la sua tradizione*, Milano, Cisalpino 2011c, 25–80.

F. Acerbi, "Pappus, Aristote et le τόπος ἀναλυόμενος", *Revue des Études Grecques* 124 (2011d), 93–113.

F. Acerbi (ed.), *Diofanto, De polygonis numeris*, Pisa – Roma, Fabrizio Serra Editore 2011e.

F. Acerbi, "The Number of Endings of the Adjective συναμφότερος", *Glotta* 88 (2012a), 1–8.

F. Acerbi, "I codici stilistici della matematica greca: dimostrazioni, procedure, algoritmi", *Quaderni Urbinati di Cultura Classica*, n. s., 101(2) (2012b), 167–214.

F. Acerbi, "Commentari, scolii e annotazioni marginali ai trattati matematici greci", *Segno e Testo* 10 (2012c), 135–216.

F. Acerbi, "Aristotle and Euclid's Postulates", *Classical Quarterly* 63 (2013a), 680–685.

F. Acerbi, "Why John Chortasmenus Sent Diophantus to the Devil", *Greek, Roman, and Byzantine Studies* 53 (2013b), 379–389.

F. Acerbi, "La concezione archimedea degli oggetti matematici", *La Matematica nella Società e nella Cultura*, 6 (2013c), 227–252.

F. Acerbi, "Traces of Menelaus' *Sphaerica* in Greek Scholia to the *Almagest*", *SCIAMVS* 16 (2015), 91–124.

F. Acerbi, "Byzantine Recensions of Greek Mathematical and Astronomical Texts: A Survey", *Estudios bizantinos* 4 (2016), 133–213.

F. Acerbi, "The Mathematical *Scholia Vetera* to *Almagest* I.10–15. With a Critical Edition of the Diagrams and an Interpretation of Their Symmetry Properties", *SCIAMVS* 18 (2017), 133–259.

F. Acerbi, "Composition and Removal of Ratios in Geometric and Logistic Texts from the Hellenistic to the Byzantine Period", in M. Sialaros (ed.), *Revolutions and Continuity in Greek Mathematics*, Berlin, Springer 2018, 131–188.

F. Acerbi, "There is no *consequentia mirabilis* in Greek mathematics", *Archive for History of Exact Sciences* 73 (2019), 217–242.

F. Acerbi, "Mathematical Generality, Letter Labels, and All That", *Phronesis* 65 (2020a), 27–75.

F. Acerbi, "Arithmetic and Logistic, Geometry and Metrology, Harmonic Theory, Optics and Mechanics", in S. Lazaris (ed.), *A Companion to Byzantine Science*, Leiden, Brill 2020b, 105–159.

F. Acerbi, "Topographie du Vat. gr. 1594", in D. Bianconi, F. Ronconi (eds.), *La «collection philosophique» face à l'histoire. Péripéties et tradition*, Spoleto, Centro Italiano di Studi sull'Alto Medioevo 2020c, 239–321.

F. Acerbi, "Interazioni fra testo, tavole e diagrammi nei manoscritti matematici e astronomici greci", in *La conoscenza scientifica nell'Alto Medioevo. Atti della LXVII Settimana di Studio, Spoleto, 25 aprile – 1 maggio 2019*, Spoleto, Centro Italiano di Studi sull'Alto Medioevo 2020d, 585–621.

F. Acerbi (ed.), *Les Prolégomènes à l'Almageste*, Pisa – Roma, Fabrizio Serra Editore 2022.

F. Acerbi, S. Martinelli Tempesta, B. Vitrac, "Gli interventi autografi di Giorgio Gemisto Pletone nel codice matematico Marc. gr. Z. 301", *Segno e Testo* 14 (2016), 411–456.

F. Acerbi, P. Riedlberger, "Uno scolio tardo-antico sulla rimozione di rapporti, fonte dello Pseudo-Domnino", *Koinonia* 38 (2014), 395–426.

F. Acerbi, N. Vinel, B. Vitrac. "Les *Prolégomènes à l'Almageste*. Une édition à partir des manuscrits les plus anciens: II. Le traité des figures isopérimétriques", *SCIAMVS* 11 (2010), 92–196.

F. Acerbi, B. Vitrac (eds.), *Héron, Metrica*, Pisa – Roma, Fabrizio Serra Editore 2014.

W.J. Aerts, *Periphrastica. An investigation into the use of εἶναι and ἔχειν as auxiliaries or pseudo-auxiliaries in Greek from Homer up to the present day*, Amsterdam, Hakkert 1965.

S. Aimar, "Modals and Copulae in Aristotle", unpublished typescript.

R.J. Allan, *The Middle Voice in Ancient Greek. A Study in Polysemy*, Amsterdam, Gieben 2003.

R.J. Allan, "The *infinitivus pro imperativo* in Ancient Greek. The Imperatival Infinitive as an Expression of Proper Procedural Action", *Mnemosyne* 63 (2010), 203–228.

G.J. Allman, *Greek Geometry from Thales to Euclid*, Dublin 1889.

S. Amigues, "Les temps de l'impératif dans les ordres de l'orateur au greffier", *Revue des Études Grecques* 90 (1977), 223–238.

M. Antonelli, J. von Plato, "Nested Indirect Proofs in Aristotle's Deductive Logic", unpublished typescript.

R.C. Archibald, *Euclid's Book on Divisions of Figures, with a Restoration Based on Woepcke's Text and on the* Practica Geometriae *of Leonardo Pisano*, Cambridge, Cambridge University Press 1915.

F. Arendt, "Eine Interpolation des Eutokios in unserem Apolloniostext", *Bibliotheca Mathematica*, III. Folge, 14 (1913–14), 97–98.

M. Asper, *Griechische Wissenschaftstexte. Formen, Funktionen, Differenzierungsgeschichten*, Stuttgart, Franz Steiner Verlag 2007

G. Aujac (ed.), *Géminos, Introduction aux Phénomènes*, Paris, Les Belles Lettres 1975.

G. Aujac (ed.), *Autolycus de Pitane, La sphère en mouvement, Levers et couchers héliaques. Testimonia*, Paris, Les Belles Lettres 1979.

G. Aujac, "Le langage formulaire dans la géométrie grecque", *Revue d'Histoire des Sciences* 37 (1984), 97–109.

G. Aujac, "Le rapport *di' isou* (*Euclide* V, définition 17): Définition, utilisation, transmission", *Historia Mathematica* 13 (1986), 370–386.

J. Avigad, E. Dean, J. Mumma, "A Formal System for Euclid's *Elements*", *The Review of Symbolic Logic* 2 (2009), 700–768.

W. Ax (ed.), *M. Tulli Ciceronis scripta quae manserunt omnia*, fasc. 46, *De Divinatione, De Fato, Timaeus*, Leipzig, B.G. Teubner 1938.

N. A. Bailey, *Thetic Constructions in Koine Greek*, Vrije Universiteit Amsterdam PhD Dissertation 2009.

E.J. Bakker, *Linguistics and Formulas in Homer. Scalarity and the Description of the Particle* per, Amsterdam – Philadelphia, John Benjamins Publishing Company 1988.

E.J. Bakker, "Boundaries, Topics, and the Structure of Discourse: An Investigation of the Ancient Greek Particle δέ", *Studies in Language* 17 (1993), 275–311.

E.J. Bakker, "Voice, Aspect and Aktionsart: Middle and Passive in Ancient Greek", in B. Fox, P.J. Hopper (eds.), *Voice. Form and Function*, Amsterdam – Philadelphia, John Benjamins Publishing Company 1994, 23–47.

S.J. Bakker, *The Noun Phrase in Ancient Greek. A Functional Analysis of the Order and Articulation of NP Constituents in Herodotus*, Leiden – Boston, Brill 2009a.

S.J. Bakker, "On the Curious Combination of the Particles γάρ and οὖν", in S.J. Bakker, G. Wakker (eds.), *Discourse Cohesion in Ancient Greek*, Leiden – Boston, Brill 2009b, 41–61.

W.F. Bakker, *The Greek Imperative. An Investigation into the Aspectual Differences between the Present and Aorist Imperatives in Greek Prayer from Homer up to the Present Day*, Amsterdam, Hakkert 1966.

A. Barker, *Scientific Method in Ptolemy's* Harmonics, Cambridge, Cambridge University Press 2000.

J. Barnes, "Logical Form and Logical Matter", in A. Alberti (ed.), *Logica, mente e persona*, Firenze, Leo S. Olschki 1990, 7–119.

J. Barnes, "'A Third Sort of Syllogism': Galen and the Logic of Relations", in R.W. Sharples (ed.), *Modern thinkers and ancient thinkers*, London, UCL Press 1993, 172–194.

J. Barnes, "What is a Disjunction?", in D. Frede, B. Inwood (eds.), *Language and Learning: Philosophy of Language in the Hellenistic Age*, Cambridge, Cambridge University Press 2005, 274–298.

J. Barnes, "Osservazioni sull'uso delle lettere nella sillogistica di Aristotele", *Elenchos* 27 (2006), 277–304.

J. Barnes, *Truth, etc.*, Oxford, Clarendon Press 2007.

O. Becker, "Eudoxos-Studien II. Warum haben die Griechen die Existenz der vierten Proportionale angenommen?", *Quellen und Studien zur Geschichte der Mathematik, Astronomie, und Physik* B2 (1932–1933), 369–387.

O. Becker, "Eudoxos-Studien IV. Das Prinzip des ausgeschlossenen Dritten in der griechischen Mathematik", *Quellen und Studien zur Geschichte der Mathematik, Astronomie, und Physik* B3 (1936a), 370–388.

O. Becker, "Zur Textgestaltung des eudemischen Berichts über die Quadratur der Möndchen durch Hippokrates von Chios", *Quellen und Studien zur Geschichte der Mathematik, Astronomie, und Physik* B3 (1936b), 411–419.

O. Becker, "Zum Text eines mathematischen Beweises im eudemischen Berichts über die Quadraturen der ‚Möndchen' durch Hippokrates von Chios bei Simplicius", *Philologus* 99 (1954–55), 313–316.

J. Beere, B. Morison, "A Mathematical Form of Knowing How: the Nature of Problems in Euclid's Geometry", unpublished typescript.

M. Beeson, "Constructive Geometry", in T. Arai *et al.* (eds.), *Proceedings of the 10th Asian Logic Conference*. Singapore, World Scientific 2010, 19–84.

K. Bentein, *Verbal Periphrasis in Ancient Greek. Have- and Be- Constructions*, Oxford, Oxford University Press 2016.

J.L. Berggren, A. Jones (transl. comm.), *Ptolemy's* Geography. *An Annotated Translation of the Theoretical Chapters*, Princeton – Oxford, Princeton University Press 2000.

J.L. Berggren, N. Sidoli, "Aristarchus's *On the Sizes and Distances of the Sun and the Moon*: Greek and Arabic Texts", *Archive for History of Exact Sciences* 61 (2007), 213–254.

J.L. Berggren, R.S.D. Thomas (transl. comm.), *Euclid's* Phaenomena. *A Translation and Study of an Hellenistic Treatise in Spherical Astronomy*, American Mathematical Society 1996.

G. Berkeley, *Principles of Human Knowledge*, London 1734.

E.W. Beth, "Kants Einteilung der Urteile in analytische und synthetische", *Algemeen Nederlands Tijdschrift voor Wijsbegeerte en Psychologie* 46 (1953/54), 253–264.

E.W. Beth, "Über Lockes ‚allgemeines Dreieck'", *Kant-Studien* 48 (1956/57), 361–380.

D.L. Blank, "Remarks on Nicanor, the Stoics and the Ancient Theory of Punctuation", *Glotta* 61 (1983), 48–67.

D. Blank, C. Atherton, "The Stoic Contribution to Traditional Grammar", in B. Inwood (ed.), *The Cambridge Companion to the Stoics*, Cambridge, Cambridge University Press 2003, 310–327.

F. Blass, A. Debrunner, *Grammatik des neutestamentlichen Griechisch*, Bearbeitet von F. Rehkopf, Göttingen, Vandenhoeck und Ruprecht 1976.

J. Blomqvist, *Greek Particles in Hellenistic Prose*, Lund, CWK Gleerup 1969.

J. Blomqvist, "Juxtaposed τε καί in Post-Classical Prose", *Hermes* 102 (1974), 170–178.

S. Bobzien, "Stoic Syllogistic", *Oxford Studies in Ancient Philosophy* 14 (1996), 133–192.

S. Bobzien, "The Stoics on Hypotheses and Hypothetical Arguments", *Phronesis* 42 (1997), 299–312.

S. Bobzien, "Logic. The 'Megarics'. The Stoics", in K. Algra, J. Barnes, J. Mansfeld, M. Schofield (eds.), *The Cambridge History of Hellenistic Philosophy*, Cambridge, Cambridge University Press 1999, 83–157.

S. Bobzien, "Peripatetic Hypothetical Syllogistic in Galen – Propositional Logic off the Rails?", *Rhizai* 2 (2004), 57–102.

S. Bobzien, "Stoic Sequent Logic and Proof Theory", *History and Philosophy of Logic* 40 (2019), 234–265.

S. Bobzien, "Demonstration and the Indemonstrability of the Stoic Indemonstrables", *Phronesis* 65 (2020), 355–378.

S. Bobzien, S. Shogry, "Stoic Logic and Multiple Generality", *Philosophers' Imprint* 31 (2020), 1–36.

Ch. Brandis (ed.), *Scholia in Aristotelem*, editio altera quam curavit O. Gigon, Berlin, Walter De Gruyter 1961.

S. Brentjes, "Two comments on Euclid's *Elements*? On the relation between the Arabic text attributed to al-Nayrīzī and the Latin text ascribed to Anaritius", *Centaurus* 43 (2001), 17–55.

A. Bronowski, *The Stoics on* Lekta. *All There Is to Say*, Oxford, Oxford University Press 2019.

J. Brunel, *L'aspect verbal et l'emploi des préverbes en Grec, particulièrement en Attique*, Paris, Klincksieck 1939.

J. Brunschwig (ed.), *Aristote, Topiques. Tome I: Livres I–IV*, Paris, Les Belles Lettres 1967.

J. Brunschwig, "Le modèle conjonctif", in Brunschwig 1978a, 59–86.

J. Brunschwig (ed.), *Les Stoïciens et leur logique*, Paris, Vrin 1978a.

J. Brunschwig, "Stoic Metaphysics", in B. Inwood (ed.), *The Cambridge Companion to the Stoics*, Cambridge, Cambridge University Press 2003, 206–232.

W. Burkert, *Lore and Science in Ancient Pythagoreanism*, Cambridge (Mass.), Harvard University Press 1972 (German original 1962).

H.L.L. Busard (ed.), *The First Latin Translation of Euclid's* Elements *Commonly Ascribed to Adelard of Bath*, Toronto, Pontifical Institute of Mediaeval Studies 1983.

H.L.L. Busard (ed.), *The Latin Translation of the Arabic Version of Euclid's* Elements *Commonly Ascribed to Gerard of Cremona*, Leiden, E.J. Brill 1984.

E. Casari, "Sulla disgiunzione nella logica megarico-stoica", in *Actes du VIIIe Congres international d'histoire des sciences*, Paris 1958, 1217–1224.

L. Castagnoli, "Il condizionale crisippeo e le sue interpretazioni moderne", *Elenchos* 25 (2004), 353–395.

V. Caston, "Something and Nothing: The Stoics on Concepts and Universals", *Oxford Studies in Ancient Philosophy* 17 (1999), 145–213.

W. Cavini, "La negazione di frase nella logica greca", in M.C. Donnini Macciò, M.S. Funghi, D. Manetti (eds.), *Studi su papiri greci di logica e di medicina*, Firenze, Leo S. Olschki 1985, 7–126.

J.-P. Changeux, A. Connes, *Matière à pensée*, Paris, Odile Jacob 1989.

P. Chantraine, *Histoire du parfait grec*, Paris, Honoré Champion 1926.

D. Charles, *Definition in Greek Philosophy*, Oxford, Oxford University Press 2010.

D. Charles, M. Peramatzis, "Aristotle on Truth-Bearers", *Oxford Studies in Ancient Philosophy* 50 (2016), 101–141.

J. Corcoran, "A Mathematical Model of Aristotle's Syllogistic", *Archiv für Geschichte der Philosophie* 55 (1973), 191–219.

P. Crivelli, "Presupposti esistenziali della negazione in Aristotele", *Annali del Dipartimento di Filosofia* 5 (1989), 45–90.

P. Crivelli, "Indefinite Propositions and Anaphora in Stoic Logic", *Phronesis* 39 (1994), 187–206.

P. Crivelli, "The Stoics on Definition", in Charles 2010, 359–423.

P. Crivelli, "Aristotle on Syllogisms from a Hypothesis", in A. Longo, D. del Forno (eds.), *Argument from Hypothesis in Ancient Philosophy*, Napoli, Bibliopolis 2011, 95

P. Crivelli, "Aristotle's Logic", in C. Shields (ed.), *The Oxford Handbook of Aristotle*, Oxford, Oxford University Press 2012, 113–149.

P. Crivelli, "Stoic Logic", unpublished typescript forthcoming in J. Klein, N. Powers (eds.), *The Oxford Handbook of Hellenistic Philosophy*, Oxford, Oxford University Press.

P. Crivelli, D. Charles, "'ΠΡΟΤΑΣΙΣ' in Aristotle's *Prior Analytics*", *Phronesis* 56 (2011), 193–203.

C. Czinczenheim (ed.), *Édition, traduction et commentaire des* Sphériques *de Théodose*, Thèse, Université de Paris IV – Sorbonne, Lille, Atelier national de reproduction des thèses 2000.

M.L. D'Ooge, F.E. Robbins, L. Ch. Karpinski (transl. comm.), *Nicomachus of Gerasa, Introduction to arithmetic*, New York, Macmillan 1926.

C. Dalimier, "Apollonios Dyscole sur la fonction des conjonctions explétives", *Revue des Études Grecques* 112 (1999), 719–730.

W. De Boer (ed.), *Galeni De propriorum animi cuiuslibet affectuum dignotione et curatione, De animi cuiuslibet peccatorum dignotione et curatione, De atra bile*, Leipzig – Berlin,1937

V. De Falco (ed.), [*Iamblichi*] *Theologoumena arithmeticae*, Leipzig, B.G. Teubner 1922.

V. De Falco, M. Krause (eds.), *Hypsikles, Die Aufgangszeiten der Gestirne*, Göttingen, Vandenhoeck und Ruprecht 1966.

I.J.F. De Jong, "Γάρ Introducing Embedded Narratives", in Rijksbaron 1997, 175–185.

A. De Morgan, *Syllabus of a Proposed System of Logic*, London 1860.

A. De Morgan, *A Budget of Paradoxes*, Chicago – London, Open Court 1915.

V. De Risi, "Euclid's Common Notions and the Theory of Equivalence", *Foundations of Science* (2020).

M. Decorps-Foulquier, *Recherches sur les* Coniques *d'Apollonios de Pergé et leurs commentateurs grecs*, Paris, Klincksieck 2000.

M. Decorps-Foulquier, M. Federspiel, R. Rashed (eds.), *Apollonius de Perge, Coniques*, 4 vol. in 7 tomes, Berlin – New York, De Gruyter 2008–10.

J.D. Denniston, *The Greek Particles*, Oxford, Clarendon Press 1954[2].

É. des Places, Études *sur quelques particules de liaison chez Platon*, Paris, Les Belles Lettres 1929.

É. des Places, "Style parlé et style oral chez les écrivains grecs", in *Mélanges Bidez*, Bruxelles 1934, 267–286.

A.M. Devine, L.D. Stephens, *Discontinuous Syntax. Hyperbaton in Greek*, Oxford, Oxford University Press 2000.

E.J. Dijksterhuis, *Archimedes*, Princeton, Princeton University Press 1987.

H. Dik, *Word Order in Ancient Greek. A Pragmatic account of Word Order Variation in Herodotus*, Amsterdam, Gieben 1995.

T. Dorandi (ed.), *Diogenes Laertius, Lives of Eminent Philosophers*, Cambridge, Cambridge University Press 2013.

K.J. Dover, *Greek Word Order*, Cambridge, Cambridge University Press 1960.

Y. Duhoux, "Études sur l'aspect verbal en grec ancien, 1: Présentation d'une méthode", *Bulletin de la Société Linguistique de Paris* 90 (1995), 241–299.

Y. Duhoux, "Grec écrit et grec parlé. Une étude contrastive des particules aux Ve-IVe siècles", in Rijksbaron 1997, 15–48.

Y. Duhoux, *Le verbe grec ancien. Éléments de morphologie et de syntaxe historiques*, Louvain-La-Neuve, Peeters 2000.

M. Duncombe, "Aristotle's Two Accounts of Relatives in *Categories* 7", *Phronesis* 60 (2015), 436–461.

I. Düring (ed.), *Die Harmonielehre des Klaudios Ptolemaios*, Göteborg, Elanders Boktryckeri Aktiebolag 1930.

I. Düring (ed.), *Porphyrios Kommentar zur Harmonielehre des Ptolemaios*, Göteborg, Elanders Boktryckeri Aktiebolag 1932.

R. Dyckhoff, "Indirect Proof and Inversions of Syllogisms", *The Bulletin of Symbolic Logic* 25 (2019), 196–207.

G. Dye, B. Vitrac, "Le *Contre les géomètres* de Sextus Empiricus: sources, cible, structure", *Phronesis* 54 (2009), 155–203.

K. Ebbinghaus, *Ein formales Modell der Syllogistik des Aristoteles*, Gottingen, Vandenhoeck & Ruprecht 1964.

D. Ebrey, "Why Are There No Conditionals in Aristotle's Logic?", *Journal of the History of Philosophy* 53 (2015), 185–205.

L. Edelstein, I.G. Kidd (ed.), *Posidonius*, 3 vol., Cambridge, Cambridge University Press 1988–99.

B. Einarson, "On Certain Mathematical Terms in Aristotle's Logic", *The American Journal of Philology* 57 (1936), 33–54; 151–172.

J.W. Engroff, *The Arabic Tradition of Euclid's* Elements, *Book V*, Cambridge (Mass.), Harvard University PhD Dissertation 1980.

J. Evans, J.L. Berggren (transl. comm.), *Geminos's* Introduction to the Phenomena. *A Translation and Study of a Hellenistic Survey of Astronomy*, Princeton – Oxford, Princeton University Press 2006.

B.M. Fanning, *Verbal Aspect in New testament Greek*, Oxford, Clarendon Press 1990.

R. Fecht (ed.), *Theodosii De Habitationibus Liber De Diebus et Noctibus Libri duo*, Berlin, Weidmann 1927.

M. Federspiel, "Sur la locution ἐφ' οὗ/ἐφ' ᾧ servant à désigner des êtres géométriques par des lettres", in J.-Y. Guillaumin (ed.), *Mathématiques dans l'Antiquité*, Saint-Étienne, Publications de l'Université de Saint-Étienne 1992, 9–25.

M. Federspiel, "Sur l'opposition *défini/indéfini* dans la langue des mathématiques grecques", *Les Études Classiques* 63 (1995), 249–293.

M. Federspiel, "Notes linguistiques et critiques sur le Livre II des *Coniques* d'Apollonius de Pergè (Première partie)", *Revue des Études Grecques* 112 (1999), 409–443.

M. Federspiel, "Notes linguistiques et critiques sur le Livre III des *Coniques* d'Apollonius de Pergè (Première partie)", *Revue des Études Grecques* 115 (2002), 110–148.

M. Federspiel, "Sur quelques effets du 'principe d'abréviation' chez Euclide", *Les Études Classiques* 71 (2003), 321–352.

M. Federspiel, "Sur les emplois et les sens de l'adverbe ἀεί dans les mathématiques grecques", *Les Études Classiques* 72 (2004), 289–311.

M. Federspiel, "Sur l'expression linguistique du rayon dans les mathématiques grecques", *Les Études Classiques* 73 (2005), 97–108.

M. Federspiel, "Sur la formation de quelques expressions de la théorie euclidienne des proportions", *Revue des Études Anciennes* 108 (2006a), 471–481.

M. Federspiel, "Sur le sens et l'emploi de la locution δι' ἴσου dans les mathématiques grecques", *Mélanges Germaine Aiujac, PALLAS* 72 (2006b), 171–185.

M. Federspiel, "Sur le sens de μεταλαμβάνειν et de μετάληψις dans les mathématiques grecques", *Les Études Classiques* 74 (2006c), 105–113.

M. Federspiel, "Les problèmes des livres grecs des *Coniques* d'Apollonius de Perge. Des propositions mathématiques en quête d'auteur", *Les Études Classiques* 76 (2008a), 321–360.

M. Federspiel, "Notes linguistiques et critiques sur le Livre III des *Coniques* d'Apollonius de Pergè (Seconde partie)", *Revue des Études Grecques* 121 (2008b), 515–546.

M. Federspiel, "Sur l'élocution de l'ecthèse dans la géométrie grecque classique", *L'Antiquité Classique* 79 (2010), 95–116.

N. Festa (ed.), *Iamblichi De communi mathematica scientia liber*, Leipzig, B.G. Teubner 1891.

K. Fine, "A Defence of Arbitrary Objects", *Proceedings of the Aristotelian Society, Supplementary Volumes* 57 (1983), 55–77.

K. Fine, *Reasoning with Arbitrary Objects*, Oxford, Basil Blackwell 1985.

M. Frede, *Die stoische Logik*, Göttingen, Vandenhoeck & Ruprecht 1974.

M. Freid, S. Unguru, *Apollonius of Perga's* Conica. *Text, Context, Subtext*, Leiden – Boston – Köln, Brill 2001.

G. Friedlein (ed.), *Anicii Manlii Torquati Severini Boethii De institutione arithmetica libri duo, De institutione musica libri quinque. Accedit geometria quae fertur Boethii*, Leipzig, B.G. Teubner 1867.

A.L. Gaffuri, "La teoria grammaticale antica sull'interpunzione dei testi greci e la prassi di alcuni codici medievali", *Aevum* 68 (1994), 95–115.

J.-L. Gardies, *Le raisonnement par l'absurde*, Paris, Presses Universitaires de France 1991.

J.-L. Gardies, "Do Mathematical Constructions Escape Logic?", *Synthèse* 134 (2003), 3–24.

M. Geymonat, "Grafia e interpunzione nell'antichità greca e latina, nella cultura bizantina e nella latinità medievale", in B. Mortara Garavelli (ed.), *Storia della punteggiatura in Europa*, Roma-Bari, Laterza 2008, 27–62.

M. Giaquinto, *Visual Thinking in Mathematics. An Epistemological Study*, Oxford, Oxford University Press 2007.

M. Giaquinto, "Visualizing in Mathematics", in P. Mancosu (ed.), *The Philosophy of Mathematical Practice*, Oxford, Oxford University Press 2008a, 22–42.

M. Giaquinto, "Cognition of Structure", in P. Mancosu (ed.), *The Philosophy of Mathematical Practice*, Oxford, Oxford University Press 2008b, 43–64.

M. Giaquinto, "Crossing Curves: A Limit to the Use of Diagrams in Proofs", *Philosophia Mathematica* 19 (2011), 281–307.

K. Gödel, "Russell's Mathematical Logic", in P.A. Schilpp (ed.), *The Philosophy of Bertrand Russell*, Evanston – Chicago, Northwestern University 1944, 123–153.

K. Gödel, "What is Cantor's Continuum Hypothesis?", in P. Benacerraf, H. Putnam (eds.), *Philosophy of mathematics. Selected readings*, 2nd ed., Cambridge, Cambridge University Press 1983, 470–485.

V. Goldschmidt, "Ὑπάρχειν et ὑφεστάναι dans la philosophie stoïcienne", *Revue des Études Grecques* 85 (1972), 331–344.

B.R. Goldstein (ed.), *The Arabic Version of Ptolemy's* Planetary Hypotheses, Philadelphia, American Philosophical Society 1967.

A. Gomez-Lobo, "Aristotle's Hypotheses and Euclid's Postulates", *Review of Metaphysics* 30 (1977), 430–439.

R. Goulet (transl. comm.), *Cléomède, Théorie élémentaire*, Paris, Vrin 1980.

R. Goulet, "La classification stoïcienne des propositions simples", in Brunschwig 1978a, 171–198.

J.-B. Gourinat, *La dialectique des Stoïciens*, Paris, Vrin 2000.

S. Gulwani, V.A. Korthikanti, A. Tiwari, "Synthetising Geometry Constructions", *ACM SIGPLAN Notices* 46 (2011), 50–61.

E. Halley (ed.), *Menelai Spæricorum libri III*, Oxonii 1758.

M. Hand, "Objectual and Substitutional Interpretations of the Quantifiers", in D. Jacquette (ed.), *Philosophy of Logic*, Amsterdam, North Holland 2007, 649–674.

R.J. Hankinson, "Galen and the Logic of Relations", in L. Schrenk (ed.), *Aristotle in Late Antiquity*, Studies in Philosophy and the History of Philosophy 27, Washington, The Catholic University of America Press 1994, 57–75.

T.L. Heath (ed.), *Aristarchus of Samos, The Ancient Copernicus*, Oxford, Clarendon Press 1913.

T.L. Heath, *Mathematics in Aristotle*, Oxford, Clarendon Press 1949.

J.L. Heiberg (ed.), *Sereni Antinoensis opuscula*, Lipsiae, B.G. Teubner 1896.

J.L. Heiberg (ed.), *Theodosius Tripolites Sphaerica*, Berlin, Weidmann 1927.

A. Hellwig, "Zur Funktion und Bedeutung der griechischen Partikeln", *Glotta* 52 (1974), 145–171.

E. Hiller (ed.), *Theoni Smyrnaei philosophi platonici Expositio rerum mathematicarum ad legendum Platonem utilium*, Leipzig, B.G. Teubner 1878.

J. Hintikka, *Logic, Language-Games and Information*, Oxford, Oxford University Press 1973.

J. Hintikka, "Aristotle's Incontinent Logician", *Ajatus* 77 (1977), 48–65.

J. Hintikka, U. Remes, *The Method of Analysis. Its Geometrical Origin and Its general Significance*, Dordrecht – Boston, Reidel 1974.

R. Hoche (ed.), *Nicomachi Geraseni Pythagorei Introductionis Arithmetica libri II*, Leipzig, B.G. Teubner 1866.

W. Hodges, "Indirect Proofs and Proofs from Assumptions", unpublished typescript.

J.P. Hogendijk, "Arabic Traces of Lost Works of Apollonius", *Archive for History of Exact Sciences* 35 (1986), 187–253.

J.P. Hogendijk, "On Euclid's Lost Porisms and Its Arabic Traces", *Bollettino di Storia delle Scienze Matematiche* 7 (1987), 93–115.

J.P. Hogendijk, "Traces of the Lost *Geometrical Elements* of Menelaus in Two Texts of al-Sijzī", *Zeitschrift für Geschichte der arabisch-islamischen Wissenschaften* 13 (2000), 129–164.

C.A. Huffman, *Archytas of Tarentum: Pythagorean, Philosopher and Mathematician King*. Cambridge, Cambridge University Press 2005.

F. Hultsch (ed.), *Pappi Alexandrini Collectionis quae supersunt*, 3 vol., Berlin, Weidmann 1876–78.

J. Humbert, *Syntaxe grecque*, Paris, Klincksieck 1960.

E. Husserl, *Logische Untersuchungen. Gesammelte Werke, Band XIX/1*, Berlin, Springer 1984.

E. Hussey, "Aristotle on Mathematical Objects", in I. Mueller (ed.), *ΠΕΡΙ ΤΩΝ ΜΑΘΗΜΑΤΩΝ*, Edmonton, Academic Printing & Publishing 1991 (= *Apeiron* 24,4), 105–133.

D.E.P. Jackson, "The Arabic Translation of a Greek Manual of Mechanics", *Islamic Quarterly* 16 (1972), 96–103.

F. Jacoby, *Die Fragmente der griechischen Historiker*, 15 vol., Berlin, Weidmann – Leiden, Brill 1923–58.

B. Jacquinod (ed.), *Études sur l'aspect verbal chez Platon*, Saint-Étienne, Publications de l'Université de Saint-Étienne 2000.

A. Jones (ed.), *Pappus of Alexandria, Book 7 of the* Collection, 2 vol., New York, Springer 1986.

A. Jones "Peripatetic and Euclidean Theories of the Visual Ray", *Physis* 31 (1994), 45–76.

A. Jones (ed.), *Astronomical Papyri from Oxyrhynchus*, Philadelphia, American Philosophical Society 1999.

A. Jones, "Pseudo-Ptolemy *De Speculis*", *SCIAMVS* 2 (2001), 145–186.

A. Jones, "Ptolemy's *Canobic Inscription* and Heliodorus' Observation Reports", *SCIAMVS* 6 (2005), 53–97.

P. Joray, "The Principle of Contradiction and Ecthesis in Aristotle's Syllogistic", *History and Philosophy of Logic* 35 (2014), 219–236.

P. Joray, "A Completed System for Robin Smith's Incomplete Ecthetic Syllogistic", *Notre Dame Journal of Formal Logic* 58 (2017), 329–342.

G. Junge, W. Thomson (eds.), *Pappus, Commentary on Book X of Euclid's Elements*, Cambridge (Mass.), Harvard University Press 1930.

Ch.H. Kahn, *The Verb 'be' in Ancient Greek*, Indianapolis – Cambridge, Hackett Publishing Company 1973.

K. Kalbfleisch (ed.), *Claudii Galeni Institutio Logica*, Leipzig, B.G. Teubner 1896.

I. Kant, *Kritik der reinen Vernunft*, 2nd ed., Riga, Hartknoch 1787.

E. Kheirandish (ed.), *The Arabic Version of Euclid's Optics*, Edited and Translated with Historical Introduction and Commentary, 2 vol., New York – Berlin – Heidelberg – Tokyo, Springer 1998.

J.C. King, "Instantial Terms, Anaphora and Arbitrary Objects", *Philosophical Studies* 61 (1991), 239–265.

W.R. Knorr, "Archimedes and the *Elements*: Proposal for a revised Chronological Ordering of the Archimedean Corpus", *Archive for History of Exact Sciences* 19 (1978), 211–290.

W.R. Knorr, "The Hyperbola-Construction in the *Conics*, Book II: Ancient Variations on a Theorem of Apollonius", *Centaurus* 25 (1982), 253–291.

W.R. Knorr, "Construction as Existence Proof in Ancient Geometry", *Ancient Philosophy* 3 (1983), 125–148.

W.R. Knorr, *The Ancient Tradition of Geometric Problems*, Boston, Birkhäuser 1986.

W.R. Knorr, "Pseudo-Euclidean Reflections in Ancient Optics: a Re-Examination of Textual Issues Pertaining to the Euclidean *Optica* and *Catoptrica*", *Physis* 31 (1994), 1–45.

W.R. Knorr, "The Wrong Text of Euclid: On Heiberg's Text and its Alternatives", *Centaurus* 38 (1996), 208–276.

M. Krause (ed.), *Die Sphärik von Menelaos aus Alexandrien in der Verbesserung von Abû Nasr Mansûr B. 'Alî B. 'Irâq*, Berlin, Weidmann 1936.

S. Kripke, "Is There a Problem about Substitutional Quantification?", in G. Evans, J. McDowell (eds.), *Truth and Meaning*, Oxford, Clarendon Press 1976, 325–419.

K.G. Kühn (ed.), *Claudii Galeni Opera Omnia*, 20 vol., Lipsiae 1821–33.

R. Kühner, F. Blass, *Ausführliche Grammatik der griechischen Sprache*, erster Teil: *Elementar- und Formenlehre*, 2 vol., 3. Auflage, Hannover, Verlag Hahnsche Buchhandlung 1890–92.

R. Kühner, B. Gerth, *Ausführliche Grammatik der griechischen Sprache*, zweiter Teil: *Satzlehre*, 2 vol., 3. Auflage, Hannover und Leipzig, Verlag Hahnsche Buchhandlung 1898–1904.

P. Kunitzsch, *Der Almagest. Die Syntaxis Mathematica des Claudius Ptolemäus in arabischer-lateinischer Überlieferung*, Wiesbaden, Harrassowitz 1974.

J. Lallot (ed.). *Apollonius Dyscole, De la construction (syntaxe)*, 2 vol., Paris, Vrin 1997.

J. Lallot (transl. comm.), *La grammaire de Denys le Thrace*, 2nd ed., Paris, CNRS Éditions 1998.

F. Lasserre, *Die Fragmente des Eudoxos von Knidos*, Berlin, De Gruyter 1966.

F. Lasserre, *De Léodamas de Thasos à Philippe d'Oponte*, Napoli, Bibliopolis 1987.

C. Lattmann "Diagrammatizing Mathematics: Some Remarks on a Revolutionary Aspect of Ancient Greek Mathematics", in M. Sialaros (ed.), *Revolutions and Continuity in Greek Mathematics*, Berlin, De Gruyter 2018, 107–129.

C. Lattmann, *Mathematische Modellierung bei Platon zwischen Thales und Euklid*, Berlin, De Gruyter 2019.

J. Lear, *Aristotle and Logical Theory*, Cambridge, Cambridge University Press 1980.

H.D.P. Lee, "Geometrical Method and Aristotle's Account of First Principles', *Classical Quarterly* 29 (1935), 113–124.

A. Lejeune (ed.), *L'Optique de Claude Ptolémée dans la version latine d'après l'arabe de l'émir Eugène de Sicile*, Leiden, Brill 1989².

E.J. Lemmon, *Beginning Logic*, London etc., Chapman & Hall 1990.

J. Locke, *An Essay Concerning Human Understanding*, London 1700.

P. Lorenzen, *Formal Logic*, Dordrecht, D. Reidel 1965.

L. Löwenheim, "On Making Indirect Proofs Direct", *Scripta Mathematica* 12 (1946), 125–139.

J. Łukasiewicz, *Aristotle's Syllogistic from the Standpoint of Modern Formal Logic*, 2nd ed., Oxford, Oxford University Press 1957.

E. Maass (ed.), *Commentariorum in Aratum reliquiae*, Berlin, Weidmann 1898.

P. Mäenpää, "From Backward Reduction to Configurational Analysis", in M. Otte, M. Panza (eds.), *Analysis and Synthesis in Mathematics*, Dordrecht, Kluwer 1997, 201–226.

P. Mäenpää, J. von Plato, "The Logic of Euclidean Construction Procedures", *Acta Philosophica Fennica* 49 (1990), 275–293.

M. Malink, "TΩI vs TΩN in *Prior Analytics* 1.1–22", *Classical Quarterly* 58 (2008), 519–536.

M. Malink, *Aristotle's Modal Syllogistic*, Cambridege (Mass.), Harvard University Press 2013.

M. Malink, "Aristotle on Principles as Elements", *Oxford Studies in Ancient Philosophy* 53 (2017), 163–213.

M. Malink, "Demonstration *by Reductio ad Impossibile* in *Posterior Analytics* 1.26", *Oxford Studies in Ancient Philosophy* 58 (2020), 91–156.

P. Mancosu, *Philosophy of Mathematics and Mathematical Practice in the Seventeenth Century*, New York – Oxford, Oxford University Press 1996.

P. Mancosu, "On the Constructivity of Proofs. A Debate Among Behmann, Bernays, Gödel, and Kaufmann", in W. Sieg, R. Sommer, C. Talcott (eds.), *Reflections on the Foundations of Mathematics*, Natick (Mass.), Peters 2002, 349–371.

K. Manders, "The Euclidean Diagram (1995)", in P. Mancosu (ed.), *The Philosophy of Mathematical Practice*, Oxford, Oxford University Press 2008a, 80–133.

K. Manders, "Diagram-Based Geometric Practice", in P. Mancosu (ed.), *The Philosophy of Mathematical Practice*, Oxford, Oxford University Press 2008b, 65–79.

K. Manitius (ed.), *Hipparchi in Arati et Eudoxi Phaenomena commentariorum libri tres*, Leipzig, B.G. Teubner 1894.

K. Manitius (ed.), *Gemini Elementa Astronomiae*, Leipzig, B.G. Teubner 1898.

K. Manitius (ed.), *Procli Diadochi Hypotyposis Astronomicarum Positionum*, Leipzig, B.G. Teubner 1909.

K. Manitius (transl. comm.), Ptolemäus, *Handbuch der Astronomie*, 2nd ed., Vorwort und Berichtigungen von O. Neugebauer, 2 vol., Leipzig, B.G. Teubner 1963.

L. Marrone, "Le questioni logiche di Crisippo (*Pherc* 307)", *Cronache Ercolanesi* 27 (1997), 83–100.

E.W. Marsden (ed.), *Greek and Roman Artillery: Technical Treatises*, Oxford, Oxford University Press 1971.

P.K. Marshall (ed.), *A. Gellii Noctes Atticae*, 2 vol., Oxford, Oxford University Press 1968.

R. Masià, "A New Reading of Archytas' Doubling of the Cube and its Implications", *Archive for History of Exact Sciences* 70 (2016), 175–204.

R. Masià, "First Steps to Automatic Processing of Ancient Greek Mathematical Texts", unpublished typescript.

B. Mates, *Stoic Logic*, Berkeley – Los Angeles, University of California Press 1961.

C.M. Mazzucchi, "Per una punteggiatura non anacronistica, e più efficace, dei testi greci", *Bollettino della Badia Greca di Grottaferrata*, n.s. 51 (1997), 129–143.

K.L. McKay, "The Use of the Ancient Greek Perfect Down to the Second Century A.D.", *Bulletin of the Institute of Classical Studies* 12 (1965), 1021.

K.L. McKay, "On the Perfect and Other Aspects in the Greek Non-Literary Papyri", *Bulletin of the Institute of Classical Studies* 27 (1980), 23–49.

K. Meiser (ed.), *Anicii Manlii Torquati Severini Boethii commentarii in librum Aristotelis περὶ ἑρμηνείας*, Leipzig, B.G. Teubner 1877.

H. Mendell, *Aristotle and the Mathematicians. Some Cross-Currents in the Fourth Century*, Stanford University PhD Dissertation 1986.

H. Mendell, "Making Sense of Aristotelian Demonstration", *Oxford Studies in Ancient Philosophy* 16 (1998), 161–225.

S. Menn, “The Stoic Theory of Categories”, *Oxford Studies in Ancient Philosophy* 17 (1999), 215–247.

S. Menn, “How Archytas Doubled the Cube”, in B, Holmes, K.-D. Fischer (eds.), *The Frontiers of Ancient Science. Essays in Honor of Heinrich von Staden*, Berlin – Boston, De Gruyter 2015, 407–435.

M. Mignucci (transl. comm.), *Aristotele, Gli Analitici Primi*, Napoli, Loffredo 1969.

M. Mignucci, “The Stoic Notion of Relatives”, in J. Barnes, M. Mignucci (eds.), *Matter and Metaphysics. Fourth Symposium Hellenisticum*, Napoli, Bibliopolis 1988, 131–221.

J. Mogenet (ed.), *Autolycus de Pitane. Histoire du texte, suivie de l'édition critique des traités de la sphère en mouvement et des levers et couchers*, Louvain, Publications Universitaires de Louvain 1950.

J. Mogenet, A. Tihon (eds.), *Le «Grand Commentaire» de Théon d'Alexandrie aux Tables Faciles de Ptolémée*, 3 vol., Città del Vaticano, Biblioteca Apostolica Vaticana 1985–99.

N. Moler, P. Suppes “Quantifier-free axioms for constructive plane geometry”, *Compositio Mathematica* 20 (1968), 143–152.

P. Monteil, *La phrase relative en Grec ancien*, Paris, Klincksieck 1963.

A.C. Moorhouse, *Studies in the Greek Negatives*, Cardiff, University of Wales Press 1959.

R. Morelon, “La version arabe du Livre des Hypothèses de Ptolémée, édition et traduction de la première partie”, *Mélanges de l'Institut dominicain d'études orientales* 21 (1993), 7–85.

A. Moreschini Quattordio, “L'uso dell'imperativo e dell'infinito in Omero e nella tradizione epigrafica”, *Studi Classici e Orientali* 19/20 (1970–71), 347–358.

B. Morison, “Logic”, in J.R. Hankinson (ed.), *The Cambridge Companion to Galen*, Cambridge, Cambridge University Press 2008, 66–115.

G.R. Morrow (transl. comm.), *Proclus, A Commentary on the First Book of Euclid's Elements*, Princeton, Princeton University Press 1970.

I. Mueller, “Aristotle on Geometric Objects”, *Archiv für Geschichte der Philosophie* 52 (1970), 156–171.

I. Mueller, “Greek Mathematics and Greek Logic”, in J. Corcoran (ed.), *Ancient Logic and its Modern Interpretations*, Dordrecht – Boston, D. Reidel 1974, 35–70.

I. Mueller, *Philosophy of Mathematics and Deductive Structure in Euclid's* Elements, Cambridge (Mass.) – London, MIT Press 1981.

I. Mueller, “On some Academic theories of mathematical objects”, *Journal of Hellenic Studies* 106 (1986), 111–120.

I. Mueller, “On the Notion of a Mathematical Starting Point in Plato, Aristotle, and Euclid”, in A.C. Bowen (ed.), *Science and Philosophy in Classical Greece*, New York – London, Garland Publishing 1991, 59–97.

Ch. Mugler, *Dictionnaire historique de la terminologie géométrique des Grecs*, Paris, Klincksieck 1958.

I. Müller (ed.), *Claudii Galeni Pergameni scripta minora*, vol. II, Leipzig, B.G. Teubner 1891.

J. Mumma, ”Proofs, pictures, and Euclid”, *Synthèse* 175 (2010), 255–287.

H. Mutschmann, J. Mau (eds.), *Sexti Empirici Opera*, 3 vol., Leipzig, B.G. Teubner 1914–58.

R. Netz, *The Shaping of Deduction in Greek Mathematics*, Cambridge, Cambridge University Press 1999.

R. Netz, “The Aristotelian Paragraph”, *Proceedings of the Cambridge Philological Society* 47 (2001), 211–232.

R. Netz, “Eudemus of Rhodes, Hippocrates of Chios and the Earliest form of a Greek Mathematical Text”, *Centaurus* 46 (2004), 243–286.

R. Neuberger-Donath, “The Obligative Infinitive in Homer and its Relationships to the Imperative”, *Folia Linguistica* 14 (1980), 65–82.

J. Noret, “Notes de ponctuation et d'accentuation byzantine”, *Byzantion* 65 (1995), 69–88.

V. Pambuccian, “Axiomatizing geometric constructions”, *Journal of Applied Logic* 6 (2008), 24–46.

F.G.H. Pang, *Revisiting Aspect and* Aktionsart. *A Corpus Approach to Koine Greek Event Typology*, Leiden – Boston, Brill 2016.

M. Panza, "The twofold role of diagrams in Euclid's plane geometry", *Synthèse* 186 (2012), 55–102.

G. Pasquali, *Filologia e storia*, Firenze, Felice Le Monnier 1998.

G. Patzig, *Aristotle's Theory of the Syllogism. A Logico-Philosophical Study of Book* A *of the* Prior Analytics, Dordrecht, D. Reidel 1968 (German original 1959).

Ch.S. Peirce, *Collected Papers*, Edited by Ch. Hartshorne and P. Weiss, 6 vol., Cambridge (Mass.), Harvard University Press 1931–35.

F.Th. Perkins, "Symmetry in Visual Recall", *The American Journal of Psychology* 44 (1932), 473–490.

R.D. Peters, *The Greek Article. A Functional Grammar of ὁ-items in the Greek New Testament with Special Emphasis on the Greek Article*, Leiden – Boston, Brill 2014.

R. Pettigrew, "Aristotle on the Subject Matter of Geometry", *Phronesis* 54 (2009), 239–260.

E. Pistelli (ed.), *Iamblichi in Nicomachi arithmeticam introductionem liber*, Leipzig, B.G. Teubner 1894.

M. Pohlenz (ed.), *Plutarchi Moralia*, vol. VI fasc. 2, Leipzig, B.G. Teubner 1958.

D.M.W. Powers, "Applications and Explanations of Zipf's Law", in D.M.W. Powers (ed.), *NeMLaP3 /CoNLL98: New Methods in Language Processing and Computational Natural Language Learning*, ACL 1998, 151–160.

P. Pritchard, *Plato's Philosophy of Mathematics*, Sankt Augustin, Academia Verlag 1995.

R. Rashed, H. Bellosta (eds.), *Apollonius de Perge, La section des droites selon des rapports*, Berlin – New York, De Gruyter 2010.

R. Rashed, A. Papadopoulos (eds.), *Menelaus' Spherics. Early Translations and al-Māhānī / al-Harawī's Version*, Berlin – New York, De Gruyter 2017.

P. Riedlberger (ed.), *Domninus of Larissa, Encheiridion and Spurious Works*, Pisa – Roma, Fabrizio Serra Editore 2013.

A. Rijksbaron, "The Greek Perfect: Subject versus Object", in Rijksbaron 2019, 39–59 (original Id., "Het Griekse perfectum: subject contra object", *Lampas* 17 (1984), 403–420.

A. Rijksbaron, *Aristotle, Verb Meaning and Functional Grammar. Towards a New Typology of States of Affairs*, Amsterdam, Gieben 1989.

A. Rijksbaron, "The Treatment of the Greek Middle Voice by the Ancient Grammarians", in H. Joly (ed.), *Actes du Colloque International "Philosophies du langage et théories linguistiques dans l'Antiquité", Grenoble 1985*, Bruxelles, Editions Ousia – Université des Sciences Sociales de Grenoble, 1986, 427–444.

A. Rijksbaron, "Adverb or Connector? The Case of καὶ … δέ", in Rijksbaron 1997a, 187–208.

A. Rijksbaron (ed.), *New Approaches to Greek Particles*, Amsterdam, Gieben 1997a.

A. Rijksbaron, *The Syntax and Semantics of the Verb in Classical Greek*, 3rd ed., Chicago – London, The University of Chicago Press 2006.

A. Rijksbaron, *Form and Function in Greek Grammar. Linguistic Contributions to the Study of Greek Literature*, Leiden – Boston, Brill 2019.

D.G. Robertson, "Chrysippus on Mathematical Objects", *Ancient Philosophy* 24 (2004), 169–191.

Ph. Roelli, *Latin as the Language of Science and Learning*, Berlin – Boston, De Gruyter 2021.

S. Rommevaux, A. Djebbar, B. Vitrac, "Remarques sur l'Histoire du Texte des *Éléments* d'Euclide", *Archive for History of Exact Sciences* 55 (2001), 221–295.

W.D. Ross (ed.), *Aristotle's Metaphysics*, Oxford, Clarendon Press 1924.

W.D. Ross (ed.), *Aristotle's Prior and Posterior Analytics*, Oxford, Clarendon Press 1949.

F. Rudio, "Der Bericht des Simplicius über die Quadraturen des Antiphon und des Hippokrates", *Bibliotheca Mathematica*, III. Folge, 3 (1902), 7–62.

C.J. Ruijgh, Review of: Ch.H. Kahn, *The Verb 'be' in Ancient Greek, Lingua* 48 (1979), 43–83.

C.J. Ruijgh, "Sur la valeur fondamentale de εἶναι: une réplique", *Mnemosyne* 37 (1984), 264–270.

C.J. Ruijgh, "L'emploi 'inceptif' du thème du présent du verbe grec. Esquisse d'une théorie de valeurs temporelles des thèmes temporels", *Mnemosyne* 38 (1985), 1–61.

C.J. Ruijgh, "La place des enclitiques dans l'ordre des mots chez Homère d'après la loi de Wackernagel", in H. Eichner, H. Rix (eds.), *Sprachwissenschaft und Philologie. Jacob Wackernagel und die Indogermanistik heute*, Wiesbaden, Ludwig Reichert Verlag 1990, 213–233.

C.J. Ruijgh, "Les valeurs temporelles des formes verbales en grec ancien", in J. Gvozdanović, Th. Janssen (eds.), *The Function of Tense in Texts*, Amsterdam, North-Holland 1991, 197–217.

C.J. Ruijgh, Review of: S. Porter, *Verbal Aspect in the Greek of the New Testament with Reference to Tense and Mood, Mnemosyne* 48 (1995), 352–366.

C.J. Ruijgh, "À propos de λάθε βιώσας: le valeur de l'impératif aoriste", *Hyperboreus* 6 (2000), 325–348.

M. Ruipérez S., *Structure du système des aspects et des temps du verbe en grec ancien. Analyse fonctionnelle synchronique*, Paris, Les Belles Lettres 1982 (Spanish original 1954).

B. Russell, "Mathematical Logic as Based on the Theory of Types", *American Journal of Mathematics* 30 (1908), 222–262.

L. Russo, "The Definitions of the Fundamental Geometric Entities Contained in Book I of Euclid's *Elements*", *Archive for History of Exact Sciences* 52 (1998), 195–219.

A.I. Sabra, "Thâbit ibn Qurra on Euclid's Parallels Postulate", *Journal of the Warburg and Courtauld Institutes* 31 (1968), 12–32.

K. Saito, "A preliminary study in the critical assessment of diagrams in Greek mathematical works," *SCIAMVS* 7 (2006), 81–144.

K. Saito, "Diagrams and Traces of Oral Teaching in Euclid's *Elements*: labels and references", *ZDM – Mathematics Education* 50 (2018), 921–936.

K. Saito, N. Sidoli, "Diagrams and arguments in ancient Greek mathematics: lessons drawn from comparisons of the manuscript diagrams with those in modern critical editions," in K. Chemla (ed.), *The History of Mathematical Proof in Ancient Traditions*, Cambridge, Cambridge University Press 2012, 135–162.

Th. Scaltsas, "Relations as Plural Predications in Plato", in A. Marmodoro, D. Yates (eds.), *The Metaphysics of Relations*, Oxford, Oxford University Press 2016.

H. Schenkl (ed.), *Epicteti Dissertationes ab Arriano digestae*, Leipzig, B.G. Teubner 1916.

M.J. Schiefsky (transl. comm.), *Hippocrates, On Ancient Medicine*, Leiden – Boston, Brill 2005.

R. Schöne (ed.), *Damianos Schrift über Optik, mit Auszügen aus Geminos*, Berlin, Reichsdruckerei 1897.

H. Sefrin-Weis (ed.), *Pappus of Alexandria: Book 4 of the Collection*, London, etc., Springer 2010.

J. Sesiano (ed.), *Books IV to VII of Diophantus'* Arithmetica, *in the Arabic Translation Attributed to Qusṭā ibn Lūqā*, New York – Heidelberg, Springer 1982.

S. Shapiro (ed.), *The Oxford Handbook of Philosophy of Mathematics and Logic*, Oxford, Oxford University Press 2005.

C.M.J. Sicking, "The Distribution of Aorist and Present Tense Stem Forms in Greek, Especially in the Imperative", *Glotta* 69 (1991), 14–43, 154–170.

C.M.J. Sicking, J.M. van Ophuijsen, *Two Studies in Attic Particle Usage: Lysias and Plato*, Leiden – New York – Köln, Brill 1993.

C.M.J. Sicking, P. Stork, *Two Studies in the Semantics of the Verb in Classical Greek*, Leiden – New York – Köln, Brill 1996.

N. Sidoli, "The Concept of *Given* in Greek Mathematics", *Archive for History of Exact Sciences* 72 (2018a), 353–402.

N. Sidoli, "Uses of construction in problems and theorems in Euclid's *Elements* I–VI", *Archive for History of Exact Sciences* 72 (2018b), 403–452.

N. Sidoli, J.L. Berggren, "The Arabic version of Ptolemy's *Planisphere* or *Flattening the Surface of the Sphere*: Text, Translation, Commentary", *SCIAMVS* 8 (2007), 37–139.

N. Sidoli, Y. Isahaya (eds.), *Thābit ibn Qurra's* Restoration of Euclid's *Data (*Kitāb Uqlīdis fī al-Muʿṭaīyāt*): Text, Translation, Commentary*, New York etc., Springer 2018.

R. Simson, *Opera quaedam reliqua*, Glasgow 1776.

T. Smiley, "What is a Syllogism?", *Journal of Philosophical Logic* 2 (1973), 136–154.

R. Smith, "What is Aristotelian Ecthesis?", *History and Philosophy of Logic* 3 (1982), 113–127.

R. Smith, "Completeness of an Ecthetic Syllogistic", *Notre Dame Journal of Formal Logic* 24 (1983), 224–232.

R. Smith (transl. comm.), *Aristotle, Prior Analytics*, Indianapolis – Cambridge, Hackett 1989.

P.T. Stevens, "Aristotle and the Koine–Notes on the Prepositions", *Classical Quarterly* 30 (1936), 204–217.

F. Stjernfelt, *Diagrammatology. An Investigation on the Borderlines of Phenomenology, Ontology, and Semiotics*, Dordrecht, Springer 2007.

P. Stork, *The Aspectual Usage of the Dynamic Infinitive in* Herodotus, Bouma's Boekhuis, Groningen 1982.

G. Striker, "Aristoteles über Syllogismen »aufgrund einer Hypothese«", *Hermes* 107 (1979), 33–50.

A. Stückelberger, G. Graßhoff (eds.), *Ptolemaios, Handbuch der Geographie*, 3 vol., Basel, Schwabe 2006.

P. Tannery, "Le fragment d'Eudème sur la quadrature des lunules", *Mémoires de la Société des sciences physiques et naturelles de Bordeaux*, 2e série, 5 (1883), 217–237, reprinted in Id., *Mémoires Scientifiques*, I (1912), 339–370.

P. Tannery, "Pour l'histoire des lignes et surfaces courbes dans l'antiquité", *Bulletin des Sciences mathématiques*, 2e série, 7 (1883), 278–291, and 8 (1884), 19–30 and 101–112, reprinted in Id., *Mémoires Scientifiques*, II (1912), 1–47.

P. Tannery, "Sur l'authenticité des axiomes d'Euclide", *Bulletin des Sciences mathématiques*, 2e série, 8 (1884), 162–175, reprinted in Id., *Mémoires Scientifiques*, II (1912), 48–63.

P. Tannery, "Aristote, *Météorologie*, Livre III, Ch. V", *Revue de philologie, de littérature et d'histoire anciennes* 9 (1886), 38–46, reprinted in Id., *Mémoires Scientifiques*, IX (1929), 51–61.

P. Tannery, "Simplicius et la quadrature du cercle", *Bibliotheca Mathematica*, III. Folge, 3 (1902), 342–349, reprinted in Id., *Mémoires Scientifiques*, III (1915), 119–130.

L. Tarán, *Speusippus of Athens. A Critical Study with a Collection of the Related Texts and Commentary*, Leiden, Brill 1981.

A. Tarski, *A Decision Method for Elementary Algebra and Geometry*, 2nd ed., Berkeley – Los Angeles, University of California Press 1951.

A. Tarski, "What is Elementary Geometry?", in L. Henkin, P. Suppes, A. Tarski (eds.), *The Axiomatic Method*, Amsterdam – New York – London, North Holland 1959, 16–29.

W. Theiler (ed. comm.), *Poseidonios, Die Fragmente*, Berlin – New York, De Gruyter 1982.

P. Thom, *The Syllogism*, München, Philosophia Verlag 1981.

A. Tihon, "Notes sur l'astronomie grecque au Ve siècle de notre ère (Marinus de Naplouse – Un commentaire au Petit Commentaire de Théon)", *Janus* 63 (1976), 167–184.

A. Tihon (ed.), *Le "Petit Commentaire" de Théon d'Alexandrie aux Tables Faciles de Ptolémée*, Città del Vaticano, Biblioteca Apostolica Vaticana 1978.

A. Tihon, R. Mercier (ed.), *Πτολεμαίου Πρόχειροι Κανόνες. Les* Tables Faciles *de Ptolémée*, 2 vol., Leuven, Peeters 2011.

M. Timpanaro Cardini (ed.), *Pitagorici. Testimonianze e frammenti*, 3 vol., Firenze, La Nuova Italia 1958–64.

K. Tittel, *De Gemini Stoici studiis mathematicis quaestiones philologae*, Lipsiae, Typis M. Hoffmanni 1895.

M.N. Tod, "The Greek Numeral Notation", *The Annual of the British School at Athens* 18 (1911–12), 98–132.

M.N. Tod, "Three Greek Numeral Systems", *The Journal of Hellenic Studies* 33 (1913), 27–34.

M.N. Tod, "Further Notes on the Greek Acrophonic Numerals", *The Annual of the British School at Athens* 28 (1926–27), 141–157.

M.N. Tod, "The Greek Acrophonic Numerals", *The Annual of the British School at Athens* 37 (1936–37), 236–258.

M.N. Tod, "The Alphabetic Numeral System in Attica", *The Annual of the British School at Athens* 45 (1950), 126–139.

M.N. Tod, "Letter-labels in Greek Inscriptions", *The Annual of the British School at Athens* 49 (1954), 1–8.

R.B. Todd (ed.), *Cleomedis Caelestia*, Leipzig, B.G. Teubner 1990.

G. Tomkowicz, S. Wagon, *The Banach-Tarski Paradox*, Cambridge, Cambridge University Press 2016.

G.J. Toomer (ed.), *Diocles, On Burning Mirrors*, Berlin – New York, Springer 1976.

G.J. Toomer (transl. comm.), *Ptolemy's Almagest*, London, Duckworth 1984.

G.J. Toomer (ed.), *Apollonius, Conics, Books V to VII. The Arabic Translation of the Lost Greek Original in the Version of the Banū Mūsā*, 2 vol., Berlin – New York, Springer 1990.

P.M.J.E. Tummers (ed.), *Anaritius' Commentary on Euclid. The Latin Translation, I–IV*, Nijmegen, Ingenium Publishers 1994.

M. Ugaglia, "Boundlessness and Iteration: On the Meaning of ἀεί in Aristotle", *Rhizai* 6 (2009), 193–213.

M. Ugaglia, "Aristotle and the Mathematical Tradition on *diastēma* and *logos*: An Analysis of *Physics* 3.3, 202a18–21", *Greek, Roman, and Byzantine Studies* 56 (2016), 49–67.

S. van der Pas, "The Normal Road to Geometry: δή in Euclid's Elements and the Mathematical Competence of His Audience", *Classical Quarterly* 64 (2014), 558–573.

E. van Emde Boas, A. Rijksbaron, L. Huitink, M. De Bakker, *The Cambridge Grammar of Classical Greek*, Cambridge, Cambridge University Press 2019.

N. Vinel (ed.), *Jamblique, In Nicomachi Arithmeticam*, Pisa – Roma, Fabrizio Serra Editore 2014.

B. Vitrac (transl. comm.), *Euclide, Les Éléments*, 4 vol., Paris, Presses Universitaires de France 1990–2001.

B. Vitrac, "Note Textuelle sur un (Problème de) Lieu Géométrique dans les *Météorologiques* d'Aristote (III.5, 375b16–376b22)", *Archive for History of Exact Sciences* 56 (2002), 239–283.

B. Vitrac, "A Propos des Démonstrations Alternatives et Autres Substitutions de Preuves Dans les *Éléments* d'Euclide", *Archive for History of Exact Sciences* 59 (2004), 1–44.

B. Vitrac, "Promenade dans les préfaces des textes mathématiques grecs anciens", in P. Radelet-de-Grave (ed.), *Liber amicorum Jean Dhombres*, Turnhout, Brepols 2008, 9–46.

B. Vitrac, A. Djebbar, "Le Livre XIV des *Éléments* d'Euclide: versions grecques et arabes", *SCIAMVS* 12 (2011), 29–158; 13 (2012), 3–156.

B. Vitrac, "La chaotique transmission de la Proposition IX.19 des *Éléments* d'Euclide (manuscrits et imprimés)", forthcoming.

J. von Plato, *Elements of Logical Reasoning*, Oxford, Oxford University Press 2013.

J. von Plato, "Infinity and Incommensurability: What Proofs in Euclid and Pythagoras?", 2015, unpublished typescript.

J. von Plato, "Aristotle's deductive logic: A proof-theoretical study", in D. Probst, P. Schuster (eds,), *Concepts of Proof in Mathematics, Philosophy and Computer Science*, Berlin, De Gruyter 2016, 323–346.

J. von Plato, *The Great Formal Machinery Works*, Princeton, Princeton University Press 2017.

G.C. Wakker, *Conditions and Conditionals. An Investigation of Ancient Greek*, Amsterdam, Gieben 1994.

H. Weidemann, "Le proposizioni modali in Aristotele, *De Interpretatione* 12 e 13", *Dianoia* 10 (2005), 27–41.

H. Weidemann, "De Interpretatione", in C. Shields (ed.), *The Oxford Handbook of Aristotle*, Oxford, Oxford University Press 2012, 81–112.

C. Wescher, *La poliorcétique des grecs*, Paris, Imprimerie impériale 1867.

B. Wilck, "Euclid's Kinds and (Their) Attributes", *History of Philosophy & Logical Analysis* 23 (2020), 362–397.

A. Willi, *The Languages of Aristophanes. Aspects of Linguistic Variation in Classical Attic Greek*, Oxford, Oxford University Press 2003.

J.G. Winter (ed.), *Papyri in the University of Michigan Collection. Miscellaneous Papyri*, Ann Arbor, University of Michigan Press 1936.

F. Woepcke, "Notice sur des traductions arabes de deux ouvrages perdus d'Euclide", *Journal Asiatique*, 4e série, 18 (1851), 217–247.

J. Würschmidt, "Die Schrift des Menelaus über die Bestimmung der Zusammensetzung von Legierungen", *Philologus* 80 (1925), 377–409.

H.G. Zeuthen, *Die Lehre von den Kegelschnitten im Altertum*, Kopenhagen, A.F. Höst & Sohn 1886.

H.G. Zeuthen, "Die geometrische Construction als 'Existenzbeweis' in der antiken Geometrie", *Mathematische Annalen* 47 (1896), 222–228.

INDICES

F. Acerbi, *The Logical Syntax of Greek Mathematics*, Sources and Studies in the History of Mathematics and Physical Sciences,
https://doi.org/10.1007/978-3-030-76959-8

INDEX NOMINUM

Before 1800

After 1800

Mentions in the bibliography are excluded

Index Fontium

Manuscripts

Papyri

Index Locorum

Simple mentions of propositions of the *Elements* are excluded; the Appendices are likewise excluded.

Index Rerum

GPSR Compliance
The European Union's (EU) General Product Safety Regulation (GPSR) is a set of rules that requires consumer products to be safe and our obligations to ensure this.

If you have any concerns about our products, you can contact us on

ProductSafety@springernature.com

In case Publisher is established outside the EU, the EU authorized representative is:

Springer Nature Customer Service Center GmbH
Europaplatz 3
69115 Heidelberg, Germany

www.ingramcontent.com/pod-product-compliance
Ingram Content Group UK Ltd.
Pitfield, Milton Keynes, MK11 3LW, UK
UKHW021008290726
14059UKWH00001BA/30

* 9 7 8 3 0 3 0 7 6 9 5 8 1 *